8.95

Hemp
PE, S, CF, Te, Am, K, Er, C, CB

DIAGNOSTIC MICROBIOLOGY

FOURTH EDITION

Copyright © 1974 by The C. V. Mosby Company

All rights reserved. No part of this book may be reproduced
in any manner without written permission of the publisher.
Previous editions copyrighted 1962, 1966, 1970
Printed in the United States of America
Distributed in Great Britain by Henry Kimpton, London

Library of Congress Cataloging in Publication Data

Bailey, William Robert, 1917-
   Diagnostic microbiology.

   Includes bibliographical references.
    1. Medical microbiology—Laboratory manuals.
2. Micro-organisms, Pathogenic—Identification.
I. Scott, Elvyn G., joint author. II. Title.
[DNLM: 1. Microbiology—Laboratory manuals.
QW25 B156d 1974]
QR46.B26 1974   616.01   74-2482
ISBN 0-8016-0420-6

TS/CB/B 9 8 7 6 5 4 3

DIAGNOSTIC MICROBIOLOGY

A textbook for the isolation and identification of pathogenic microorganisms

W. ROBERT BAILEY, Ph.D.

Professor of Health Sciences and Biology,
Coordinator of Allied Health Science Programs,
University of Delaware,
Newark, Delaware

ELVYN G. SCOTT, M.S., M.T. (ASCP)

Microbiologist-in-Charge,
Section of Microbiology, Department of Pathology,
Wilmington Medical Center,
Wilmington, Delaware

FOURTH EDITION

with 92 illustrations

THE C. V. MOSBY COMPANY
SAINT LOUIS   1974

PREFACE *to fourth edition*

Like many other scientific endeavors in the second half of the twentieth century, microbiology has undergone progressive change that necessitates the updating and revising of textbooks, especially those that deal with the pragmatic aspects of the subject. In this new edition of *Diagnostic Microbiology* we have attempted to bring the latest applied literature in the field to the student and the laboratorian, with the defined purpose of meeting the professional needs of both.

In view of the fact that there is some disparity between taxonomists and clinical microbiologists concerning the classification of the bacteria, we have decided to delete the former Part I of the text because the new classification, which will appear in the eighth edition of *Bergey's Manual of Determinative Bacteriology,* and which would have been contained in that part, will serve no useful purpose for either students of diagnostic microbiology or clinical personnel.

The reader will find that the text has been substantially revised in many areas. The material in the first three parts of the book includes the latest methods and new terminologies, especially in reference to blood culturing and anaerobic methods. New material will be found in the section on identification of pathogenic bacteria re-lating to the streptococci, including the enterococci, the pseudomonads, mycoplasma T strains, the neisseriae, and the extensively revised group of gram-negative enteric bacteria. There is also new substance and some regrouping in those chapters on the aerobes and anaerobes. The spirochetes have been expanded and the viruses and rickettsiae updated.

Two new chapters have been added— one on quality control and the other on intestinal parasites of clinical significance, the latter being fully contributed by Dr. Vichazelhu Iralu of the University of Pennsylvania. New material is to be found also on antibiotic susceptibility testing and on fluorescent antibody procedures.

We would like to acknowledge enthusiastically the assistance rendered and the guidance offered by many of our associates, including Albert Balows, William Ewing, Sydney Finegold, Charles Hall, William Holloway, George Kubica, William Martin, Vera Sutter, and John Washington. We are much indebted.

To our wives, Pamela and Lois, we continue to express our thanks for their indulgence and generous help during the preparatory months.

W. ROBERT BAILEY
ELVYN G. SCOTT

PREFACE *to first edition*

Diagnostic Microbiology is the first edition of a new series and not a revision of the former publication *Diagnostic Bacteriology,* the latest edition of which we revised (1958). This new title derives in part from the fact that the new volume includes microorganisms other than bacteria. The reader will note, for example, that the former Society of American Bacteriologists has now become the American Society for Microbiology.

Since this book is designed to be used as a reference text in medical bacteriology laboratories and as a textbook for courses in diagnostic bacteriology at the college level, the material has been consolidated and placed in separate parts and chapters. The selected sequence will be commensurate with the needs of both the diagnostician and the student.

For purposes of orientation in taxonomy and ready reference, an outline of bacterial classification has been included. For the student beginning diagnostic work, some pertinent background information is presented on the cultivation of microorganisms, the microscopic examination of microorganisms, and the proper methods for collecting and handling specimens.

A number of chapters include recommended procedures for the cultivation of both the common and the rare pathogens isolated from clinical material and should serve to familiarize the microbiologist with the wide variety of pathogens that may be encountered. An additional chapter has been devoted to the methods employed in the microbiological examination of surgical tissue and autopsy material.

To effect further consolidation of the book's content, one part has been devoted to a series of chapters which cover the various groups of bacteria of medical importance—their taxonomic position, general characteristics, and procedures for their identification. The chapter on the enteric bacteria introduces the new classification of the family Enterobacteriaceae, outlines the group biochemical characteristics, and discusses the serological aspects. The chapter on the mycobacteria includes a discussion of the increasingly important unclassified (anonymous) acid-fast bacilli, giving the methods for their identification, certain cytochemical tests, and animal inoculation procedures.

The chapter on laboratory diagnosis of viral and rickettsial diseases includes a guide to the collection of specimens and offers recommendations for the appropriate time of collection. In the chapter on laboratory diagnosis of systemic mycotic infections, the biochemical approach in identifying the pathogenic fungi is brought to the reader's attention.

The remainder of the book includes prescribed tests for the susceptibility of bacteria to antibiotics, serological procedures on microorganisms and patients' sera, and a technical section on culture media, stains, reagents, and tests, each in alphabetical sequence.

We would like to express our gratitude to Mrs. Isabelle Schaub and to Sister Marie Judith for committing the continuation of the original publication, *Diagnostic Bacte-* *riology,* to our care and responsibility. We also asknowledge the many kindnesses extended by a number of microbiologists and clinicians in permitting the use of published and unpublished materials.

W. ROBERT BAILEY
ELVYN G. SCOTT

CONTENTS

PART ONE

Laboratory methods

1 General requirements for cultivation of microorganisms

Great advances relative to the nutritional requirements of bacteria and other microorganisms have been realized during the current century. Consequently, with the exception of a few fastidious forms, most pathogens can be cultivated in the laboratory on artificial media. Emphasis should be placed, therefore, on the proper preparation and selection of culture media in order to isolate and grow the various microbes. By the same token, the need for quality control should be stressed. A working definition of a **culture medium** that will assist the student of diagnostic microbiology may be expressed as follows: that which contains the essential nutrients in the proper concentration, an adequate amount of salt and an adequate supply of water, is free of inhibitory substances for the organism to be cultivated, is of the desired consistency, has the proper reaction (pH) for the metabolism of that organism, and is sterile. It is obvious that such a definition applies to the preparation of inanimate media because a different set of circumstances is compulsory when one is dealing with obligate parasites.

Additional requirements, such as those generated by the temperature and oxygen relationships of the culture and by the synthetic disability of certain fastidious pathogens, have to be met also.

Heterotrophic microorganisms, the group to which the pathogens belong, exhibit a wide variety of needs. Despite this, however, numerous dehydrated media that are stable and eminently suitable for use in a diagnostic laboratory are available on the market. This ready supply minimizes the preparator's concern for deterioration of prepared media and also reduces storage needs. It is recommended that media preparators observe the directions given on the container labels.

PREPARATION OF MEDIA

Clean, detergent-free glassware and equipment are essential to good media preparation. The screw cap, the metal closure, and the plastic plug have virtually replaced the cotton plug in tubes of media, but personnel are cautioned concerning the exclusive use of the push-on metal closure. With tubes of media in storage over a long period, air contamination is likely to occur. Tubes cooling after sterilization take in air—the most likely means of initial contamination.

If a culture medium is to be prepared from its ingredients, the latter should be weighed out accurately in a suitable container and the required amount of water added, preferably in two portions. Adding approximately half of the required amount to dissolve the ingredients first, and then adding the remainder, is a desirable procedure. If it is an agar-containing medium, heat will be necessary to dissolve the agar. Heating on an open flame is **not recom-**

mended, although it is often practiced. The use of a boiling water bath or steam bath is preferred. Complete dissolution of the ingredients of the medium in the required amount of water before sterilization gives a consistent and homogeneous product. Flasks of agar-containing media should be mixed after sterilization. This promotes a uniform consistency for pouring plates.

Media may be **liquid** (broth) or **solid** (containing agar). Sometimes a semisolid consistency is desired, in which case a low concentration of agar is used. Culture media may be also **synthetic,** in which all ingredients are known, or **nonsynthetic,** in which the exact chemical composition is unknown.

The inclusion of small amounts of a carbohydrate, such as glucose, in media is often recommended for the enhancement of growth in routine plating or broth media. The presence of such a carbohydrate in some instances can lead to a product of fermentation that may alter the characteristic appearance of a reaction produced by a microorganism. This is exemplified later.

Dehydrated media or media prepared according to accepted formulae are now available for the cultivation of anaerobic bacteria—which often precludes the necessity for cultivation under strict anaerobic conditions in an anaerobic jar. Thioglycollate broth or agar is an excellent example of this type. In the preparation of this medium (which contains peptone, an amino acid, and other ingredients), sodium thioglycollate is added as a reducing agent. This substance possesses sulfhydryl groups (SH−), which tie up molecular oxygen and prevent the formation and accumulation of hydrogen peroxide in the medium.

The cultivation of catalase-negative microorganisms, such as the clostridia, which are unable to break down this toxic substance, becomes possible under these conditions. The addition of thioglycollate or other reducing agents, for example,

cysteine or glutathione, to culture media such as nutrient gelatin, milk, and others renders them suitable for anaerobic or microaerophilic cultivation.

Through the addition of **glucose** to thioglycollate medium the growth of undesirable gram-negative bacteria may be suppressed in a mixed culture from which the isolation of anaerobes is being attempted.

STERILIZATION OF MEDIA

Sterilization may be effected in various ways: by **heat, filtration,** or **chemical** methods. The method of choice will depend on the medium, its consistency, and its labile constitutents. Moist heat rather than dry heat is employed in the sterilization of culture media, and the method of application will vary with the type of medium.

Moist heat

Steam under pressure. The usual application for most media is steam under pressure in an autoclave where temperatures in excess of 100° C. are obtained. The reader is undoubtedly familiar with the principle of the autoclave, and this will not be discussed at length here; yet one important feature about autoclave sterilization should always receive the attention of the operator. In the normal procedure, where one uses 15 pounds of steam for 15 minutes, the autoclave chamber should be flushed **free of air** before the outlet valve is closed. A temperature of 105° C. on the chamber thermometer is used as an index of complete live steam content. When this temperature is reached, the outlet valve may be closed and the pressure allowed to build up to the required level of 15 pounds to attain generally a temperature of 121° C. Automation, however, takes care of such operational needs.

The time of exposure to this temperature and pressure may be allowed to exceed 15 minutes if large volumes of material are being sterilized. It is not recommended that an exposure of more than 30 minutes be used, because overheating can cause a breakdown of nutrient constituents.

A lower pressure may be necessary at times, such as in the heat sterilization of certain carbohydrate solutions. A pressure of 10 to 12 pounds for 10 to 15 minutes will reduce the possibility of hydrolysis.

For the convenience of the reader the temperatures of the autoclave that correspond to the various live steam pressures (above atmospheric) are given in Table 1-1.

It is strongly recommended that the efficiency of the autoclave be checked at regular intervals. This may be conveniently done by exposing filter paper strips, impregnated with spores of the thermophile *Bacillus stearothermophilus* and an appropriate amount of dehydrated culture medium with an indicator.[2]

The paper strips,* contained in small envelopes, are inserted in the center of a basket of tubes of media or other articles to be tested. The basket or article is then placed near the bottom at the front of the autoclave chamber and the usual sterilizing cycle carried out. When the cycle is complete, the load is removed from the sterilizer and the sporestrip envelope is sent to the laboratory.

In the microbiology laboratory, the sporestrips are individually removed, using aseptic technique, and placed directly into tubes containing 12 to 15 ml. of sterile distilled water. After dissolution of

the medium on the strips, the tubes are gently shaken and placed at 55° C. and then observed daily for several days. An unexposed sporestrip is processed in a like manner as a control with each sterilization check.

The control is examined after appropriate incubation. This should reveal a change in the color of the indicator showing that acid (yellow if bromcresol purple is used) has been produced through fermentation of the glucose. If the remainder of the tubes are of the same appearance, this would indicate that sterilization has not been effected. The same reaction should be observed also with the unheated control strip.

Successful sterilization is indicated by the unchanged appearance of the heated tubes after 7 days' incubation at 55° C. The spores of *B. stearothermophilus* are destroyed when exposed to 121° C. for 15 minutes.

Recently a form of test kit more convenient for use in the microbiology laboratory has been introduced.* It consists of a sealed glass ampule containing a standardized spore suspension of *B. stearothermophilus,* culture medium, and indicator. The ampule is exposed, incubated, and read as previously described for the sporestrip; an unheated ampule is also included as a positive control.

These ampules are only for professional use and are used just once. Because they contain live cultures they should be han-

*Kilit Sporestrips No. 1, Baltimore Biological Laboratory, Division of BioQuest, Cockeysville, Md.

*Kilit ampule, Baltimore Biological Laboratory, Cockeysville, Md.

Table 1-1. Autoclave temperatures corresponding to steam pressures in the chamber

PRESSURE (POUNDS)	TEMPERATURE (° C.)	PRESSURE (POUNDS)	TEMPERATURE (° C.)	PRESSURE (POUNDS)	TEMPERATURE (° C.)
1	102.3	10	115.6	15	121.3
3	105.7	11	116.8	16	122.4
5	108.8	12	118.0	17	123.3
7	111.7	13	119.1	18	124.3
9	114.3	14	120.2	20	126.2

dled with care to prevent breakage. Each ampule is destroyed after using, preferably by incineration; unused ampules are stored in a refrigerator at 2° to 10° C.

A third but less reliable alternative is to use adhesive tape on which the word "sterile" is printed invisibly. The word becomes visible if the autoclaving is efficient. The tape may be placed on any suitable container to be autoclaved.

Flowing steam. The flowing steam procedure represents another application of moist heat and may be employed in the sterilization of materials that cannot withstand the elevated temperatures of an autoclave. **Fractional sterilization,** or **tyndallization,** introduced by John Tyndall in 1877, is a procedure involving the use of flowing steam in an Arnold sterilizer.

Material to be so sterilized is exposed for 30 minutes on 3 successive days. After the first and second days the material is placed at room temperature to permit any viable spores present to germinate. Vegetative bacteria are destroyed at flowing steam temperature, whereas their endospores are resistant. This procedure is used for media such as milk containing an indicator and others that may be precipitated or changed chemically by the normal autoclave treatment.

Inspissation. A third type of moist heat application is the process known as **inspissation,** or thickening through evaporation. This is used in the sterilization of high protein-containing media that cannot withstand the high temperatures of the autoclave. The procedure causes coagulation of the material without greatly altering the substance and appearance. Materials such as the Lowenstein-Jensen egg medium, the Loeffler serum medium, and the Dorset egg medium are inspissated.

Modern autoclaves are equipped to allow inspissation procedures. If the Arnold sterilizer or a regular inspissator is used, the tubes containing the medium are placed in a slanted position and exposed to a temperature of 75° to 80° C. for 2 hours on 3 successive days. Precautions should be taken during such treatment to prevent advanced dehydration of the medium.

Manually operated autoclaves may also be used successfully. In this situation, the tubes of medium are placed in the autoclave chamber in a slanted position in a rack with adequate spacing. The tubes may be closed with a screw cap, loosely fitted initially, when Lowenstein, Loeffler, and Dorset egg media are being prepared. For operation, the exhaust valve of the autoclave is first closed and then the door shut tightly. This will trap air in the chamber. The steam is turned on, and the chamber will contain an air-steam mixture. The pressure is then raised to 15 pounds and should be rigidly maintained for 10 minutes. The temperature ranges between 85° and 90° C. After this period, through manipulation of the steam valve and the exhaust valve, the air-steam mixture is replaced with live steam while the pressure is kept constant at 15 pounds. When the temperature reaches 105° C., the chamber contains only live steam. The outlet valve is then closed, and an additional 15 minutes at 15 pounds is allowed in order to effect complete sterilization. During the final phase the chamber temperature rises to 121° C. At the end of this period the pressure should be permitted to subside very slowly. This is achieved by closing the steam valve and keeping the outlet valve tightly closed. The chamber temperature should drop below 60° C. before opening the door. When the tubes of media are cool, their caps should be tightened.

Filtration

Certain materials cannot tolerate the high temperatures used in heat sterilization procedures without deterioration; thus other methods must be devised. Materials such as urea, certain carbohydrate solutions, serum, plasma, ascitic fluid, and some others must be filter sterilized. Filters made of sintered Pyrex, compressed asbestos, or membranes are employed. These are placed in sidearm flasks, and the entire assembly of filter and flask is sterilized in the autoclave. A test tube of appropriate length may be placed around and under the delivery tube of the filter and sterilized with the unit when only a small quantity of filtrate is to be required. Negative pressure (suction) is applied to draw or positive pressure is applied to force the materials through the filter. Millipore* and other membrane filters are used in preference to the Seitz asbestos filter for some materials because of the high adsorption capacity of the latter type. The Swinney filter,† consisting of a filter attachment affixed to a hypodermic syringe and needle, is fast and simple for small quantities of material. The nonsterile material is taken up in the syringe, the sterile filter attachment is affixed, and the material is expressed by positive pressure (exerted by the plunger) into a sterile container. Disposable sterile plastic Millipore filters are now being widely used.

Chemical methods

In addition to or in lieu of some of the foregoing, chemical methods may be used. These normally comprise two main types: (1) the use of chemical additives to solutions or (2) treatment with gases. The latter applies primarily to the sterilization of thermolabile plastic ware, such as Petri dishes, pipets, and syringes, and for this ethylene oxide may be used. The efficiency of such sterilization may be determined by the use of commercially prepared sporestrips.‡ As this is not directly associated with sterilization of media, it will not be discussed here. The use of chemical compounds, such as thymol, a crystalline phenol, as additives to concentrated thermolabile solutions is quite appropriate, however. For example,

---

*See manual by Millipore Filter Corp., Bedford, Mass.
†Millipore Filter Corp., Bedford, Mass.
‡Attest Biological Indicators, 3 M Co., St. Paul, Minn.

to sterilize a 20-times normal concentration of urea or a 20% solution of carbohydrate, approximately 1 gm. of thymol per 100 ml. of medium would be added and allowed to stand at room temperature for 24 hours. The dilution subsequently made for use would serve to nullify any bactericidal effects of the thymol.

SELECTION OF PROPER MEDIA

The multiplicity of available media increases annually. Selection of the proper media for specific purposes requires judgment by an experienced person, but for the most part, commercially available media, used in nearly all laboratories today, carry with them a recommendation based on the broad experience of others in the field. A very important facet in medium selection is the purpose for which the medium is intended. Laboratory personnel are advised to keep their selections to a minimum to avoid duplication of purpose.

Culture media, by virtue of their ingredients, may be selected for general and for specific purposes. The reader will find various terms ascribed to media throughout the literature. For example, terms such as **enrichment, enriched, selective,** and **differential** are very common as applied to media and will be found in various sections of the book. These terms are indicative of purpose. Usually the media are special, but for general use a good basic nutrient medium is needed in the preparation of broth and agar media. The choice of such a medium is governed partly by tradition and partly by the experience of the users. In the selection of a basic medium, however, it is generally understood that one which is glucose free will tend to give more consistent and reliable results. It is known that the presence of small amounts of this carbohydrate in a medium will tend to enhance growth; yet fermentation of glucose can result in a pH damaging to acid-sensitive organisms. This has been recognized in the cultivation of pneumococci and beta hemolytic streptococci. The presence of the carbohydrate in a blood agar base medium can lead to the appearance of a green zone around colonies of streptococci, making differentiation between alpha and beta hemolysis extremely difficult.

Trypticase or tryptic soy broth, a pancreatic digest of casein and soybean peptone, is a widely used general medium, and it will support the growth of many fastidious organisms without further enrichment. Because of its glucose content, however, one should be aware of its limitations in certain areas of application.

Thioglycollate medium, now available in a modified form containing 0.06% glucose, will support the growth of many aerobes, microaerophiles, and anaerobes.

In the ensuing chapters of this book, which discuss the isolation and cultivation of various pathogenic microorganisms, recommendations relative to the selection and use of media will be made. Chapter 39 gives the formulae and methods of preparation for many of the media in use at the present time.

Valuable information on the choice of media and the nutritional requirements of microorganisms may be gained from the references given at the end of this chapter.[1,3-6]

STORAGE OF MEDIA

Because of the heavy work schedule with which most diagnostic laboratories are faced today, bulk preparation of media is often necessary. The average medium should be stored in the refrigerator to avoid deterioration and dehydration. Plating media used for the cultivation of certain fastidious organisms that are sensitive to drying should be kept airtight and refrigerated. Some tubed media, especially those with tightly fitting screw caps,

paraffined corks, or rubber stoppers, may be stored for relatively long periods at room temperature. Plates of media that can be readily sealed are available commercially* from several companies and have certain advantages. Plates may also be stored in plastic bags closed with "twisters." Such plated media are in use in small laboratories, in physicians' offices, and also in many of the larger laboratories for the cultivation of certain bacteria such as gonococcus, meningococcus, and beta hemolytic streptococci, which are very sensitive to dehydration.

Refrigeration facilities will obviously determine the storage capacity. The laboratory worker is cautioned about the use of media that have been just removed from refrigeration. Plated and tubed media should be allowed to warm to room temperature before use.

---

*Baltimore Biological Laboratory, Cockeysville, Md.; Hyland Laboratories, Los Angeles; Case Laboratories, Inc., Chicago; Scott Laboratories, Fiskeville, R. I.; and others.

REFERENCES

1. BBL manual of products and laboratory procedures, ed. 5, Cockeysville, Md., 1973, Bio-Quest, Division of Becton, Dickinson & Co.
2. Brewer, J. H., and McLaughlin, C. B.: Dehydrated sterilizer controls containing bacterial spores and culture media, J. Pharm. Sci. **50:**171-172, 1961.
3. Difco manual of dehydrated culture media and reagents for microbiological and clinical laboratory procedures, ed. 9, Detroit, 1953, Difco Laboratories.
4. Difco supplementary literature, Difco Laboratories, Detroit, May 1972.
5. Porter, J. R.: Bacterial chemistry and physiology, New York, 1946, John Wiley & Sons, Inc.
6. Snell, E. E.: Bacterial nutrition-chemical factors. In Werkman, C. H., and Wilson, P. W., editors: Bacterial physiology, New York, 1951, Academic Press, Inc.

2 Optical methods in specimen examination

SPECIMEN PREPARATION

Specimens received in a laboratory for microbiological examination can vary widely. They may consist of clinical material, such as blood, urine, pus, cerebrospinal fluid, or microbial cultures isolated from such material. In his examination the laboratory worker or student should consider the nature and possible content of the specimen and be cognizant of the difficulty in cultivating and staining certain microorganisms. Special methods are often needed.

There are two ways in which microbial isolates or microorganisms in certain clinical material may be examined under a microscope—in the **living** state or in the **fixed** state.

In the living state

It is often necessary to examine certain microorganisms in the living state because they are not readily stained or because they cannot be easily cultivated. Also, when they are examined in the living state, morphology is less distorted and motility and other characteristics may be observed readily.

Two methods (the wet-mount and the hanging-drop techniques) are generally used for such examination.

Wet-mount method. In the wet-mount procedure there are two approaches, the choice being determined by the nature of the specimen: (1) place a loopful of the liquid clinical specimen or culture on a clean glass slide and cover with a coverglass; (2) place a loopful of clean water on the slide and in it emulsify some nonliquid clinical material or agar culture and add a coverglass. In either case, to reduce evaporation of the liquid, the coverglass may be ringed with petrolatum. The preparation may then be examined, either by bright-field or by dark-field microscopy. The latter method may be essential with such microorganisms as the treponeme of syphilis, which is very difficult to stain. This procedure may also be adapted to a wet-mount staining procedure for certain organisms.

Hanging-drop method. A second procedure used in the examination of organisms in the living state is the hanging-drop method. Loopfuls of the specimen may be prepared as with the wet-mount method but on a thin coverglass instead of the slide. The coverglass is then inverted over the concave area of a hollow-ground or well slide to provide the hanging drop. The coverglass can be sealed as in the wet-mount method with petrolatum or a small amount of immersion oil to reduce evaporation. Such a preparation is examined by bright-field microscopy.

In the fixed state

Bacteria and other microbes are generally examined in the fixed state, and they are more readily observed when they are stained. By drying and fixing the organisms on a glass slide, various staining procedures may be carried out. The choice of procedure will be determined by the nature of the specimen or the desired result. Staining procedures for bacteria and other microorganisms are discussed in Chapter 40.

MICROSCOPES—THEIR USE AND CARE

It is assumed that the reader understands the elements of microscopy and is generally familiar with the parts of a microscope. For this reason, discussion of the instrument will be restricted to a few points that could bear emphasis. The lenses of a microscope should be kept clean and free of grease and dust. In using the average compound light microscope equipped with a condenser, only the flat plane or face of the mirror is used. The light source selected should be one that gives uniform illumination, such as a frosted bulb, a fluorescent light, or a special microscope lamp. Most of the new light microscopes have built-in lamps. For routine bacteriological examination, the condenser of the microscope should be kept almost fully racked up. This position usually provides optimal and uniform illumination of the field. The user of a monocular microscope is strongly advised to keep both eyes open in order to reduce eyestrain. Since many laboratories today are equipped with a variety of microscopes, several instruments will be briefly discussed here with some specific advice and information offered. All light microscopes should be kept covered when not in use, and full attention should be given to protection of the objective lenses. Immersion oil should never be left on the objective when the instrument is not to be used for some time. The nosepiece should be so rotated that the low-power objective is in position before the microscope is put away. Repeated use of a solvent such as xylol to remove oil from a lens is discouraged, since this will tend to loosen the mounting cement around the lens. **Clean lens paper only** should be used for wiping the objective free of oil. Other commercial

tissue is too rough and will scratch the lens.

Bright-field microscope

The bright-field microscope is used for the majority of routine examinations where stainable microorganisms are to be observed. In bright-field microscopy the organisms appear dark against a bright background.

Dark-field microscope

In contrast to the effect in bright-field microscopy, organisms appear bright against a dark background under the dark-field microscope. This appearance may be produced either by an opaque stop inserted below the condenser or by a stop built into the condenser. This stop permits only peripheral rays of light to enter the condenser. These rays pass through the specimen at such an angle that the field appears unilluminated. Any particles, such as microorganisms in the field, reflect the light and appear bright in the optical system. This type of microscope is particularly adaptable to the examination of microorganisms that are difficult to stain. The treponeme of syphilis in the exudate from a lesion or in other material may be thus examined. Examination is carried out with the living organisms in a wet-mount preparation. Motility is readily determined for other bacteria as well.

In using the conventional dark-field microscope it is essential to use immersion oil on the top of the condenser as well as on the coverglass of the wet-mount preparation to minimize light refraction.

Phase microscope

The phase microscope is an instrument that permits relatively accurate observation of (1) bacteria in tissue sections, (2) parasites in various types of clinical material, (3) Negri or inclusion bodies in virus-infected material, and (4) histological preparations. A halo of light produced by light passing from the source through an annular diaphragm initiates the effect obtained. Elementary physics reveals that rays of light that pass through an area composed of materials varying in refractive index will emerge out of phase and give a pattern of bright and dark relief. Objects such as bacteria and inclusion bodies, being of a different refractive index from surrounding tissue, will consequently show up in such a microscope, whereas in normal bright-field microscopy these objects would have to be stained to be observed.

Ultraviolet microscope

The use of ultraviolet light instead of visible white light in microscopy allows for greater resolution. This is due to the shorter wavelength of ultraviolet light. As a result, one can obtain a magnification two to three times higher than that which is possible with visible light. Because of the invisiblity of ultraviolet light a photographic plate is used for recording the image, and since glass is impervious to such light, quartz rather than glass lenses are employed.

Fluorescence microscope

Fluorescence microscopy has become very popular in a number of laboratories because of its application to the diagnosis of disease. The light source employed is ultraviolet. The principle involved here is that certain fluorescent chemical complexes have the property of absorbing ultraviolet light and emitting rays of visible light. Microorganisms can be treated with a fluorescent dye or a dye-antibody complex that causes them to fluoresce and become readily distinguishable in a mixed population.

The fluorescent antibody technique has both advantages and disadvantages in diagnostic microbiology. Among the former are the time saved by eliminating lengthy cultural study, the sensitivity of the test, the lack of interference from contaminating microorganisms if their staining reactions are known, and the detection of nonviable pathogenic organisms. The disadvantages of the technique are the same that are experienced with most serological tests: the cross-reactions that inevitably develop between species. Anyone, therefore, planning to use the fluorescent antibody technique should be aware of such limitations. Successful results may now be obtained with Group A streptococci, enteropathogenic *Escherichia coli,* and other bacteria and in the diagnosis of rabies. The technique has further application in the detection of treponemal antibody (FTA). The reader's attention is directed to Chapter 37, where immunofluorescence is discussed in further detail. The value of this procedure in diagnostic parasitology and mycology is discussed in Public Health Service Publication No. 729.[1]

Electron microscope

The electron microscope has become essential to various fields of research. In the biological sciences, virology presently makes the greatest use of this, but the instrument has also become an invaluable piece of equipment to cytologists.

An electron source is used in lieu of a light source, and magnets are used instead of lenses. Only nonliving material may be examined, and specimens require special preparation. In the more modern instruments a direct magnification of 200,000 diameters may be obtained.

The image may be observed on a fluorescent screen through a window at the base of the evacuated column, or a photographic plate may be inserted to record the image. The result, however, is not a photograph, since photons are not used. The coined term "electronmicrograph" has been in use for some years.

Great advances in the refinement of techniques used in specimen preparation have been made in recent years, and much knowledge has now been amassed relative to the interpretation of electronmicrographic images.

REFERENCE

1. Cherry, W. B., et al.: Fluorescent antibody techniques in the diagnosis of communicable diseases, Public Health Service Pub. No. 729, Washington, D. C., 1960, U. S. Government Printing Office.

3 General principles in staining procedures

Bacteria and other microorganisms are usually transparent, and this makes the study of morphological detail difficult when they are examined in the natural state. The early methods of fixing and staining initiated by Paul Ehrlich and Robert Koch allowed the microbiologist to distinguish many structural features not formerly seen.

Unless some specific morphological feature, dependent on the age of the culture, is to be demonstrated, the microbiologist is advised to use a young culture for routine staining procedures, as old cells lose their affinity for most dyes. Unless one is dealing with organisms that have an unusually long generation time, a 24-hour culture is expected to yield favorable results.

PREPARATION OF A SMEAR

In routine staining procedures the first step is the preparation of a **smear**. A loopful of liquid culture or fluid specimen is spread over an appropriate area of a clean glass slide to make a film (Fig. 3-1, D), or a section of a colony taken with a needle from a plate or growth from a slant culture is emulsified in a loopful of clean water and spread over the required area (Fig. 3-1, D and E). For best results the film should be homogeneous.

In either event the mixing and spreading should never be done vigorously; such treatment is potentially hazardous when working with pathological material be-

cause it may create aerosols. In addition, harsh treatment will tend to destroy characteristic arrangements of cells such as chains and clusters.

A smear ideally should always be permitted to dry in the air and then, if necessary, be heat fixed by passing gently through a Bunsen flame to promote adherence of the cells to the slide. Overheating causes distortions and should be avoided. The slide should then be allowed to cool.

When a large number of cultures are to be examined by a routine staining procedure, such as the Gram stain, several small smears can be made on the same slide in areas appropriately marked with a glass-marking pencil (Fig. 3-1, F).

STAINING OF BACTERIA
Simple stain

1. After fixation of the smear, draw a vertical line on the glass slide about 1 inch from the left or right end, using a wax pencil (Fig. 3-1, D). This will keep the stain away from the fingers when holding the slide.
2. Flood the smeared area with the stain to be used (crystal violet, fuchsin, methylene blue, or safranin).
3. Allow to react for the appropriate time, 30 seconds to 3 minutes, depending on the stain.
4. Wash off with a gentle stream of cool water. Avoid having the water fall

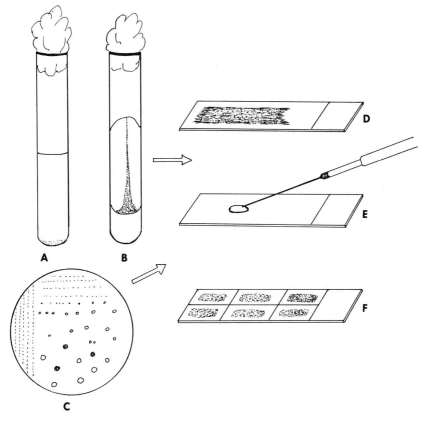

Fig. 3-1. Preparation of bacterial smears. **A** and **B,** Broth and agar slant cultures. **C,** Colonies on agar plate. **D,** Smear from liquid culture. **E,** Loopful of water for suspension of microorganisms. **F,** Multiple smear.

directly on the smear with any force.

5. Blot dry between sheets of bibulous paper and examine under the oil-immersion lens.

6. It is good practice to dispose of blotting paper used in this simple procedure and in many others involving pathogenic microorganisms. The slides, if they are not to be retained for reference, should be placed in disinfectant.

Special staining procedures
Gram stain

The Gram stain ranks among the most important stains for bacteria. Devised by Hans Christian Gram, a Dane, in 1884, it allows one to distinguish broadly between various bacteria that may exhibit a similar morphology. There are various modifications of the technique; the one described in Chapter 40 has been found reliable, even in the hands of a beginner.

On microscopic examination of a gram-stained smear containing a mixed bacterial flora, the differential features of the method will become apparent. Many bacteria will have retained the violet-iodine combination and stain **purple** (gram positive), others will have been stained **red** by the safranin (gram negative). Thus, in using this procedure not only are form, size, and other structural details made visible, but the microorganisms present can be grouped into gram-positive and gram-negative types by their reactions. This is an important diagnostic tool in subsequent identification procedures.

Probably the difference in the staining reaction between gram-positive and gram-negative bacteria can be attributed to their different chemical compositions. Gram-negative cell walls have a higher lipid[2] content than gram-positive cell walls, and although a crystal violet–iodine complex is formed in both kinds of cell, the alcohol removes the lipid from the gram-negative cells, thereby increasing the cell permeability and resulting in the loss of the dye complex. The complex is retained, however, in the gram-positive cells, in which dehydration by the alcohol causes a decrease in permeability.

Acid-fast stain

Another differential staining technique that depends upon the chemical composition of the bacterial cell is the acid-fast stain, which is used in staining tubercle bacilli and other mycobacteria. Since these microorganisms are difficult to stain with the ordinary dyes, basic dyes are used in the presence of controlled amounts of acid and are generally applied with heat. Once stained, the tubercle bacillus is resistant to subsequent treatment with **acid** alcohol, whereas most other bacteria are decolorized. For convenience in viewing, a counterstain of contrasting color is usually applied. The method most commonly employed is that of Ziehl and Neelsen, which is described as it is used at the Center for Disease Control in Atlanta.[1] This procedure is outlined in Chapter 40.

Acid-fast organisms are difficult to stain and, once stained, are not easily decolorized. Many theories have been introduced to account for this property of acid fastness. The most acceptable concept is that acid fastness is determined by the selective permeability of the cytoplasmic membrane. The brilliance of the red color is due to the retention of the dye, carbolfuchsin, in solution within the cell. Should the cell be mechanically disrupted, its acid-fast property is lost. The fact that this may not be a precise explanation of the reaction does not invalidate the usefulness of the technique for detecting fundamental differences in bacterial genera.

Capsule stain

The capsule of bacteria does not have the same affinity for dyes as do other cell components, and this necessitates the use of special staining procedures. Some are designed to stain the cell and its background but not the capsule, so that the surrounding envelope is seen by contrast, as in the Anthony method. Other procedures will produce a differential staining effect, wherein the capsule will take a counterstain, as in the Muir method. Another procedure involves the negative stain principle in which the capsule shows up as a clear halo against a dark background, as in the India ink method.

The various procedures are given in Chapter 40.

Flagella stain

Flagella are fragile, heat-labile appendages and require careful handling and staining by special procedures in order to observe them. Success in staining depends on the freshness and effectiveness of a mordant. Since flagella are beyond the resolving power of the ordinary light microscope, their size has to be increased in order to bring them into view. This may be accomplished by coating their surfaces with a precipitate from an unstable colloidal suspension (the mordant). This precipitate then serves as a layer of stainable material, and the threadlike structure of the flagellum shows up when the appropriate stain is applied.

Various methods may be used, but the authors have found the Gray method very satisfactory. The procedure may be found in Chapter 40.

Staining of metachromatic granules

Various methods for staining metachromatic granules have been devised. Some are preferred to others. We have found that two stains yield highly satisfactory results. One of these is the methylene blue stain and the other is Albert's stain. Both procedures are outlined in Chapter 40.

Spore stain

Spores are relatively resistant to physical and chemical agents and are not easily stained. In routine staining procedures such as the Gram stain they appear as unstained refractile bodies. Heat is usually required during a special procedure for staining in order to promote penetration of the stain into the spores.

The Dorner method, or a modification thereof, gives good results. The procedure is given in Chapter 40.

Relief staining

Relief staining does not stain microbial cells but produces an opaque or darkened background against which the cells stand out in sharp relief as white or light objects. It simulates dark-field microscopy and allows a rapid study of morphology.

Staining of spirochetes

On the whole, spirochetes have a poor affinity for the standard dyes and usually require special stains, such as the silver impregnation method. The Fontana stain usually gives good results. The procedure may be found in standard texts.

Staining of rickettsiae

Although rickettsiae are gram negative, they do not respond well to the Gram stain and are more successfully stained by the Giemsa or Castañeda methods. These procedures may be found in Chapter 40.

Staining of yeasts and fungi

Fixed smears of yeast can be stained quite readily with crystal violet or methylene blue when these dyes are applied for 30 to 60 seconds. Wet mounts of yeast can be stained effectively with methylene blue or Gram's iodine. Cells can be emulsified in a drop of either stain and covered with a coverglass. Alternatively, a loopful or small drop of the dye may be placed at the edge of the coverglass, covering an unstained suspension (wet mount), and the dye will spread rapidly beneath it. Lactophenol cotton blue is excellent for staining fungi. The procedure is described in Chapter 40.

MOUNTING OF STAINED SMEARS

Microbiologists often want to develop or add to a slide collection for reference and teaching purposes, as well as for demonstrations. For such purposes, permanent slides may be prepared by a relatively simple procedure.

If a slide has been examined under oil, blot it with bibulous paper, **do not wipe,** and place a drop of Canada balsam or permount on a desirable area of the smear. Place a clean cover glass carefully over this to avoid trapping air bubbles. With the butt end of an inoculating loop or needle, press on the cover glass to force the mounting material to the sides of the cover glass to completely cover the area beneath.

Place the slides in a horizontal position away from dust to permit proper setting of the balsam. When the slides are completely dry, wipe away excess dye with cheesecloth dampened with xylene, and cut away excess balsam with a razor blade. Some persons prefer to add a border of nail polish or other appropriate substance to the cover glass as a protective device. Most permanently mounted slides will last for years without deterioration.

REFERENCES

1. Kubica, G. P., and Vestal, A. L.:Tuberculosis-laboratory methods in diagnosis, Washington, D. C., 1959, U. S. Government Printing Office.
2. Salton, M. R. J.: The bacterial cell wall, Amsterdam, 1964, Elsevier Press, Inc.

4 Methods of obtaining pure cultures

The importance of **pure cultures** cannot be overemphasized in diagnostic microbiology. They are essential to the study of the following: colony characteristics, biochemical properties, morphology, staining reaction, immunological reactions, and the susceptibility of a microbial species to antimicrobial agents.

In general, pure cultures are best obtained by using solid media, either in streak plates or in pour plates. Both methods are useful, and each has certain advantages.

STREAK PLATE METHODS

The streak plate, if properly performed, is probably the most practical and most useful method for obtaining discrete colonies and pure cultures. Streaking methods may vary from one laboratory to another, but the following techniques may be used for inoculating any type of agar plate. These plates may be prepared in advance, in any desired quantity, and stored at room temperature or in a refrigerator until the time of inoculation. The agar surface should be free of beads of condensation water before the streaking is done. The plate may be dried by inverting it and propping it on the lid (Fig. 4-1, *A*).

Method 1

Method 1 (Fig. 4-1, *B*), although designed for broth cultures, may also be used for cultures from agar plates or agar slants.

Should either of the latter be used, a reduced amount of inoculum is recommended and the overlapping should be moderated.

1. Place a loopful of the inoculum near the periphery of the plate as indicated.
2. With the loop, spread the inoculum over the top quarter of the plate's surface.
3. Stab the loop into the agar several times and continue streaking, overlapping the previous streak as shown.
4. Stab the loop as before and continue streaking.
5. Flame the loop, allow it to cool, overlap the previous streak, and complete the streaking.
6. Lift the loop and streak the center of the plate with zigzag motions.

Discrete colonies should be found in the central portion of the plate, whereas additional information on hemolytic activity may be gained from the effect of a reduced oxygen tension on the organisms **stabbed** into the agar. For example, beta hemolytic streptococci may appear to be alpha hemolytic on the surface.

Method 2

The second method of streaking (Fig. 4-1, *C*) may be used either for cultures growing on solid media (colonies or slanted growth) or for heavy broth cultures. The technique is primarily designed, however, for the former.

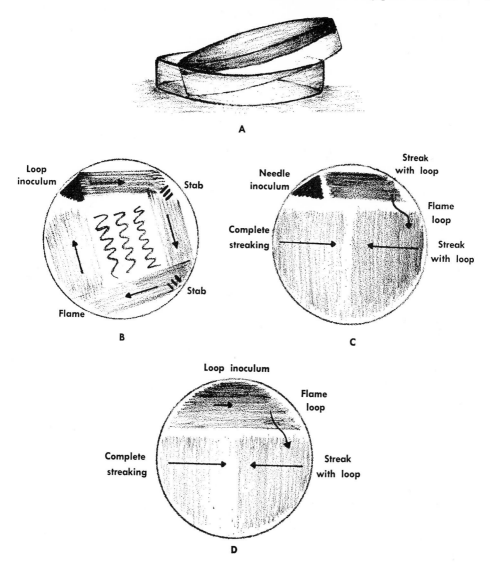

Fig. 4-1. Streak plate technique. **A,** Drying a plate of agar in the laboratory. **B,** Method 1. **C,** Method 2. **D,** Method 3.

1. Select the desired colony from a crowded plate or pick up growth from a slant with a needle. Streak this carefully on a restricted area of the plate as indicated. Flame the needle and put it away.
2. With a sterile cool loop make one light sweep through the needle-inoculated area and streak the top quarter of the plate's surface with close parallel strokes. Flame the loop and **allow it to cool.**
3. Turn the plate at right angles, make one light sweep with the loop through the lower portion of the second streaked area, and streak, as before, approximately one half of the remaining portion. Do not overlap any of the previously streaked areas.
4. Turn the plate 180 degrees and streak the remainder of the plate with the same loop, avoiding any areas previously streaked.

5. Discrete colonies should appear in these lower portions of the plate.

Method 3

Method 3 (Fig. 4-1, *D*) constitutes the simple technique and is used largely for liquid cultures, but it may also be used for cultures from solid media.

1. Place a loopful of the inoculum near the periphery of the plate and cover approximately one fourth of the plate with close parallel streaks. Flame the loop and allow it to cool.
2. Make one light sweep through the lower portion of this streaked area, turn the plate at right angles, and streak approximately one half of the remaining portion, without overlapping previous streaks.
3. Turn the plate 180 degrees and streak the remainder of the plate, avoiding previously streaked areas.
4. The appearance of this plate will resemble that obtained in method 2.

POUR PLATE METHODS

Pour plates are generally used in the laboratory as a means of determining the approximate number of viable organisms in a liquid such as water, milk, urine, or broth culture, and data are expressed as the number of colony-forming units per milliliter of the substance examined. Pour plates are also used to determine the hemolytic activity of deep colonies of some bacteria, such as the streptococci, by using an agar medium containing blood. The pour plate also lends itself to pure culture study when the components of a mixed culture are to be separated. Differences are recognized by size, shape, and color of the colonies.

The pour plate method consists of the preparation of serial dilutions of the specimen in sterile water or saline from which a prescribed volume may be pipetted into tubes of melted agar medium, respectively, or directly into sterile Petri dishes. The former are then poured into Petri dishes and allowed to set, whereas melted agar medium is added to the latter. After a suitable incubation period the colonies may be examined as surface or subsurface colonies for differences or counted, depending on the exercise. The melted agar medium used should not exceed 50° C., and the dilutions should be adequately mixed with the medium. Additional tests on isolates may be carried out as desired.

The usefulness of the pour plate depends to a great extent on the number of colonies developing on the plate. This is particularly important when studying organisms for their hemolytic activity or where colonial characteristics aid in the separation of various types. The greatest accuracy in counting is obtained with plates that contain between 30 and 300 colonies. In excess of the higher figure, counting is difficult, differential characteristics are not readily distinguishable, and colony selection without contamination is almost impossible.

A procedure that may be followed in counting overcrowded plates in order to estimate numbers present involves the use of a **counting plate.** The agar plate may be placed on the counting plate, which is ruled off in square centimeters. In this way, five or more square-centimeter areas may be counted, the average count calculated, and this figure then multiplied by 62.5 to give the estimated number of colonies per plate. The average Petri dish (inside diameter, 90 mm.) has an area of 62.5 sq. cm.

If a specimen is suspected of containing a large number of organisms, an overcrowded pour plate may be avoided by using the following procedure, which is recommended for hemolytic streptococci:

1. Examine a gram-stained smear of the original material: if more than several organisms per microscopic field are seen, proceed to step 2.
2. Inoculate 8 to 10 ml. of broth, sterile buffered saline, or water with one loopful of the original material.
3. Transfer one loopful of this dilution

to a tube of melted agar at about 45° C.

4. Using a sterile pipet, add about 0.5 to 1 ml. of sterile defibrinated blood to the inoculated tube and twirl between the palms of the hands to thoroughly mix the contents; then pour into a sterile Petri dish. Alternatively, the medium and inoculum may be combined and poured into the dish containing blood and the entire contents mixed by rotating the plate with the lid in place.

5. Allow the medium to gel, then incubate the plate in an inverted position (lid on bottom). This prevents collection of condensation on the agar surface. Unless the surface is dry, it will be difficult to obtain discrete surface colonies.

USE OF ENRICHMENT MEDIA

In the examination of fecal material for the presence of pathogenic bacteria, it is frequently necessary to use an **enrichment** medium (to be differentiated from an enriched medium, which contains a nutritive supplement). These media, because of their chemical composition, will inhibit or kill off the normal intestinal flora, such as the coliforms (commensals), and will permit salmonellae and some shigellae, which may be present only in small numbers, to grow almost unrestrictedly, thus enriching the population of such forms. After incubation, subcultures from these media must be made to solid plating media in order to obtain isolated colonies for further study. Tetrathionate broth and selenite broth are examples of enrichment media and are used primarily in enteric bacteriology.

USE OF DIFFERENTIAL AND SELECTIVE MEDIA

Differential and selective media, by virtue of their chemical compositions, will characterize certain bacterial genera by their distinctive colonial appearances in culture. Differential media, such as eosin–methylene blue agar (EMB) and MacCon-

key agar, contain lactose and a dye or indicator in the decolorized state. Bacteria that ferment the lactose with the production of acid or aldehyde will produce red colonies or colonies with a metallic sheen, depending on the medium. This distinguishes lactose-fermenting from lactose-nonfermenting organisms. The latter may be the pathogens.

Selective media are complex plating media that not only serve to differentiate among certain genera but also are highly selective in their action on other organisms. Such media as Salmonella-Shigella agar, desoxycholate citrate agar, and bismuth sulfite agar will inhibit the growth of the majority of coliform bacilli, along with many strains of *Proteus* (those developing are prevented from swarming), and permit selective isolation of enteric pathogens. The coliform colonies that do develop are readily differentiated from lactose-nonfermenting organisms by color and opacity.

Phenylethyl alcohol agar is a selective medium for the isolation of gram-positive cocci in specimens or cultures contaminated with gram-negative organisms, particularly *Proteus* species. Although many gram-negative rods form visible colonies on this medium, spreading or swarming does not occur.* On this medium, colonies of gram-positive cocci are similar to those produced on ordinary media, whereas the gram-negative organisms that do grow will produce very small colonies. To avoid contamination when selecting colonies, only those that are well separated from the small forms should be used.

Infusion agar containing potassium tellurite and either serum or blood may be used as a selective plating medium for the isolation of members of the genus *Corynebacterium,* particularly in the laboratory diagnosis of diphtheria. On this medium staphylococci and streptococci are usually inhibited.

---

*Swarming may also be inhibited by increasing the agar content of the medium to 5%.

Pathogenic staphylococci that do develop appear as black colonies on tellurite medium and as golden yellow colonies with yellow halos on mannitol salt agar, a medium containing a high salt concentration (Chapter 16).

USE OF MEDIA CONTAINING ANTIBIOTICS

Media containing antibiotics are selectively inhibitory and may be effectively used for isolating pathogenic species from a mixed population. Sabouraud dextrose agar containing cycloheximide and chloramphenicol will support the growth of dermatophytes and most fungi causing systemic mycoses, although it will greatly inhibit the majority of saprophytic fungi and bacteria found in clinical specimens. *Staphylococcus, Haemophilus,* and *Neisseria* species are inhibited on blood agar media containing neomycin and polymyxin, whereas Group A hemolytic streptococci from cultures taken of patients with suspected streptococcal disease will grow out readily.

An enriched chocolate agar medium containing ristocetin and polymyxin B has proved effective for the selective isolation of *Neisseria* and *Haemophilus* species from clinical material.* The antibiotics suppress saprophytic neisseriae, *Mima* species, and *Pseudomonas* species.

OTHER METHODS

The isolation of species of the genus *Clostridium* from clinical material is frequently complicated by the presence of rapidly growing facultative organisms such as coliforms, *Pseudomonas* species, and spreading species of *Proteus.* Anaerobic blood agar plates may be overgrown by these organisms, and the isolation of clostridia becomes a difficult task. If the anaerobe produces spores in thioglycollate

---

*Thayer-Martin selective medium, Difco Laboratories, Detroit; Baltimore Biological Laboratory, Cockeysville, Md. (available also with vancomycin, colistin, and nystatin).

medium (used in the primary culturing of wound specimens), the heat resistance of these spores ("heat shock" method) may be applied in the following manner:

1. Inoculate a fresh tube of thioglycollate medium[1] with 0.1 ml. of the original culture or specimen containing the suspected species of *Clostridium.*
2. Heat this tube at approximately 80° C. for 15 to 30 minutes. This will kill all vegetative forms but will not destroy spores, if present.
3. Incubate the heated culture for 24 to 48 hours.
4. Examine daily by making gram-stained smears: if gram-positive sporulating bacilli are found, proceed to step 5.
5. Inoculate two blood agar plates, incubating one aerobically and one anaerobically.

Clostridium perfringens rarely can be recovered successfully by this heating procedure, however, since it fails to sporulate in most media. A method that will aid in the successful isolation of this organism from mixed cultures is as follows:

1. Incubate the thioglycollate medium containing the original mixed culture for a minimum of 48 to 72 hours.
2. Streak several blood agar plates serially from this culture and incubate aerobically and anaerobically.
3. It will be found occasionally that the plates are overgrown with gram-negative organisms. In most cases, however, these will be fewer in number, and isolated colonies of suspected *Clostridium perfringens* will be readily obtained.
4. Transfer such colonies to tubes of thioglycollate medium for subsequent pure culture study.

Other methods for the isolation of pure cultures, based on physical and chemical alterations, may be used. For example, variation of incubation temperatures, changes in the relative acidity or alkalinity of the medium, or variations in oxygen tension or gas concentration may be employed for the isolation of specific microorganisms. Isolation of certain pathogenic bacteria from clinical material frequently may be accomplished more readily by animal inoculation. The techniques of these various methods will be more fully described in subsequent chapters. For further information, however, the reader may refer to the manual[2] included in the references.

REFERENCES

1. Brewer, J. H.: Clear liquid mediums for the aerobic cultivation of anaerobes, J.A.M.A. **115:**598-600, 1940.
2. Blair, J. E., Lennette, E. H., and Truant, J. P., editors: Manual of clinical microbiology, Bethesda, Md., 1970, American Society for Microbiology.

PART TWO

Recommended procedures with clinical specimens

5 Collection and handling of specimens for microbiological examination

Generally, a report from the bacteriological laboratory can indicate only what has been found by microscopic and cultural examination. An etiological diagnosis is thus confirmed or denied. Failure to isolate the causative organism, however, is not necessarily the fault of inadequate technical methods—it is frequently the result of faulty collecting technique. On a busy hospital ward the collection of specimens is too often relegated to personnel who have neither the knowledge or understanding of the requirements and consequences of such procedures. The microbiologist may also deserve criticism on occasion. Negligence on his part to provide adequate supplies or proper instructions may well result in poorly collected samples. The following are **general considerations** regarding the collection of material for culture. Specific instructions for the handling of a variety of specimens will be given in subsequent chapters.

Whenever possible, specimens should be obtained **before antibiotics or other antimicrobial agents have been administered.** Often, a purulent cerebrospinal fluid will reveal no bacterial pathogens on smear or culture when an antibiotic has been given within the previous 24 hours. A patient with salmonellosis will invariably show a negative stool culture if the specimen has been collected while the patient was receiving suppressive antibacterial therapy, only to reveal a positive culture several days after therapy has been terminated. If the culture has been taken after initiation of antibacterial therapy, the laboratory should be informed, so that specific counteractive measures, such as adding penicillinase or merely diluting the specimen, may be carried out.

It is axiomatic that material should be collected where the suspected organism is **most likely to be found** and with as little external contamination as possible. This is particularly true of draining lesions containing coagulase-positive staphylococci. Not infrequently the phage type of the primary infecting strain is found to be different from that of a surface-contact strain that may grow out.

Another factor contributing to the successful isolation of the causative agent is the **stage of the disease** at which the specimen is collected for culture. Enteric pathogens are present in much greater numbers during the **acute** or diarrheal stage of intestinal infections, and they are more likely to be isolated at that time. Viruses responsible for causing meningoencephalitis are isolated from cerebrospinal fluid with greater frequency when the fluid is obtained during the **onset** of the disease rather than at a time when the symptoms of acute illness have subsided.

There are occasions when the patient must contribute actively in the collection of a specimen, such as with a morning sputum sample. He should have full instructions, and his cooperation should be encouraged by the ward attendant. Too often, a container is placed on the patient's bedside table, with the only instructions being to "spit in this cup when you cough anything up."

Specimens should be of a **quantity sufficient** to permit complete examination and should be placed in sterile containers that preclude subsequent contamination of patient, nurse, or ward messenger. A serious danger to the laboratory worker, as well as to all others involved, is the soiled outer surface of a sputum container or a leaking stool sample. The hazard of spread of disease by inadequately trained nonprofessional workers is frequently overlooked. Its control requires continued education and constant vigilance by those in responsible and supervisory positions.

Provision must be made for the **prompt delivery** of specimens to the laboratory if the subsequent results of analysis are to have etiological validity. It is difficult, for example, to isolate *Shigella* from a fecal specimen that has remained on the hospital ward too long, permitting overgrowth by commensal organisms and an increasing death rate of the shigellae. In some instances it may be necessary to take culture media and other equipment to the patient's bedside to ensure prompt inoculation of the specimen. This is, of course, an unusual circumstance and requires prior arrangements with the laboratory.

Although not a function of specimen collection, it is an essential prerequisite that **the laboratory be given sufficient clinical information** to guide the microbiologist in his selection of suitable media and appropriate techniques. Likewise, it is important for the clinician to appreciate the **limitations and potentials** of the bacteriological laboratory—to realize that a negative report does not exclude the correctness of his diagnosis. As so aptly noted by Joan Stokes: "It is illogical to use a test for diagnosis of a particular infection and **at the same time** use the clinical findings to gauge its reliability."[9] It is essential that close cooperation and frequent consultation between the clinician, nurse, and microbiologist be the rule rather than the exception.

SPECIMEN CONTAINERS AND THEIR TRANSPORT

In the microbiological examination of clinical specimens it is essential that the container bearing the specimen does not contribute its own microbial flora. Furthermore, the original flora should neither multiply nor decrease because of prolonged incubation on the ward or prolonged refrigeration in the laboratory refrigerator. In other words, **a sterile container should be used and the specimen be plated as soon as possible.** Although these are not hard-and-fast rules, any deviation should be the responsibility of the microbiologist.

A variety of containers* have been devised for collecting bacteriological specimens. Many of these can be used repeatedly, after proper sterilization and cleaning, whereas others must be incinerated after use. Apart from the sterile Pyrex glass Petri dish and its modern counterpart, the presterilized and disposable plastic dish, the most useful and most used piece of collecting equipment is the cotton-, calcium alginate–, or polyester-tipped wooden applicator stick. These swabs are best prepared by autoclaving the wooden applicator sticks in Sorensen buffer, pH 7.2, for 5 minutes, drying them, and then tipping them with polyester batting,† calcium alginate,‡ or long-fibered medicinal cotton. One combination packaged as a sterile outfit consists of a Pyrex test tube (20 by 150 mm.), either cotton plugged or with a stainless-steel or disposable plastic cap,§ containing the ap-

---

*Falcon Plastics, Division of BioQUEST, Oxnard, Calif.
†Dacron polyester filling, half-pound bags, Sears, Roebuck and Co.
‡Colab Laboratories, Inc., Chicago Heights, Ill.
§Morton culture tube closure from Scientific Products, Division of American Hospital Supply Corp., Evanston, Ill.

plicator stick, and also a small tube (10 by 75 mm.) with several drops of thioglycollate broth (not needed with a polyester tip). This may be used for the collection of material from the throat, nose, eye, ear, from wound and operative sites, from urogenital orifices, and from the rectum, but is not recommended for optimal recovery of anaerobes (see below). The swab, inoculated with material from the patient, is placed in the inner broth tube to prevent drying out, and the whole outfit is properly labeled and promptly sent to the laboratory.

A recent innovation is a sterile disposable culture unit (Culturette)* consisting of a plastic tube containing a sterile polyester-tipped swab and a small glass ampule of modified[1] Stuart's holding medium. The unit is removed from its sterile envelope, and the swab is used to collect the specimen. It is then returned to the tube, the ampule is crushed, and the swab is forced into the released holding medium. This will provide sufficient moisture for storage up to 72 hours at room temperature.

Certain workers[7] believe that some cotton used for applicators may contain fatty acids which may be detrimental to microbial growth. An excellent substitute may be found in calcium alginate wool, derived from alginic acid, a natural plant product.[4] This silky-fibered material dissolves readily in certain solutions that are compatible with bacterial preservation (such as dilute Ringer solution with sodium hexametaphosphate) to form a soluble sodium alginate. Calcium alginate–tipped wooden applicators or flexible aluminum nasopharyngeal swabs, as well as the citrate or hexametaphosphate diluents are available commercially.†

A modification of the cotton applicator is the substitution of 28-gauge Nichrome or thin aluminum wire in place of the

wooden stick. Because of its flexibility, the wire applicator is recommended for collecting specimens from the nasopharynx. In order to ensure that the small amount of cotton will adhere to the wire during passage through the nasal tract, the end of the wire must be bent over and some collodion applied before the wrapping with cotton; otherwise, the outfit remains the same as that previously described. A commercially prepared wire swab also is available.*

A variety of transport media have been devised for prolonging the survival of microorganisms when a significant delay occurs between collection and culturing. Stuart and others[10,11] advocated a medium that has proved effective in preserving the viability of pathogenic agents in clinical material. The medium consists of buffered semisolid agar devoid of nutrients and containing sodium thioglycollate as a reducing agent and is used in conjunction with cotton swabs. Stuart's medium† maintains a favorable pH and prevents both dehydration of secretions during transport as well as oxidation and enzymatic self-destruction of the pathogen present.

A preliminary report on a new transport medium* by Cary and Blair[3] indicates that with initial fecal specimens, salmonellae and shigellae can be recovered for as long as 49 days, Vibrio (cholerae) comma for 22 days, and Pasteurella (Yersinia) pestis for at least 75 days.

The use of polyester-tipped swabs for delayed recovery of Group A streptococci from throat cultures will be discussed in Chapter 8. A report by Hosty and co-workers[6] has indicated that the incorporation of a small amount of silica gel in the glass tube containing the polyester swab will further maintain the viability of Group A streptococci in throat swabs for as long as 3 days before plating.

*Culturette from Scientific Products, Division of American Hospital Supply Corp., Evanston, Ill.
†Colab Laboratories, Inc., Chicago Heights, Ill.

*Colab Laboratories, Inc., Chicago Heights, Ill.; Falcon No. 2047, 2050, Oxnard, Calif.
†Baltimore Biological Laboratory, Cockeysville, Md.

The collection of specimens for anaerobic culturing poses a special problem in that the conventional methods previously described will not lead to optimal recovery of these air-intolerant microorganisms. Finegold and associates[12] have so aptly pointed out that a crucial factor in the final success of anaerobic culturing is the transport of clinical specimens: the lethal effect of atmospheric oxygen must be nullified until the specimen has been processed anaerobically in the laboratory. These workers recommend the use of a double-stoppered collection tube, gassed out with oxygen-free CO_2 or N_2*; the specimen (pus, body fluid or other liquid material) is injected through the rubber stopper while avoiding the introduction of air. In the laboratory, the reverse procedure is carried out and the material inoculated to prereduced media and incubated under anaerobic conditions. If only a swab specimen can be obtained, it should be collected on a swab which has been prepared in a "gassed out" tube,* and then transferred to a rubber-stoppered tube half filled with prereduced and anaerobically sterilized transport medium, such as the Carey and Blair medium previously described. Further details in anaerobic culture techniques are described in Chapters 23 and 24.

Containers for special purposes, such as those used for the collection of 24-hour urine samples, can usually be devised from equipment found around the hospital; or they may be purchased from a surgical supply house.

Although most pathogenic microorganisms are not greatly affected by small changes in temperature, they are generally susceptible to drying out, particularly when on cotton applicator sticks. However, some bacteria, such as the meningococcus in cerebrospinal fluid, are quite sensitive to low temperatures and require immediate culturing. On the other hand,

clinical material likely to contain abundant microbial flora may in most instances be held at 5° C. in a refrigerator for several hours before culturing if it cannot be processed right away. This is particularly true with such specimens as urine, feces, and sputum samples and material on swab sticks taken from a variety of sources, with the exception of wound cultures that may contain oxygen-sensitive anaerobes. These are best kept at room temperature until inoculation. Not only will refrigeration preserve the viability of most pathogens, it will also prevent overgrowth of commensal organisms, increased numbers of which could make the isolation of a significant microbe more difficult. Objectively, however, the sooner a parasite leaving the sheltered environment of its host is transferred to an appropriate artificial culture medium the better are the chances of its survival and subsequent multiplication.

There may be occasions when it is necessary to submit specimens to a reference laboratory in a distant city, thus requiring transportation by mail or express. If, for example, there is no viral diagnostic service available in the immediate area, it may become necessary to ship specimens in the frozen state to preserve their viability. This is especially true of virus-containing material, such as cerebrospinal fluid, throat and rectal swabs, stools, and tissue, which should be frozen immediately and shipped in Dry Ice (obtained from a local ice cream plant). Whole blood is **not** frozen. Rather, the serum is separated and sent in a sterile tube. Freezing and thawing of blood would result in extensive hemolysis, rendering the serum unsatisfactory for serological testing.

On the other hand, if a culture slant of an isolated organism is to be sent to a reference laboratory (state or public health laboratory), it is not necessary to send it in the frozen state; in fact, this could be harmful to the culture for reasons previously cited.

In most instances microbiological specimens can be satisfactorily shipped

---

*Available commercially from Scott Laboratories, Inc., Fiskeville, R. I.

through the mails, provided that **special precautions** are taken against breakage and subsequent contamination of the mailing container. Postal regulations require that cultures on solid media be transported in a stout glass container closed with a leak-proof rubber or paraffined cork stopper. This glass container is then completely wrapped in absorbent cotton or other suitable absorbent packing material and placed in a cylindrical sheet metal box with soldered joints and a metal screw cover with a rubber or felt washer.* The metal container is enclosed in a closely fitting (to avoid shifting) wooden or papier-mâché box or tube of sufficient strength to resist rough handling in the mail. A label affixed to the outside of the container should carry these words: "Specimen for bacteriological examination; this package shall be packaged with letter mail."†

Double mailing containers should always be used when the microbiologist considers the specimen to be of a hazardous nature and when he thinks it may constitute a definite danger to the person handling the container. This includes clinical specimens to be examined for the presence of tubercle bacilli, enteric pathogens, respiratory pathogens, cultures for identification, and so forth.

For the shipment of fecal specimens containing salmonellae or shigellae over long distances, the filter paper method[2] may also be employed. In collecting the specimen by this technique, fresh fecal material must be spread fairly thinly over a strip of filter paper or clean blotting paper and allowed to dry at room temperature. Using forceps, the smeared strip is then folded inward from the ends in such a way that the fecal material is covered. The folded specimen may then be inserted

in a plastic envelope (polyethylene is recommended) and placed in a container to conform with postal regulations. It is possible to ship a large number of such specimens in one container via the mail. The pathogens are unaffected by this treatment, whereas the normal intestinal flora dies off. On receipt at the laboratory the paper specimen may be cut into three pieces. One piece is placed in physiological saline for suspension and direct plating, and the remaining pieces are placed in selenite and tetrathionate broths, respectively, for enrichment and subsequent plating.

When it is necessary to ship fecal specimens in the unfrozen state, or when such specimens must be held for some time before culturing, it is recommended that they be placed in a preservative solution. The buffered glycerol-saline solution of Sachs[8] has proved satisfactory. (Its final pH should be 7.4; it should be discarded if it becomes acid.) Approximately 1 gm. of feces is emulsified in not more than 10 ml. of preservative, and the preserved specimen is shipped in a heavy glass container (universal type) with a screw cap and contained in the regular double mailing container just described.

HANDLING OF SPECIMENS IN THE LABORATORY

In previous sections the importance of a properly collected specimen was stressed and the responsibility of personnel in its collection was indicated. In the following section the subsequent handling of the specimen in the laboratory will be considered. In addition, the responsibility of the laboratory worker will be pointed out.

Since it is not always practical for many specimens to be inoculated as soon as they arrive in the laboratory, refrigeration at 4° to 6° C. offers a safe and dependable method of storing many clinical samples until they can be conveniently handled. However, some may require immediate plating, whereas others must be immediately frozen.

---

*Whole assembly (interior metal container and cardboard mailing tube), Arthur H. Thomas Co., Philadelphia.

†Information from the U. S. Postal Service indicates that specimens must have first class postage affixed in order to be packed with letter mail.

The length of time of refrigeration varies with the type of specimen: swabs from wounds (except for anaerobic cultures), the urogenital tract, throat, and rectum and samples of feces or sputum can be refrigerated for 2 to 3 hours without appreciable loss of pathogens. Urine specimens for culture may be refrigerated for 24 to 48 hours without affecting the bacterial flora (except the tubercle bacillus, which may be adversely affected by the urine); on the other hand, cloudy cerebrospinal fluid from a patient with purulent meningitis should be examined **at once.**

Specimens submitted for the isolation of virus should be frozen immediately; this is particularly true of spinal fluid, where only a few cells may be present in which viruses grow and remain viable. Specimens of clotted blood for virus serology may be refrigerated but, for the reasons given previously, never frozen.

Gastric washings and resected lung tissue submitted for culture of *Mycobacterium tuberculosis* should be processed soon after delivery, since tubercle bacilli may die rapidly in either type of specimen.

Pieces of hair or scrapings from the skin and nails submitted for the isolation of fungi may be kept at room temperature and protected from dust for several days before inoculation. On the other hand, sputum, bronchial secretions, bone marrow, and purulent material from patients suspected of having systemic fungal infection should be inoculated to appropriate media as soon as possible, especially when a diagnosis of histoplasmosis is considered.

REFERENCES

1. Amies, C. R.: A modified formula for the preparation of Stuart's transport medium, Canad. J. Public Health, **58:**296-300, 1967.
2. Bailey, W. R., and Bynoe, E. T.: The "filter paper" method for collecting and transporting stools to the laboratory for enteric bacteriological examination, Canad. J. Public Health **44:**468-475, 1953.
3. Cary, S. G., and Blair, E. B.: New transport medium for shipment of clinical specimens, J. Bact. **88:**96-98, 1964.
4. Forney, J. E., editor: Collection, handling, and shipment of microbiological specimens, Public Health Serv. Public. No. 976, Washington, D. C., Nov. 1968, U. S. Government Printing Office.
5. Higgins, M.: A comparison of the recovery rate of organisms from cotton-wool and calcium alginate wool swabs, Ministry Health Public Lab. Serv. Bull. 43, 1950.
6. Hosty, T. S., Johnson, M. B., Freear, M. A., Gaddy, R. E., and Hunter, F. R.: Evaluation of the efficiency of four different types of swabs in the recovery of group A streptococci, Health Lab. Sci. **1:**163-169, 1964.
7. Pollock, M. R.: Unsaturated fatty acids in cotton plugs, Nature **161:**853, 1948.
8. Sachs, A.: Difficulties associated with bacteriological diagnosis of bacillary dysentery, J. Roy. Army Med. Corps **73:**235-239, 1939.
9. Stokes, E. J.: Clinical bacteriology, London, 1960, Edward Arnold (Publishers) Ltd.
10. Stuart, R. D.: The diagnosis and control of gonorrhea by bacteriological cultures, Glasgow Med. J. **27:**131-142, 1946.
11. Stuart, R. D., Tosach, S. R., and Patsula, T. M.: The problem of transport of specimens for culture of gonococci, Canad. J. Public Health **45:**73-83, 1954.
12. Sutter, V. L., Attebery, H. R., Rosenblatt, J. E., Bricknell, K. S., and Finegold, S. M.: Anaerobic bacteriology manual, Los Angeles, 1972, Dept. Contin. Educ. Health Sciences, School of Med., UCLA.

PART THREE

Cultivation of pathogenic microorganisms from clinical material

6 Microorganisms encountered in the blood

The organisms most likely to be found in blood cultures include the following:

Staphylococci (coagulase positive, coagulase negative)
Coliform bacilli and related enteric organisms
Serratia marcescens
Alpha and beta hemolytic streptococci
Pneumococci
Enterococci
Haemophilus influenzae
Clostridium perfringens and related organisms
Pseudomonas species
Proteus species
Bacteriodes species and related anaerobes
Anaerobic cocci
Neisseria meningitidis
Salmonella species
Brucella species
Francisella tularensis
Listeria monocytogenes
Acinetobacter calcoaceticus (Herellea vaginicola)
Streptobacillus moniliformis
Leptospira species
Vibrio fetus and related vibrios
Opportunistic fungi such as *Candida, Torulopsis,* and *Nocardia* species, *Blastomyces dermatitidis, Histoplasma capsulatum*

BLOOD CULTURE—GENERAL CONSIDERATIONS

Blood for culture is probably the most important single specimen submitted to the microbiology laboratory for examination. The presence of living microorganisms in the patient's blood almost always reflects active and possibly spreading infection in the tissues. The prognosis of such a bacteremia or septicemia may well depend on its prompt recognition by bacteriological means. Subsequent initiation of specific therapy based upon laboratory findings may well prove to be lifesaving. Likewise, a negative culture would prove helpful in ruling out microbial etiology.

It is important to know which infections are likely to show a positive blood culture, and thus further reading on bacteremias is recommended.[7,23] A **transient bacteremia** frequently occurs during the course of many diseases, including pneumococcal pneumonia, bacterial meningitis, urinary tract infection, typhoid fever, and generalized salmonella infections. Wound infections caused by a beta hemolytic streptococcus, by *Staphylococcus aureus,* and by *Bacteroides* species, infections of the gallbladder and biliary tract, osteomeylitis, peritonitis, and puerperal sepsis show a transient positive blood culture, which is also a frequent concomitant of operative manipulation in chronically infected areas, as in instrumentation of the urinary tract. It is generally unrewarding, however, to attempt to isolate the causative organism by culturing the blood in cases of tetanus, diphtheria, shigellosis, or tuberculosis.

Although a mild transitory bacteremia is a frequent finding in many infectious diseases, the persistent, continuous, or recurrent type of bacteremia, when present, is indicative of a more serious condition. When the classical syndrome of a **septicemia** (chills, fever, prostration) due to the presence of actively multiplying bacteria and their toxins in the bloodstream, is found, one rarely encounters difficulty in isolating the causative organism. However, when localizing signs of infection are notably absent and a persistent positive blood culture is demonstrated, the possibility of a serious intravalvular infection, such as bacterial endocarditis, must be considered.

In some diseases the probability of obtaining a positive blood culture depends on the **stage of the disease** at which the culture is made. For example, bacteria can be cultured from the blood during the first several days of illness in typhoid fever,* tularemia, plague, and anthrax, but the organisms disappear from the blood during the later course of the disease. Thus, the early recovery of these bacteria from the blood is important, since it may be the only reliable means of making a diagnosis available to the clinician.

Perhaps the most challenging systemic diseases from the clinical and bacteriological points of view are **subacute bacterial endocarditis** (SBE), and **brucellosis.** Confirmation of their diagnoses is almost entirely dependent on the isolation of the organisms from the blood. Frequently, the causative bacteria are present only in small numbers. At other times organisms may appear in large numbers as "showers" to be cleared from the circulation by natural filtering mechanisms, only to reappear several hours later.

Futhermore, isolation of the causative organism may be complicated by the fact that the patient received prior antibiotic therapy. This may not have been at a drug level high enough to eradicate the microorganism but one that, when carried over in the blood sample, may prove sufficient to inhibit growth in the blood culture bottle.† For this reason it is recommended that a **minimal dilution** of 1 volume of

blood in 10 volumes of broth be made as a safeguard against such an eventuality. This dilution also aids in counteracting the bactericidal effect of normal serum.

Since there is usually a lag period of 1 to 2 hours between the time of entrance of the bacteria into the circulation and the subsequent chill,[7] blood for cultures should be drawn, ideally, **before** the expected temperature rise. The frequency with which blood samples are taken for culture is also important. In patients with suspected subacute bacterial endocarditis who have not received any antibacterial agents, a total of three to four cultures (samples being taken at convenient intervals during a 24- to 48-hour period) should be adequate to establish the diagnosis in most cases. This method is to be preferred to the procedure of taking culture material at 2-hour intervals around the clock.

Austrian[5] believes that after appropriate antibacterial therapy, it is incumbent on the attending physician to obtain a negative blood culture in order to document the absence of bacteremia. In patients with subacute bacterial endocarditis he recommends a blood culture at the termination of therapy and again 2 weeks later, the period during which those patients who will most likely suffer a bacteriological relapse are observed to do so.

In **brucellosis,** positive blood cultures are usually obtained during exacerbation of symptoms and elevation of temperature, although even during the acute phase, isolation of the organism is not regularly successful. Occasionally, culture of the bone marrow has yielded positive results when peripheral blood cultures have been sterile in cases of brucellosis, subacute bacterial endocarditis, histoplasmosis, and—after the first week—in typhoid fever. Routine bacterial and fungal cultures are recommended concurrently when **bone marrow biopsies** are performed for other diagnostic purposes.

A still more difficult problem for the bacteriologist is the generalized infection caused by a **rarely isolated** organism that

---

*If negative, repeat weekly for several weeks, if patient continues to be febrile.[20]

†Aberrant forms of bacteria may occur in blood cultures of patients receiving antibiotic therapy.[40]

requires special methods for its isolation. The causative agents of diseases such as leptospirosis, rat-bite fever, listeriosis, and infection by pasteurellae require complex culture media and, in some instances, animal injection for their isolation. Some of these special procedures will be discussed in a later section.

Blood cultures should be made **immediately** in all cases of unexplained postoperative shock, particularly in those following genitourinary tract manipulation. A single negative blood culture should never be depended on to eliminate the possibility of a bacteremia, because a single specimen may be sterile even though the bloodstream as a whole is infected. A minimum of three cultures taken at half-hour intervals is recommended. Blood cultures should be made on all patients with an **unexplained fever** of more than several days' duration. Chills and fever in patients with urinary tract infections, infected burns, rapidly progressing tissue infections, postoperative wound sepsis, or with indwelling venous catheters make them immediate candidates for having blood cultures taken. Debilitated patients who develop chills and fever while undergoing prolonged therapy with antibiotics, corticosteroids, immunosupressives, antimetabolites, or parenteral hyperalimentation also should have blood drawn for culture.

The diagnosis of a bacteremia can be made only by growing out pathogenic agents from culture media suitably inoculated with adequate amounts of the patient's blood. In the bacteriological examination of the blood, several important principles must be considered:

1. An acceptable blood collection technique, which minimizes chances of contamination, is mandatory.
2. Enriched and osmotically stabilized **aerobic** and **anaerobic** culture media must be utilized, and conditions of incubation must be carried out to provide optimal conditions for bacterial growth.
3. Results of presumptively positive

findings must be reported **without delay** to the attending physician.

ROLE OF ANTICOAGULANTS AND OSMOTIC STABILIZERS IN BLOOD CULTURING

As Rosner[45] so aptly observed, the most important attribute of an effective blood culture system is its ability to support rapid growth of a wide variety of microorganisms so that their prompt recovery and identification can be accomplished.

One of these elements would be a means of preventing the blood sample from clotting in the culture bottle, since bacteria become entrapped in clotted blood.[64] Various anticoagulants have been recommended; the most effective one to date appears to be sodium polyanethol sulfonate (Liquoid), a synthetic polyanionic agent. First reported by Von Haebler and Miles[58] in 1938 as a useful anticoagulant in blood culturing, it is now apparent that Liquoid* in a concentration of 0.03% will inhibit clotting, neutralize the bactericidal effect of human serum,[42] prevent phagocytosis,[46] and inactivate certain antibiotics, such as streptomycin, kanamycin, gentamicin, and polymyxin B.[55,56] Furthermore, Liquoid is stable at autoclave temperatures. It is not recommended, however, in agar pour plate cultures or membrane filter culture systems,[25] and may be inhibitory to some strains of streptococci, neisseriae, and *Haemophilus influenzae.*

Sodium citrate in a concentration of 0.5% to 1.0% is another commonly used anticoagulant. Some workers feel it may be inhibitory to gram-positive cocci, but Huddleson's work[36] indicates that this effect has been noted only at a pH near 6.0. Our own unpublished results of blood culturing, using sodium citrate in a 0.5% concentration in tryptic broth, showed an increased recovery of gram-positive cocci in this medium over the recovery of gram-positive cocci in thioglycollate medium

---

*Available in a 5% sterile aqueous solution as Grobax from Roche Diagnostics, Nutley, N. J.

with Liquoid. Perhaps optimal results could be obtained by utilizing paired culture flasks containing citrate or Liquoid in appropriate concentration.

Another factor in the successful recovery of small numbers of circulating microorganisms in the blood is the use of an osmotic stabilizer, such as 10% to 30% sucrose.[46] This is especially true of gram-negative rod bacteremias in patients receiving antibiotic therapy, where low levels of bacteria (10 colonies per milliliter or less) can be expected.[24] The protective effect of the sucrose on organisms having undergone cell wall damage (L forms) may be in preventing osmotic pressure changes,[46] thus creating a more favorable environment for cell growth. Most microorganisms with intact cell walls also will grow well in enriched hypertonic media.[23]

In summary, it seems that Liquoid protects the bacterial cell from antagonistic factors, while the sucrose protects it from osmotic pressure changes, permitting rapid recovery and growth. Each agent, incorporated separately or together in broth culture bottles, is strongly recommended.

BLOOD CULTURE PROCEDURES

When blood for culture is drawn, scrupulous care must be exercised in preparing the site for venipuncture. A 2% solution of tincture of iodine is applied to the skin over the area of the selected vein by means of several saturated cotton-tipped applicator sticks. The iodine is allowed to dry and then is removed with sponges saturated with 80% isopropyl alcohol. The entire process is repeated, a tourniquet is applied, and 10 to 12 ml. of blood is withdrawn by a sterile 21-gauge needle on a 20-ml. syringe. The blood is transferred immediately to a small sterile tube or bottle fitted with a rubber stopper and containing 3.4 ml. of 0.35% Liquoid in sterile physiologic saline.* It is then

mixed well and promptly transported to the laboratory, where it is inoculated to media as soon as possible. If it is necessary to hold the specimen for a short interval, the blood may be refrigerated for not more than 1 to 2 hours. For culture, about 5 ml. of the blood is transferred to a bottle containing 100 ml. of thioglycollate medium, and the remainder is distributed in a bottle containing 100 ml. of trypticase soy broth or tryptic soy broth, using aseptic technique throughout. The practice of making agar pour plates for colony counts, and as a means of recognizing contamination, is of questionable value and is not recommended as a routine procedure. A few bacteria in the bloodstream have the same clinical significance as do a large number, and it does not always follow that bacteria originally present in the blood will grow out in the pour plate as well as in the broth flask, although some workers report a greater recovery of streptococci by this method.

A method that has for a number of years been used successfully in the laboratory of the Delaware Division, Wilmington Medical Center, is also recommended.[48] By use of a modified Castañeda[13] bottle containing an agar slant and 75 ml. of trypticase soy broth and a similar bottle containing thioglycollate medium with 10% sucrose and Liquoid, 10 ml. of blood is distributed between the two vessels. (See p. 381 for preparation.) The partial vacuum in the bottles facilitates the inoculation. Similar blood culture outfits* recently have been introduced commercially and have proved satisfactory for use in small laboratories with limited personnel.

---

*B-D Vacutainer tube No. S3208 XF308, Becton, Dickinson & Co., Rutherford, N. J.

---

*Among sources of supply are the following: BBL-BioQuest, Cockeysville, Md.; Becton, Dickinson & Co., Rutherford, N.J.; Difco Laboratories, Detroit, Mich.; Lederle Laboratories, Pearl River, N. Y.; Hyland Laboratories, Los Angeles; Case Laboratories, Chicago; Pfizer Diagnostics, Flushing, N.Y.; and others. Many of these bottles contain added carbon dioxide and a vacuum, useful in isolating microaerophiles and capnophiles.

Generally, thioglycollate medium is satisfactory for the growth of most anaerobic organisms from the blood, especially if the bottle has been reheated and cooled prior to introduction of the blood sample and has been incubated with loosened cap in an anaerobic jar. However, the use of a commercially available prereduced medium* in a bottle with a gas-tight stopper that permits needle puncture is recommended as the most convenient and efficient system for recovery in bacteremia caused by anaerobes.[52] It should be noted that most so-called anaerobic media sold commercially are unable to support the growth of strict anaerobes because they have only part of the atmosphere replaced with carbon dioxide.

Since conventional methods for recovering microorganisms from the blood may require several days for growth and identification, a rapid method using membrane filters was introduced. This procedure proved cumbersome and time-consuming and has recently been modified[50] by lysing the blood instead of agglomerating the red cells prior to filtration. This technique requires less time (5 minutes) and a smaller blood sample (10 ml.), making it practical for routine clinical usage.

If the patient is receiving penicillin therapy at the time the blood is drawn for culture, the concentration of the antibiotic in the blood may be high enough to inhibit the growth of susceptible bacteria in the sample. In order to recover these organisms, it may be necessary to add penicillinase,† an enzyme produced by certain *Bacillus* species, to the blood culture, thereby nullifying the effect of the penicillin. Some workers, however, do not recommend the addition of penicillinase to the culture on the theory that it may increase the chance of contamination. Furthermore, if the blood is sufficiently diluted in the broth medium (1:10 to 1:30), it is thought that the inhibitory effect of the penicillin (as well as complement, natural antibody, and other antibacterial components of serum) is reduced to a level of no consequence.[4]

Hamburger and co-workers[31] have reported that magnesium ion will antagonize the antistaphylococcal action of tetracycline and recommended the inclusion of 100 mg. of magnesium sulfate per milliliter of blood in the blood cultures of patients receiving tetracycline antibiotics. The addition of cephalosporinase also has been recommended in treating blood from patients receiving cephalothin.[15] As previously noted, it has been reported that the addition of Liquoid to blood broth cultures will effectively neutralize the activity of streptomycin, kanamycin, gentamicin, and polymyxin.

By the use of the aforementioned techniques and media, most of the common pathogens can be readily cultivated. The latter include the aerobic and anaerobic streptococci, staphylococci, pneumococci, meningococci, coliform bacilli, and members of the genera *Salmonella, Haemophilus, Bacteroides, Clostridium, Acinetobacter, Pseudomonas, Alcaligenes,* and *Candida.* **Special methods** or special media are necessary for the isolation of *Actinomyces, Brucella, Pasteurella, Leptospira, Listeria, Vibrio,* and *Spirillum.* These will be described in subsequent sections.

Incubation and examination of cultures

Blood cultures are incubated aerobically at 35° to 37° C. (except where otherwise indicated) and are examined daily throughout the first week and several times a week thereafter for evidence of growth. When a culture is **positive,** the **broth** may assume one of several appearances. For one type of growth characteristic of gram-negative rods, the medium above the red cell layer becomes uniform-

---

*Brain Heart Infustion supplemented (with or without Liquoid), prereduced and anaerobically sterilized, Scott Laboratories, Fiskeville, R. I.: also from McGaw Laboratories, Glendale, Calif.
†Baltimore Biological Laboratory, Cockeysville, Md.; Difco Laboratories, Detroit.

ly turbid, and gas bubbles caused by fermentation of glucose may be present. When pneumococci or meningococci are present, a similar although less distinct turbidity is produced, usually accompanied by a greenish tint in the medium. Initially, streptococci may grow only in thioglycollate medium, appearing as "cotton ball" colonies on top of the sedimented red cells, whereas the upper layer of broth may remain clear. This commonly occurs in blood cultures from patients with subacute bacterial endocarditis caused by microaerophilic alpha hemolytic streptococci. Careful observation is thus all-important in this suspected diagnosis. If beta hemolytic streptococci or other hemolytic organisms are present, marked hemolysis of the blood may be observed, along with turbidity. Pathogenic species of staphylococci will grow out readily, and they frequently produce a large jellylike coagulum throughout the broth due to coagulase production. Grossly visible *Staphylococcus* colonies will also develop in the medium and impart turbidity. If the culture is contaminated with *Bacillus* species or a saphrophytic fungus, the broth becomes hemolyzed and a fairly thick pellicle forms on its surface. *Clostridium* species grow readily in the thioglycollate medium, producing hemolysis, an unpleasant odor, and gas under pressure. *Bacteroides* species produce less gas, but a foul odor, in this medium. Growth of *Haemophilus influenzae* does not generally produce any change in the medium and may be detected only by subculturing to chocolate agar.

In the **agar slant** bottle, bacterial growth generally appears either as large or small discrete colonies or as confluent growth on the slant, with cloudiness in the broth. Colonies of pneumococci are small, translucent, and difficult to recognize, as are colonies of Group A streptococci. Staphylococcus colonies, on the other hand, are quite large, opaque, and easily seen, as are those of coliform bacilli, yeasts, and so forth. Growth of clostridia generally appears between the glass side wall and the

agar slant and produces large bubbles of gas. *Proteus* and *Pseudomonas** colonies generally appear as a translucent film on the agar slant. Otherwise, growth is similar to that previously described.

A **negative** blood culture usually remains clear, but it may develop cloudiness with continued incubation, sometimes caused by shaking the bottle before making smears, which in this instance will show no bacteria. After prolonged incubation (10 days or more) the thioglycollate bottle may develop a turbidity from fibrin or agar particles. This occurs usually above the red cell layer, and it may suggest bacterial growth. In either case turbidity alone is not a reliable guide; a gram-stained smear should be examined to determine the presence of microorganisms.

When growth appears on **any** medium, the culture should be inoculated to a blood agar plate and a tube of thioglycollate broth, and a gram-stained smear should be examined. If gram-negative rods are seen in the original bottles, an eosin–methylene blue (EMB) or a MacConkey agar plate should be streaked also. These will permit rapid preliminary identification if the organism is a coliform. The gram-negative rod may prove, however, to be *Haemophilus influenzae,* in which case it will not grow out on regular blood agar or on EMB agar, necessitating the use of chocolate agar. Also, *Bacteroides* species may be present, which require **anaerobic** conditions for isolation. In the interest of rapid diagnosis, therefore, it is expedient to use a sufficient variety of media for subculturing a positive blood culture.

Failure to hold cultures long enough or to examine them frequently enough may result in a delay in recognizing the presence of pathogenic microorganisms or may even result in their loss of viability. Frequent openings of the culture bottles,

---

*Pseudomonas and yeasts may require 3 to 4 days to grow out and are best recovered in an **air-vented** bottle.

on the other hand, may also result in contaminating the contents. The following schedule is offered as a desirable **routine** for the examination of regular blood cultures. Special methods for handling the less common pathogen will be described subsequently.

Broth culture procedure

1. After 18 to 24 hours' incubation, shake the flask well and subculture a loopful of the blood broth to a blood agar plate and to a chocolate agar plate. At the same time prepare a thin smear of the blood broth and examine by Gram stain. Reincubate the blood culture.

2. Incubate both plates in a candle jar and examine for growth in 24 and 48 hours. Discard thereafter if negative.

3. After 14 days' incubation* repeat the foregoing procedure. Discard the broth flask if the smear is negative and the plates show no growth after 48 hours.

Thioglycollate culture procedure

1. After 18 to 24 hours' incubation make a gram-stained smear and subculture to a blood agar plate and a tube of thioglycollate medium. Discard the plate and broth tube after 48 hours if there is no growth. Reincubate the thioglycollate bottle.

2. After 14 days' incubation repeat step 1. Discard the thioglycollate bottle if the smear is negative and the subcultures show no growth after 48 hours.

3. If the smear shows the presence of bacteria and the aerobic subculture is negative, inoculate blood agar and chocolate agar plates and incubate **anaerobically.** Also inoculate a fresh tube of thioglycollate medium from the original blood culture bottle. This

anaerobic incubation is often required for the primary isolation of microaerophilic streptococci as well as *Bacteroides* species and *Streptobacillus moniliformis.*

The most frequently isolated skin contaminants include microaerophilic diphtheroid bacilli, which usually grow out only in the thioglycollate medium. These organisms are strongly catalase-positive, nonmotile, gram-positive rods of characteristic morphology, which may be confused morphologically with *Listeria monocytogenes,* a distinct pathogen from which they must be differentiated. The reader is referred to Chapter 27 for a discussion of the biochemical reactions of the latter organism.

Conclusion

If all cultures, subcultures, and gram-stained smears are negative, the blood culture may be reported as **"No growth, aerobic or anaerobic, after 14 days' incubation."**

It is strongly recommended that a stock culture be made, on appropriate media,* of all significant organisms isolated from blood cultures. This is important for subsequent antibiotic susceptibility testing by the tube dilution technique or for more extensive studies, such as serological tests, animal inoculation, or referral to a diagnostic center.

PNEUMOCOCCAL BACTEREMIA

Pneumococcal bacteremia can be demonstrated in about 25% of adult patients with pneumococcal pneumonia,[6] less frequently in infants and children.[10] In many cases, the blood culture may be the **only** specimen from which pneumococci are recovered; it therefore provides the most useful and readily available procedure for confirming a clinical impression of bac-

---

*After reviewing his records of 40,000 blood cultures, Ellner[21] elected to discard cultures that showed no growth after 7 days, since none in the series became positive after 5 days.

*CTA medium, Baltimore Biological Laboratory, Cockeysville, Md., is recommended. The culture, after overnight incubation, may be stored in the refrigerator.

teremia or meningitis and should be obtained from any patient with a suspected pneumococcal infection.

Although the procedure previously outlined is a satisfactory one for isolating pneumococci from the blood, certain precautions should be observed in the handling of these cultures. After overnight incubation, gram-stained smears of the broth should be made and examined regardless of the appearance of the flask. The pneumococci, if present, will have grown out in this period, and longer incubation may result in a loss of the organism through autolysis in a fluid medium.* Pneumococci will appear as gram-positive cocci in chains resembling streptococci; some may show the typical lancet-shaped diplococcus with evidence of a capsule.

Whether or not organisms are seen in the smear of the blood broth, a blood agar streak plate should be made. This demonstrates the presence of pneumococci in numbers too small to be seen in the smear. The blood plate is incubated in a candle jar for 24 to 48 hours and then examined for characteristic translucent flat colonies surrounded by a zone of green discoloration. Further methods for the identification of the pneumococcus may be found in Chapter 18.

ISOLATION OF NEISSERIA AND HAEMOPHILUS

The isolation of *Neisseria meningitidis* and *Haemophilus influenzae* from the blood culture of all patients with purulent meningitis should be attempted. Through use of the procedures previously described, both organisms may be recovered after 18 to 24 hours' incubation† by subculturing from the broth bottle* to a **chocolate agar** plate and following with incu-

bation for 18 to 24 hours in a candle jar. If this subculture shows no growth, the procedure is repeated after incubating the original broth flask for an additional 5 to 10 days. Gram-stained smears of *Neisseria* and *Haemophilus* are difficult to interpret because of the presence of gram-negative debris and a heavy red background. Consequently it is a good practice to subculture the broth and thioglycollate flasks routinely when these organisms are suspected, regardless of the appearance of the smears.

Attention is called to a *Haemophilus* species, *H. aprophilus,* which is rarely isolated from blood cultures of patients with bacterial endocarditis.[63] The organism is a gram-negative, nonmotile rod that does not require V factor but does require X factor and an increased carbon dioxide tension for its growth.

SEPTICEMIA AND BACTEREMIC SHOCK CAUSED BY GRAM-NEGATIVE RODS

The association between severe and frequently fatal septicemia caused by gram-negative enteric bacilli, and confinement in the hospital is becoming increasingly apparent.[18,19] In the early antibiotic era the primary concern was for sepsis caused by *Staphylococcus aureus;* in the past decade, however, many institutions are reporting a great increase in infections caused by *E. coli* and members of the *Klebsiella-Enterobacter-Serratia* group and *Proteus* and *Pseudomonas* species.[54]

Factors that have played a primary role in initiating these infections include the following:

1. Those compromising the resistance of the host, such as the use of adrenal corticosteroids, immunosuppressive agents, and antimetabolite drugs
2. The widespread utilization of broad-spectrum antimicrobial agents, which may upset the endogenous intestinal flora, with emergence of resistant strains
3. Procedures producing new artificial avenues of infections, such as the use

---

*Pneumococci, neisseriae, and *Haemophilus* do not always grow in the thioglycollate medium. This medium is recommended, however, for the possible isolation of other etiological agents of pneumonia and bacterial meningitis.

†*N. gonorrhoeae* may require 7 or more days.

of indwelling bladder and venous catheters, tracheostomies, and mechanical ventilators

4. More extensive surgical procedures
5. The increasing number of elderly, chronically debilitated patients admitted to the hospital
6. Nosocomial infections from contaminated IV fluids*

Since an attendant mortality of 45% to 50% accompanies these gram-negative septicemias,[34] the importance of an early diagnosis and the initiation of specific therapy cannot be overemphasized. Therefore, it is strongly recommended that any patient exhibiting signs of **unexplained fever** or **shock,** particularly after gastrointestinal surgery or manipulative urological procedures, should have a blood culture taken immediately.

LIMULUS TEST FOR ENDOTOXEMIA

Despite the fact that appropriate antibiotic therapy is presently available, the mortality in patients with gram-negative bacillary sepsis is still significantly high, especially when shock is present.[9] In theory, the early and rapid detection of the etiologic agent from the blood and the determination of its susceptibility to antibiotic agents would improve this statistic, although other factors, such as age and the presence of underlying disease play a great role in the outcome.

Many factors, including low levels of circulating organisms, the bactericidal power of normal serum factors, or the presence of antimicrobial drugs in the sample of blood,[23] adversely affect rapid detection of bacteremia. Since the symptoms of gram-negative bacillary sepsis—fever, shock, intravascular coagulation—have been attributed to the presence of circulating endotoxin in the blood, a test for the rapid detection of this substance would offer confirmation of

suspected sepsis before results of a blood culture were available. Such a test, based on the gelation of a lysate derived from the blood cells of the horseshoe crab, *Limulus polyphemus,* by minute amounts of endotoxin, has been described.[38] Various authors have reported on the usefulness of the test,[12,39] particularly for early detection of endotoxemia and differentiation between gram-negative and gram-positive septicemias. Some, however, feel that the results in detection of endotoxemia are unrewarding, and suggest that the limulus test would be more usefully employed in the predetection of hazardous amounts of endotoxin in pharmaceuticals to be administered parenterally. Others indicate that it may also be useful in assaying endotoxin in the liver, spleen, or other organs, or in body fluids such as urine, cerebrospinal fluid, and so forth.*

Commercial preparations of limulus lysate are not generally available, since standardization has not been entirely satisfactory. The interested reader is advised to await FDA release of an approved reagent at a later date.

ANAEROBIC BACTEREMIA

An increasing interest, by both the microbiologist and the clinician, in the recovery of anaerobes from clinical material has resulted in numerous reports of the isolation of these bacteria from the blood. In a retrospective study of bacteremias at the Mayo Clinic involving over 3,000 positive blood cultures, Washington[60] found anaerobes in 11% of these, representing 20% of the patients with bacteremia. Sullivan and co-workers,[51] while studying bacteremia in 300 patients following urological procedures, found a 31% positive incidence in those having transurethral resections. Although enterococci and *Klebsiella* were isolated most frequently, there was

---

*See special supplement to vol. 20, no. 9, Morbidity and mortality reports, Atlanta, 1971, Center for Disease Control.

---

*From the Round Table, Session 173: *Limulus* lysate test for the detection of endotoxin, Seventy-third Annual Meeting, American Society for Microbiology, Miami, 1973.

also a relatively large number of anaerobic cocci, *Bacteroides,* and so forth recovered. The authors reported the highest isolation rate in osmotically stabilized broth containing Liquoid; a membrane filter system also was useful for detecting multiple organism bacteremia (26% in this series).

In papers from the Center for Disease Control, Felner and Dowell,[22] and Alpern and Dowell[3] have reported the isolation of such anaerobes as *Bacteroides fragilis,* fusobacteria, clostridia (nonhistotoxic), peptococci, and propionibacteria from the blood of patients with symptoms of endocarditis, gram-negative rod septicemia, or neonatal sepsis. Alpern and Dowell[2] also studied 27 patients with documented *Clostridium septicum* septicemia associated with malignant disease. They concluded that *C. septicum* bacteremia is more common than previously suspected, especially in patients with evidence of an altered integrity of their intestinal walls, and stressed the importance of early bacteriological diagnosis, since prompt initiation of antibiotic therapy may be lifesaving. Only by employing good anaerobic techniques and multiple blood samplings can these results be achieved.

INFECTIONS CAUSED BY SPECIES OF BRUCELLA

The diagnosis of *Brucella* infections in man can be made with certainty only by the isolation of the causative organism. *Brucella* is most often found in the bloodstream, particularly in the febrile period, although it is not recovered as readily as the more commonly isolated pneumococcus, streptococcus, or staphlyococcus. Two reasons have been advanced for this: (1) the relatively complex nutritional requirements of brucellae and (2) their presence in the blood only in small numbers because of their intracellular proclivities. The first difficulty has been overcome in recent years by the introduction of commercially available culture media such as tryptose, trypticase soy, and Albimi liquid and solid media,* which will readily support the growth of small inocula. The problem of the small numbers of brucellae can be resolved partly by culturing large quantities of blood or by obtaining numerous cultures over a period of several days. Bone marrow cultures have also proved useful in some instances when blood cultures have been negative.

A simple method for cultivating brucellae and permitting examination of cultures by inoculation of agar has been described by Ruiz-Castañeda[13] and modified by Scott.[48] This method utilizes a bottle containing trypticase soy broth and trypticase agar slant, which permits direct observation of colonies developing from the broth culture.† The bottle is under a partial vacuum and is closed with a thin rubber diaphragm in a screw cap. When blood is to be cultured for brucellae, the bottle is taken to the patient's bedside. Immediately after obtaining the specimen by venipuncture, 5 to 10 ml. of blood is injected directly into the bottle through the previously sterilized (by iodine and alcohol) rubber diaphragm, using the same syringe and needle. The blood broth is mixed well to prevent clotting, and the agar slant is inoculated by tilting the bottle. The bottle is then labeled, the rubber seal pierced with a sterile 18-gauge needle containing a wisp of cotton in the hub, and the bottle is incubated in a candle jar *(Br. abortus* requires increased carbon dioxide for initial growth) in an upright position at 36° C.

The culture is examined at regular intervals by careful inspection of the agar slant for discrete colonies or confluent growth. The later may not be readily apparent. If no growth is observed, the bottle is gently shaken and tilted so that the blood broth flows over the agar slant again, and is then replaced in the candle jar. This pro-

---

*Difco Laboratories, Detroit; Baltimore Biological Laboratory, Cockeysville, Md.; Albimi Laboratories, Brooklyn.
†Baltimore Biological Laboratory, Cockeysville, Md.

cedure is repeated every third day, and the culture is held for 30 days before it is discarded as negative. This method reduces the possibility of contaminating the culture or infecting the worker by repeated opening and subculturing of the bottle.

If growth occurs and consists of small coccoid or short bacillary gram-negative organisms, it may be identified as a species of *Brucella* by agglutination with type-specific antisera, hydrogen sulfide production, and susceptibility to certain dyes. These characteristics are described in Chapter 22.

INFECTIONS CAUSED BY SPECIES OF FRANCISELLA, PASTEURELLA, AND YERSINIA

The organisms comprising this group are nonmotile* gram-negative rods that usually exhibit polar staining. They are biochemically inactive and moderate to strictly fastidious in their growth requirements. They are primarily involved in diseases of lower animals, where they cause a highly fatal type of hemorrhagic septicemia; but closely related species can cause bubonic plague and tularemia in humans.

The classification of the genus *Pasteurella* has recently been altered in accordance with more recent information on the characteristics of these bacteria. *Pasteurella tularensis* is now classified as *Francisella tularensis; Past. pestis* and *Past. pseudotuberculosis* are now in the genus *Yersinia,* which now includes *Y. enterocolitica.*

Tularemia, caused by *Francisella tularensis,* is a naturally occurring disease in animals such as wild rodents, rabbits, and deer. Human beings contract the disease accidentally as a result of handling infected animals (chiefly rabbits) or from the bites of infected deerflies or ticks. The organism will not grow on ordinary media, differing in this respect from *Yersinia pestis,* the causative agent of plague. It may be cultured on a coagulated egg me-

dium, or cystine-glucose blood agar, or in heart infusion broth containing glucose, cystine, and hemoglobin. *F. tularensis* is difficult to cultivate from infected material, but successful isolation from the blood, pleural fluid, or pus from unopened skin lesions may be achieved by the following procedure:

1. Inoculate six blood cystine-glucose agar slants and one plain agar control slant with 0.5 to 1 ml. of the patient's blood, freshly drawn by venipuncture. Distribute the blood over the slants before it clots and incubate the slants on their sides aerobically at 36° C. Pus or pleural fluid should be similarly spread over the slants.
2. Observe the slants daily for the presence of growth. Minute droplike colonies may appear in 3 to 5 days but can take as long as 10 days to develop.
3. Examine a stained smear of the colony for small gram-negative coccobacilli. If present, the culture may be identified by specific agglutination.
4. Do not report the culture as negative before 3 weeks of incubation.

Note: Great caution is mandatory in handling cultures!

Guinea pigs can be infected without difficulty, and subcutaneous or intraperitoneal injection of a heavy saline suspension of blood or ground-up tissue is the preferred procedure when only small numbers of organisms may be present. Animals usually survive 5 to 7 days but may die in 2 days; they show enlarged caseous lymph nodes, enlarged spleens, and necrotic foci in the liver and spleen. Accidental laboratory infection is an ever-present hazard; animal injection, therefore, should not be attempted unless the laboratory worker has been previously immunized against tularemia.

Y. pestis infection of man is most commonly of the **bubonic plague** type, and the bacilli may be cultivated from material obtained from buboes, the spleen, or blood.

Y. pseudotuberculosis is motile at 25° C.

In the pneumonic type the sputum is most likely to yield the organism. Blood cultures taken during the first 3 days of bubonic disease are positive in most cases. Blood specimens are first cultured in broth and then subcultured to blood agar plates after several days' incubation. On this medium after 48 hours the colonies are small (1 mm. in diameter), glistening, and transparent, with round granular centers and a notched margin. Cultures are identified by biochemical reactions and specific agglutination, although specific bacteriophage lysis is considered more reliable.

Other members of this group that may infect man are *Past. multocida* and *Y. pseudotuberculosis.* Both are primarily animal pathogens, and man is an accidental host. Approximately one half of the human cases of *Past. multocida* infection result from the bites or scratches of dogs or cats; the remainder are either pulmonary infections resulting from probable animal contact or from miscellaneous causes.[35] Bacteremia is apparently a rare occurrence, although a case of endocarditis caused by *Pasteurella,* nov. sp., following a cat bite was reported.[29] *Y. pseudotuberculosis* infections are very rare in human subjects and can be fatal in a fulminating septic type but more commonly are seen as a nonfatal mesenteric lymphadenitis.

Past. multocida grows readily on blood or chocolate agar plates, giving rise to large raised colonies, which show a tendency to become confluent. These organisms also show a great susceptibility to penicillin when it is used in disc form on the culture plate. Further descriptive bacteriology may be found in Chapter 22.

BACTEREMIA CAUSED BY LEPTOSPIRA SPECIES

Human leptospiral infection results principally from contact with an environment that has been contaminated by pathogenic leptospires from a urinary shedder or other reservoir hosts. The organisms enter the body through the abraded skin or the mucosal surfaces of the mouth, nasopharynx, conjunctivae or genitalia[57]; their principal sources are contaminated waters, soils, vegetation, food stuffs, and so on. Occupational exposure is a primary factor in leptospirosis: the disease is most frequently associated with such people as farm workers, veterinarians, livestock handlers, abbatoir attendants, and sewer workers. The disease is also associated with recreational pursuits such as swimming and camping.

The laboratory diagnosis of leptospirosis depends on isolating the organism from the blood or demonstrating a rise in antibody titer in the serum. Leptospirae may be recovered from the blood and occasionally the spinal fluid during the first week of the initial febrile illness, before circulating antibodies appear. Cultural methods or the inoculation of young guinea pigs or weanling hamsters may be employed. The organism is readily cultivated on relatively simple media enriched with rabbit serum, such as Fletcher's or Stuart's medium.* Blood cultures are made at the patient's bedside or in the laboratory, using venous blood directly from the syringe or defibrinated blood in a sterile bottle. The semisolid medium is dispensed in screw-capped tubes or rubber-stoppered vaccine bottles, and a minimal inoculum of blood is introduced directly into the bottle by puncturing the rubber diaphragm at the patient's bedside. For further details on the isolation and identification of *Leptospira,* the reader is referred to Alexander's excellent chapter in the American Society for Microbiology manual.[1]

RAT-BITE FEVER CAUSED BY SPIRILLUM MINUS AND STREPTOBACILLUS MONILIFORMIS

There are two kinds of human disease that may follow the bite of an infected rat: one, known as **sodoku** in Japan, is caused by a spiral organism presently classified as *Spirillum minus;* the other, more common

---

*Difco Laboratories, Detroit.

to the United States, is caused by a highly pleomorphic organism, *Streptobacillus moniliformis.* These diseases present similar symptomatology, including a primary ulcer, regional lymphadenopathy, fever, and malaise in addition to a characteristic skin rash. The latter infection may also be accompanied by a severe type of multiple arthritis.

A definitive diagnosis of rat-bite fever depends on the demonstration of the causative organism in the blood, lesion exudate, joint fluid, and so forth. In sodoku the presence of spirilla can be demonstrated by injecting 1 to 2 ml. of the patient's whole blood intraperitoneally into several white mice or guinea pigs known to be free from naturally occurring spirilla. Mouse peritoneal exudate or the supernate of defibrinated guinea pig blood is obtained and examined weekly for a period up to 4 weeks, either under dark-field microscopy or by special stains, for the presence of spirilla.

In rat-bite fever caused by *Streptobacillus,* the blood is cultured in the routine manner, making sure that a thioglycollate bottle is used. After several days of incubation at 36° C. the organism grows out as "cotton balls" on the surface of the red cell layer in the thioglycollate flask. Whether or not growth occurs, the thioglycollate and blood mixture should be subcultured to fresh tubes of the same medium, using a thin sterile capillary pipet for the transfer. When growth is observed, Gram stains and Wayson stains are made and examined for the presence of pleomorphic gram-negative rods. Coccobacillary, rod-shaped, branching cells and large swollen club-shaped cells are seen; special methods may also reveal the presence of L forms of *Streptobacillus.* These are fragile, filterable forms lacking a rigid cell wall; they require a complex medium for growth and are capable of reverting to the parent bacterial cell. Agglutinins are formed against the *Streptobacillus,* and a titer of 1:80 may be of diagnostic significance.

INFECTIONS CAUSED BY LISTERIA MONOCYTOGENES

Listeriosis* is a bacterial infection of man and animals that is being recognized with increasing frequency, probably because of an increased awareness of its existence. The causative organism, *Listeria monocytogenes,* was first isolated from rabbits in 1926, and the first human case was reported in 1929. Although the disease is common in domestic animals, the natural habitat of the organism or its epidemiology or epizoology has not been clearly defined. The German authority, Seeliger,[49] suspects that, because the organism is well equipped to survive in the soil, it "has a biological reservoir outside of animals and man." It is interesting to note that a recent paper[61] reports the isolation of *L. monocytogenes* from vegetation or soil taken from 11 out of 12 farms and from 6 out of 7 nonagricultural (suburban residential) sites.

Listeriosis is primarily an infection of the infant and the newborn baby, in whom it causes a purulent meningitis or meningoencephalitis, or a granulomatous septicemia acquired in utero or during delivery. In the adult it may cause abortion or an aseptic type of mononucleosis or an acute septicemia. The mortality for human cases in the United States from 1933 to 1966 was 42%.

L. monocytogenes is most frequently isolated from the blood and cerebrospinal fluid of human patients, although it may be recovered from meconium, vaginal and cervical secretions, discharge from the eye, or tissue collected at autopsy (particularly from the brain, liver, and spleen).†
A procedure for isolating *L. monocyto-*

---

*The reader is referred to an excellent monograph on the subject by Gray and Killinger,[28] and also the CDC report on human listeriosis in the United States, 1967-1969.[11]

†Directions for culturing these specimens may be found in Blair, J. E., Lennette, E. H., and Truant, J. P., editors: Manual of Clinical Microbiology, Baltimore, Md., 1970, American Society for Microbiology, Chapter 11.

genes from the blood (Seeliger and Cherry[49]) is as follows:

1. After careful disinfection of the skin, obtain 15 ml. of blood by venipuncture, using a sterile syringe and needle.
2. Immediately inoculate 5 ml. into each of two flasks containing 100 ml. of a good fluid medium such as trypticase or tryptic soy broth, or brain heart infusion broth; mix well to prevent clotting.
3. Incubate one flask in a candle jar or CO_2 incubator at 35° C. Inspect for growth and subculture to sheep blood agar at 2-day intervals for at least 1 week, incubating plates in CO_2 for 48 hours before discarding.
4. Refrigerate the second flask at 4° C., and if no growth occurs from the 35° C. culture, subculture from the refrigerated flask to blood agar at weekly intervals for 4 weeks.
5. Transfer 5 ml. of blood to a sterile tube, allow to clot and separate serum, inactivate at 56° C. for 30 minutes, and store in freezer for subsequent serological testing.

Agglutination tests on acute- and convalescent-phase sera may be carried out, although most children under the age of 10 years, as well as many adults, have normal antibody titers to *Listeria.* Titers below 1:160 should not be considered positive.

INFECTIONS CAUSED BY VIBRIO FETUS AND RELATED SPECIES

Vibrio fetus was first recognized as a causative agent of contagious abortion in cattle by McFaydean and Stockman in England in 1909. It was not until 1948, however, that the first human case was reported in the United States by Ward.[59] Up to 1970, *V. fetus* had been isolated from human sources in at least 74 instances,[8] in many cases from the blood. Although the organism has been found in the mouth and in the female genital tract, the epidemiology in humans remains un-

clear, although a bovine origin is suspected. Human infection is probably more common than is generally believed.

Since there are no distinguishing clinical features of human vibriosis, save an undulating fever of unknown origin, signs of sepsis, and dysentery in persons who may have had contact with infected domestic animals or in those who have had acute thrombophlebitis or recent dental extractions, a definitive diagnosis must depend on the isolation and identification of *V. fetus* or related species from the blood or other body sites.

V. fetus is readily isolated from blood cultured by the same techniques as those used in cases of suspected brucellosis. The organism will grow well in tryptose phosphate or trypticase soy broth and in thioglycollate medium. Since the bacterium is a microaerophile, primary incubation in 5% to 10% carbon dioxide is essential. Strains may tend to die out on repeated subculture; aliquots of the original blood cultures frozen at −70° C. may prolong survival of the vibrios. See references for a description of the cultural characteristics of these microorganisms.

POLYMICROBIAL BACTEREMIA

The finding of two different bacterial species in a blood culture is reportedly an uncommon occurrence and may cause concern as to the possibility of contamination. It has been the recent experience of some workers, however, that such findings are not unusual, particularly in patients with gastrointestinal, hepatic, or urinary tract abnormalities, or with hematologic disorders.[22,32,33] The incidence of two or more isolates from a single culture is generally reported as 6% to 10%; the organisms implicated include the gram-negative enteric bacilli, bacteroides, clostridia, and the gram-positive cocci. In these instances, the key to diagnosis and definitive antibacterial therapy rests with the bacteriology laboratory; as Elizabeth King[37] so aptly pointed out: "If people really looked, we would find that mixed infections are not as rare as previously supposed. Many workers consider any organism isolated from blood or spinal fluid to be in pure culture."

ENDOCARDITIS CAUSED BY COAGULASE-NEGATIVE STAPHYLOCOCCI

Although coagulase-negative staphylococci are generally considered as contaminants when isolated from blood cultures, they can be responsible for a serious illness, such as bacterial endocarditis. Until about 1955, their increased incidence in this disease was reported to be around 3%, but a study by Geraci and co-workers[27] in 1968 indicated an incidence of 10% in bacterial endocarditis. The etiological significance of coagulase-negative staphylococci from two or more blood cultures, therefore, should not be overlooked.

BLOOD CULTURES ON NEWBORN INFANTS

Multiple blood cultures (3 or more at 30-minute intervals) are preferred for the diagnosis of bacteremias in adults; in newborn infants, however, this becomes impractical for obvious reasons. Recently, a retrospective study of single blood cultures taken on infants less than 28 days old was reported by Franciosi and Favara.[26] Approximately 1 to 3 ml. of heparinized blood was collected in a bottle containing 5 ml. of trypticase soy broth and agar slant; of 56 patients with positive cultures, 28 were considered to have septicemia, and a single culture was considered satisfactory for diagnosis. Since clinical symptoms of neonatal septicemia frequently are minimal, with the possibility of rapid deterioration and death within hours after birth, this method can be lifesaving in the newborn.

INFECTIONS CAUSED BY L FORMS OF BACTERIA

Cell-wall defective forms of microorganisms, or L forms, have been isolated from the blood of patients treated with antibiotic agents, such as penicillin or methicillin, which are antagonistic to cell-wall synthesis.[40] These fragile forms either do not survive or fail to grow in conventional isotonic blood culture media, and the culture remains sterile. However, in an osmotically stabilized medium such as the previously described hypertonic sucrose broth with Liquoid, these aberrant forms may revert to their parent organism and subsequently be recovered and identified by conventional means.

The interested reader is referred to an excellent compilation of papers edited by Guze[30] on the subject of L forms and similar types.

INFECTIONS CAUSED BY FUNGI (FUNGEMIA)

The incidence of blood stream invasion by fungi, particularly members of the genus *Candida,* has increased significantly. This infection, once considered a rarity, is now a frequent complication in patients requiring therapy with adrenal hormones, cytotoxic agents, irradiation, or multiple antibiotics. Fungemia has also been reported more recently in patients after prolonged venous catheterization[47] or parenteral hyperalimentation[16] and as a postoperative complication of renal transplants or cardiac valve prostheses.[14] Candidal endocarditis in heroin addicts is also an increasing problem.[41]

The diagnosis of fungemia depends primarily on isolation of the organism from blood culture, the results of which are not always readily available. Species of *Candida,* particularly *C. albicans* and *C. tropicalis,* will grow in the trypticase broth–agar slant bottle previously described (p. 36) within 1 to 2 days, especially if the bottle is vented to the air. Other yeasts, however, such as *C. parapsilosis, C. guilliermondii,* and *Torulopsis glabrata* may require 1 to 2 weeks' incubation before growing out. In our limited experience, thioglycollate medium does not yield as many positive yeast recoveries as does the trypticase broth. It should be noted that not all patients with disseminated candidiasis show a positive blood culture, especially if they are receiving antifungal therapy for their disease.[53]

Sternal marrow or the buffy coat of venous blood may be examined for fungi by methods similar to those used in hematological studies, using the Giemsa stain or some other preferred bone marrow stain. Cultures should be made of the aspirated material on Sabouraud dextrose agar and brain heart infusion agar without added antibiotics. For best results this is done by transferring 0.5 to 1.0 ml. of the specimen directly to several slants of the media at the patient's bedside, using the syringe and needle with which the marrow aspirate was obtained.* These media are incubated at room temperature and at 36° C. and examined for the presence of growth at frequent invervals for several weeks. If actinomycosis is suspected, tubes of thioglycollate medium must also be inoculated; the brain heart infusion slant should be incubated **anaerobically.**

Appropriate methods for the identification of isolates are found in Chapter 32.

NONCULTURAL TECHNIQUES FOR DETECTION OF BACTEREMIA

A special method for early detection of bacteremias using a radioisotope has recently been introduced.[17] The method is based on the ability of an organism to release CO_2 from ^{14}C-labeled glucose or other labeled substrate, and the $^{14}CO_2$ formed is detected by sampling the atmosphere above the blood culture by radiometric means. Early reports indicate that the method is very sensitive and can often detect the presence of microorganisms in the blood as early as 5 to 8 hours after inoculation.[44] This procedure, it should be noted, offers no clue as to the probable identity of the organism; consequently,

conventional procedures must take over thereafter.

Another method makes use of gas chromatography to detect metabolic end products resulting from bacterial growth in blood cultures after a short incubation period. In a limited study, however, it was shown that certain end products, such as volatile fatty acids, were detected, but never before the time that the blood culture was positive by gross inspection.[23] The procedure requires further study for its applicability in the clinical laboratory.

REFERENCES

1. Alexander, A. D.: Leptospira. In Blair, J. E., Lennette, E. H., and Truant, J. P., editors: Manual of clinical microbiology, Bethesda, 1970, American Society for Microbiology.
2. Alpern, R. J., and Dowell, V. R., Jr.: *Clostridium septicum* infections and malignancy, J.A.M.A. **209:**385-388, 1969.
3. Alpern, R. J., and Dowell, V. R., Jr.: Nonhistotoxic clostridial bacteremia, Amer. J. Clin. Path. **55:**717-722, 1971.
4. Anderson, T. G.: Personal communication, 1960.
5. Austrian, R.: The role of the microbiological laboratory in the management of bacterial infections, Med. Clin. N. Amer. **50:**1419-1432, 1966.
6. Austrian, R.: Current status of bacterial pneumonia with especial reference to pneumococcal infection, J. Clin. Path. **21** (suppl. 2), 1968.
7. Bennett, I.: Bacteremia, Veterans Admin. Tech. Bull. TB 10/104, Washington, D. C., 1954, U. S. Government Printing Office.
8. Bokkenheuser, V.: *Vibrio fetus* infection in man, Amer. J. Epidem. **91:**400-409, 1970.
9. Bryant, R. E., Hood, A. F., Hood, C. E., and Koenig, M. G.: Factors affecting mortality of gram-negative rod bacteremia, Arch. Intern. Med. **127:**120-128, 1971.
10. Burke, J. P., Klein, J. O., Gezon, H. M., and Finland, M.: Pneumococcal bacteremia, Amer. J. Dis. Child. **121:**353-359, 1971.
11. Busch, L. A.: Human listeriosis in the United States, 1967-1969, J. Infect. Dis. **123:**328-332, 1971.
12. Cardis, D. T., Reinhold, R. B., Woodruff, P. W. H., and Fine, J.: Endotoxemia in man, Lancet **1:**1381-1386, 1972.
13. Castañeda-Ruiz, M.: Practical method for routine blood cultures in brucellosis, Proc. Soc. Exp. Biol. Med. **64:**114-115, 1947.
14. Chandhuri, M. R.: Fungal endocarditis after valve replacements, J. Thorac. Cardiovasc. Surg. **60:**207-214, 1970.
15. Chang, T. W., and Weinstein, L.: Use of cepha-

*This procedure is also recommended in culturing blood for fungi, especially for prompt isolation of *Candida* species[41]; also *Histoplasma capsulatum* in disseminated disease. Inoculate slants directly with 1 to 2 ml. of blood.

losporinase in blood cultures, J. Bact. **90**:830, 1965.

16. Curry, C. R., and Quiz, P. G.: Fungal septicemia in patients receiving parenteral hyperalimentation, N. Eng. J. Med. **285**:1221-1225, 1971.

17. DeBlanc, H. J., de Land, F., and Wagner, H. N., Jr.: Automated radiometric detection of bacteria in 2,967 blood cultures, Appl. Microbiol. **22**:846-849, 1971.

18. Dodson, W. H.: *Serratia marcescens* septicemia, Arch. Intern. Med. **121**:145-150, 1968.

19. Edmonson, E. B., and Sanford, J. P.: The Klebsiella-Enterobacter (Aerobacter)-Serratia group; a clinical and bacteriological evaluation, Medicine **46**:323-340, 1967.

20. Edwards, P. R., and Ewing, W. H.: Identification of Enterobacteriaceae, ed. 3, Minneapolis, 1972, Burgess Publishing Co.

21. Ellner, P. D.: System for inoculation of blood in the laboratory, Appl. Microbiol. **16**:1892-1894, 1968.

22. Felner, J. M., and Dowell, V. R., Jr.: Anaerobic bacterial endocarditis, N. Eng. J. Med. **283**:1188-1192, 1970.

23. Finegold, S. M.: Early detection of bacteremia. In Sonnenwirth, A. C., editor: Septicemia: laboratory and clinical aspects, Springfield, Ill., 1973, Charles C Thomas, Publisher.

24. Finegold, S. M., White, M. L., Ziment, I., and Winn, W. R.: Rapid diagnosis of bacteremia, Appl. Microbiol. **18**:458-463, 1969.

25. Finegold, S. M., Ziment, I., White, M. L., Winn, W. R., and Carter, W. T.: Evaluation of polyanethol sulfonate (Liquoid) in blood cultures, Antimicrob. Agents Chemother. **7**:692-696, 1967.

26. Franciosi, R. A., and Favara, B. E.: A single blood culture for confirmation of the diagnosis of neonatal septicemia, Amer. J. Clin. Pathol. **57**:215-219, 1972.

27. Geraci, J. E., Hanson, K. C., and Giuliani, E. R.: Endocarditis caused by coagulase-negative staphylococci, Mayo Clin. Proc. **43**:420-434, 1968.

28. Gray, M. L., and Killinger, A. H.: *Listeria monocytogenes* and listeric infections, Bact. Rev. **30**:309-382, 1966.

29. Gump, D. W., and Holden, R. A.: Endocarditis caused by a new species of *Pasteurella,* Ann. Intern. Med. **76**:275-278, 1972.

30. Guze, L. B., editor: Microbial protoplasts, spheroplasts and L-forms, Baltimore, 1968, The William & Wilkins Co.

31. Hamburger, M., Carleton, J., and Harcourt, M.: Reversal of the antistaphylococcal activity of tetracycline by magnesium, Antibiot. Chemother. (N. Y.) **7**:274-278, 1957.

32. Hermans, P. E., and Washington, J. A. II: Polymicrobial bacteremia, Ann. Intern. Med. **73**:387-392, 1970.

33. Hochstein, H. D., Kirkham, W. R., and Young, V. M.: Recovery of more than one organism in

septicemias, N. Eng. J. Med. **273**:468-474, 1965.

34. Holloway, W. J., and Scott, E. G.: Gram-negative rod septicemia, Delaware Med. J. **40**:181-185, 1968.

35. Holloway, W. J., Scott, E. G., and Adams, Y. B.: *Pasteurella multocida* infection in man, Amer. J. Clin. Path. **51**:705-708, 1968.

36. Huddleson, I. F.: Growth of bacteria in blood, Bull. W. H. O. **21**:187-199, 1959.

37. King, E. O.: Personal communication, 1965.

38. Levin, J., and Bang, F. B.: The role of exotoxin in the extracellular coagulation of Limulus blood, Bull. Johns Hopkins Hosp. **115**:265-274, 1964.

39. Levin, J., Poore, T. E., Zauber, N. P., and Oser, R. S.: Detection of endotoxin in the blood of patients with sepsis due to gram-negative bacteria, N. Eng. J. Med. **283**:1313-1316, 1970.

40. Lorian, V., and Waluslka, A.: Blood cultures showing aberrant forms of bacteria, Amer. J. Clin. Path. **57**:406-409, 1972.

41. Louria, D. B., Hensle, T., and Rose, J.: The major medical complications of heroin addiction, Ann. Intern. Med. **67**:1-22, 1967.

42. Lowrance, B. L., and Traub, W. H.: Inactivation of the bactericidal activity of human serum by Liquoid (sodium polyanetholsulfonate), Appl. Microbiol. **17**:839-842, 1969.

43. Rammell, C. G.: Inhibition by citrate of the growth of coagulase-positive staphylococci, J. Bact. **84**:1123-1125, 1962.

44. Randall, E. L.: Comparison of a radiometric method with conventional methods for detection of bacteremia, paper M 43, Bact. Proc., p. 87, 1972.

45. Rosner, R.: Effect of various anticoagulants and no anticoagulant on ability to isolate bacteria directly from parallel clinical blood specimens, Amer. J. Clin. Path. **49**:216-219, 1968.

46. Rosner, R.: A quantitative evaluation of three blood culture systems, Amer. J. Clin. Path. **57**:220-227, 1972.

47. Salter, W., and Zinneman, H. H.: Bacteremia and candida septicemia, Minn. Med. **50**:1489-1499, 1967.

48. Scott, E. G.: A practical blood culture procedure, J. Lab. Clin. Med. **21**:290-294, 1951.

49. Seeliger, H. R. P., and Cherry, W. B.: Human listeriosis, Washington, D. C., 1957, U. S. Government Printing Office.

50. Sullivan, N. M., Sutter, V. L., Attebery, H. R., and Finegold, S. M.: Clinical evaluation of improved blood culture procedures, paper M 46, Bact. Proc., p. 87, 1972.

51. Sullivan, N. M., Sutter, V. L., Attebery, H. R., and Finegold, S. M.: Bacteremia after genital tract manipulation: bacteriological aspects and evaluation of various blood culture systems, Appl. Microbiol. **23**:1101-1106, 1972.

52. Sutter, V. L., Attebery, H. R., Rosenblatt, J. E., Bricknell, K. S., and Finegold, S. M.: Anaerobic

bacteriology manual, Los Angeles, 1972, Dept. Contin. Educ. Health Sciences, School of Med., UCLA.

53. Taschdjian, C. L., Kazinn, P. J., and Toni, E. F.: Opportunistic yeast infections with special reference to candidiasis, Ann. N. Y. Acad. Sci. **174:**606-622, 1970.

54. Thoburn, R., Fekety, F. R., Cluff, L. E., and Melvin, V. B.: Infections acquired by hospitalized patients, Arch. Intern. Med. **121:**1-10, 1968.

55. Traub, W. H.: Antagonism of polymyxin B and kanamycin sulfate by Liquoid (Sodium polyanetholsulfonate) *in vitro,* Experientia **25:**206-207, 1969.

56. Traub, W. H., and Lawrence, B. L.: Media-dependent antagonism of gentamicin sulfate by Liquoid (sodium polyanetholsulfonate), Experientia **25:**1184-1185, 1969.

57. Turner, L. H.: Leptospirosis, Brit. Med. J. **1:**231-235, 1969.

58. Von Haebler, T., and Miles, A. A.: The action of sodium polyanethol sulfonate (Liquoid) on blood cultures, J. Path. Bact. **46:**245-252, 1938.

59. Ward, B. R.: The apparent involvement of *Vibrio fetus* in an infection in man, J. Bact. **55:**113-114, 1948.

60. Washington, J. A. II: Comparison of two commercially available media for detection of bacteremia, Appl. Microbiol. **22:**604-607, 1971.

61. Welshimer, H. J., and Donker-Voet, J.: *Listeria monocytogenes* in nature, Appl. Microbiol. **21:**516-519, 1971.

62. Winn, W. R., White, M. L., Carter, W. T., Miller, A. B., and Finegold, S. M.: Rapid diagnosis of bacteremia with quantitative differential-membrane filtration culture, J.A.M.A. **197:**539-548, 1966.

63. Witorsch, P. and Gordon, P.: *Hemophilus aphrophilus* endocarditis, Ann. Intern. Med. **60:**957-961, 1964.

64. Wright, H. D.: The bacteriology of subacute infective endocarditis, J. Path. Bact. **28:**541-578, 1925.

7 Microorganisms encountered in contaminated blood and plasma from the blood bank

Of the many complications of blood transfusion, including incompatibility due to mismatched blood and serum hepatitis, those caused by bacterially contaminated blood are usually the most dangerous to the patient. Not only will severe reactions (chills, fever, and a rapid fall in blood pressure) ensue immediately, but peripheral collapse may occur within an hour after the transfusion of massively contaminated blood, and the patient frequently may succumb. It is important, therefore that the bacteriology laboratory offer a procedure for the detection of contamination in preserved blood and its products **before** they are administered to a patient.

The organisms most frequently encountered in fatal transfusion reactions comprise a small group of **psychrophilic gram-negative rods** (from donor's skin) that have the ability to multiply and elaborate endotoxins in blood stored at refrigerator temperatures.[3] These organisms and their optimal temperatures for growth are the following:

1. *Pseudomonas* species (other than *Pseudomonas aeruginosa*), which grow slowly at 4° C., rapidly at room temperature (22° to 25° C.), and slowly or not at all at 36° C.
2. *Citrobacter* species, which grow best at room temperature and at 36° C.
3. Other late lactose-fermenting members of the family Enterobacteriaceae, which will multiply at all three temperatures.
4. *Achromobacter* species, which grow best at room temperature and at 36° C.

These gram-negative organisms are generally found in the air, soil, water, and dust and in the intestinal tract of man and animals. They can be transferred readily from the latter site to the skin of man, where they may become a part of the normal transient flora of that tissue. These organisms will multiply in the refrigerated blood, autolyze, and release **endotoxins** that may produce a toxic effect when administered in a blood transfusion. The subsequent reaction that follows is the result of this phenomenon, rather than the effect produced by the introduction of the bacteria into the circulatory system or their multiplication within the tissues of the patient.

Contamination of the blood may be caused in various ways: (1) by the use of improperly sterilized equipment and reagents for the collection, (2) by some flaw in aseptic technique during collection, and most frequently (3) by the inadequate cleansing and disinfection of the skin over the site of venipuncture. For procedures directed against obviating these difficulties, the reader is referred to a manual on technical methods and procedures issued by the American Association of Blood Banks.

RECOMMENDED PROCEDURES FOR TESTING STERILITY OF BLOOD PLASMA AND BANK BLOOD
Sterility test procedure

The following methods for carrying out sterility tests are those recommended by the American Association of Blood Banks, and are included here with its permission[1]:

   a. Sterility tests shall be performed at regular intervals and not less than once monthly. Blood intended for transfusion shall not be tested by a method that entails entering the container. A record shall be kept of the results of sterility tests.

   b. Technic of sterility testing: Each month at least one container of normal-appearing blood shall be tested within the 18th to 24th day after collection. The test should be performed by inoculating 10 ml. of blood into ten times the volume of fluid thioglycollate medium in a manner that introduces blood throughout the medium without destroying the anaerobic conditions in the lower portions of the broth. The sample should be incubated for 10 days at 30° to 32° C., or separate samples should be incubated at 18° to 20° C. and 35° to 37° C., respectively, and examined every work day for evidence of microbial growth. A subculture of 1 ml. inoculated into 10 ml. of thioglycollate medium should be made on the third, fourth, or fifth day. The subculture should be incubated for 10 days as described above and examined daily for evidence of microbial growth.

Culturing plastic packs

Only when blood or plasma is collected in an open system—that is, where the blood container is entered—is it necessary to carry out a periodic sterility check.* After a transfusion reaction, however, it is strongly recommended that the following procedure, described by Walter and associates,[4] be carried out in the culturing of plastic blood packs and plasma packs. In order to secure a representative sample, the bag should be turned gently end over end to mix the contents; the integral donor tube should also be stripped several times. A 5-inch segment is then sealed with a dielectric sealer and cut free in the center of the seal. Both ends of the segment are then dipped in 70% ethyl alcohol, flamed, and with a pair of scissors similarly flame

---

*Public Health Service, Section 73.3004, amended March 25, 1972, Federal Register vol. 37, no. 59.

sterilized, one seal is cut off. This cut end then is held over a tube of thioglycollate medium, the top seal nicked to admit air, and the sample (approximately 1 ml.) allowed to run in.

Cultures should be made in duplicate when possible and held for at least 10 days at room temperature and at 35° to 37° C. At the end of the incubation period, a loopful of the culture is streaked on a blood agar plate or placed in another tube of thioglycollate medium and held at the preceding temperatures for 24 hours to confirm sterility. Stained smears of the original cultures also may be examined.

Culture of donor arm bleeding site

In addition to the periodic sterility checks of blood and the equipment and solutions used in its collection, the culture of the venipuncture site before and after the arm "prep" is also recommended as a quality control measure. The technic described by Jorness[2] may be used:

1. A cotton applicator moistened in 1 ml. of pH 7.5 sterile phosphate buffer (see p. 397 for preparation) is rubbed vigorously over a 1–square inch area of the venipuncture site for 1 minute. This is done in duplicate, before and after the "prep."

2. The applicator is then twirled in a tube (15 to 20 ml.) of melted and cooled (45° C.) soybean-casein digest agar and squeezed out against the inside of the tube.

3. Approximately 1 ml. of sterile defibrinated sheep blood is added; the contents are mixed well by inversion and are poured into a sterile Petri dish.

4. The plates are examined after 48 hours' incubation at 36° C., and colony counts are made, together with the identification of significant isolants, if indicated.

5. The number of bacteria per square inch of skin should be reduced almost to zero, if an effective degerming has been achieved.

SIGNS OF CONTAMINATION

Contamination of a bottle of blood often may be suspected from gross examination of the sample. The following changes should be looked for:

1. Change in the color of blood to a bluish black
2. Zone of hemolysis in the lowest portion of the supernatant plasma at the cell-plasma interface
3. Appearance of a large quantity of free hemoglobin in plasma when shaken
4. Appearance of small fibrin clots in a bottle, which had formerly appeared satisfactory, due to the utilization of citrate by the organisms previously described

The transfusion of contaminated blood is followed (usually after the first 50 ml.) by a sudden onset of chills, fever, hypotension, and shock. This may progress, usually fatally, to convulsions and coma. A "stat" Gram stain of blood from the blood container may reveal the presence of organisms; a blood culture from the patient also should be obtained. Prompt diagnosis and initiation of specific therapy may be lifesaving.

REFERENCES

1. American Association of Blood Banks: Technical methods and procedures, ed. 5, Chicago, 1970, The Association.
2. Joress, S. M.: A study of disinfection of the skin, Ann. Surg. **155:**296-304, 1962.
3. Koch, M. L.: Bacteriologic problems in blood banking, Bull. Amer. Assoc. Blood Banks **12:**397-401, 1959.
4. Walter, C. W., Kundsin, R. B., and Button, L. N.: New technic for detection of bacterial contamination in a blood bank using plastic equipment, New Eng. J. Med. **257:**364-369, 1957.

8 Microorganisms encountered in respiratory tract infections

The pathogens most likely to be found in the upper and lower respiratory tract include the following:

Group A beta hemolytic streptococci
Corynebacterium diphtheriae
Streptococcus pneumoniae
Neisseria meningitidis
Bordetella pertussis
Haemophilus influenzae
Mycobacterium tuberculosis, other mycobacteria
Fungi including *Candida* species, *Histoplasma capsulatum, Coccidioides immitis,* and others
Klebsiella pneumoniae, other coliform bacilli
Pseudomonas aeruginosa, other pseudomonads

THROAT AND NASOPHARYNGEAL CULTURES

Throat cultures and nasopharyngeal cultures are important as an aid in the diagnosis of certain infections, such as streptococcal sore throat, diphtheria, or candidal infections of the mouth (thrush); in establishing the focus of infection in diseases such as scarlet fever, rheumatic fever, and acute hemorrhagic glomerulonephritis; and also in the detection of the carrier state of organisms such as beta hemolytic streptococci, meningococci, *Staphylococcus aureus,* and the diphtheria bacillus.

To obtain the best results it is important that material for a throat culture be obtained **before** antimicriobial therapy and taken in a proper manner. A satisfactory method is as follows:

1. Use a sterile throat culture outfit such as a commercially available disposable unit* that contains a polyester-tipped applicator and an ampule of modified Stuart transport medium.
2. With the patient's tongue depressed and the throat well exposed and illuminated, rub the swab firmly over the back of his throat, both tonsils or tonsillar fossae, and any areas of inflammation, exudation, or ulceration. Care should be taken to avoid touching the tongue or lips with the swab.
3. Replace the swab in the inner tube, crush the ampule, force the swab into the released holding medium, and send it to the laboratory immediately, where it should be plated as soon as possible. If it is necessary to hold the culture for more than an hour, it should be refrigerated until plates can be streaked.

If diphtheria is suspected, two swabs should be taken; one should be inoculated to Loeffler medium immediately and the other handled as described in step 3.

Nasopharyngeal cultures are recommended when attempting the isolation of pneumococci, meningococci, or *Haemophilus influenzae,* because these organisms are found more commonly in the nasopharynx than in the nose or throat. Nasopharyngeal swabs are essential for the recovery of *Neisseria meningitidis* from suspected carriers or *Bordetella pertussis*

---

*Culturette, from Scientific Products, Evanston, Ill.

from suspected cases of whooping cough. They are most useful in culturing specimens from infants and small children, where sputum specimens are not readily obtainable. Such cultures are best taken on cotton-tipped nichrome, or stainless steel wire applicators (B. & S. 28 gauge)* in a sterile tube containing a few drops of broth. With the patient's head firmly held, the wire swab is gently inserted (without force) through the nose to the posterior nasopharynx, allowed to remain there for a few seconds, and then deftly withdrawn. The wire swab may also be bent at a right angle near the cotton tip (bending against the inside of the sterile tube container) and then inserted through the mouth and behind the uvula and soft palate into the nasopharynx. Care must be taken to avoid mouth and throat contamination of the swab. In either case the inoculated swab is returned to the inner tube containing the broth and thus transported to the laboratory.

Anterior nares cultures are occasionally required, particularly for the study of staphylococcal or streptococcal nasal carriers. They are taken by introducing a cotton swab moistened with broth about 1 inch into the nares, gently rotating the swab, and returning it to the broth tube previously described.

ROUTINE PROCEDURES, INCLUDING ISOLATION METHODS FOR HEMOLYTIC STREPTOCOCCI

The throat or nasopharyngeal swab is used for the direct inoculation of blood agar plates, and several methods are available. Since the routine throat culture is taken chiefly to detect the presence of **beta hemolytic streptococci**, the latter's signifi-

cance will be stressed in the following procedures:

1. If polyester-tipped* swabs are used (as is preferred), direct inoculation of a **moist sheep blood agar** plate (without added glucose, which inhibits characteristic hemolysis) is recommended. Roll the swab firmly over a small portion of the agar surface and streak by using a sterile wire loop. Spread the inoculum over the remainder of the plate to obtain well-isolated colonies. There are many techniques for spreading the inoculum; one is shown in Fig. 4-1. Make a few stabs into the agar with the loop for observation of **subsurface hemolysis** by strains producing beta-type hemolysis under reduced oxygen tension. This procedure may be used on either a freshly made culture or on one in which the polyester swab has remained in its original container for several days (must be refrigerated).

2. If an applicator immersed in broth has been used, twirl the swab vigorously against the side of the broth tube to obtain as much material in suspension as possible. It has been shown[3] that the flora of such a throat swab can be determined better by inoculating a broth suspension than by directly streaking the swab on a blood agar plate. By using a sterile loop, transfer a loopful of the suspension to a blood agar plate and streak as described previously.

3. If a **pour plate** is desired, the cotton-tipped applicator is placed in 2 ml. of trypticase soy broth and the suspension prepared as previously described.

    (a) Transfer a loopful of this suspension to a 20 by 150 mm. screw-

---

*These are cut in 8-inch lengths, and a loop is made at one end to act as a handle. The tip is tightly bent over itself, dipped in collodion, and carefully covered with a tuft of fine eye cotton. An excellent substitute is the calcium alginate swab on a soft flexible aluminum wire, Colab Laboratories, Inc., Chicago Heights, Ill.; Falcon No. 2047, 2050, Oxnard, Calif.

---

*This fiber appears to prolong the survival of beta hemolytic streptococci from throat swabs and may be mailed in an envelope to the laboratory without noticeable loss of streptococci.[9]

capped tube containing approximately 15 ml. of melted and cooled (45° to 50° C.) trypticase soy agar.

(b) Pipette 0.8 ml. of sterile defibrinated sheep blood into the agar and mix the entire contents of the tube by inversion. The sheep blood may be conveniently stored in 1 ml. amounts in small sterile tubes in a refrigerator, where it will remain stable for 7 to 10 days.

(c) Flame the lip of the tube and pour the agar into a sterile Petri dish; allow to cool and solidify.

Pour plates are useful in revealing certain strains of Group A streptococci that may appear to be nonhemolytic when growing on the surface of blood agar plates because these strains produce streptolysin O only, which is inactivated by atmospheric oxygen. Most strains of Group A streptococci, however, show beta hemolysis around surface colonies due to their production of streptolysin S, which is oxygen stable.

All throat and nasopharyngeal culture plates are incubated at 35° to 37° C. in a candle jar (or 10% carbon dioxide incubator) for 18 to 24 hours. Some strains of Group A streptococci and other pathogens grow slowly under ordinary atmospheric conditions. The use of deep agar pour plates or incubation in an atmosphere of reduced oxygen tension will aid in the recovery of these fastidious strains.

The results obtained from throat cultures on blood agar depend to an extent on the **type of blood** used in preparing the plates. For reasons described in Chapter 17, sheep blood agar plates are recommended for these cultures (horse blood is also acceptable); human blood, either fresh or outdated bank blood, should be used only when sheep blood is not available.

When examining throat and nasopharyngeal cultures for the presence of members of the genus *Haemophilus*, rab-

bit or horse blood agar should be used, since sheep blood, as well as some human blood, contains substances inhibitory to these organisms. This is particularly true for *Haemophilus influenzae.*

READING OF PLATES

Blood agar plates inoculated with throat cultures are sometimes confusing to the inexperienced bacteriologist. A knowledge of the **normal flora,** of the organisms that may be significant, and of the colony characteristics of both groups will aid in solving the problem. The organisms that may be encountered in the **normal** throat, in order of their predominance, are as follows:

Alpha hemolytic streptococci
Neisseria catarrhalis; other normal throat species of *Neisseria,* including pigmented forms
Coagulase-negative staphylococci; occasionally *Staphylococcus aureus*
H. haemolyticus (may be confused with beta hemolytic streptococci); *H. influenzae*
Pneumococci
Nonhemolytic (gamma) streptococci
Diphtheroid bacilli
Coliform bacilli (particularly after penicillin therapy)
Yeasts, including *Candida albicans*
Beta hemolytic streptococci other than Group A

Microorganisms that may be encountered in an **infected** throat are the following:

Beta hemolytic streptococci of Group A; occasionally Groups B, C, and G.
Corynebacterium diphtheriae
Bordetella pertussis
Meningococcus
A predominance of *Staphylococcus aureus,* coliform bacilli, pneumococci, *H. influenzae, C. albicans,* and so forth

Since Group A streptococci are isolated more readily from throat swabs than from other infected areas, it is for the detection of these organisms that most throat cultures are taken. Therefore, either the blood agar streak or pour plate should be carefully examined for the presence of colonies of beta hemolytic streptococci after overnight incubation. A description of these colonies will be detailed in Chapter 17.

Any colony suggestive of **beta hemolytic streptococci** should be subcultured by stabbing with a straight needle and streaked on a sector of a sheep blood agar plate or inoculated to a tube of blood broth. This will usually yield a pure culture; if it does not, the process should be repeated. Further identification of these colonies may be carried out by use of the bacitracin disc sensitivity test and confirmed, if required, by serological grouping or the fluorescent antibody procedure. These procedures are also described in Chapter 17.

Although **pneumococci** (*Streptococcus pneumoniae*) occur in small numbers in many normal throats, an increase in their number, especially in nasopharyngeal specimens, may be of etiological significance. Since some patients, particularly of the pediatric age group, may not be able to produce sputum, the laboratory confirmation of a diagnosis of pneumococcal pneumonia may rest on the isolation of pneumococci from throat or nasopharyngeal cultures. Further identification of pneumococci may be carried out by testing their susceptibility to Optochin (ethyl hydrocupreine hydrochloride) by a disc procedure or by testing their solubility in bile (desoxycholate). Refer to Chapter 18 for a description of these tests.

Normal throat cultures show a predominance of alpha hemolytic streptococci and commensal neisseriae. If a culture shows a predominance or increase of other organisms that normally occur only in small numbers—such as pneumococci, *H. influenzae,* or *Staph. aureus*—it is significant in our opinion and should be reported. A convenient method of reporting throat cultures is indicated by the following examples: (1) **normal flora;** (2) **marked increase in beta hemolytic streptococci, probably Group A** (if susceptible to the bacitracin disc); (3) **normal flora with slight increase in** *H. influenzae;* and (4) **pure culture of coagulase-positive** *Staph. aureus.*

BACTERIOLOGICAL DIAGNOSIS OF PERTUSSIS

The diagnosis of **whooping cough** is confirmed by the isolation of *Bordetella pertussis* from the respiratory secretions. Nasopharyngeal swabs are used to obtain material for culture, and a special medium, Bordet-Gengou agar containing 15% to 20% sheep blood, is used for cultivating the organism.

Nasopharyngeal swabs are obtained in the following manner: while the child's head is immobilized, the special wire applicator (see discussion of throat and nasopharyngeal cultures elsewhere in this chapter) is carefully passed through the nasal aperture along the nasal floor until it touches the posterior pharyngeal wall. It is left there for several induced coughs before withdrawal, whereupon it is immediately streaked upon the surface of the Bordet-Gengou medium, and the inoculum is spread over the plate with a sterile loop or spreader. Penicillin (0.25 to 0.5 unit per milliliter) may be incorporated in the medium at the time of preparation; this inhibits the growth of penicillin-sensitive organisms normally present in the nasopharynx, usually without affecting the growth of pertussis organisms. The plates are incubated aerobically at 35° to 37° C. for 4 or 5 days and are examined with a hand lens for the presence of typical colonies resembling mercury droplets surrounded by a hemolytic zone. *Bord. pertussis* is identified by staining reaction and agglutination with specific antiserum, as indicated in Chapter 22.

The yield of positive isolations from clinical cases of pertussis varies from 20% to 98%. This variation is apparently related to the procedures used, the stage of the disease, and the prior administration of antibiotics. By the use of the fluorescent antibody (FA) technique on pernasal smears, Whitaker and associates[17] were able to confirm the diagnosis in 94% of pertussis cases during the first week of illness. Holwerda and Eldering[7] employed the FA technique in direct staining of early growth from Bordet-Gengou plates* inoculated with pernasal swabs and were able thereby to identify *Bord. pertussis* one day earlier. At the present time the FA technique promises to become a useful aid in diagnosis; however, it should be used in conjunction with conventional cultural procedures and should not be used alone.

BACTERIOLOGICAL DIAGNOSIS OF DIPHTHERIA

The diagnosis of **diphtheria** is a clinical problem, and the responsibility should rest primarily on the attending physician. Often he will attempt to place the responsibility upon the bacteriologist and may expect a definitive diagnosis within 15 minutes, based on the examination of a direct smear of the lesion. Although the appearance of diphtheria bacilli in stained smears from lesions and membranes is highly characteristic, they cannot be identified by morphology alone, since other nonpathogenic diphtheria-like bacilli may be indistinguishable from *Corynebacterium diphtheriae.* The chance of error in either direction is too great when diagnosis is based exclusively upon microscopic examination; such a reading should be accepted only as **presumptive** evidence of infection. The identity and type of the organism in question is obtained **only** after isolating toxigenic *C. diphtheriae* in pure culture from the patient. The exact determination of the presence of these organisms requires a careful cultural and microscopic study of several days' duration followed by the performance of a virulence test in suitable animals or by an in vitro test.

In primary isolation, *C. diphtheriae* is best cultivated on appropriate media, such as Loeffler medium (3 parts animal serum and 1 part dextrose broth, coagulated as slants) or modified Pai medium, where growth occurs after 18 to 24 hours' incuba-

tion at 36° C. Differential and selective media such as cystine or chocolate-tellurite agar also give a high proportion of recoveries and should be included in primary inoculation; however, the characteristic morphology of the diphtheria bacillus on chocolate-tellurite agar is not as distinctive as that from Loeffler or Pai medium. These media are described in Chapter 39.

Material received in the laboratory for diphtheria culture is usually sent on swabs taken from the **throat** and **nasopharynx.** These are promptly inoculated to the surface of a Loeffler or Pai slant and streaked upon cystine or chocolate-tellurite agar plates. Since severe throat infections caused by beta hemolytic streptococci may produce lesions simulating diphtheria, it is recommended that sheep blood agar plates be inoculated also and examined for Group A streptococci, along with the Loeffler slant and the tellurite plate. After the media are inoculated, a smear can be prepared by rolling the swab on a slide, fixing it, and staining it with methylene blue stain. The slide should be examined for the presence of irregularly stained pleomorphic bacilli suggestive of diphtheria and for the presence of the spiral and fusiform organisms found in Vincent's infection, which also may resemble the diphtheritic lesion in the throat.

The inoculated media are incubated at 35° to 37° C. On a Loeffler slant that has been incubated for a short period (2 to 8 hours), *C. diphtheriae* will frequently outgrow other organisms present and will exhibit a characteristic morphology when stained with methylene blue. The bacilli are deeply stained and appear septated, with wedge-shaped ends and tapered points. The cells are often at angles to each other, as in the letters V and Y, and do not palisade as is seen with some of the diphtheroid bacilli.

If organisms resembling diphtheria bacilli are found in smears from the Loeffler slant but not from the tellurite plate, the growth from the slant is suspended in 1

---

*This appears to be the chief value of the FA technique in diagnosing pertussis.

to 2 ml. of sterile broth and a loopful of this is streaked on a fresh tellurite plate in order to isolate the organism. All plates should be held and examined for 48 hours before being discarded as negative.

The application of FA procedures in the detection of *C. diphtheriae* in known or suspected cases of clinical diphtheria has been used as a rapid presumptive diagnostic procedure. The conjugate used cross-reacts with toxigenic strains of *C. ulcerans* (rarely found in the United States), and does not differentiate between toxigenic and nontoxigenic strains of *C. diphtheriae.* It is recommended that the procedure is preferably used in conjunction with conventional culturing procedures in order to achieve the greatest degree of sensitivity and specificity.

LABORATORY DIAGNOSIS OF VINCENT'S INFECTION (FUSOSPIROCHETAL DISEASE)

Vincent's stomatitis is a pseudomembranous or ulcerative infection of the gums, mouth, or pharynx. It is thought to be caused by a gram-negative spirochete, *Borrelia vincentii,* and a straight or slightly curved, anaerobic gram-negative rod with sharply pointed ends, *Fusobacterium fusiforme,* in symbiotic association with other oral microorganisms. A reduction in local tissue resistance is generally a precursor of the infection.

The pseudomembrane present on tonsillar lesions in the anginal form of this infection has been mistaken for a similar lesion in diphtheria, and it is important that they be differentiated. Fusospirochetal organisms can be readily demonstrated by staining smears with crystal violet (that used in the Gram stain) for 1 minute and looking for large numbers of the characteristic forms previously described, along with numerous pus cells. In acute infections the spirochetal form predominates; in chronic cases the fusiforms are more preponderant. Since these organisms are often present in the normal mouth and gums, their presence in smears must be correlated with the clinical findings to be of diagnostic significance.

NASOPHARYNGEAL CULTURES FOR MENINGOCOCCI

Nasopharyngeal cultures are of value in demonstrating the presence of *Neisseria meningitidis* in cases of suspected meningococcemia and meningococcal meningitis, and in the detection of meningococcal carriers. The specimen may be collected through the mouth by passing a wire swab under and beyond the uvula to the posterior wall of the nasopharynx, or through the nostril by passing the wire swab along the floor of the nasal passage to the posterior wall. In either case it is desirable to secure a bit of the mucus by gently twirling the swab while in place against the nasopharyngeal wall. It should be noted that *N. meningitidis* can be recovered from healthy people; carrier rates among population groups experiencing a meningitis outbreak may rise as high as 80%. Under these circumstances it is doubtful that a single positive culture is significant. A **negative** culture from a patient convalescing from meningococcal meningitis who may be in contact with younger susceptible persons upon returning home is perhaps more meaningful information. *N. meningitidis* is extremely sensitive to cold, dehydration, unfavorable pH, or the inhibitory activity of other microflora; therefore, culture material should be inoculated as soon as possible.

Meningococci may be isolated on a moist blood agar or chocolate agar plate used in routine throat and nasopharyngeal cultures, where they appear after 18 to 24 hours' incubation in a candle jar as small, gray, mucoid, nonhemolytic colonies of gram-negative diplococci.*

Like the gonococcus, the meningococcus is **oxidase positive;** this provides a helpful test that can be used for detecting the presence of these colonies in a mixed

---

*Thayer-Martin agar medium has proved extremely useful in carrier surveys.

culture. It must be noted, however, that the nonpathogenic neisseriae of the oropharynx also will give a positive oxidase test. In order to identify *N. meningitidis*, therefore, it is necessary to carry out a slide agglutination test with a polyvalent antiserum* or a capsular swelling (quellung) reaction test. Fermentation reactions with several carbohydrates are also essential to identification. Refer to Chapter 19 for further information.

THROAT CULTURES FOR SPECIES OF CANDIDA

Although the skin is a frequent site of infection by the yeastlike organism *Candida albicans*, an intraoral infection called **thrush** is sometimes seen, particularly in infants and neonatal patients. Material for a culture is taken from the whitish, loosely adherent membrane attached to the inner cheek, palate, or other portions of the oral mucosa, using the technique described for throat cultures. On receipt in the laboratory, the swab is inoculated to slants of Sabouraud dextrose agar, with and without added antibiotics (see Chapter 32), and incubated both at room temperature and at 36° C. Primary growth of *Candida* occurs in 2 to 3 days as a yeastlike colony with a characteristic fermentative odor. Identification of *C. albicans* depends on the demonstration of typical chlamydospores on a special medium, and other tests. The procedure is described in Chapter 32.

SPUTUM CULTURES

Bacterial pneumonia, pulmonary tuberculosis, and chronic bronchitis constitute a most important group of human diseases. Since specific treatment frequently depends upon a bacteriological diagnosis, the prompt and accurate examination of a properly collected sputum specimen by smear, culture, and antibiotic susceptibility testing becomes imperative. This is true with pneumococcal pneumonia, and

particularly in pneumonia caused by *Klebsiella pneumoniae* and *Staphylococcus aureus*, both of which are more rapidly fatal than that caused by *Streptococcus pneumoniae*.[1]

Furthermore, less common bacterial species, including the gram-negative enteric bacilli, *Pseudomonas aeruginosa**, and *Haemophilus influenzae*, are being reported in hospital-associated pneumonia of the debilitated patient.[4] These also may cause pneumonia,[15] thereby complicating conditions such as hematological disorders, virus infections, alcoholism, and pregnancy,[16] and only proper bacteriological examination of the sputum will reveal their presence.

Sputum specimens are also submitted from patients with bronchiectasis, lung abscess, suspected pulmonary mycotic infections, pulmonary tuberculosis, or infection with *Mycoplasma pneumoniae*. The laboratory diagnosis of these diseases is sufficiently complex to warrant special sections in this text, to which the interested reader is referred.

Recovery of an etiologic agent from sputum depends not only upon the laboratory methods used but also upon the **care taken** in securing the specimen. Too often, the culturing of unsuitable material results in misleading information for the clinician, the true infecting agent having been missed entirely. It requires no special courage on the part of the microbiologist to discard a specimen that is obviously saliva as being **unsatisfactory** for bacteriological examination.

The collection of sputum for culture also requires the cooperation of the patient and should include instructions to obtain material from a deep cough (tracheobronchial sputum), which is expectorated directly into a sterile Petri dish or other satisfac-

---

*Difco Laboratories, Detroit.

*These organisms are most frequently involved in patients requiring mechanical ventilation, including reservoir nebulizers, resuscitators, and tracheostomy equipment.

tory container.* The volume of the specimen need not be large—1 to 3 ml. of purulent or mucopurulent material will be sufficient for most examinations except tuberculosis culturing. On receipt in the laboratory the specimen should be examined without delay, especially if histoplasmosis is suspected. However, refrigeration of the specimen for not more than 1 to 3 hours is satisfactory for the recovery of most pathogens if immediate culturing is not feasible.

The organisms most frequently associated with **acute** bacterial infections of the lower respiratory tract are the following:

Pneumococcus (*Streptococcus pneumoniae*)
Klebsiella pneumoniae
Haemophilus influenzae
Staphylococcus aureus
Coliform bacilli, *Pseudomonas, Proteus* species
Mycoplasma pneumoniae

With the exception of *Mycobacterium tuberculosis,* it is generally possible to recover the important pulmonary pathogens from the upper respiratory tract of apparently healthy people. The recovery of these organisms in small numbers, therefore, is not always sufficient evidence of their etiological role in a particular infection. In acute bacterial pneumonia, however, the pathogen is generally present in large numbers and in pure culture.

EXAMINATION OF GRAM-STAINED SMEARS

An early presumptive diagnosis of bacterial pneumonia can often be made by examining a direct smear of sputum properly stained by the Gram method. The presence of many encapsulated, lancet-shaped cocci occurring singly, in pairs, or in short chains, along with pus cells, revealed by a capsule stain, may suggest a pneumococcal infection in typical cases. Likewise,

a stained preparation showing great numbers of gram-positive cocci in clusters is characteristic of staphylococcal pneumonia. But a more important reason for examining sputum smears, in our opinion, is for the detection of numerous short, fat, encapsulated gram-negative rods, suggestive of *Klebsiella* infection.

However, the **presumptive** aspect of these findings should be emphasized— only by the isolation of pneumococci, staphylococci, pseudomonads, or klebsiellae from cultures of sputum, blood, or pleural fluid can such an impression be confirmed.

CULTURE FOR PNEUMOCOCCUS

Since *Streptococcus pneumoniae* continues to be the major cause of bacterial pneumonia, the examination of a fresh sputum specimen is the first step in the diagnostic procedure.

Blood-tinged (rusty) mucopurulent sputum freshly obtained from patients with lobar pneumonia may be handled in the following manner:

1. Select purulent or bloody flecks and prepare direct smears as follows: transfer a loopful to a clean slide, press another slide over it, squeeze the slides together, and then pull them apart. Flame both slides as soon as they are dry. Stain one of them by the Gram method and one by the acid-fast technique.
2. Streak a sheep blood agar plate* and inoculate a tube of thioglycollate medium with small loopfuls of a **selected** portion of the specimen (not saliva!). Place the blood plate in a candle jar;† incubate both the jar and the thioglycollate culture at 35° to 37° C.

---

*The use of superheated hypertonic saline aerosols for sputum induction is recommended when the cough is not productive. Proper decontamination of the equipment must be carried out when using this equipment.[5]

---

*The worker is well advised to streak serially two or more blood agar plates and a MacConkey or eosin–methylene blue (EMB) plate (if *Klebsiella* is suspected) to obtain isolated colonies when the bacterial population of the sample is unusually high.
†Austrian[2] has shown that a small percentage of pneumococci require 3% to 10% CO_2 for recovery on agar plates at the time of primary inoculation.

for 18 to 24 hours. Examine the cultures daily.

3. If the sputum is unusually tenacious, transfer a mucopurulent mass to a Petri dish containing sterile physiological saline and swish it around to remove the excess saliva. Transfer to a clean Petri dish, add a small volume of saline, and repeatedly aspirate and expel the specimen in a sterile syringe to emulsify it. **Exercise caution.** Use a loopful of this suspension to inoculate the media previously indicated.

4. If after incubation of the blood plate, small, shiny, greenish, transparent, flat colonies with depressed centers and raised edges are seen, they are most likely colonies of pneumococci. These may be cultured and further identified by determining their solubility in bile salts and their susceptibility to Optochin. These tests are described in Chapter 18.

5. Although pneumococci usually predominate in sputum cultures from patients with early lobar pneumonia, the blood plate may also show a large number of colonies of the usual throat flora, consisting of alpha hemolytic streptococci, nonpathogenic neisseriae, and so forth. Before discarding such a plate as showing only normal throat flora, it is recommended that a search for pneumococcus colonies be made while using a **hand magnifying lens** (3× to 5×). If these colonies are present, an attempt should be made to recover them in pure culture by fishing to sectors of a blood agar plate.

6. If pneumococcus colonies are not found on the primary plate, transfer a loopful of the well-mixed thioglycollate culture to another blood agar plate and incubate as described. Generally, only alpha hemolytic and nonhemolytic streptococci will be recovered, but pneumococci will be isolated occasionally, even when they are not present on the original plate.

QUANTITATIVE SPUTUM CULTURE

Quantitative sputum culturing, by means of liquifying the specimen with a mucolytic agent, homogenizing it, and plating tenfold serial dilutions of it on appropriate solid culture media,[10] has had numerous proponents for its adoption in the clinical laboratory. The technique, however, is cumbersome and time consuming, requires a considerable number of plates, and appears to offer few advantages over routine conventional methods. However, in the susceptible or compromised host, it offers an important means of uncovering and monitoring superinfections resulting from endogenous or iatrogenic induction. It can provide microbiological guidance to the clinician in directing therapy to the major pathogen, or by indicating the emergence of a potential pathogen of different antibiotic susceptibility from the one for which the therapy was initiated.[11] These results may not become apparent by conventional techniques.

TRANSTRACHEAL ASPIRATION IN PNEUMONIA

Since severe and sometimes fatal necrotizing pneumonia caused by gram-negative enteric bacilli and other unusual pathogens (including anaerobes) is occuring more frequently,[13] the reliability of a routine sputum culture (contaminated by mouth flora) has been questioned. Pecora,[12] in 1959, proposed transtracheal aspiration, in which a small-gauge catheter is threaded into the trachea through a needle introduced at the cricothyroid ligament, as an alternative procedure for obtaining material from the lower respiratory tract, uncontaminated by oropharyngeal microorganisms. The technique has proved safe and the results bacteriologically reliable,[8] yielding a high number of pulmonary pathogens, and is recommended in patients with pneumonia who are unable to raise a satisfactory sputum specimen.[6] Results of gram-stained smears of tracheal aspirates correlate better with their corresponding cultures than do smears of expectorated sputum from the same patient.

CULTURE FOR OTHER PATHOGENS

Sputum from acute or chronic pulmonary infections caused by agents other than pneumococci can vary from a thick tenacious specimen that reveals species of *Klebsiella* on culture to a foul-smelling purulent sputum from a patient with a lung abscess, from which species of *Bacteroides* and other anaerobes may be recovered. As a rule, the isolation and identification of klebsiellae pose no problem, although serological identification by means of capsular swelling tests may require the aid of a reference laboratory.* The anaerobes are more difficult to recover and identify; they should be looked for especially in the foul-smelling sputum from patients with lung abscesses, bronchiectasis, and empyema. Transtracheal aspirates from these patients should be streaked on two **freshly prepared** blood agar plates and inoculated to two tubes of enriched thioglycollate medium. One set of media is incubated in an **anaerobic jar** and subsequently handled as described in Chapter 13; the other set is handled as described for pneumococcus isolation.

The examination of both stained and unstained wet smears prepared directly from the sputum may be informative, particularly when fungus elements can be demonstrated. Dark-field illumination may also be helpful in revealing the presence of spirilla and spirochetes. See Chapter 26 for a description of anaerobic culture procedures.

It is frequently difficult to evaluate the bacterial flora of **bronchial aspirations** because these specimens are usually contaminated with upper respiratory secretions. Using special collection techniques, however, cultures of the lower respiratory tract of normal subjects are generally sterile, so that recovery of potential respiratory pathogens in these instances becomes clinically significant. It should also be noted that cultures of sputum or pharyngeal secretions generally do not accurately reflect the bacterial flora of the bronchi.[14] Routine cultures, as well as those for pathogenic fungi and tubercle bacilli, should be set up on all bronchoscopic samples.

The etiological significance of the **pathogenic fungi** in pulmonary disease has been recognized only within the last two decades. Formerly these organisms were considered exotic parasites, and special techniques for their isolation were rarely employed. In the light of present knowledge, however, it becomes a necessity for the modern diagnostic microbiology laboratory to be able to isolate and at least recognize the agents of candidiasis, histoplasmosis, cryptococcosis, sporotrichosis, and the opportunistic fungi, such as *Aspergillus, Nocardia,* and *Mucor* species. Because of the complexities of the techniques involved, a special section on medical mycology has been included in Chapter 32 of this text.

Since the introduction and widespread use of the broad-spectrum antibiotics, the laboratory rarely receives tonsillar or adenoidal tissue or swabs from the mastoid region for culture. The swab, if submitted, is handled by the methods described for throat cultures. **Tissue specimens,** however, must be minced with sterile scissors and ground with sterile alundum, using a tissue grinder made for this purpose (Chapter 32). The resulting suspension is streaked on a sheep blood agar plate, and a loopful is inoculated to a tube of enriched thioglycollate medium. A blood agar pour plate, prepared from a broth dilution of the original suspension, is also recommended. The media are incubated and examined as described for throat cultures.

---

*Some antisera are available from Difco Laboratories, Detroit, and Baltimore Biological Laboratory, Cockeysville, Md.

Special methods for anaerobes, fungi, and mycobacteria are described in their respective chapters.

REFERENCES

1. Austrian, R.: The role of the microbiology laboratory in the management of bacterial infections, Med. Clin. N. Amer. **50:**1419-1432, 1966.
2. Austrian, R., and Collins, P.: Importance of carbon dioxide in the isolation of pneumococci, J. Bacteriol. **92:**1281-1284, 1966.
3. Committee on Acute Respiratory Diseases: Problems in determining the bacterial flora of the pharynx, Proc. Soc. Exp. Biol. Med. **69:**45-52, 1948.
4. Feingold, D. M.: Hospital-acquired infections, New Eng. J. Med. **283:**1384-1391, 1970.
5. French, M. L. V., Dunlop, S. G., and Wentzler, T. F.: Contamination of sputum induction equipment during patient usage, Appl. Microbiol. **21:**899-902, 1971.
6. Hahn, H. H. and Beaty, H. N.: Transtracheal aspiration in the evaluation of patients with pneumonia, Ann. Intern. Med. **72:**183-187, 1970.
7. Holwerda, J., and Eldering, G.: Culture and fluorescent-antibody methods in diagnosis of whooping cough, J. Bact. **86:**449-451, 1963.
8. Kalinske, R. W., Parker, R. H., Brandt, D., and Hoeprich, P. D.: Diagnostic usefulness and safety of transtracheal aspiration, New Eng. J. Med. **276:**604-608, 1967.
9. Lattimer, A. D., Siegel, A. C., and De Celles, J.: Evaluation of the recovery of beta hemolytic streptococci from two mail-in methods, Amer. J. Public Health **53:**1594-1602, 1963.
10. Louria, D. B.: Uses of quantitative analyses of bacterial populations in sputum, J.A.M.A. **182:**1082-1086, 1962.
11. Louria, D. B.: The nature of opportunistic infections. In: Symposium on opportunistic pathogens, sponsored by Eastern Pennsylvania Branch American Society for Microbiology et al, Nov. 2-3, 1972, Philadelphia.
12. Pecora, D. V.: A method of securing uncontaminated tracheal secretions for bacterial examination, J. Thorac. Surg. **37:**653, 1959.
13. Pierce, A. K., Edmonson, B., McGee, G., Ketchersid, J., Loudon, R. G., and Sanford, J. P.: An analysis of factors predisposing to gram-negative bacillary necrotizing pneumonia, Amer. Rev. Resp. Dis. **94:**309-315, 1966.
14. Potter, R. T., Rotman, F., Fernandez, F., McNeill, T. M., and Chamberland, J. M.: The bacteriology of the lower respiratory tract; bronchoscopic study of 100 clinical cases, Amer. Rev. Resp. Dis. **97:**1051-1061, 1968.
15. Tillotson, J. R., and Lerner, A. M.: Pneumonia caused by *Escherichia coli* and *Pseudomonas,* Antimicrob. Agents Chemother. **6:**198-201, 1966.
16. Turck, M.: Current therapy of bacterial pneumonias, Med. Clin. N. Amer. **51:**541-548, 1967.
17. Whitaker, J. A., Donaldson, P., and Nelson, J. D.: Diagnosis of pertussis by the fluorescent-antibody method, New Eng. J. Med. **263:**850-851, 1960.

9 Microorganisms encountered in the gastrointestinal tract

Members of the family Enterobacteriaceae make up a large part of the aerobic microflora of the intestinal tract of man. These include the intestinal commensals (the coliforms and species of *Proteus*), as well as the enteric pathogens of the *Salmonella* and *Shigella* genera and related species, intestinal streptococci (enterococci), species of *Bacteroides* (probably the predominant microorganisms in the normal stool), clostridia, various yeasts (including *Candida albicans*), and occasionally pathogenic staphylococci. The vibrio of cholera may be isolated from cases of this disease, and rarely the necessity arises for attempting to isolate *Mycobacterium tuberculosis* from fecal material.

At least 65 species of virus have been recovered from the intestinal tract of man, and the number is constantly increasing. These viruses comprise mainly three groups: the polioviruses, the Coxsackie viruses, and the echoviruses* (enteric cytopathogenic human orphan). All require special techniques for their recovery. Rectal swabs and fecal specimens for this purpose must be maintained in a frozen state at all times prior to attempting recovery.

---

*Although not considered an enterovirus, the virus of infectious hepatitis is present in the feces.

ISOLATION OF SALMONELLA AND SHIGELLA

Salmonella and *Shigella* are present in the stool in appreciable numbers only during the **acute stage** (first 3 days) of a diarrheal disease; therefore, fecal specimens should be obtained within this period whenever possible. Bits of bloody mucus or epithelium should be selected from the freshly voided specimen for inoculation to appropriate media. In **chronic** dysentery, however, the immediate culturing of material obtained during **proctoscopic** examination offers the best means of obtaining positive *Shigella* isolations. When stool specimens are not readily obtainable in outbreaks of enteric disease, **rectal swabs** afford the most practical method of securing material for culture, although they should not be relied on for maximum recovery. Rectal swabs are of less value than fecal specimens in the examination of convalescent patients or in carrier surveys. In all the aforementioned situations it is well to remember that rapid proliferation of normal bacterial commensals occurs and that certain forms, particularly *Shigella,* may die off rather rapidly after the specimen is collected. These samples, therefore, should be **inoculated to appropriate culture media soon after collection.** Whenever possible, **multiple** stool specimens should be examined; numerous investigations have demonstrated the value of this procedure.[5]

A great variety of culture media have been devised for the isolation of salmonellae and shigellae from fecal specimens. These may be divided into several general groups (discussed in more detail in Chapters 20 and 39) without any sharp division among groups.

1. **Differential nonselective media,** such as eosin–methylene blue (EMB) agar, MacConkey agar, or Leifson desoxycholate agar, contain certain carbohydrates, indicators, and chemicals that are inhibitory to many gram-positive bacteria. Differentiation of enteric bacteria is achieved through the incorporation of lactose (and sucrose in some brands of EMB agar), since the organisms that attack lactose will form colored colonies, whereas those that do not ferment lactose will appear as colorless colonies. Among the latter are found the salmonellae and shigellae, as discussed in Chapter 4.

2. **Differentially moderately selective media,** such as Salmonella-Shigella (SS) agar, Hektoen enteric agar (HEA), and desoxycholate citrate agar, and highly selective media, such as Kauffmann brilliant green agar and Wilson-Blair bismuth sulfite agar, are complex combinations of nutrients and chemicals that serve to inhibit many coliforms, especially strains of *Escherichia,* as well as *Proteus,* but permit the growth of most *Salmonella* and many *Shigella* organisms isolated in the United States. These pathogens, along with slow or lactose-nonfermenting organisms, generally form colorless colonies, although some may appear as black or greenish colonies on certain media (Chapter 20). The lactose-fermenting organisms, which are not inhibited, grow as pink, orange, or red colonies on media other than bismuth sulfite agar. Coliforms, if they develop, may appear as small black, brown, or greenish colonies on bismuth sulfite agar.

3. **Selective enrichment media,** such as Leifson selenite broth, Mueller tetrathionate broth, or Hajna GN broth, incorporate nutrients and such chemicals as selenium salts, tetrathionate (by oxidation of thiosulfate through addition of iodine just prior to use), or desoxycholate and citrate. Such agents inhibit the growth of gram-positive organisms and temporarily (12 to 18 hours) limit the growth of coliforms and *Proteus,* while encouraging the multiplication of salmonellae and shigellae.

It is recognized that selective plating media used for isolating enteric pathogens, by virtue of their inhibitory action on coliforms, may also inhibit less hardy pathogens. Thus, whereas media containing brilliant green, bismuth sulfite or bile salts and citrate may be excellent for isolation of salmonellae, their value may be limited in the recovery of some *Shigella* species.

The plating medium, xylose lysine desoxycholate (XLD) agar,* as modified by Taylor[7] is recommended by him for general application, but with particular reference to *Shigella* and *Providencia.* The differentiation of the various species is based upon xylose fermentation, lysine decarboxylation, and hydrogen sulfide production. In a comparison of the recovery efficiency of various enrichment broths and plating media in a series of more than 1,000 fecal specimens, Taylor and Schelhart[8] concluded that the most successful combination utilized GN or selenite-F enrichment medium, followed by plating on XLD agar. A further study by Rollender and associates[6] indicated a definite superiority of XLD agar over MacConkey agar in the recovery of salmonellae and shigellae by direct inoculation of stools and urine specimens. All of these liquid media must be subcultured to differential or selective plating media in order to demonstrate colonies of enteric pathogens.

From the foregoing discussion it should be most apparent that **no single medium can be used for all purposes;** a variety of plating and enrichment media will result in a higher number of positive isolations than will be obtained when only one me-

---

*Difco Laboratories, Detroit; Baltimore Biological Laboratory, Cockeysville, Md.

dium is used. The following procedures are recommended for culturing stool and rectal specimens for the presence of *Salmonella* and *Shigella*[1]:

1. Using a cotton-tipped swab, **heavily** inoculate an SS agar, Hektoen enteric agar (HEA), brilliant green agar, and desoxycholate citrate agar or Mac-Conkey agar plate with fresh material. Streak with the swab or a wire spreader to provide good distribution of isolated colonies over a major portion of each plate (Chapter 4). A suspension of fecal material in enrichment medium also may be used to inoculate these plates (see step 4).

2. At the same time **lightly** inoculate an EMB, XLD, or desoxycholate agar plate and streak as in step 1.

3. Since bismuth sulfite agar is the best medium at present for the isolation of typhoid bacilli, its use is recommended whenever typhoid fever or the carrier state is suspected. **Heavily** streak a bismuth sulfite agar plate as described here. In addition, make two pour plates, using about 0.1 ml. and 5 ml. of a heavy fecal suspension in broth mixed with the melted and cooled (45° C.) agar medium.

4. Also inoculate heavily a tube of selenite enrichment and GN broths with the fecal specimen. After overnight incubation, transfer several loopfuls from these broth media as follows: selenite to EMB or MacConkey, and GN to XLD plate.

5. Incubate all media at 36° C. for 18 to 24 hours and for 48 hours if indicated (bismuth sulfite plates usually require a longer period). Examine the plates in a good light for suspicious colonies.

Salmonellae and shigellae produce typical **colorless** colonies on these media, except on XLD agar, when they appear as **red** colonies, sometimes with black centers. On HEA, however, coliforms appear as salmon to orange in color, whereas salmonellae and shigellae appear bluish green. *Salmonella typhi* forms **black,** opaque colonies on bismuth sulfite agar. It should be kept in mind that atypical colonies may appear, especially on selective media. On such plates great care must be exercised in picking colonies, since microscopic growth of other inhibited bacteria may be present on or near the ones selected, resulting in mixed cultures. Refer to Chapter 20 and discussion of the Enterobacteriaceae for further identification procedures.

ISOLATION OF ENTEROPATHOGENIC ESCHERICHIA COLI

It is now widely accepted that certain serological types of *Escherichia coli* are responsible for outbreaks and sporadic cases of infantile diarrhea, particularly in the newborn. Results of earlier investigations of this disease were inconclusive because only biochemical methods of identification had been used; it awaited the introduction, by Kauffmann and associates, of definitive typing methods and the establishment of an antigenic schema by which these bacteria could be classified.

Specimens should be collected early in the course of the illness and before any antibiotics have been administered. They may be obtained from freshly soiled diapers or by means of a rectal swab and must be cultured without delay. Since enteropathogenic* *E. coli* organisms are inhibited by the selective media used for the isolation of salmonellae and shigellae (SS agar, desoxycholate citrate agar, and so forth), it is necessary to utilize the less inhibitory media, such as EMB or MacConkey agar, to recover them. Blood agar plates also should be inoculated because pure cultures are easier to obtain on this medium. It may be necessary to inoculate other media, however, in order not to overlook the presence of salmonellae and shigellae.

---

*"Enteropathogenic" is used to differentiate these strains of *E. coli* from those responsible for urinary tract infections, appendicitis, peritonitis, septicemia, and so forth.

After overnight incubation, the plates are examined for colonies of coliforms, and at least three *Escherichia*-like colonies are picked from each plate and suspended individually in a drop of polyvalent *E. coli* OB antiserum on a microscope slide that has been appropriately marked. The slide is rotated and examined for agglutination, using a dark background and suitable illumination. A rapid clumping is presumptive evidence of group identification. For a more complete description of further identification procedures, see Chapter 20.

ISOLATION OF PATHOGENIC STAPHYLOCOCCI

It has been well demonstrated that the administration of broad-spectrum antibiotics to patients may result in a disturbance of the normal microflora of the lower intestinal tract. This may result in the development of large numbers of coagulase-positive staphylococci in this area and induce a syndrome characterized by fever, abdominal pain, and a fulminating diarrhea.

Through the use of a selective medium containing a high concentration of sodium chloride, such as Staphylococcus Medium 110* or mannitol salt agar,* pathogenic staphylococci can be isolated from the stool. By using this technique the carrier rate of coagulase-positive staphylococci found in normal adults may be shown to be as high as 15% to 20%.[2] This figure may be higher in infants or in patients hospitalized for more than several days or in those who have been given antibiotics. A diagnosis of staphylococcal enteritis, based only upon the demonstration of coagulase-positive staphylococci in the stool, therefore, is questionable in light of these findings.

On the other hand, our own experience parallels that of Hinton and associates.[2] When coagulase-positive staphylococci are isolated in **large numbers** from a stool

specimen by employing a medium such as phenylethanol blood agar, and a Gram stain of a direct smear of the fecal specimen reveals **large numbers** of gram-positive cocci in clusters, we believe that this is a laboratory report of considerable clinical significance and that it should be so interpreted.

ISOLATION OF VIBRIO CHOLERAE (COMMA)

At the present time, Asiatic cholera is confined chiefly to India, Pakistan, Burma, Nepal, and China. Infections persisting in small numbers of human cases (apparently there is no chronic carrier state) between epidemics keep the disease smoldering in these areas. This, then, spreads geographically (generally westward), producing yearly epidemics. Only infrequently is a laboratory in the continental United States called upon to attempt isolation and culture of this microorganism.

Cholera vibrios are present in enormous numbers in the characteristic **rice water stool** produced by the profuse diarrhea in the early stages of the disease. At times the vibrios may be demonstrated in direct smears prepared from flecks of mucus from the stool, but prompt culture of the specimen on taurocholate gelatin agar or TCBS agar yields good results. After overnight incubation, colonies of *Vibrio cholerae* appear small, smooth, and translucent, distinct from the rough coliform colonies. SS and EMB media inhibit the growth of the organism. *V. cholerae* is discussed in more detail in Chapter 20.

ISOLATION OF MYCOBACTERIUM TUBERCULOSIS FROM THE STOOL

The examination of fecal specimens for *Mycobacterium tuberculosis* is not routine and is of questionable value when carried out. Tubercule bacilli isolated from intestinal contents probably reflect the presence of pulmonary tuberculosis, because sputum is often swallowed. The method recommended is desribed by Kubica and Dye[4] and is outlined in Chapter 28.

---

*Difco Laboratories, Detroit; Baltimore Biological Laboratory, Cockeysville, Md.

ISOLATION OF CANDIDA ALBICANS FROM THE STOOL

Although the yeastlike fungus *Candida albicans* (and other species of *Candida*) can be found normally on the skin, in the mouth, and in the intestinal tract of man, it is frequently responsible for a severe type of intestinal disturbance. Intensive use of antibiotics, particularly the tetracyclines, in the treatment of bacterial infection apparently leads to the suppression of a fair portion of the normal intestinal bacterial flora, with consequent overgrowth and infection by *Candida*.

Microscopic examination of direct smears of fecal material aids in the differentiation of the saprophytic and pathogenic phase of *C. albicans*. The presence of **mycelial elements and yeast forms** suggests the invasive or pathogenic phase of *C. albicans*. When direct smears reveal only the yeast form, the fungus is probably present as a saprophyte.[3]

C. albicans can be readily isolated on Sabouraud dextrose agar containing antibiotics. Methods for isolating and identifying this organism may be found in Chapter 32.

REFERENCES

1. Edwards, P. R., and Ewing, W. H.: Identification of enterobacteriaceae, ed. 3, Minneapolis, 1972, Burgess Publishing Co.
2. Hinton, N. A., Taggart, J. G., and Orr, J. H.: The significance of the isolation of coagulase-positive staphylococci from stool, Amer. J. Clin. Path. **33:**505-510, 1960.
3. Kozinn, P. J., and Taschdjian, C. L.: Enteric candidiasis, Pediatrics **30:**71-85, 1962.
4. Kubica, G. P., and Dye, W. E.: Laboratory methods for clinical and public health mycobacteriology, Public Health Service Pub. No. 1547, Washington, D. C., 1967, U. S. Government Printing Office.
5. McCall, C. E., Martin, W. T., and Baring, J. R.: Efficiency of cultures of rectal swabs and fecal specimens in detecting *Salmonella* carriers; correlation with numbers of salmonellas isolated, J. Hyg. **64:**261-269, 1966.
6. Rollender, W., Beckford, O., Kostroff, B., and Belsky, R.: Comparison of xylose lysine deoxycholate agar and MacConkey agar for the isolation of salmonella and shigella from clinical specimens, paper M 238, Bact. Proc., p. 106, 1968.
7. Taylor, W. I.: Isolation of shigellae; xylose lysine agars; new media for isolation of enteric pathogens. Amer. J. Clin. Path. **44:**471-475. 1965.
8. Taylor, W. I., and Schelhart, D.: Isolation of *Shigella*. VI. Performance of media with stool specimens, paper M 242, Bact. Proc., p. 107, 1968.

10 Microorganisms encountered in the urinary tract

Bacterial infections of the urinary tract affect patients of all age groups and both sexes, and they vary in severity from an unsuspected infection to a condition of severe systemic disease. The clinical diagnosis of pyelonephritis is frequently overlooked because of the absence of urinary tract symptoms and pyuria.[6] The correlation of a bacteriuria with unsuspected active pyelonephritis at autopsy has been demonstrated.[10] The role of the indwelling catheter in the development of a bacteriuria, frequently accompanied by a gram-negative rod bacteremia, has also been revealed.[4] The demonstration of bacteria by appropriate cultural methods is therefore the only reliable means of making a specific diagnosis.

Urine is ordinarily an excellent culture medium for the multiplication of the common pathogens of the urinary tract, and when bacteria are deposited in the urine, they tend to increase greatly, exceeding 1,000,000 per milliliter. Since specimens of urine, either clean voided or catheterized, are frequently contaminated on collection, the recovery of organisms, even known pathogens, does not necessarily establish the diagnosis of a urinary tract infection. Studies by Kass, Sanford, MacDonald, Beeson, and others, summarized in an extensive bibliography and review of this topic,[15] have indicated that bacterial counts on fresh, voided urine from **infected** patients show more than 100,000 (10^5) or-

ganisms per milliliter,* whereas specimens from noninfected or normal persons may be sterile or contain up to 1,000 organisms per milliliter. It should be pointed out, however, that bacterial counts of less than 10^5 per milliliter may occur in patients who are receiving specific antibacterial therapy or in patients who are excessively hydrated, with a consequent dilution of urine.[1] Therefore, to evaluate the clinical significance of a **positive** urine culture, it is strongly recommended that some means of estimating the number of organisms present in a specimen be part of a routine urine culture.

As will be noted from the following, the bacterial flora of **normal** voided urine differs from that of infected specimens. The organisms listed in the first group are those that are most frequently encountered in normal urine.

Staphylococci, coagulase negative
Diphtheroid bacilli
Coliform bacilli
Enterococci
Proteus species
Lactobacilli
Alpha hemolytic and beta hemolytic streptococci
Saprophytic yeasts
Bacillus species

The flora of **infected** urine, as a rule, will include one or more of the following:

---

*This figure is based on first voided urine in the morning; specimens collected later may give lower counts, due to hydration.

Escherichia coli, Klebsiella-Enterobacter-Serratia division

Proteus mirabilis and other *Proteus* species, *Providencia* species

Pseudomonas aeruginosa and other *Pseudomonas* species

Enterococci *(Streptococcus faecalis)*

Staphylococci, coagulase positive and coagulase negative

Alcaligenes species

Acinetobacter (Herellea) species

Haemophilus species (probably *Corynebacterium vaginalis*)

Candida albicans, Torulopsis glabrata, other yeasts

Beta hemolytic streptococci, usually Groups B and D

Neisseria gonorrhoeae

Mycobacterium tuberculosis, other mycobacteria

Salmonella and *Shigella* species

COLLECTION OF SPECIMENS

The distal portion of the human urethra and the perineum are normally colonized with bacteria, particularly in the female. These organisms will readily contaminate a normally sterile urine upon voiding, but contamination can be prevented by using the "clean catch" technique. In this the periurethral area (tip of penis, labial folds, vulva) is carefully cleansed with two separate washes with plain soap and water or a mild detergent, and well rinsed with warmed sterile water to remove the detergent, with the glans penis or labial folds retracted. The urethra is then flushed by passage of the first portion of the voiding, which is discarded. The subsequent urine, voided directly into a sterile container, is used for culturing and colony counting.

It is recommended that two successive clean-voided midstream specimens be collected in order to approach a 95% confidence level when using a bacterial count of 10^5 per milliliter as an index of significant bacteriuria.[8]

In order to increase the accuracy of localizing urinary tract infections, Stamey and colleagues[18] have recommended the use of suprapubic needle aspiration of the female bladder and a technique of dividing voided urine from the male into urethral, midstream (bladder), and postprostatic massage cultures. In the former procedure, both urethral and vaginal contamination is eliminated; in the latter, localization of a lower urinary tract infection to the urethra or prostate is facilitated.

Since urine generally will support the growth of most of the urinary pathogens as well as do routine broth media, it is absolutely essential for culture purposes that urine be processed within an hour of collection or **stored in a refrigerator** at 4° C. until it can be cultured. Studies have indicated that such specimens may be kept in the refrigerator for 5 to 10 days or more without significantly reducing their bacterial content.[14] If specimens are received at various times during a laboratory work day, they may be placed in a refrigerator as received, then set up together at some designated hour in the afternoon.

CULTURE PROCEDURES

A common practice in a number of laboratories has been to centrifuge urine specimens and inoculate either liquid or solid media with the sediment or to inoculate the urine directly into a tube of broth. Although occasional situations may arise that require these methods of handling, the general practice may well be questioned. Centrifugation will concentrate contaminants as well as pathogens; the introduction of only a few contaminating bacteria by directly inoculating broth will invariably result in a positive culture. For these reasons it is strongly recommended that one of the following **quantitative** methods be used for culturing urine specimens.

Calibrated loop–direct streak method

The calibrated loop–direct streak method, first reported by Hoeprich[5] in 1960, makes use of a calibrated bacteriological loop commonly employed in dairy bacteriology to inoculate and streak plates of standard or differential culture media. After proper incubation the number of colonies present is estimated and reported as a measure of the degree of bacteriuria.

The method is used most successfully as a **screening procedure;** its confidence level, when compared with actual plate counts, is highest when counts of 10^5 (and greater) bacteria per milliliter are observed. If a more precise count is indicated, conventional plate count procedures (see next section) should be employed.

1. Using a flame-sterilized and **cooled** 4-mm. platinum loop,* delivering 0.01 ml., deposit one **vertical** loopful of a well-mixed, uncentrifuged urine specimen on (a) a blood agar plate and (b) an EMB or a MacConkey agar plate.
2. Streak both plates by passing the loop through the inoculum and downward to the lower edge of the plate in a T pattern from the inoculum site (Mayo technique).
3. Incubate both plates overnight at 35° to 37° C. and read the following morning. Examine and count the colonies on both plates; estimate the total count from the blood plate and gram-negative bacterial count from the EMB or MacConkey plate. In each case the colony count is multiplied by 100 (0.01 ml. used) to give an estimate of the number per milliliter of urine. One hundred colonies would represent 10^4 actual cells present, for example.

After determining the plate count, proceed with the identification of the organisms present and determine their susceptibility to antibiotics as described in subsequent sections.

4. If no growth occurs in 24 hours, hold the plates for another day and, if still negative, report as: **"No growth after 48 hours."**

Pour plate method

Although the pour plate method has been criticized as not entirely reflecting the true bacterial count of a urine speci-

men (clumps of organisms can give rise to single colonies),[16] it remains the most accurate procedure yet devised for measuring the degree of bacteriuria. The procedure is as follows:

1. Prepare three tenfold dilutions of well-mixed urine in sterile distilled water in sterile screw-capped tubes as follows:

1:10 (10^{-1}) = 1.0 ml. urine + 9.0 ml. water
1:100 (10^{-2}) = 1.0 ml. of 1:10 dilution + 9.0 ml. water
1:1000 (10^{-3}) = 1.0 ml. of 1:100 dilution + 9.0 ml. water

Use separate 1-ml. serological pipets for each dilution, with adequate shaking between dilutions.

2. Pipet 1 ml. of each of the dilutions into appropriately labeled, sterile Petri dishes.
3. Overlay each dilution with 15 to 20 ml. of melted and cooled (50° C.) nutrient or infusion agar; mix well by carefully swirling the dishes.
4. When solidified, invert the dishes and incubate overnight at 35° to 37° C.
5. Using a Quebec colony counter, enumerate the number of colonies in the plate yielding 30 to 300 colonies; multiply by the dilution to obtain the total number of bacteria per milliliter of urine.

Other methods

Recently attention has been given to chemical tests for the rapid detection of bacteriuria. These include the reduction of triphenyltetrazolium chloride (TTC) by metabolizing bacteria,[17] the Griess nitrite test, based on the rapid reduction of nitrate by members of the Enterobacteriaceae and staphylococci, and the catalase test,[13] in which rapid gas production from urea reacting with hydrogen peroxide indicates bacteriuria. These tests, however, have been considered unsatisfactory due to substantial numbers of false-negative results; commercial test kits based on these principles likewise have shown correspondingly equivocal results.

---

*Platinum-rhodium inoculating loop, Cat. No. 7012-F 15, Arthur H. Thomas Co., Philadelphia.

In order to avoid the rather tedious pour plate procedures described previously, a number of screening **culture** tests for bacteriuria have been proposed. In one such method (dip-slide,[2,3] tube[12] or spoon[11]), an agar-coated vehicle is dipped into a freshly voided midstream urine specimen; and after overnight incubation, the colonial growth is counted or compared with density photographs of known colony counts and the significance of the culture determined.* A similar procedure utilizes a filter paper strip[9] that is dipped into the urine specimen, then transferred to the surface of an agar plate. After incubation, the number of colonies growing in the inoculum area are counted and their significance evaluated. Multiple specimens can be cultured on a single agar plate.

Unfortunately, all of these procedures possess a common disadvantage: that bacterial contamination in the collection of the specimen may produce a false-positive indication of significant bacteriuria. We are inclined to agree with Taylor[19] that all of these methods—chemical or dip-culture—lack the accuracy of methodology required in a clinical microbiology laboratory and have their principal value as a screening procedure for bacteriuria in a doctor's office. A duplicate urine specimen held in the refrigerator could be submitted for conventional culture and antibiotic sensitivity testing if the screening test proved to be positive.

Interpretation of colony counts

On the basis of the colony count estimate, report the degree of bacteriuria as a plate count of the following:

1. Less than 1,000 (10^3) bacteria per milliliter
2. Between 1,000 and 100,000 (10^3 to 10^5) bacteria per milliliter
3. Greater than 100,000 (10^5) bacteria per milliliter

Generally, counts of less than 10^3 are suggestive of contamination, whereas urines containing between 10^3 and 10^5 organisms are suspected of infection, and counts of 10^5 and greater are indicative of infection, although no single concentration of bacteria distinguishes infected from contaminated specimens. Counts in infected patients may be low when the rate of urine flow is rapid, or the patient is receiving suppressive therapy, or occasionally when the urine pH is less than 5 and the specific gravity of the urine is less than 1.003.[16]

Examination of direct smears

The examination of a direct smear of fresh **uncentrifuged** urine has long been used as a screening device to distinguish between true bacteriuria and contamination.[7] It is done by making a smear of the well-mixed urine with a 4-mm. (0.01-ml.) bacteriological loop, allowing the smear to air dry, fixing it with heat, and performing a Gram stain. A positive smear is one showing at least several bacteria in the majority of oil-immersion fields.[8]

If one uses the point of 10^5 organisms as representing a significant colony count, the correlation of the direct smear has been variously reported as 75% to 95%.[5,7,14,16] If lower counts are compared to the direct smear (counts between 10^4 and 10^5), the accuracy falls to around 50%.[14]

REFERENCES

1. Clapp, M. P., and Grossman, A.: The quantitative evaluation of bacteriuria and pyuria, Amer. J. Med. Sci. **248:**158-163, 1964.
2. Cohen, S. N., and Kass, E. H.: A simple method for quantitative urine culture, New Eng. J. Med. **277:**176-180, 1969.
3. Grettman, D., and Naylor, G. R. E.: Dip-slide aid to quantitative urine culture in general practice, Brit. Med. J. **3:**343-345, 1967.
4. Hodgin, U. G., and Samford, J. P.: Gram-negative rod bacteremia, Amer. J. Med. **39:**952-960, 1965.

*Commercial kits utilizing these techniques may be obtained from Smith, Kline & French Laboratories, Philadelphia; Flow Laboratories, Rockville, Md.; Abbott Laboratories, S. Pasedena, Calif.; Ayerst Laboratories, New York; BioQuest, Cockeysville, Md.; and others.

5. Hoeprich, P. D.: Culture of the urine, J. Lab. Clin. Med. **56:**899-907, 1960.

6. Kaitz, A. L., and Williams, E. J.: Bacteriuria and urinary-tract infection in hospitalized patients, New Eng. J. Med. **262:**425-430, 1960.

7. Kass, E. H.: Asymptomatic infections of the urinary tract, Trans. Assoc. Amer. Physicians **69:**56-64, 1956.

8. Kass, E. H.: Pyelonephritis and bacteriuria; a major problem in preventive medicine, Ann. Intern. Med. **56:**46-53, 1962.

9. Leigh, D. A., and Williams, J. D.: Method for the detection of significant bacteriuria in large groups of patients, J. Clin. Path. **17:**498-503, 1964.

10. MacDonald, R. A., Levitin, H., Mallory, G. K., and Kass, E. H.: Relation between pyelonephritis and bacterial counts in urine; autopsy study, New Eng. J. Med. **256:**915-922, 1957.

11. Mackay, J. P., and Sandys, G. H.: Laboratory diagnosis of infections of urinary tract in general practice by means of dip-inoculation transport medium, Brit. Med. J. **2:**1286-1288, 1965.

12. Mackay-Scollay, E. M.: A simple quantitative and qualitative microbiological screening test for bacteriuria, J. Clin. Path. **22:**651-653, 1969.

13. Montgomerie, J. Z., Kalmanson, G. M., and Guze, L. B.: Use of catalase test to detect significant bacteriuria, Amer. J. Med. Sci. **251:**184-187, 1966.

14. Mou, T. W., and Feldman, H. A.: The enumeration and preservation of bacteria in urine, Amer. J. Clin. Path. **35:**572-575, 1961.

15. Prother, G. C., and Sears, B. R.: In defense of the urethral catheter, J. Urol. **83:**337-344, 1960.

16. Pryles, C. V.: The diagnosis of urinary tract infection, Pediatrics **26:**441-451, 1960.

17. Simmons, N. A., and Williams, J. D.: A simple test for significant bacteriuria, Lancet **1:**1377-1378, 1962.

18. Stamey, T. A., Gavan, D. E., and Palmer, J. M.: The localization and treatment of urinary tract infections; the role of bactericidal urine levels as opposed to serum levels, Medicine **44:**1-36, 1965.

19. Taylor, W. I.: Rapid detection of bacteria in urine, Lab. Med. **2:**29, 1971.

11 Microorganisms encountered in the genital tract

The microbial flora of the normal human genital tract consists chiefly of nonpigmented staphylococci and gram-positive rods. In normal females the vaginal flora varies considerably with the pH of the secretions and the amount of glycogen present in the epithelium; these factors in turn depend on ovarian function. In most instances, however, microaerophilic lactobacilli (the Doederlein bacillus groups[9]) predominate, together with gram-negative enteric bacilli, species of *Bacteroides*, enterococci, species of *Haemophilus (Corynebacterium)*, and coagulase-negative staphylococci.

The normal cervix is generally sterile or contains only a few bacteria, probably because of its relatively alkaline reaction. The organisms present are identical with those found in the upper vagina.

The bacterial flora of the vulva is a mixture of organisms present on the skin of this area, including the acid-fast saprophyte *Mycobacterium smegmatis,* and other bacteria descending from the vagina.

The organisms most frequently isolated from the genital tract are as follows:

Coliform bacilli, enterococci, and other intestinal commensals, including *Bacteroides* species and other anaerobes
Lactobacilli
Haemophilus species *(Corynebacterium vaginale)*
Trichomonas vaginalis
Candida albicans, other *Candida* species, and saprophytic yeasts

Beta hemolytic streptococci of Groups B and D
Nonpathogenic mycobacteria
Anaerobic streptococci (peptostreptococci)
Neisseria gonorrhoeae
Treponema pallidum
Haemophilus ducreyi
Mycobacterium tuberculosis
Mycoplasma species

ISOLATION OF THE GONOCOCCUS (NEISSERIA GONORRHOEAE)

The laboratory diagnosis of **gonorrhea** depends upon the demonstration of intracellular diplococci in smears and upon the isolation and identification of *Neisseria gonorrhoeae* by culture procedures. As a rule, gonococci may be found readily in smears of pus from **acute** infections, particularly in the male. In **chronic** infections, especially of the female, the value of the smear findings decreases; cultural methods generally yield a higher percentage of positive results.

Collection of specimens and primary inoculation of media

In the **female,** the best site to obtain a culture is the **cervix,** and this should be collected with care by an experienced professional. A sterile bivalve speculum is moistened with warm water (the usual lubricants contain antibacterial substances that may be lethal to gonococci) and inserted, and the cervical mucus plug is removed with a cotton ball and forceps. A sterile cotton or polyester-tipped applicator is then inserted into the endocervical canal, moved from side to side, allowed to remain for a few seconds, then removed and immediately inoculated to a Thayer-Martin chocolate agar plate (directions for preparation in Chapter 39) or a Transgrow bottle* (this must be inoculated in an upright position to prevent escape of contained CO_2 gas). The Thayer-Martin plate is streaked in a Z pattern with the swab and thereafter cross-streaked with a sterile bacteriological loop, preferably in the clinic; the Transgrow bottle is inoculated over the entire agar surface, working from the bottom up, after moistening the swab with excess moisture contained in the bottle. The Thayer-Martin plate should be placed in a candle jar (approximately 3% CO_2) within 1 hour and incubated at 35° to 36° C. for 20 hours before examining. The Transgrow bottle, used primarily for convenience in a physician's office, VD clinic, or small laboratory, should be incubated at 35° to 36° C. overnight before shipping; this provides sufficient growth for survival during prolonged transport. It is then sent to the diagnostic facility by mail or other convenient means, avoiding any marked temperature changes.

When the cervical culture is negative, and especially as a follow-up of treatment efficacy, the **anal culture** is recommended.† This specimen is easily obtained without using an anoscope, by carefully inserting a cotton-tipped applicator approximately 1 inch into the anal canal, and moving it from side to side to obtain material from the crypts. If the swab shows the presence of fecal material, it is discarded and another swab is used. The swab is inoculated to Thayer-Martin medium as previously described.

In special situations where a cervical specimen is not indicated, for example, in children or hysterectomized patients, a urethral or vaginal culture may be substituted.

In the **male** with a purulent urethral exudate, the examination of a gram-stained direct smear (made by **rolling** the swab over the slide, to preserve cell morphology), is ususally sufficient to confirm a clinical diagnosis of gonorrhea. However, since an appreciable number of males may be asymptomatic and the stained smear probably negative, a **urethral culture** is recommended. This is readily obtained by gently inserting a nasopharyngeal applicator (Falcon No. 2050) several centimeters into the urethra, removing it carefully, and immediately inoculating a Thayer-Martin plate or Transgrow bottle as previously described. In homosexuals, anal and pharyngeal specimens should also be obtained.

Occasionally, material from other sites may be submitted for gonococcal culture; these include throat swabs, freshly voided urine, joint fluid, swabs from the eye, and so forth. The swabs and sediments from centrifuged fluids should be immediately inoculated to Thayer-Martin medium and handled as indicated above.

Direct smears for gonococci

In order to obtain the highest percentage of positive findings in cases of suspected gonorrhea, both smears and cultures should be made. In the case of purulent material, slides should be prepared by the examining physician at the time the cultures are taken. Such smears should be carefully prepared by **rolling** the swab over the slide rather than by rubbing it on. This distributes the pus cells into layers, thereby permitting accurate observation of intracellular organisms. The practice of submitting moist material stuck between two slides is condemned for obvious reasons. Smears of urine and other fluid specimens should be prepared from the centrifuged sediment of these materials.

---

*Available from Baltimore Biological Laboratory, Cockeysville, Md.; Difco Laboratories, Detroit; shelf life at room temperature in excess of 3 months.
†Criteria and techniques for the diagnosis of gonorrhea: Venereal Disease Branch, Center for Disease Control, Atlanta, 1972.

Slides must be carefully stained by the Gram method and examined for the presence of characteristic **gram-negative diplococci** with their adjacent sides flattened (coffee-bean shaped). These organisms are generally found inside pus cells in acute gonorrhea, but in very early infections or in chronic gonorrhea they may be found extracellularly only, and frequently as single cocci. It should also be noted that recent administration of a specific antimicrobial agent such as penicillin may either eradicate or alter the morphology and staining reaction of *N. gonorrhoeae.* Care should be taken to distinguish gonococci from bipolar-staining, gram-negative coccobacilli, *Moraxella osloensis (Mima polymorpha* var. *oxidans),* which have been mistaken for neisseriae.[21] Biochemical tests must be carried out to differentiate them with certainty. Positive smears should be reported as "**intracellular (extracellular) gram-negative diplococci resembling gonococci were found; many (few) pus cells present.**"

Culture media

Many different and complex media have been introduced for the isolation of the gonococcus, but excellent results may be obtained by using the medium introduced by Thayer and Martin [22] in 1964. The original formula, an enriched chocolate agar medium containing the antibiotics ristocetin and polymyxin B, was recommended for the isolation of *Neisseria gonorrhoeae* and *N. meningitidis.* The authors reported that the medium showed an inhibitory action against other neisseriae and most species of Mimeae, and it suppressed *Pseudomonas* and *Proteus* species. The removal of one of the antibiotics from the market necessitated a suitable substitute, and the above authors reported the successful use of vancomycin, sodium colistimethate, and nystatin[23] (V-C-N inhibitor*). Comparison of the new medium with the original formula showed comparable growth of *N. gonorrhoeae* from both male and female patients, along with a greater inhibition of staphylococci and saprophytic neisseriae.

A further improvement in Thayer-Martin chocolate agar was the incorporation of a chemically defined supplement* that resulted in as good as or better recovery of gonococci than was achieved with previous media.[16]

Recently a modification of the Thayer-Martin medium—Transgrow—was introduced, prepared according to the formulation of Martin and Lester.[17] Recommended for transport and growth of pathogenic neisseriae (while maintaining their viability for at least 48 hours at room temperature), the medium supresses contaminating organisms in a manner similar to the Thayer-Martin formula. The agar content of Transgrow has been increased to 2% and the glucose content to 0.25%; the medium also contains trimethoprim lactate, which is inhibitary to *Proteus* species. It is available commercially,† packaged in a tightly closed, flat bottle under a partial CO_2 atmosphere, and has a shelf life of 3 to 4 months. Transgrow medium appears to be superior to other transport media for maintaining the viability of pathogenic neisseriae and can be mailed to the reference laboratory after overnight incubation without appreciable loss of gonococci. The procedure for inoculation of Transgrow bottles has been described previously. Comparative studies with Thayer-Martin medium by CDC's Veneral Disease Research Laboratory have indicated good correlation, both in the laboratory and in field trials.[18]

Handling of specimens in the laboratory

Ideally, specimens for the isolation of gonococci will be submitted on Thayer-

---

*Available from Baltimore Biological Laboratory, Cockeysville, Md.; Difco Laboratories, Detroit.

---

*Available from Baltimore Biological Laboratory, Cockeysville, Md.; Difco Laboratories, Detroit.
†Baltimore Biological Laboratory, Cockeysville, Md.; Difco Laboratories, Detroit.

Martin plates (or in Transgrow bottles) that have been properly inoculated previously and incubated in candle jars overnight. In a hospital situation, however, the specimen for culture is usually received on a cotton or polyester swab in holding medium such as the Culturette,* having been obtained from the patient a short time before. Occasionally, one may also receive a tube of purulent material from an aspirated joint, a freshly voided urine sample, or other specimens. In any case, the specimen (or sediment following centrifugation) should be inoculated **immediately** to a freshly prepared Thayer-Martin plate (brought to room temperature), using the Z and cross-streaking procedure previously described. Swab specimens should not be left at room temperature unless in Transgrow or other transport media, and **under no circumstances** in a 36° C. incubator—gonococci either die off or are rapidly overgrown by commensal bacteria. However, the swab may be refrigerated at 4° to 6° C. for several hours before inoculation, without a noticeable reduction in recovery of gonococci (not true of meningococci, which are notably cold sensitive).

All cultures, including previously inoculated Thayer-Martin plates (or Transgrow bottles received with loosened caps), are placed in a CO_2 incubator (5% to 10% CO_2) or in a candle jar with a tight-fitting lid,† containing a short, thick, smokeless candle affixed to a glass slide that is lighted inside the jar before closing with the lid. This will generate approximately 3% CO_2 before extinguishing itself.

The plates or bottles are incubated at 35° to 36° C.‡ overnight and examined for growth; cultures showing no growth are returned to the incubator for a total period of not less than 48 hours.* Typical growth of *N. gonorrhoeae* appears as translucent, mucoid, raised colonies of varying size; in Transgrow bottles, colonies may or may not appear typical. The **oxidase test** is used to verify the presence of gonococcus colonies, and along with the Gram stain of those colonies reacting positively, serves to presumptively identify *N. gonorrhoeae* from a urogenital site. These procedures and further identification tests by carbohydrate fermentation and FA may be found in Chapter 19.

Isolation of gonococci from synovial fluid

An interesting paper from Holmes and colleagues[7] at the USPH Hospital in Seattle reports on the recovery of *N. gonorrhoeae* from "sterile" synovial fluid in gonococcal arthritis by means of a medium developed for the propagation of L forms of those organisms in vitro. The medium was a modification of that introduced by Bohnhoff and Page[1] for *N. meningitidis* L forms and consisted of trypticase soy broth supplemented with 10% sucrose and 1.25% agar, to which was added 20% inactivated horse serum after sterilization. After inoculation, the surface of the medium was overlaid with 0.5 ml. of the same medium containing one half the agar concentration, and following 72 hours' incubation, oxidase-positive, reverting L form–type colonies developed. On subculture to chocolate agar, tiny colonies that were identified as *N. gonorrhoeae* by FA staining and by carbohydrate fermentation grew out. No growth occurred on regular chocolate agar or Thayer-Martin agar inoculated at the same time, after 96 hours' incubation.

The authors hypothesized that the gonococci may have existed in the synovial fluid in a cell wall–deficient, osmotically fragile state, in which they are not usually isolated by conventional techniques. We agree that a further trial of this medium in patients with suspected gonococcal arthritis is certainly warranted.

---

*Available from Scientific Products, Evanston, Ill.
†Available from Erno Products Co., 65 N. 2nd St., Philadelphia 19106; EC jar (clear) 120 mm. mouth opening #C-3118, with metal screw cap lid. This will hold approximately twelve 90-mm. plates and a candle.
‡The incubator should be adjusted to this temperature, since many strains of *N. gonorrhoeae* will not grow well at 37° C.

---

*Nearly one half of positive cultures require this incubation period.

MISCELLANEOUS VENEREAL DISEASES
Syphilis

Although full discussion is not intended to be within the scope of this text, mention should be made of the serological diagnosis of **syphilis** and the methods most frequently used in making this diagnosis. These tests are basically of two types: the **nontreponemal** antibody tests, such as the VDRL and rapid plasma reagin (RPR) flocculation tests, and the Kolmer complement fixation test; or the **treponemal** antibody tests, such as the Treponema pallidum immobilization (TPI) and fluorescent antibody-absorption (FTA-ABS) tests. Further discussion of these may be found in Chapter 29.

Mention should be made of **dark-field** microscopy as an aid to the diagnosis of early syphilis. An ample drop of tissue fluid expressed from a primary lesion is placed on a coverglass, which is inverted over a glass slide and pressed to make a thin film. This preparation is then examined microscopically, using oil immersion and a properly adjusted dark-field illumination. The presence of motile treponemes of characteristic morphology in a typical early lesion establishes the diagnosis.

A fluorescein-labeled *Treponema pallidum* conjugate has recently become available* for the direct identification of *T. pallidum* by the immunofluorescent technique. Material from the lesion is applied to a slide and air dried; the conjugate is then added, rinsed off, and dried; a cover slip is applied, and the preparation is examined by fluorescence microscopy. Preparations may be made, dried, and examined later, eliminating the necessity of immediate dark-field examination.

Chancroid

Chancroid is a soft chancre of the genitalia of venereal origin, caused by *Haemophilus ducreyi* (Ducrey bacillus), an ex-

tremely small gram-negative rod occurring singly or in small clumps.

H. ducreyi can be demonstrated by the following procedure:[2]

Ten ml. of the patient's blood is distributed in 5-ml. amounts in sterile screw-capped tubes. After clotting has occurred, the serum is removed, transferred aseptically to another sterile tube and inactivated in a 56° C. water bath for 30 minutes. After cooling, the serum is inoculated with material obtained from the undermined border of the genital lesion (previously cleansed with saline-moistened gauze) and incubated at 35° C. for 48 hours. A gram-stained smear of the growth is then examined for the presence of coccobacillary gram-negative organisms in tangled chains or in long parallel strands ("schools of fish"), which are diagnostic of *H. ducreyi*.

Granuloma inguinale

Granuloma inguinale is an infection of the genital region, characterized by a slowly progressive ulceration and caused by *Calymmatobacterium granulomatis*. This organism appears within the cytoplasm of large mononuclear cells as a small (1 to 2 μ), plump, heavily encapsulated coccobacillus and is best observed in impression smears of cleansed tissue obtained by punch biopsy at the edge of the lesion, stained with the Wright stain. A diligent search of smears prepared on several occasions may be necessary in order to find the organisms.

Lymphogranuloma venereum (LGV)

Lymphogranuloma venereum is caused by one of a large group of gram-negative intracellular parasites responsible for some important human pathogens, including psittacosis and trachoma. Once considered to be large viruses, they are now classified as **chlamydiae,** and may have been derived from gram-negative bacteria.[10] The diagnosis is made by demonstrating a skin hypersensitivity to an antigen prepared from infected yolk sacs (Lygranum) and a fourfold rise in complement fixation titer

---

*Available from Baltimore Biological Laboratory, Cockeysville, Md., Catalog No. 40806.

on paired (acute and convalescent) sera, using a similar antigen. A serological test for syphilis (STS) should also be done on these patients in order to rule out syphilis.

TRICHOMONAD INFECTION

The most practical method of confirming a diagnosis of trichomonad infection is by the demonstration of **actively motile** flagellates in a saline suspension of vaginal or urethral discharges. The preparation must be examined shortly after collecting, or the characteristic motility and morphology may not be apparent. Occasionally, cultures of vaginal specimens, semen, urine, and other materials will reveal trichomonads when wet smears are negative. The Kupferberg[14] medium is recommended for this purpose, with the addition of 5% human serum and antibiotic (chloramphenicol, 1 mg. per milliliter of medium). The base medium may be obtained in dehydrated form.* This method may be more convenient in some situations, since cultures are examined after 2 to 4 days' incubation, rather than immediately.

CANDIDA INFECTION

The yeastlike fungus *Candida albicans* is a frequent cause of vaginitis, and it can readily be identified by the examination of gram-stained smears and growth on Sabouraud dextrose agar slants. Methods are described in Chapter 32.

ISOLATION OF HAEMOPHILUS SPECIES (PROBABLY CORYNEBACTERIUM VAGINALE)

The bacteriological examination of cases of nonspecific urethritis and vaginitis frequently fails to provide the clinician with any specific agent against which to direct therapy. Gardner and Dukes[5] reported the isolation of a small, pleomorphic, gram-negative bacillus from many of their cases of nonspecific vaginitis, and they named it *Haemophilus vaginalis*. The organism

grew out on sheep blood agar plates that had been incubated for 24 to 48 hours in a candle jar as minute colorless, transparent, pinpoint colonies, best seen by oblique illumination, and frequently surrounded by a clear zone of hemolysis. The characteristic "puff ball" colony appeared in thioglycollate medium enriched with ascitic fluid, and subcultures of these on sheep blood plates produced the colonies just described.

Zinnemann and Turner[24] have reported the successful cultivation of *H. vaginalis* on a medium containing neither hemin (X factor), nicotinamide adenine dinucleotide (V factor), nor other coenzymes required by members of the genus *Haemophilus,* and have proposed reclassification of this organism as *Corynebacterium vaginale.* Their work has been amply confirmed by Dunkelberg and coworkers[4] at the Venereal Disease Research Laboratory, CDC, who also suggest that the bacterium be reclassified as *C. vaginale.*

Corynebacterium (Haemophilus) vaginale is apparently associated with vaginitis in the pregnant female,[15] although symptoms are generally mild or absent; the organism has also been implicated in puerperal pyrexia, nongonococcal urethritis of the male, and other genitourinary infections.[4]

NONGONOCOCCAL URETHRITIS ASSOCIATED WITH T STRAINS OF MYCOPLASMAS

The microorganisms of the *Mycoplasma* group, often called PPLO, are the smallest free-living organisms that can be cultivated on artificial media. They lack a rigid cell wall, require protein-enriched media for growth, and exhibit a characteristic "fried egg" colony when growing on solid media.

Currently, seven mycoplasmas of human origin have been described.[3] Of these, *M. hominis, M. fermentans,* and T strains are found in the human genitourinary tract. T-strain mycoplasmas have been associated with from 60% to

---

*Baltimore Biological Laboratory, Cockeysville, Md.; Difco Laboratories, Detroit.

over 90% of cases of nongonococcal urethritis in males, whereas their natural occurrence in a similar group of normal controls ranged from 21% to 26%, as reported by Shepard.[19] These findings suggested that T strains of mycoplasmas played an important etiological role in this genital infection. A urease color test medium for their detection in clinical material also has been introduced,[20] which has proved sensitive and specific for the recovery and identification of T strains from clinical material (Chapter 39, No. 55e).

MICROBIAL AGENTS IN HUMAN ABORTION

Listeria monocytogenes is most frequently associated in humans with septic perinatal infections in the female (some with bacteremia) and purulent meningitis, meningoencephalitis, or miliary granulomatosis in the neonate. The mother may exhibit flulike symptoms with a low-grade fever in the last trimester of her pregnancy and subsequently may abort or deliver a stillborn baby.[11] Postpartum cultures from the vagina, cervix, or urine may be positive for *L. monocytogenes* from a few days to several weeks, although the patient may appear asymptomatic.[6] In uncontaminated material (blood, cerebrospinal fluid), *L. monocytogenes* can be readily isolated by conventional techniques, but contaminated specimens or tissue may require selective media and cold enrichment procedures. Refer to Chapter 27 for a further description of these methods.

Although the place of *Vibrio fetus* in causing abortion in cattle is well established, its role in human abortion is not conclusive. In describing 17 human infections caused by *V. fetus*, Elizabeth King[12] of the Communicable Disease Center noted that three of these patients (from the French literature) had accompanying problems of pregnancy. Hood and Todd[8] of Charity Hospital in New Orleans reported what is apparently the first human case of *V. fetus* infection involving

pregnancy in the Western Hemisphere. The bacterium was isolated from the placenta of the mother and the brain of the aborted fetus.

A recent report on the recovery of mycoplasmas from fetal membranes associated with spontaneous abortions and premature births suggested to Kundsin and Driscoll[13] that mycoplasmas may be a cause of human reproductive failure. Both T strains and *Mycoplasma hominis* were recovered from prenatal cervical cultures; in general, when compared to controls, there seemed to be a greater tendency to premature labor in women harboring both of these mycoplasmas. The authors hypothesized that reported success in treating reproductive failures with tetracycline antibiotics may be due to unintentional eradication of the mycoplasmas.

It is apparent that the clinical microbiologist should be aware of all these agents and their role in human abortion and, through close communication with the clinician, offer microbiological aid in their early detection.

REFERENCES

1. Bohnhoff, M., and Page, M. I.: Experimental infection with parent and L-phase variants of *Neisseria meningitidis*, J. Bact. **95:**2070-2077, 1968.
2. Borchardt, K. A., and Hoke, A. W.: Simplified laboratory technique for diagnosis of chancroid, Arch. Dermat. **102:**188-192, 1970.
3. Crawford, Y. E.: Mycoplasma. In Blair, J. E., Lennette, E. H., and Truant, J. P., editors: Manual of clinical microbiology, Bethesda, Md., 1970, American Society for Microbiology.
4. Dunkelberg, W. E., Skaggs, R., and Kellogg, Jr., D. S.: A study and new description of *Corynebacterium vaginale (Haemophilus vaginalis)*, Amer. J. Clin. Path. **53:**370-377, 1970.
5. Gardner, H. L., and Dukes, C. D.: *Haemophilus vaginalis* vaginitis, Amer. J. Obstet. Gynec. **69:**962-976, 1955.
6. Gray, M. L., and Killinger, A. H.: *Listeria monocytogenes* and listeric infections, Bact. Rev. **30:**309-382, 1966.
7. Holmes, K. K., Gutman, L. T., Belding, M. E., and Turck, M.: Recovery of *Neisseria gonorrhoeae* from "sterile" synovial fluid in gonococcal arthritis, N. Eng. J. Med. **284:**318-320, 1971.
8. Hood, M., and Todd, J. M.: *Vibrio fetus;* a cause of human abortion, Amer. J. Obstet. Gynec. **80:**506-511, 1960.

9. Hunter, C. A., Jr., Long, K. R., and Schumacher, R. R.: A study of Doderlein's vaginal bacillus, Ann. N. Y. Acad. Sci. **83:**217-226, 1959.

10. Jawetz, E., Schacter, J., and Hanna, L.: Psitacosis–lymphogranuloma venereum–trachoma group of agents. In Blair, J. E., Lennette, E. H., and Truant, J. P., editors: Manual of clinical microbiology, Bethesda, Md., 1970, American Society for Microbiology.

11. Kelly, C. S., and Gibson, J. L.: Listeriosis as a cause of fetal wastage. Obstet. Gyn. **40:**91-97, 1972.

12. King, E. O.: Human infection with *Vibrio fetus* and a closely related vibrio, J. Infect. Dis. **101:**119-128, 1957.

13. Kundsin, R. B., and Driscoll, S. G.: Mycoplasmas and human reproductive failure, Surg., Gynec. Obstet. **131:**89-92, 1970.

14. Kupferberg, A. B., Johnson, G., and Sprince, H.: Nutritional requirements of *Trichomonas vaginalis,* Proc. Soc. Exp. Biol. Med. **67:**304-308, 1948.

15. Lewis, J. F., O'Brien, S. M., Ural, U. M., and Burke, T.: *Corynebacterium vaginale* vaginitis in pregnant women, Amer. J. Clin. Path. **56:**580-583, 1971.

16. Martin, J. E., Jr., Billings, T. E., Hackney, J. F., and Thayer, J. D.: Primary isolation of *N. gonorrhoeae* with a new commercial medium, Public Health Rep. **82:**361-363, 1967.

17. Martin, J. E. and Lester, A.: Transgrow, a medium for transport and growth of *Neisseria gonorrhoeae* and *Neisseria meningitidis,* HSMHA Health Reports **86:**30-33, 1971.

18. Printz, D. W., VD Branch, Center for Diseae Control, Atlanta. Personal communication, 1972.

19. Shepard, M. C.: Nongonococcal urethritis associated with human strains of "T" mycoplasmas, J.A.M.A. **211:**1335-1340, 1970.

20. Shepard, M. C., and Lunceford, C. D.: Urease color test medium U-9 for the detection and identification of "T" mycoplasmas in clinical material, Appl. Microbiol. **20:**539-543, 1970.

21. Svihus, R. H., Lucero, E. M., Mikolajczyk, R. J., and Carter, E. E.: Gonorrhea-like syndrome caused by penicillin-resistant Mimeae, J.A.M.A. **177:**121-124, 1961.

22. Thayer, J. D., and Martin, J. E., Jr.: A selective medium for the cultivation of *N. gonorrhoeae* and *N. meningitidis,* Public Health Rep. **79:**49-57, 1964.

23. Thayer, J. D., and Martin, J. E., Jr.: Improved medium selective for cultivation of *N. gonorrhoeae* and *N. meningitidis,* Public Health Rep. **81:**559-562, 1966.

24. Zinnemann, K., and Turner, G. C.: The taxonomic position of *"Haemophilus vaginalis" (Corynebacterium vaginale),* J. Path. Bact. **85:**213-219, 1963.

12 Microorganisms encountered in cerebrospinal fluid

Bacteriological examination of the spinal fluid is an essential step in the diagnosis of any case of suspected meningitis. The specimen must be collected under sterile conditions and transported to the laboratory without delay. If a viral etiology is suspected, a portion of the fluid must be immediately frozen for subsequent attempts at isolation of the virus.

Acute bacterial meningitis is an infection of the meninges—the membranes covering the brain and spinal cord—and is caused by a variety of gram-positive and gram-negative microorganisms, predominately *Haemophilus influenzae, Neisseria meningitidis,* and *Streptococcus pneumoniae.* Bacterial meningitis also may be secondary to infections in other parts of the body, and rarely, *Salmonella* species, coliform bacilli, staphylococci, or mycobacteria may be recovered; in neonatal meningitis, *Escherichia coli* is the most frequent organism isolated, along with Group B beta hemolytic streptococci, an increasingly important pathogen, and *Listeria monocytogenes.*

In bacterial meningitis the cerebrospinal fluid is usually **purulent,** with an increased white cell count (generally greater than 1,000 per cubic millimeter), a predominance of polymorphonuclear cells, and a reduced concentration of spinal fluid glucose. On the other hand, in meningitis caused by the tubercule bacillus or by nonbacterial agents, such as viruses, fungi, or protozoa, the fluid is usually nonpurulent, with a low cell count of the mononuclear type and a normal or moderately reduced glucose content. However, in early bacterial meningitis the cell count may be low and without a multinuclear response, whereas in early tuberculous or viral meningitis polymorphonuclear cells may predominate.

Since there are no dependable signs that will enable the clinician to differentiate the meningitides clinically, rapid and accurate means of identifying the etiological agents must be available in the laboratory. It is therefore strongly recommended that a complete microbiological work-up, including smear and culture, be carried out on all cerebrospinal fluid specimens from cases of suspected meningitis, whether the fluid be clear or cloudy.

Generally, in a case of suspected meningitis the fluid is submitted for chemical and cytological examination as well as for culturing. Since the amount of fluid provided is usually small, it is suggested that the specimen be centrifuged for 15 minutes (except where *Cryptococcus neoformans* is suspected) at 2,500 r.p.m. as soon as it is received. The supernatant fluid is then removed with a sterile capillary pipet to another tube for chemical or serological studies, leaving the sediment and one or two drops of fluid for culture procedures. In this way the entire specimen may be concentrated by centrifugation, obviating its division into several different tubes for the other examinations.

Purulent (cloudy) fluids should be examined **immediately** by a gram-stained smear, followed by appropriate culturing procedures. In these acute forms of meningitis the etiological agent frequently can be demonstrated in stained films of the specimen. In some cases, as in *Haemophilus influenzae* and pneumococcus infections, **the organism can be identified immediately by a capsular swelling (quellung) test with specific antiserum.*** Refer to Chapters 18 and 35 for procedures.

The following organisms (in order of their frequency) are isolated from cerebrospinal fluid:

> *Haemophilus influenzae,* type b (infants and children)
> *Neisseria meningitidis* (meningococcus), most common in Great Britain
> *Streptococcus pneumoniae* (pneumococcus)
> *Mycobacterium tuberculosis*
> Staphylococci, streptococci (including enterococci)
> *Cryptococcus neoformans*
> Coliform bacilli,
> *Pseudomonas* and *Proteus* species, *Edwardsiella tarda*
> *Bacteroides* species
> *Listeria monocytogenes*
> *Acinetobacter calcoaceticus*
> *Leptospira* species
> Viruses, fungi, and other agents

ROUTINE CULTURE FOR COMMONLY ISOLATED PATHOGENS

For the cultivation of *Haemophilus influenzae,* meningococci, pneumococci, streptococci, and other gram-positive cocci, and also the aerobic and anaerobic gram-negative bacilli from cerebrospinal fluid, the following procedures are recommended (keep specimens at 36° C. while awaiting examination):

1. Using a loopful of the fluid or sediment obtained by centrifugation, inoculate the following media: (a) a blood agar plate, (b) a chocolate agar plate, (c) a tube of enriched thioglycollate medium and other anaerobic media, if indicated.

2. Make a thin smear for gram staining and if appropriate, an India ink preparation (described in Chapter 40) for encapsulated *Cryptococcus neoformans.**

3. To the sediment remaining in the tube add approximately 5 ml. of dextrose ascitic fluid semisolid agar (see Chapter 39). This is done by flaming the mouths of both tubes, cooling, and pouring the semisolid medium over the remaining sediment. In our experience the diagnosis of a bacterial meningitis has been established frequently by demonstrating growth in this semisolid medium when streak plates and smears have been negative.

4. Incubate the plates in a candle jar and the fluid media under ordinary conditions at 35° to 36° C. Examine after overnight incubation (and daily thereafter) for the presence of growth. The semisolid medium will become turbid and turn yellow (due to the production of acidity that is indicated by the phenol red present) when growth has occurred. Make a gram-stained smear of this medium and subculture to appropriate solid media after reading the smear. It is important to include antibiotic sensitivity discs on these plates for obvious reasons.

5. Hold all cultures for 72 hours (5 to 7 days when antibiotics have been given before culture was obtained) before discarding as negative.

6. Identify all organisms isolated by the methods described in subsequent sections on the various genera.

---

*Available from Difco Laboratories, Detroit; Hyland Laboratories, Los Angeles. Antisera prepared by Erna Lund, Statens Serum-institut, Copenhagen, Denmark, are also recommended. Particularly useful is **Omni-serum,** a polyvalent pneumococcus antiserum that gives capsular swelling reactions with all eighty-two types.

*Gram-positive and gram-negative artifacts resembling yeast cells have been observed in the spinal fluid of patients who have had recent myelograms.[1]

In all cases of purulent meningitis an attempt should be made to detect the **primary focus** of infection. For this purpose, blood for culture should be taken and cultures also should be made from the nasopharynx, middle ear, or other sites **before** antibacterial therapy is initiated. Petechial rash is common in meningococcal meningitis; the organism can sometimes be demonstrated by smear and culture of the petechiae.

HAEMOPHILUS INFLUENZAE MENINGITIS

Haemophilus influenzae type **b,** is the most frequent cause of acute purulent meningitis in the small child and only rarely the cause in the adult (probably due to acquisition of *H. influenzae* antibodies). The organism enters by way of the respiratory tract, where it produces a nasopharyngitis, sinusitis, or middle ear infection. It may reach the bloodstream from these sites and be carried to the meninges, where it produces the characteristic symptomatology of acute meningitis.

When morphologically typical, short, slender gram-negative rods appear in sufficient numbers in the cerebrospinal fluid (either uncentrifuged or in the sediment), a direct capsular swelling procedure may be carried out using type **b** rabbit antiserum (Difco, Hyland).

The initial culture procedures outlined in the previous section are satisfactory for the isolation of *H. influenzae;* further cultural details will be found in Chapter 22.

MYCOBACTERIUM TUBERCULOSIS AS A CAUSE OF MENINGITIS

Cerebrospinal fluid from patients with tuberculosis meningitis contains at best only a few recoverable tubercle bacilli. Acid-fast stains of the sediment, therefore, rarely reveal their presence. The chance of recovering the organisms increases in proportion to the volume of spinal fluid submitted for examination. About 10 ml. are required for an adequate culture.

Although cultural procedures are pre-ferred for the recovery of tubercule bacilli from cerebrospinal fluid, the clinical significance of finding any acid-fast bacilli by smear is obvious. Because of this urgency to discover an etiological agent, two methods for demonstrating mycobacteria in stained smears are included:

1. Preparations may be made from the pellicle or coagulum ("fibrin web") that forms in some specimens. This formation takes place when the spinal fluid is allowed to stand undisturbed at room temperature or in a refrigerator overnight. Carefully remove the formed coagulum with a **new** loop (a rough, old wire loop is inadequate, since it balls up the material and makes removal difficult). Crush the coagulum by pressure between an albumin-coated slide and a coverglass. Air dry, flame, and then stain the resulting films by the acid-fast or fluorochrome method.

2. Acid-fast bacilli occasionally can be demonstrated in the sediment obtained by centrifuging the cerebrospinal fluid at 3,000 r.p.m. for 15 to 30 minutes. Since this method requires the use of the entire sediment and is not always reliable, it is recommended that the sediment be used instead for inoculation of culture media or guinea pig injection. Techniques for these procedures will be found in Chapter 28 on *Mycobacterium tuberculosis.*

LISTERIA MONOCYTOGENES AS A CAUSE OF MENINGITIS

In recent years attention has been focused on the role of *Listeria monocytogenes* in acute meningitis and meningoencephalitis in human beings. In a study of more than 420 cases of human listeriosis in the United States, Gray and Killinger[10] found that over 80% occurred in neonates or in adults over 40 years of age who had meningitis or meningoencephalitis. Less than 7% were perinatal infections, with only an influenzalike illness of the mother, but frequently there resulted premature delivery of a stillbirth or acutely ill infant. The organism is also occasionally recovered from the blood and spinal fluid of debilitated patients or those undergoing chemotherapy.[15] The organism is probably too often discarded in the laboratory as a contaminating diphtheroid when isolated from cerebrospinal or subdural fluid, blood, or other clinical specimens; a high index of suspicion on the part of the microbiologist is the best aid to its recovery. The reader is referred to Chapter 27 on *Listeria,* where identification procedures may be found.

CRYPTOCOCCUS NEOFORMANS AS A CAUSE OF MENINGITIS

Cryptococcus neoformans, a yeastlike fungus, has been responsible for a number of meningeal infections in man, some of which have been diagnosed erroneously as brain tumors or as tuberculosis meningitis. Budding yeast cells with abundant capsules occur in masses in granulomatous tissue; the cerebrospinal fluid, although generally not grossly purulent, may contain numbers of yeast cells. These are sometimes mistaken for lymphocytes in spinal fluid cell counts, since cryptococci resemble them in size and shape and the correct diagnosis may be overlooked unless cultures are made. The organism is heavily encapsulated, however, and can be identified readily by mixing a loopful of sediment from the centrifuged spinal fluid with a loopful of India ink on a slide, covering with a thin coverglass, scanning under the high dry objective, and confirming under oil immersion.* Typically, large, clear hyaline **capsules** are seen surrounding the yeast forms, some of which show single buds. Since the number of cryptococci in cerebrospinal fluid may be of the magnitude of only one cell per 15

to 20 ml. of fluid, Utz[19] has recommended that the entire specimen be inoculated directly to a series of culture slants. He has frequently made an etiological diagnosis only after inoculating 40 to 50 ml. of fluid collected during encephalography. Further identification procedures will be found in Chapter 32.

LEPTOSPIRA AS A CAUSE OF MENINGITIS

Human **leptospirosis** is a disease of protean manifestations, varying from a mild and inapparent infection to a severe and sometimes fatal illness with deep jaundice and profound prostration. It is transmitted by contact—either directly or indirectly—with the urine of carriers. The central nervous system may be involved in many cases of severe infection, and leptospirae may be recovered from the spinal fluid. This is particularly common in infections caused by *Leptospira icterohaemorrhagiae, L. canicola,* and *L. pomona.* In a report by Heath and co-workers[11] of 483 cases of human leptospirosis in the United States, the central nervous system was the organ system most frequently involved (235 cases, or 68%). Because of the diversity of clinical impressions, however, the authors concluded that many human infections, particularly of a mild type, remain undiagnosed in the United States.

Recovery of the organism (see description of techniques, Chapters 6 and 29) from the cerebrospinal fluid or demonstration of a rise in specific antibody may lead to a definitive diagnosis. All suspected cultures should be confirmed by a leptospirosis reference laboratory, such as the Center for Disease Control (CDC).

MIMA POLYMORPHA (ACINETOBACTER CALCOACETICUS) AS A CAUSE OF MENINGITIS

A disease simulating meningococcal meningitis, both clinically and bacteriologically, may be produced by a gram-negative coccobacillus, *Mima polymorpha,*[16] first described by de Bord in 1948[5] and presently classified as *Acinetobacter*

---

*Some authors believe that centrifugation may injure the fragile cryptococcal cells and recommend inoculating the uncentrifuged specimen directly to appropriate culture slants.

calcoaceticus. This organism appears coccoid on solid media and shows both bacillary and coccal forms in liquid media. It stains gram negative, frequently shows bipolar staining, is nonmotile, and does not reduce nitrates.

Moraxella osloensis, formerly classified as a member of de Bord's Mimeae *(M. polymorpha* var. *oxidans)* is similar in morphology and gram reaction to *Neisseria meningitidis* and is also oxidase positive. It has been incriminated in meningitis[20] and is best characterized by biochemical reactions, as described in Chapter 21.

PASTEURELLA MULTOCIDA AS A CAUSE OF MENINGITIS

The pleomorphic gram-negative coccobacillus *Pasteurella multocida* has been implicated in central nervous system infections, including meningitis.[13] Because of its appearance in stained films of spinal fluid, it has led to incorrect diagnoses of meningococcal, *Haemophilus influenzae,* or *M. polymorpha* types of meningitis. Unlike most of the gram-negative rods, *Past. multocida* is very susceptible to penicillin in vitro (either by the disc technique or by tube dilution). If such an organism is isolated, if it grows poorly, or not at all, on EMB agar or MacConkey agar, and if it produces no change in the butt of a TSI agar slant within 24 hours but does produce an acid reaction throughout the medium in 48 to 72 hours, the presence of *Past. multocida* should be considered. This may be confirmed biochemically and serologically by submitting the culture to a reference laboratory. See Chapter 22 for a further description of the organism.

FLAVOBACTERIUM SPECIES AS A CAUSE OF MENINGITIS

Flavobacterium species are gram-negative bacilli whose natural habitat is the soil and water. They have also been isolated from a variety of other sources, including hospital sink traps, nursery equipment, and water supplies.

A report by Brody and others[2] recorded two outbreaks of meningitis among newborn infants in hospital nurseries in which a gram-negative bacillus was isolated from the spinal fluid of 17 of the 19 infants; there were 15 deaths. The organism, first described in 1944, was studied by Elizabeth King, who found the organisms to be thin, nonmotile, gram-negative rods, proteolytic and nitrite negative, producing a small amount of indole (tested with Ehrlich reagent) and fermenting carbohydrates in nutrient broth after some delay. After overnight incubation at 36° C. a lavender-green discoloration is seen on blood agar. Another nursery outbreak was described by Cabrera and Davis,[3] who recorded 14 clinical cases of neonatal meningitis, with 10 deaths. The source of this nosocomial epidemic appeared to be a faulty sink trap in the premature nursery, and the organism isolated was similar to that described by King.

Because the organisms satisfy the requirements for inclusion in the genus *Flavobacterium,* King suggested the name *Flavobacterium meningosepticum* for them. For a more complete biochemical and serological characterization the reader is referred to King's paper.[14]

MENINGITIS CAUSED BY GRAM-NEGATIVE ENTERIC BACTERIA

Meningeal infections caused by coliform bacilli, species of *Pseudomonas* or *Proteus,* and other gram-negative enteric organisms are not very common, although an increase in hospital-associated infections may occur. In a series of 294 cases of bacterial meningitis studied at the Mayo Clinic[7,9] only 23 (7.7%) were caused by these facultative gram-negative bacilli. In 9 of the 23 cases *Escherichia coli* was the responsible pathogen, causing death in 3 out of 4 infected newborn infants, who appeared to have an unexplained and great susceptibility to this organism; other bacterial agents included *Klebsiella, Pseudomonas, Proteus,* and *Alcaligenes.* In 21 of the patients an underlying disease process was present; the infection was controlled in only 8 of the 23 patients with meningitis.

MENINGITIS DUE TO EDWARDSIELLA

A recent report by Sonnenwirth and Kallus[17] details the first case of meningitis due to *Edwardsiella tarda,* a recently described member of the family Enterobacteriaceae. Variously named "Asakusa group," "Bartholomew group" and "biotype 1483-59," the genus *Edwardsiella* (with a single species, *E. tarda*) was proposed by Ewing and colleagues[8] in 1965.

The organism is characterized by its marked production of hydrogen sulfide, a negative urease and beta-galactosidase (ONPG) reaction, and the fermentation of glucose and maltose but not lactose and mannitol. The reader is referred to the references cited for a further description of *E. tarda.*[6]

MENINGITIS CAUSED BY MORE THAN ONE BACTERIUM

In the series previously cited,[7] cultures of the cerebrospinal fluid of 40 patients (13.6%) revealed two or more organisms isolated simultaneously or subsequently during the course of the illness. These included *Streptococcus faecalis,* coagulase-negative staphylococci, *Staphylococcus aureus,* gram-negative enteric bacilli, and so forth in various combinations. In some instances the clinical significance of the additional organism was not readily determined.

In a similar study of 534 infants and children conducted at St. Louis Children's Hospital,[12] 20 patients (3.7%) were found to show two different bacterial species in initial cerebrospinal fluid cultures. *Haemophilus influenzae,* combined variously with *N. meningitidis, S. pneumoniae, E. coli, Staph. aureus, A. aerogenes (Klebsiella),* and *Enterobacter* occurred most frequently; the meningococcus or pneumococcus was also accompanied by other bacterial species. The authors pointed out that those having the responsibility for reading cerebrospinal fluid cultures should be aware of the possibility of the occurrence of simultaneous mixed bacterial meningeal infections.

ASEPTIC MENINGITIS

Aseptic or nonbacterial meningitis is a clinical syndrome rather than a disease of specific etiology. It is generally associated with a viral agent, although approximately one fourth of the cases may remain undiagnosed.

The disease is characterized by rapid onset with fever and symptoms referable to the central nervous system. In the majority of patients with viral encephalitides, the spinal fluid cell count is increased, with lymphocytes predominating (polymorphonuclear cells may appear at the onset). The protein content is generally elevated, while the spinal fluid glucose level is usually normal, an important diagnostic differential from bacterial meningitis, in which the glucose level is depressed.

Aseptic meningitis may be caused by a variety of viral agents, including Coxsackie Group B (occasionally A) viruses, poliovirus, lymphocytic choriomeningitis and arthropod-borne encephalitis viruses, enteroviruses, mumps virus, and herpes virus. The multiplicity of etiological agents requires a variety of laboratory examinations available only at a diagnostic virology laboratory. These include virus isolation by animal injection or by inoculation of embryonated eggs or tissue cell cultures from specimens of stool, throat washings, and cerebrospinal fluid, as well as complement fixation and neutralization tests on acute and convalescent sera. Since the only contribution of the bacteriology laboratory in such cases may be the responsibility for the proper collection and transportation of the necessary clinical materials, these techniques are considered more fully in Chapter 31.

Aseptic meningitis may also occur after measles, mumps, vaccinia, chickenpox, and smallpox. It may be caused by meningeal irritation, silent brain abscess, toxins, neoplasms, and allergens.[18]

EFFECT OF PRIOR ANTIBACTERIAL THERAPY ON RECOVERY OF BACTERIAL PATHOGENS

In a series of 310 cases of bacterial meningitis, Dalton and Allison[4] reported that approximately one half of the patients had received antibacterial drugs before admission to the hospital. This partial treatment reduced the recovery rate of the etiological agent by approximately 30%, particularly in the isolation of *Neisseria meningitidis, Streptococcus pneumoniae,* and a miscellaneous group.

There was also a reduction in the number of positive spinal fluid smears in the treated group, especially in meningococcal meningitis. Gram-positive organisms also tended to appear gram negative. The authors recommended caution in interpreting smears from these treated cases.

REFERENCES

1. Bartlett, R. C.: In Summary report, 1968, Chicago, Commission on Continuing Education (American Society of Clinical Pathologists).
2. Brody, B. H., Moore, H., and King, E. O.: Meningitis caused by an unclassified gram-negative bacterium in newborn infants, J. Dis. Child. **96:**1-5, 1958.
3. Cabrera, H. A., and Davis, G. H.: Epidemic meningitis of the newborn caused by flavobacteria, J. Dis. Child. **101:**289-295, 1961.
4. Dalton, H. P., and Allison, M. J.: Modification of laboratory results by partial treatment of bacterial meningitis, Amer. J. Clin. Path. **49:**410-413, 1968.
5. De Bord, G. G.: *Mima polymorpha* in meningitis, J. Bact. **55:**764-765, 1948.
6. Edwards, P. R., and Ewing, W. H.: Identification of Enterobacteriaceae, ed. 3, Minneapolis, 1972, Burgess Publishing Co.
7. Eigler, J. O. C., Wellman, W. E., Rooke, E. D., Keith, H. M., and Svien, H. J.: Bacterial meningitis. I. General review (294 cases), Proc. Staff Meet. Mayo Clin. **36:**357-365, 1961.
8. Ewing, W. H., McWhorter, A. C., Escobar, M. R., and Lubin, A. M.: *Edwardsiella,* a new genus of Enterobacteriaceae based on a new species, *E. tarda,* Int. Bull. Bact. Nomenclat. Taxon. **149:**33-38, 1965.
9. Gorman, C. A., Wellman, W. E., and Eigler, J. O. C.: Bacterial meningitis. II. Infections caused by certain gram-negative enteric organisms, Proc. Staff Meet. Mayo Clin. **37:**703-712, 1962.
10. Gray, M. L., and Killinger, A. H.: *Listeria monocytogenes* and listeric infections, Bact. Rev. **30:**309-382, 1966.
11. Heath, C. W., Jr., Alexander, J. D., and Galton, M. M.: Leptospirosis in the United States, New Eng. J. Med. **273:**912-922, 1965.
12. Herweg, J. C., Middelkamp, J. N., and Hartmann, A. F., Sr.: Simultaneous mixed bacterial meningitis in children, J. Pediat. **63:**76-83, 1963.
13. Hubbert, W. T., and Rosen, M. N.: *Pasteurella multocida* infections. II. *Pasteurella multocida* infection in man unrelated to animal bite, Amer. J. Public Health **60:**1109-1116, 1970.
14. King, E. O.: Studies on a group of previously unclassified bacteria associated with meningitis in infants, Amer. J. Clin. Path. **31:**241-247, 1959.
15. Louria, D. B., Blevins, A., and Armstrong, D.: *Listeria* infections, Ann. N. Y. Acad. Sci. **174:**545-551, 1970.
16. Olaffson, M., Lee, Y. C., and Abernathy, T. J.: *Mima polymorpha* meningitis; report of a case and review of the literature, New Eng. J. Med. **258:**465-470, 1958.
17. Sonnenwirth, A. C., and Kallus, B. A.: Meningitis due to *Edwardsiella tarda;* first report of meningitis caused by *E. tarda,* Amer. J. Clin. Path. **49:**92-95, 1968.
18. Swartz, M. N., and Dodge, P. R.: Bacterial meningitis; a review of selected aspects, New Eng. J. Med. **272:**898-902, 1965.
19. Utz, J. P.: Recognition and current management of the systemic mycoses, Med. Clin. N. Amer. **51:**519-527, 1967.
20. Waite, C. L., and Kline, A. H.: *Mima polymorpha* meningitis; report of case and review of the literature, J. Dis. Child. **98:**379-384, 1959.

13 Microorganisms encountered in wounds; anaerobic procedures

Although the microbial flora of infected wounds frequently is so varied, a group designated as **organisms most frequently isolated from wounds** would include the following:

> *Staphylococcus aureus*
> *Streptococcus pyogenes*
> Coliform bacilli (from the lower half of body)
> *Bacteroides* species and other anaerobic nonsporing gram-negative and gram-positive rods
> *Proteus* species
> *Pseudomonas* species
> *Clostridium* species
> Anaerobic cocci *(Peptococcus, Peptostreptococcus)*
> Enterococci

Another group, which might be labeled **organisms rarely isolated from wounds,** includes the following:

> *Clostridium tetani*
> *Francisella tularensis, Past. multocida*
> *Mycobacterium tuberculosis, M. marinum,* and other mycobacteria
> *Corynebacterium diphtheriae*
> *Bacillus anthracis*
> Systemic fungi *(Sporotrichum, Actinomyces, Nocardia,* and so forth)
> *Erysipelothrix insidiosa*

Since **anaerobic** microorganisms are the predominant microflora of humans and are constantly present in the intestinal tract, upper respiratory tract, and genitourinary tract, it is not unexpected to find them invading both usual and unusual anatomical sites, giving rise to severe and often fatal infections. This is particularly true when the host defenses, either natu-rally or artificially, have been so altered as to permit an overgrowth of these organisms. Therefore, it seems that anaerobes deserve more attention than they have been given and it behooves the microbiologist and technologist to familiarize themselves with the techniques for the isolation and identification of these indigenous bacteria.

Although anaerobic procedures are not more difficult to carry out than those used in aerobic bacteriology, only a strict adherence to basic principles and a degree of patience will ensure successful recovery of these pathogens. These principles include the following:*

1. **The selection of specimens most likely to contain clinically significant anaerobes.** As suggested by Sutter and co-workers,[19] there are certain bacteriological clues that should prompt the microbiologist to carry out anaerobic culturing of selected clinical material. Some of these specimens include:

    a. Pus from any deep wound or aspirated abscess, especially if associated with a **foul or fetid** odor, or containing "sulfur granules"

    b. Necrotic tissue or debrided material from suspected gas gangrene

---

*The reader is referred to several excellent anaerobic laboratory manuals,[8,9,19] for a more complete discussion of the subject.

c. Material from infections close to a mucous membrane

d. Uterine cultures from postabortal sepsis (avoiding vaginal contamination)

e. Material from abscesses of the brain, lung, or liver or from intra-abdominal, perirectal, subphrenic, or other sites

f. Aspirated fluids from normally sterile or obviously infected sites, including blood, peritoneal, pleural, synovial, or amniotic fluids

g. Food suspected of causing botulism or food poisoning

2. **The avoidance of normal flora in collection.** Since anaerobic organisms are part of the normal flora of body sites, such as the skin, oropharynx, intestinal tract, or external genitalia, the following are **not** routinely cultured anaerobically:

a. Swabs from the throat, nose, eye, ear, decubiti, superficial wounds, urethra, vagina, cervix, or rectum

b. Expectorated sputum, bronchial secretions, voided urine, feces, and gastric contents

Furthermore, all other specimens should be collected to minimize contamination by normal anaerobic microflora, and **collection by needle aspiration, rather than swab culture, is recommended.** This is particularly true in the following clinical situations, as recommended by Finegold[19]:

a. Pus from a closed abscess

b. Pleural fluid by thoracentesis

c. Urine by suprapubic bladder aspiration

d. Pulmonary secretions by transtracheal aspiration

e. Uterine secretions or sinus tract material by insertion of an intravenous type of plastic catheter through a decontaminated area and aspiration with a syringe

The air should be expelled from the syringe and its contents injected directly into a "gassed out" sterile tube, described below. If a swab **must** be used, a two-tube system has been recommended,[19] one tube containing the swab in oxygen-free CO_2, the other containing a prereduced and anaerobically sterilized (PRAS) transport medium, such as Cary and Blair semisolid medium.* After collection, the swab should be inoculated to appropriate culture media as soon as possible, or held at room temperature **(not refrigerated).** Some authors recommend placing the swab in a "gassed out" tube containing a few drops of salt solution.[9]

3. **The use of a "gassed out" collection tube.** As indicated previously, once a clinical specimen has left a body site where the Eh (oxidation-reduction potential) may be as low or lower than minus 250 millivolts,[9] any anaerobes present must be protected from the toxic effect of atmospheric oxygen until the specimen is properly set up anaerobically. For this reason the use of a "gassed out" collection tube is strongly recommended, since it will provide an anaerobic environment during transport to the laboratory. The outfit consists of a tube with a recessed rubber stopper and screw cap; it is prereduced by being flushed out with oxygen-free CO_2, then sterilized by autoclaving. The tube can be prepared by hand, with a modified Hungate method,[2] but although expensive, it is more readily available commercially.† Aspirated fluid, swabs of pus or portions of tissue may be collected in these tubes; if the stopper is removed, the tube should be held upright (to prevent loss of CO_2) and the specimen introduced rapidly to avoid as little exposure to air as possible.

4. **The use of freshly prepared or prereduced anaerobically sterilized (PRAS) culture media.** For the primary inoculation of specimens for anaerobic culturing, at least several types of solid media are required and should be used within several hours of pouring. Since this may require the preparation of these media on a daily

---

*Available commercially from Scott Laboratories, Fiskeville, R. I.; Hyland Laboratories, Costa Mesa, Calif.

†Anaport, Scott Laboratories, Fiskeville, R. I.

basis, which may not be practical for many laboratories, several alternates are suggested:

  a. Prereduced culture media (but not plating media) may be obtained commercially from several sources.*
  b. Freshly poured plates may be stored in anaerobic jars kept at room temperature, to be opened and used as needed.
  c. Plating media may be stored for not more than 72 hours in a relatively air-tight cabinet (for example, an unused incubator) that is fed with a constant stream of CO_2 (0.5 liters per minute)—probably the best procedure—suggested by Martin at the Mayo Clinic.[14] Growth characteristics of representative anaerobes indicated no adverse effects of storage in such apparatus.

Under no circumstances should plating media be refrigerated—cold temperatures increase the absorption of atmospheric oxygen.

  5. **The provision of a proper anaerobic environment.** Numerous methods have been introduced for providing an anaerobic environment, including the PRAS rolltube method of Hungate,[10] the anaerobic chamber[18] or glove box,[1] and the anaerobic jar.[4,7] It appears that the use of the anaerobic jar with a catalyst and hydrogen or nitrogen gas plus 5% CO_2 offers the most satisfactory and practical method for achieving an anaerobic environment in the clinical laboratory, and this will be discussed further in the next section. Other methods, including techniques that depend on displacement of oxygen by inert gases alone, by chemical means, or by cultivation in a vacuum are not recommended in routine clinical practice.

*Scott Laboratories, Fiskeville, R. I.; Hyland Laboratories, Costa Mesa, Calif.; McGaw Laboratories, Glendale, Calif. Recently, Robbin Reducible Media plates have become available from Scott Laboratories, in a variety of formulations. They contain cysteine hydrochloride and palladium chloride as reducing agents and may be reduced by being placed in an anaerobic jar for 24 hours before use.

METHODS OF OBTAINING ANAEROBIOSIS
Anaerobic jar method

One of the most satisfactory methods for the cultivation of anaerobes is the use of the **anaerobic jar,** a tightly sealed container in which the oxygen is completely eliminated by various means, including a catalyst. This principle was first applied by Laidlaw[12] in 1915 and later adapted by McIntosh and Fildes.[13] Many modifications of the McIntosh and Fildes jar have been introduced, including the Brewer,* GasPak,* and Torbal jar,* and are recommended for routine use.

The Brewer jar

The Brewer[7] modification of the Brown jar has for some time proved to be an essential piece of laboratory equipment. In it the oxygen is removed by means of an electrically heated platinized asbestos catalyst with the electric connection outside the jar (eliminating the danger of explosion). The reaction chamber within the lid is shielded by a heavy wire mesh screen. An H_2-CO_2-N_2 gas mixture is recommended, and is employed as follows[8]: the jar is loaded with the media to be incubated, along with an indicator for anaerobiosis (see following section), and an evenly rolled length of Plasticine with ends joined is placed on the jar rim. The jar lid is pressed down on the Plasticine and clamped in place, and the whole assembly is placed in a metal safety shield.

After evacuating the jar to approximately 30 cm. of mercury (by water or vacuum pump), the jar is slowly filled with a gas mixture containing 10% hydrogen, 10% carbon dioxide, and 80% nitrogen.† This evacuation and filling procedure is repeated two more times, the outlet tubing is

*Baltimore Biological Laboratory, Cockeysville, Md.; Torsion Balance Co., Clifton, N. J.; Baird & Tatlock, Ltd., London, England.
†These gases are available commercially in a range of cylinder sizes from Matheson Co., Inc., Joliet, Ill., East Rutherford, N. J., or Newark, Calif., along with the necessary regulators, gauges, valves, and so forth. Consultation with the hospital's oxygen therapy staff is suggested in setting up such an assembly.

clamped, the jar is disconnected from the vacuum-gas assembly, and the electrical element is connected and allowed to heat for 10 minutes. The jar is then disconnected and placed in the incubator.

The GasPak jar

The recent introduction by Brewer and Allgeier[3,4,5] of the GasPak anaerobic jar* for both 100-mm. and 150-mm. plates, a disposable hydrogen and carbon dioxide generator envelope,* and a disposable anaerobic indicator* makes possible the most practical system for the cultivation of anaerobes that is available to any microbiological laboratory. The polycarbonate plastic anaerobic jar, used with the disposable hydrogen generator, has no external connections, thereby eliminating the need for vacuum pumps, gas tanks, manometers, and so forth. It uses a room-temperature catalyst (palladium-coated alumina pellets), which obviates the need for an electrical connection to heat the catalyst.

To use this system the inoculated media are placed in the jar, along with one hydrogen generator envelope with a top corner cut off and a methylene blue anaerobic indicator.[6] After 10 ml. of water is introduced with a pipet into the envelope, the cover (containing the catalyst in a screened reaction chamber in the lid) is immediately placed in position and the clamps are applied and **screwed only hand tight** (rubber attachments on the clamp provide release of excessive pressure).

Hydrogen is generated in the jar by the following reaction:

$$Mg + ZnCl_2 + 2H_2O \xrightarrow{NaCl} MgCl_2 + Zn(OH)_2 + H_2 \uparrow$$

The hydrogen reacts with the oxygen in the presence of the catalyst:

$$2H_2 + O_2 \longrightarrow 2H_2O$$

An anaerobic atmosphere is thereby produced.

As this environment is achieved, condensed water will appear as a visible mist or fog on the inner wall of the jar, and the lid over the catalyst chamber will become warm. If this does not occur within 25 minutes, either the catalyst needs replacing* or the lid was not secured properly. After overnight incubation the methylene blue indicator should appear colorless and the jar will be under a slight positive pressure.

It should be pointed out that the disposable hydrogen generator may also be used in the Torbal jar or Brewer jar. (See the manufacturer's directions for such use.)

Since hydrogen is an **explosive** gas, every precaution must be taken to prevent a laboratory accident when using it:

1. Any open flame in the vicinity must be extinguished.
2. All jars must be inspected for cracks and discarded if faulty.
3. The metal screen inside a Brewer jar lid must be intact.
4. A wooden or metal safety shield should be used whenever the catalyst is activated electrically (heat).

Other methods

Other anaerobic methods mentioned previously include the use of PRAS culture media and roll tubes, introduced by Hungate,[10] with modifications by Holdeman and Moore[9] at the Virginia Polytechnic Institute's Anaerobe Laboratory, and the anaerobic glove box[1] or chamber.[18]

In the Hungate technique, a closed tube of PRAS medium is inoculated through the rubber stopper by needle and syringe; in the VPI technique, an open tube of PRAS medium is inoculated in the presence of a continuous stream of O_2-free gas introduced by a flame-sterilized cannula inserted in the tube.

The anaerobic chamber techniques uti-

---

*Baltimore Biological Laboratory, Cockeysville, Md.

*This should be done after each use, since excess moisture and hydrogen sulfide (from H_2S-producing organisms) will inactivate the catalyst. Workers at the VPI Anaerobe Laboratory[9] have suggested that the catalyst may be reactivated by being heated in a 160° C. drying oven for 2 hours and subsequently stored in an airtight container with desiccant, such as a discarded antibiotic disc cartridge container.

lize a plastic glove box or rigid chamber with attached and sealed gloves, which are used to manipulate the material inside this work area. The chamber is kept continuously anaerobic by a catalyst and hydrogen gas, and material is passed in and out through an interchange, a rigid appurtenance attached by a gas-tight seal to the chamber.

Several authors[11,16] recently have compared the three anaerobic systems—GasPak, glove box and roll tube—for their effectiveness in the isolation of anaerobic organisms from clinical material, and have concluded that recovery of most anaerobes was comparable in all three systems, but that the **GasPak** method was as effective as the other more complex methodologies. The interested reader is referred to the papers cited for further details.

INDICATORS OF ANAEROBIOSIS

The use of an **indicator** for oxidation-reduction potential is essential in anaerobic culturing. Among those available, the original Fildes and McIntosh methylene blue indicator is recommended (see Chapter 41 for preparation). The indicator is prepared freshly each time by mixing equal parts of the solutions of methylene blue, glucose, and sodium hydroxide in a test tube and boiling until colorless (the methylene blue is reduced to its leuco base). This tube is immediately placed into the previously loaded jar, and the jar is then sealed and charged by the methods described. If anaerobic conditions were secured and maintained throughout the incubation period, the indicator solution will have remained **colorless.** Should the indicator turn blue, showing that oxygen was absorbed, anaerobiosis was not achieved or maintained. A disposable indicator also may be used; it is available commercially* in a sealed envelope which is opened and prepared at the time of use.

Louis D. S. Smith[17] recommends the

inclusion of a culture of a strict and fastidious anaerobe, such as *Clostridium haemolyticum,* as a monitor for the adequacy of one's anaerobic media and methods; if isolated colonies of this organism are obtained by the procedures used, they are satisfactory for all known anaerobic pathogens.

INOCULATION OF CULTURES

A variety of liquid and solid culture media are available for primary inoculation of clinical specimens, including various selective media, and many are available commercially.* At the Wilmington Medical Center, we have adopted the recommendations of Dr. William Martin at the Mayo Clinic,[15] and routinely inoculate the following media **as soon as possible** after collection (methods previously described).

1. If the specimen is received on a cotton swab, it is placed directly into two tubes of enriched thioglycollate medium† (THIO) and gently rotated (avoiding agitation). Optional: freshly boiled and cooled chopped meat-glucose medium (CMG)‡

2. The swab is removed and used to inoculate the following **freshly prepared** solid media (may be stored anaerobically in a GasPak jar at room temperature)‡:

   2 trypticase soy blood agar (BA)
   1 brucella blood agar with added vitamin K (BRBA)
   1 phenylethanol blood agar (PEA)
   1 kanamycin-vancomycin blood agar (KVBA) with added vitamin K and hemin

   The plates are streaked to secure isolated colonies, with a platinum-iridium loop (**not** nichrome, which oxidizes the inoculum).

3. If fluid material is submitted, it is inoculated with a capillary pipet; one

---

---

*Scott Laboratories, Fiskeville, R. I., Hyland Laboratories, Costa Mesa, Calif.
†Freshly prepared, or boiled (10 minutes) and cooled, after which 10% horse or rabbit serum is added. Store at room temperature—not in refrigerator!
‡Preparation of these media is given in Chapter 39.

drop is deposited on each of the plates, and several drops are introduced to the bottom of a tube of thioglycollate medium, with as little agitation as possible.

4. If tissue is received, it is promptly transferred to a sterile tissue grinder and ground with sterile alundum or sand and thioglycollate, avoiding aeration. The resulting homogenate is inoculated as a fluid specimen (step 3).

5. If clostridia are suspected, an egg yolk agar plate (EYA)* is also inoculated. Optional: Nagler egg yolk agar antitoxin plate.*

EXAMINATION OF DIRECT FILMS

It is important to prepare and examine a gram-stained film of the original specimen **after** inoculating the media; there is usually sufficient residual material remaining on the swab.

The type and approximate number of organisms present are noted, as well as the presence, shape, and location of spores(? clostridia), branching gram-positive elements (? actinomyces), gram-positive cocci in pairs or chains (? staphylococci, streptococci), and gram-negative rods with round or pointed ends (? coliforms, fusobacteria). Pleomorphism and irregularity of staining are also seen with anaerobes.

A preliminary report of these findings should be submitted to the attending physician without delay, since this may aid in the selection of appropriate antimicrobial agents.

INCUBATION OF CULTURES

Incubate all inoculated media at 35° to 36° C. in the following manner (based on one culture):

1. One BA plate (No. 1) in a candle jar (approximately 3% CO_2), and one THIO in air. These are to be used for routine "aerobic" culture.

2. One BA plate (No. 2) in candle jar, to be used for comparison with anaerobic plates.

3. The KVBA, BRBA, PEA (and EYA, if used) plates in a GasPak jar with H_2-CO_2 generator, anaerobic indicator, and a paper towel in the bottom of the jar to absorb excess moisture formed during incubation.*

4. One THIO, with cap loosened (along with a tube of CMG, if used), in a separate GasPak jar, along with a generator and an indicator.†

5. The first BA plate and THIO in air are examined after overnight incubation, then subcultured; isolants are identified, and antibiotic susceptibility tests are set up, as required by aerobic culture methods delineated in subsequent chapters. The second BA plate (No. 2) is held for later comparison with the anaerobic plates.

6. The anaerobic THIO (and CMG) tube is examined after 18 to 24 hours; if no growth occurs, it is reincubated for an additional 24 hours.

7. **All of the anaerobic plates are incubated for a minimum of 48 hours** (slow-growing organisms may require 3 to 5 days); if the jar is opened sooner, some fastidious strains may cease to grow, even if reincubated anaerobically.

In emergency situations, as in suspected gas gangrene, duplicate sets of plates, including a Nagler egg yolk agar antitoxin plate,‡ may be inspected after 12 to 14 hours' incubation, while the second set may be held longer, if indicated. This permits the early identification of *Clostridium perfringens,* confirming a presumptive positive direct Gram stain.

---

*Preparation of these media is described in Chapter 39.

*Note that a fresh charge of reactivated catalyst should be used each time a jar is loaded (p. 93).
†Described in Chapter 39.
‡This is useful as a "back-up" tube and may be examined when turbid only when there is no growth on the primary plate (step 3) or there has been failure of anaerobiosis after incubation.

EXAMINATION OF CULTURES

After appropriate incubation, remove the KVBA, BRBA, and PEA (and EYA, if used) plates from the GasPak jar, emptying only one anaerobic jar at a time to avoid undue exposure to air. Also remove the second BA plate from the CO_2 jar, and examine and compare the growths on each plate, using a hand lens. Describe and record the colony types observed on a work sheet; those types present on the anaerobic plates, but **not** on the CO_2 plate are presumed to be anaerobes, and are handled as described below:

1. Each different colony type is picked and inoculated to the following:

   > A one-fourth sector of BRBA, to be incubated anerobically
   > A one-fourth sector of BRBA, to be incubated under CO_2
   > A tube of enriched thioglycollate medium (see footnote p. 94)

   The anaerobic BRBA and THIO (cap loosened) plates are incubated for 48 hours, and the CO_2 plate for 18 to 24 hours. While the above subculture procedures are being completed, a number of these inoculated plates can be safely held at room temperature in a glass jar under a continuous stream of CO_2, until they are set up anaerobically.* The loose-fitting jar top is fabricated of $1/2$ inch thick Plexiglas and contains a small drill hole to permit escape of the CO_2, which is introduced to the bottom of the jar at a flow rate producing a steady stream of bubbles through a water bottle.†

2. The primary KVBA and BRBA plates are also examined under ultraviolet light‡ for the presence of colonies of *Bacteroides melaninogenicus,* which characteristically exhibit a brick-red fluorescence. Usually, this is sufficient to identify *B. melaninogenicus;* however, other biochemical tests are described in Chapter 23, if further species identification is required.

3. After overnight incubation, the CO_2 jar plate is examined; if growth occurs, the subculture was probably not an anaerobe,* and further workup on that isolant is discontinued.

4. After 48 hours' incubation, the anaerobic subcultures are examined for growth, hemolysis, and colonial morphology. If in **pure culture,** a Gram stain (and a repeat THIO subculture if insufficient growth has occurred in THIO [step 1]) is made, examined, and recorded on the work sheet.

5. The growth in pure culture of the THIO is also gram stained, examined, and the results noted, with particular attention paid to gram reaction, cellular morphology, and arrangement of cells. These observations, along with those made in step 4 will determine the biochemical and other tests to be carried out for definitive identification of the anaerobic isolant. These procedures are described in Chapter 23.

6. The EYA plate is examined for the presence of lecithinase activity, which is indicated by the formation of an **opaque zone** in the medium around the growth. This plate can also be used for the detection of catalase by exposing it to air for 30 minutes and then dropping 3% hydrogen peroxide on the colony.[19] If subcultures are required, they should be made before exposing the plate to air.

From the results of the above observations, the group to which the isolant belongs is determined, and further biochemical tests are then carried out for definitive identification. These procedures are described in the following sections under

---

*Many isolants may survive in the air for a period of 8 hours or more (Tally, F. P. et al., paper M60, Abstr. Ann. Meeting, ASM, 1973).
†Procedure adopted from Martin.[14]
‡"Blak-Ray" UV lamp and viewbor, model UVL 56, from Ultra-Violet Products, San Gabriel, Calif.

---

*Some species of *Clostridium* are aerotolerant.

specific headings, such as micrococci, streptococci, Bacteroidaceae, anaerobic sporeformers, and so forth.

REFERENCES

1. Aranski, A., Syed, A., Kenney, E. B., and Freter, R.: Isolation of anaerobic bacteria from human gingiva and mouse cecum by means of a simplified glove box procedure, Appl. Microbiol. **17:**568-576, 1969.
2. Attebery, H. R., and Finegold, S. M.: Combined screw-cap and rubber-stopper closure for Hungate tubes (pre-reduced, anaerobically sterilized roll tubes and liquid media), Appl. Micorbiol. **18:**558-561, 1969.
3. Brewer, J. H., and Allgeier, D. L.: Disposable hydrogen generator, Science **147:**1033-1034, 1965.
4. Brewer, J. H., and Allgeier, D. L.: Safe self-contained carbon dioxide–hydrogen anaerobic system, Appl. Microbiol. **14:**985-988, 1966.
5. Brewer, J. H., and Allgeier, D. L.: A disposable anaerobic system designed for field and laboratory use, Appl. Microbiol. **16:**848-850, 1968.
6. Brewer, J. H., Allgeier, D. L., and McLaughlin, C. B.: Improved anaerobic indicator, Appl. Microbiol. **14:**135-136, 1966.
7. Brewer, J. H., and Brown, J. H.: A method for utilizing illuminating gas in the Brown, Fildes and McIntosh or other anaerobe jars of the Laidlaw principle, J. Lab. Clin. Med. **23:**870-874, 1938.
8. Dowell, V. R., Jr., and Hawkins, T. M.: Laboratory methods in anaerobic bacteriology, Public Health Service Pub. No. 1803, Washington, D. C., 1968, U. S. Government Printing Office.
9. Holdeman, L. V., and Moore, W. E. C., editors: Anaerobe laboratory manual, Blacksburg, Va., 1972, Virginia Polytechnic Institute and State University.
10. Hungate, R. E.: A roll tube method for cultivation of strict anaerobes. In: Methods in microbiology, vol. 3 B, New York, 1969, Academic Press, Inc.
11. Killgore, G. E., Starr, S. E., Del Bene, V. E., Whaley, D. N., and Dowell, V. R., Jr.: Comparison of three anaerobic systems for the isolation of anaerobic bacteria from clinical specimens, paper M 86, Abstr. Annual Mtg., American Society for Microbiology, 1972.
12. Laidlaw, P. P.: Some simple anaerobic methods, Brit. Med. J. **1:**497, 1915.
13. McIntosh, J., and Fildes, P.: Cultivation of anaerobes, Lancet **1:**768, 1916.
14. Martin, W. J.: Practical method for isolation of anaerobic bacteria in the clinical laboratory, Appl. Microbiol. **22:**1168-1171, 1971.
15. Martin, W. J.: Personal communication, 1973.
16. Rosenblatt, J. E., Fallon, A. M., and Finegold, S. M.: Recovery of anaerobes from clinical specimens, Appl. Microbiol. **25:**77-85, 1973.
17. Smith, L. D. S., and Holdeman, L. V.: The pathogenic anaerobic bacteria, Springfield, Ill., 1968, Charles C Thomas, Publisher.
18. Socransky, S., MacDonald, J. B. and Sawyer, S.: The cultivation of *Treponema microdentium* as surface colonies, Arch. Oral Biol. **1:**171-172, 1959.
19. Sutter, V. L., Attebery, H. R., Rosenblatt, J. E., Bricknell, K. S., and Finegold, S. M.: Anaerobic bacteriology manual, Los Angeles, 1972, Dept. Contin. Educ. Health Sci., Univ. Extens., School Med., UCLA.

14 Microorganisms encountered in the eyes, ears, mastoid region and sinuses, the teeth, and effusions

EYE CULTURES

Because of the constant washing activity of tears and their antibacterial constituents, the number of organisms recovered from cultures of many eye infections may be relatively low. Unless the clinical specimen is obviously purulent, it is recommended that relatively large inocula and a variety of media be used to ensure recovery of an etiological agent.

The following organisms are **most frequently isolated** from infections of the eye:

Pathogens

> *Staphylococcus aureus*
> *Haemophilus* species
> *Streptococcus pneumoniae*
> *Neisseria gonorrhoeae*
> Alpha and beta hemolytic streptococci
> *Moraxella lacunata*
> *Herellea vaginicola (Acinetobacter calcoaceticus)*
> Coliform bacilli
> *Pseudomonas aeruginosa,* other enteric bacilli
> *Corynebacterium diphtheriae*
> Viruses, fungi, other agents

Nonpathogens, or opportunists

> *Corynebacterium xerosis,* other diphtheroid bacilli
> Coagulase-negative staphylococci
> *Sarcina,* other micrococci
> Saprophytic fungi

If a diagnosis of purulent conjunctivitis has been made, the purulent material is collected on a sterile cotton swab or surgical instrument (before the local application of antibiotics, irrigating solutions, or other medications) from the surface of the lower cul-de-sac and inner canthus of the eye. This purulent material should be inoculated immediately to blood agar and chocolate agar plates, which should be incubated in a candle jar, and to a tube of thioglycollate broth. All media should be held for at least 48 hours, and any resulting growth should be identified by the appropriate methods.* One should be aware of the possibility of gonococcal infection in the eye of the newborn infant, and the oxidase test may be carried out on the chocolate agar plate to detect colonies of neisseriae. Sabouraud dextrose and brain-heart blood agar slants should be inoculated if a mycotic infection is suspected. Whenever possible, a **gram-stained smear** of the purulent material present also should be examined. Frequently, the results may give sufficient information to the physician to confirm a clinical diagnosis and serve as a guide to proper therapy.

Scrapings of the conjunctivae for the presence of eosinophils or of ulcerative lesions of the cornea for demonstrating viral inclusion bodies are preferably taken by the opthalmologist. These scrapings are transferred to a glass slide, dried, and stained by the Wright-Giemsa method.

---

* *Pseudomonas aeruginosa* and *Strep. pneumoniae* may cause a severe and damaging corneal infection, and should be reported without delay to the physician.

Moraxella lacunata (Morax-Axenfeld bacillus) is a short, thick, gram-negative diplobacillus that causes a subacute or chronic catarrhal conjunctivitis, which is particularly severe in the outer angles of the eye. The organism is best cultivated on Loeffler medium, where it causes characteristic pitting and proteolysis of the medium, although nonproteolytic strains are being isolated with some frequency. Identification methods are described in Chapter 21.

Haemophilus aegyptius (Koch-Weeks bacillus) is a small gram-negative bacillus, morphologically resembling *H. influenzae,* and is the causative agent of an acute epidemic conjunctivitis commonly called **pink eye.** The organism grows well on blood agar or chocolate agar plates. (See Chapter 22 for cultural characteristics.)

Corynebacterium diphtheriae may cause a pseudomembranous conjunctivitis, and the organism must be demonstrated by culture and proved toxigenic before a diagnosis of diphtheria can be made. The organism may be readily cultivated on Loeffler or Pai medium. Their identifying characteristics and tests for virulence are described in Chapter 27.

EAR CULTURES

The following organisms are encountered **most frequently** from cultures of the ear:

Pathogens

> *Pseudomonas aeruginosa*
> *Staphylococcus aureus*
> *Proteus* species
> Alpha and beta hemolytic streptococci
> *Streptococcus pneumoniae*
> *Haemophilus influenzae*
> Coliform and other enteric bacilli
> *Aspergillus fumigatus, Candida albicans,* and other fungi

Nonpathogens, or opportunists

> Coagulase-negative staphylococci
> Diphtheroids
> *Gaffkya tetragena*
> *Bacillus* species
> Saprophytic fungi

The following organisms are **rarely** occurring pathogens in such cultures:

Corynebacterium diphtheriae
Actinomyces israelii
Mycobacterium tuberculosis, other mycobacteria
Mycoplasma pneumoniae

Material from the ear, especially that obtained after perforation of the eardrum, is best collected by the otolaryngologist, using sterile equipment and a sterile cotton or polyester swab. Discharges from the ear in chronic otitis media usually reveal the presence of pseudomonads and *Proteus* species. Acute or subacute otitis usually yields pyogenic cocci. In external otitis the external ear should be cleansed with a 1:1,000 aqueous solution of benzylkonium chloride or other detergent in order to free the skin of contaminating bacterial flora before a culture is taken, if the results are to be of clinical significance. Otherwise, a variety of nonpathogenic bacteria and saprophytic fungi will be recovered. Specimens are cultured as described in the preceding section, with the addition of phenylethanol agar, to recover other organisms in the presence of overgrowing *Proteus* species.

MASTOID AND ACCESSORY SINUS CULTURES

The widespread use of antimicrobial agents in the treatment of acute infections of the middle ear (otitis media) has resulted in a significant decrease in the incidence of acute mastoiditis, an infection of the mastoid process and surrounding structure. The offending organisms, usually originating from a suppurative otitis, generally consist of alpha and beta hemolytic streptococci and pneumococci. Occasionally, an infection from type 3 pneumococcus may progress to severe meningitis and septicemia.

Cultures from the mastoid region are generally taken on a cotton or polyester swab (before antibiotic therapy) and are handled in the laboratory as any other wound culture would be. Pyogenic cocci are usually isolated; but with the widespread use of antibiotics, enteric gram-negative bacilli are being isolated with increasing frequency.

Swabs from the paranasal sinuses are normally sterile, or they may contain a few saprophytes picked up during collection. An acute suppurative sinusitis, however, may follow a common cold or occur after infected water is forced into the nose while swimming or diving. The most frequent isolates in this infection include the streptococci, staphylococci, and pneumococci; *Bacteroides* and *Haemophilus influenzae* are occasionally isolated and may give rise to serious complications. The aerobic and anaerobic methods previously described for wound cultures (Chapter 13) are generally satisfactory.

CULTURES OF TEETH

The bacterial flora of the normal mouth is made up of a wide variety of microorganisms, including bacteroides, streptococci, filamentous rods, neisseriae, and spirochetes, in addition to lactobacilli and yeasts. This flora appears to be affected by dietary factors, antibiotic therapy, and the presence of certain pathological conditions of the teeth and gums.

The role of spirochetes and fusiform bacilli in Vincent's infection has already been discussed (p. 59). For a further discussion of the role of microorganisms in dental caries and other infections of the tissue, the reader is referred to textbooks on dental pathology.[1,4] It should be noted, however, that as a result of studies with germ-free animals, certain strains of acidogenic streptococci, such as *Streptococcus mutans*,* have been implicated in the initiation of dental caries,[2] thereby casting doubt on the lactobacilli as the primary etiological agent.

Effective methods and culture media are available for the bacteriological examination of extracted teeth, tooth sockets, and the periapical region through the root canal of the tooth in situ. Specimens are obtained by the dental surgeon, using a rigidly aseptic technique and sterile equipment. The apex of an extracted tooth is cut off with a pair of cutting forceps, and the apical fragment is transferred directly to a tube of enriched medium, such as brain-heart infusion with *p*-aminobenzoic acid* (penicillinase if pencillin has been used) and 10% ascitic fluid or animal serum, and 0.1% agar. The medium is held at 36° C. for several days and observed for **growth.** Any indication of growth (increase in turbidity) is confirmed by examining a gram-stained smear, which also serves to guide in the selection of appropriate aerobic and anaerobic media to be used in subculturing. Cultures from sockets, obtained either with a sterile curet or cotton applicator, are handled in a similar manner.

In culturing root canals, the following method[4] is recommended, using strict asepsis throughout. After a sterile field is established, the seal and previous dressing are removed and discarded. The canal is cleansed of any residual medicament by flushing with about 1 ml. of sterile water and is dried by inserting a fresh absorbent point with a wiping motion, then removing and discarding it. Another fresh sterile point is then inserted into the apex and allowed to remain in place for about 1 to 2 minutes, then removed. If the tip appears to be moist with exudate or blood, it is dropped into a tube of the culture medium previously indicated. If the point appears to be dry upon removal, it is discarded and a fresh point, aseptically moistened with the culture medium, is inserted into the root canal, left for several minutes, and cultured as before. After 48 hours' incubation at 36° C. the culture tube is examined for the presence of **growth,** which is evidenced by any increase in turbidity, especially around the tip or surface of the absorbent point. Generally, negative cultures are kept for 1 week, and two consecutive negative cultures are obtained before filling the root canal.

---

*S. *mutans* has also been incriminated in endocarditis.[3]

*Difco Laboratories, Detroit; Baltimore Biological Laboratory, Cockeysville, Md.

CULTURES OF EFFUSIONS

All fluids suspected of exudative origin should be examined bacteriologically. Clear or slightly cloudy specimens should be centrifuged at 2,500 r.p.m. for 30 minutes, the supernatant fluid removed aseptically, and the sediment examined by means of smears and cultures. Specimens that are grossly purulent should be examined directly by gram staining thin films and immediately examining the stained films microscopically. All specimens should also be inoculated to routine media, including blood agar plates, enriched thioglycollate broth, and isolation media for *Mycobacterium tuberculosis* and systemic fungi if indicated. If the effusion source suggests a fastidious flora, the use of special media may be required.

Synovial, or joint, fluid usually obtained in only small amounts, should also be examined for the presence of gonococci by inoculating chocolate agar plates, as described in Chapter 11. **Pleural effusions** occur most frequently in patients with pulmonary tuberculosis; the fluid should be centrifuged and the sediment used for inoculation of tuberculosis culture media or for guinea pig injection. In **empyema** the fluid is generally purulent or seropurulent and may yield pneumococci, streptococci, coagulase-positive staphylococci, or *Haemophilus influenzae* on culture. On occasion, especially in ruptured abscesses in the lung, a variety of anaerobes, including *Bacteroides* species, are recovered frequently. Effusions from patients with **peritonitis** vary in character from the thin cloudy fluid found in tuberculous peritonitis to the foul-smelling purulent specimen from a case of colon bacillus peritonitis. In general, the methods for handling these specimens are those described for routine culturing of specimens from wounds (Chapter 13) and should include inoculation of both **aerobic** and **anaerobic media.**

REFERENCES

1. Grossman, L. I.: Endodontic practice, ed. 7, Philadelphia, 1970, Lea & Febiger.
2. Jordan, H. V.: Bacteriological aspects of experimental dental caries, Ann. N. Y. Acad. Sci. **131:**905-912, 1965.
3. Lockwood, W. R., et al.: *Streptococcus mutans* endocarditis, Ann. Intern. Med. **80:**369-370, 1974.
4. Nolte, W. A., editor: Oral microbiology, ed. 2, St. Louis, 1973, The C. V. Mosby Co.

15 Microorganisms encountered in material removed at operation and necropsy

EXAMINATION OF MATERIAL OBTAINED AT OPERATION

Although largely a neglected function of the bacteriology laboratory in the past, the current availability of specific antibacterial agents has made the microbiological examination of surgical tissue an essential adjunct to the histopathological diagnosis.

Likewise, the increasing use of more refined techniques in postmortem bacteriology has contributed to disclosing a possible etiology of an infectious process, as well as providing a means of evaluating the effectiveness of prior antibiotic therapy.

Collection of specimen

In order to carry out a proper microbiological examination of excised tissue, a thorough search for aerobic and anaerobic microorganisms, acid-fast bacilli, brucellae, fungi, viruses, and other pathogenic agents must be made. An **adequate** specimen, therefore, is a prerequisite; the surgeon or pathologist must assume the responsibility for obtaining sufficient material at the time of operation.

It is also necessary that a specimen bottle of **sufficient size** be available to the operator. A sterile, wide-mouthed, screwcapped bottle with a neoprene liner in the lid or a suitable sterile plastic container is recommended. This receptacle should be conveniently placed on the instrument table at the time of surgery so that the op-

erator may deposit material directly into the jar. This prevents the possibility of accidentally fixing the tissue with formaldehyde or other germicidal chemical, and it also avoids any possible contamination by the surgical pathologist in his examination.

In the collection of material from chronic draining sinuses and ulcers, appropriate sterile equipment should be available to the examining physician. This may include curets, scissors, syringes, needles, medications, dressings, containers, and so forth. With the aid or these instruments, deep curettage of the sinus tract may be carried out, and it should include the procurement of a portion of the wall of the tract. Finegold's method[6] is recommended for collecting material for **anaerobic** culturing from sinus tracts or draining wounds. An intravenous plastic catheter is introduced into the tract as deeply as possible after appropriate decontamination of the skin opening, and material is aspirated with a syringe. This is immediately transferred to a "gassed out" tube (p. 91) and submitted promptly to the laboratory for incubation. These methods are preferred to that of collecting a specimen on a swab stick, since such specimens are usually taken rather gingerly and on culturing are likely to yield only members of the normal skin flora.

Material from an ulcer should contain tissue from its base as well as the edge of the lesion. Closed and fluctuant abscesses should be aspirated, using a 15-gauge needle (to secure any dead tissue and debris that may be present) and a large-caliber syringe, and the material should be transferred to a "gassed out" tube. Inspection for the presence of **granules** in the aspirated pus and a Gram stain should be a part of every examination, since this may be the first clue to an infection caused by actinomycetes or certain fungi.

In some instances, **contaminated** material may be submitted for bacteriological examination. Such specimens as tonsils, autopsy tissue, or similar material may be cauterized or blanched by immersing in boiling water for 5 to 10 seconds to reduce surface contamination. The specimen may then be dissected with sterile instruments to permit culturing of the **center** of the specimen, which will not be affected by the heating.

All surgical specimens intended for microbiological examination should be divided by the operator, using sterile instruments; one half is submitted for histological examination, and the other half is sent to the bacteriology laboratory. A detailed clinical history should accompany the histological specimen in order to guide the pathologist in directing subsequent microbiological studies. Since viable organisms may be few in number in tissue, especially in old chronic lesions, and since they may be irregularly distributed throughout the tissue, it is desirable to secure multiple specimens when the lesion is large enough.[9]

It is well to bear in mind that surgical specimens differ from other clinical material in that they are frequently obtained at considerable risk and expense to the patient. Furthermore, a specimen may represent the entire pathological process. It is obvious, therefore, that supplementary specimens cannot be obtained with the ease with which similar specimens of blood, urine, or feces can be secured. It is strongly recommended that a portion of the tissue moistened in sterile physiological saline be **refrigerated or frozen** for subsequent studies should preliminary or routine examination prove unproductive.

The routine culturing of any biopsy specimen for fungi has been recommended by Utz.[7] He has obtained positive cultures from necrotic tissue, liver, kidney, prostate, epididymis, testis, muscle, skin, brain, synovium, spleen, and so forth, as well as from ulcers of the nose, mouth, epiglottis, and larynx.

Preparation of tissues for culture

In the important step of preparing tissues for culture, it is not enough to merely scrape material from the surface of the specimen with a bacteriological loop. One must thoroughly **grind** the tissue into a fine suspension, since only a few organisms may be present in the whole sample.*

To prepare this suspension the tissue should first be finely minced with sterile scissors and transferred to a sterile tissue grinder, of which several types are available. A small one is illustrated in Weed's[9] paper on the isolation of fungi from tissue; we have used one produced locally,† which has proved satisfactory. The tissue is ground to a pasty consistency (10% to 20% suspension), using sterile sand and sterile saline or broth (preferred). After settling, the supernatant fluid is transferred to another small tube by use of a sterile capillary pipet. It is then inoculated to culture media, as discussed later, or injected into animals in the same manner as with any other biological fluid. Intraperitoneal or intramuscular injection of guinea pigs and intraperitoneal injection of several mice is recommended. These are examined daily and are cultured at the time of death.

---

*Yeastlike fungi, such as *Cryptococcus neoformans,* may be macerated and rendered nonviable by grinding. It is therefore recommended that **minced** tissue be inoculated directly to fungus media.

†Ten Broeck tissue grinder, small size, heavy-walled Pyrex glass, Bellco Glass, Inc., Vineland, N. J.

Histological examination

Although a discussion on histological techniques is beyond the intended scope of this text, it should be pointed out that a thorough histological examination of fixed tissue is an essential part of any pathological diagnosis. The microscopic study of carefully selected and sectioned tissue, stained by both routine and special methods (such as the Gram, acid-fast, Gomori, Gridley, periodic acid–Schiff methods), will serve two useful purposes. First, it will demonstrate the histopathology of the lesion—whether it is of a neoplastic or inflammatory nature. If the lesion proves to be of noninflammatory origin, a bacteriological study need not be carried out. Second, if the appearance suggests infection, it may serve as a guide to the selection of appropriate culture media and isolation techniques.

Selection of culture media

In the selection of culture media, one must decide not only whether the tissue contains a mixed or contaminated flora, indicating the need for selective isolation media, but also whether it contains **fastidious pathogens** that may require special media.

No general rule can be laid down regarding the particular kinds of media to use; the choice will depend largely on the organism sought or suspected. Enriched media, such as blood agar, chocolate agar, heart infusion broth, anaerobic media, enriched thioglycollate medium, and media for the primary isolation of mycobacteria and fungi, must be considered. Incubation in both a candle jar and an anaerobic jar should be carried out. This has been discussed in earlier chapters.

Isolation of some fastidious pathogens

Although the techniques for isolation of fastidious pathogens are presented in other parts of the text, there are certain facts relative to the recovery of these microorganisms that are pertinent to this section. In the isolation of **pathogenic fungi,** for example, the following facts are important:

1. Tissue for mycological examination should be minced, rather than ground, and transferred with the knife blade directly to fungus media slants (see footnote, p. 103).
2. In severe, disseminated histoplasmosis, cultures of lymph nodes and needle biopsies of the liver or bone marrow are likely to yield *Histoplasma capsulatum.*
3. *Actinomyces* species are microaerophilic or anaerobic, require enriched media, are inhibited by antibiotics, and grow best at 36° C. in an anaerobic jar.
4. *Nocardia* species are aerobic, will grow on simple media, such as Sabouraud agar, and are inhibited by antibiotics.
5. Most of the other fungi, with the exception of cryptococci, can be isolated on media containing antibiotics. Some require enriched media and may grow at both room and incubator temperatures.

It is apparent that no one medium, no specified temperature of incubation, and no single standard technique in handling will be suitable for all fungi. Massive inoculation of the tissue should be made to multiple tubes of both simple and enriched media and incubated at both temperatures for at least 4 weeks before discarding. Specific recommendations will be found in Chapter 32.

Acid-fast stains of surgical specimens may be positive in less than one half of those cases subsequently proved to contain pathogenic mycobacteria.[10] A negative stain is thus of little value in ruling out mycobacteria, and a positive smear does not always denote tuberculosis; therefore, bacteriological studies must be made. These methods will be further considered in Chapter 28.

The isolation of brucellae from contaminated surgical specimens can be accomplished only by the use of enriched media containing antibiotics, since routine cultures will generally be overgrown by common pathogens, such as *Staphylococcus aureus.* A fresh meat–extract agar with added glucose* and 5% animal blood has proved satisfactory for the isolation of *Brucella* species from surgical specimens.[8] The addition of antibiotics increased the usefulness of the medium for isolating brucellae from clinical material containing a mixed flora.

Cultures of most of these fastidious pathogens should be incubated for 3 to 6 weeks before discarding. Many of the organisms isolated from chronic infections—including *Brucella, Histoplasma, Coccidioides,* and mycobacteria—require long incubation periods before growth becomes apparent. During the first day or so after primary inoculation, the medium should be examined for overgrowth by *Proteus,* pseudomonads, or other contaminants. It is recommended that a portion of the original tissue suspension be stored in a refrigerator until the cultures are obviously free of contaminants. If possible, a portion of the original tissue may be frozen; it may serve as a guarantee against loss by breakage or overgrowth on media.

Final identification methods should be carried out on all isolates whenever possible. Special procedures involving virulence tests by animal injection, serological analysis, and final identification may not be feasible for the small laboratory. In such cases pure cultures of these isolates should be sent to a reference laboratory for further study. Means are available for the handling and rapid transport of the specimens, as described in Chapter 5.

EXAMINATION OF MATERIAL OBTAINED AT AUTOPSY

Although it is true that postmortem invasion of the bloodstream and organs by commensal organisms can occur, it has become apparent that invasion is not as rapid as was formerly believed. Although earlier studies interpreted the high incidence of positive autopsy cultures as evidence of either antemortem infection or agonal invasion of tissue, O'Toole and co-workers[4] indicate that contamination of tissue by the environment or personnel at the autopsy may be more significant. In a report by Minckler and co-workers[3] on experience gained by culturing 475 tissues obtained surgically or by a sterile autopsy technique, it was suggested that a "normal bacterial flora" of the internal tissues of humans was not an impossibility. This would have a significant implication in the homotransplantation of organs, particularly in recipients whose immune mechanisms have been altered by irradiation or immunosuppressive agents. An evaluation of the results of autopsy blood cultures by Wood and associates[11] indicated a good correlation with the results of antemortem cultures or with anatomical data derived at autopsy. A higher degree of correlation was obtained when the blood for culture was obtained within 18 hours of death. Prior antibiotic therapy apparently does not appreciably reduce the recovery of pathogens from this source. The microscopic examination of direct smears stained by the Gram, Ziehl-Neelsen, or other methods, along with a thorough evaluation by culturing procedures of blood, tissue, or swabs from infected areas, may reveal significant data related to the cause of death.

Blood for culture is collected with a sterile syringe and a 14- or 15-gauge needle after searing the atrium or ventricle of the heart with a soldering iron. From 5 to 10 ml. of blood are inoculated to broth and thioglycollate bottles in the manner described for blood cultures.* If blood is not

---

*Brucella agar (Albimi Laboratories, Flushing, N. Y.) and trypticase soy agar (Baltimore Biological Laboratory, Cockeysville, Md.) also can be used.

*Silver and Sonnenwirth (Amer. J. Clin. Path. **52:**433-437, 1969) indicate that heart blood obtained through the closed chest and before bowel manipulation gave more significant clinical results than with the usual methods.

secured by using this approach, puncture of the pulmonary artery or auricle after searing is recommended. The specimen is inoculated as previously described. Small fragments of blood clots may appear, but these do not interfere with the culturing if large-bore needles are used. Some workers[3] believe that cultures of aspirated splenic pulp are of greater value in confirming a diagnosis of bacteremia or septicemia than is a blood culture.

Cultures of other tissues may be obtained by searing the surface with a soldering iron or hot spatula and passing a swab or fine-pointed pipet through the seared area to an unheated and uncontaminated region.

Another method of obtaining autopsy culture material that compared favorably with results obtained by a sterile autopsy technique has been described by de Jongh and associates.[2] In this procedure the tissue surface is seared to dryness with a heated steel spatula, and a 1-cc. cube of tissue is excised from the center of the seared area using sterile forceps and scissors. Tissue blocks thus obtained are minced and ground as described previously and inoculated to appropriate aerobic and anaerobic media, as perhaps indicated by a Gram stain of the homogenate.

If bacterial endocarditis is suspected, a small portion of the friable vegetation is removed with sterile scissors and forceps and submitted to the laboratory in a Petri dish. Here it is gently washed in at least three changes of sterile saline solution and then ground with sterile sand and broth in a tissue grinder, as described previously. A smear is prepared of this suspension, gram stained, and examined; a portion also is inoculated to blood agar and chocolate agar plates and enriched thioglycollate medium. The plates are incubated in a candle jar at 36° C. for 48 hours and examined for the presence of growth. If a delay has occurred in obtaining the specimen (more than 4 hours post mortem), with probable overgrowth of contaminants, it is well to streak the suspension on a phenyl-ethyl alcohol blood agar plate and on other media. This is done in order to isolate alpha hemolytic streptococci and other gram-positive cocci that may be overgrown by *Proteus* and other gram-negative bacteria on the other plates. If so indicated, anaerobic subcultures of the thioglycollate medium may be carried out.

Collection of autopsy specimens for virus isolation*

An effort should be made to obtain postmortem specimens in all fatal cases of central nervous system disease of suspected viral etiology, especially if antemortem studies were not carried out. Using sterile precautions at the time of autopsy, the whole brain is removed and refrigerated but **not** frozen; a 3-inch segment of the descending colon is tied off at both ends; and aliquots of various other tissues are obtained and placed in sealed sterile containers and refrigerated. A blood specimen also should be collected by cardiac puncture; the serum should be separated and then refrigerated. In a fatal case of respiratory disease, lung tissue and a tracheobronchial swab should be collected and refrigerated until shipment; a blood specimen also should be obtained. All specimens thus obtained for virus isolation should be transported **without delay** to the nearest virus reference laboratory, after making the proper arrangements by telephone for their shipment and handling at the laboratory on receipt.

Procedures for isolation of Listeria monocytogenes from tissue

The tissues of all fetuses, premature infants, and young babies coming to necropsy as a result of an infectious process should be cultured for listeriae for reasons given previously (see discussion of blood cultures). Specimens of the brain, liver, and spleen are most likely to contain the

---

*Abstracted from the instruction pamphlet by the New Jersey State Department of Health, Division of Laboratories, Dr. Martin Goldfield, Director.

organism. The isolation procedure is that recommended by Cherry of the Center for Disease Control in Atlanta.[5]

1. Prepare a suspension of the specimen that has been removed aseptically by grinding 2 to 5 gm. of the tissue in a tissue grinder with a few grams of sand and 20 ml. of sterile distilled water or broth (avoid saline).
2. Using a broken-tipped pipet, transfer 5 to 10 ml. of the suspension to each of two flasks of infusion broth.
3. Incubate one flask at 36° C. for 24 hours, and inoculate a drop of this to a blood agar plate and a tellurite blood agar plate. Incubate these in a candle jar for at least 48 hours along with the original broth flask.
4. Store the second flask in a refrigerator at 4° C. If the 36° C. subcultures (step 3) are unsuccessful, subculture material from the refrigerated flask at weekly intervals for at least 1 month. It is also recommended that swabs and similar material be handled in like manner.
5. *Listeria monocytogenes* is identified in the manner described in Chapter 27.

If the material is likely to be contaminated, as in the case of autopsy specimens, feces, and so forth, the following method is recommended[1]:

1. Grind the tissue with sand and 10 to 20 ml. of tryptose broth in a tissue grinder.
2. Inoculate the suspension into two tubes of broth (20 by 150 mm., screw-capped test tubes containing 9 ml. tryptose broth), one of which is incubated at 36° C. for 18 to 24 hours and the other at 4° C.
3. After incubation, inoculate a plate of modified McBride medium with the 36° C. culture (Chapter 39) and incubate at 36° C. for 24 hours in a candle jar, along with the original broth tube.
4. Examine the plate with a binocular scanning microscope (10 to 15×) or

hand lens and oblique (40-degree angle from mirror to substage), transmitted light. *Listeria* colonies are slightly raised, **blue green** in color, with a finely textured surface.

5. Pick suspected colonies and stab into 0.1% dextrose semisolid agar.
6. **Motile forms,** which appear as diffuse growths throughout the semisolid agar, with a maximum growth zone just below its surface, are subjected to confirmatory studies as possible listeriae, described previously (Chapter 11).
7. If no listeriae are obtained from the 36° C. cultures, repeat the process using the 4° C. cultures. Subculture these at weekly intervals for 3 months while the original broth culture is held in a refrigerator, since substances present in the tissue may inhibit the growth of the listeriae.

REFERENCES

1. Bearns, R. E., and Girard, K. F.: On the isolation of *Listeria monocytogenes* from biological specimens, Amer. J. Med. Techn. **25**:120-126, 1959.
2. De Jongh, D. S., Loftis, J. W., Green, G. S., Shively, J. A., and Minckler, T. M.: Postmortem bacteriology—a practical method for routine use, Amer. J. Clin. Path. **49**:424-428, 1968.
3. Minckler, T. M., Newall, G. R., O'Toole, W. F., Niwayama, G., and Levine, P. H.: Microbiology experience in collection of human tissue, Amer. J. Clin. Path. **45**:85-92, 1966.
4. O'Toole, W. F., Saxena, H. M. K., Golden, A., and Ritts, R. E.: Studies of postmortem microbiology using sterile autopsy technique, Arch. Path. (Chicago) **80**:540-547, 1965.
5. Seeliger, H. P. R., and Cherry W. B.: Human listeriosis; its nature and diagnosis, Department of Health, Education, and Welfare, Washington, D. C., 1957, U. S. Government Printing Office.
6. Sutter, V. L., Atteberg, H. R., Rosenblatt, J. E., Bricknell, K. S., and Finegold, S. M.: Anaerobic bacteriology manual, Los Angeles, 1972, Dept. of Continuing Education, Health Sciences, UCLA.
7. Utz, J. P.: Recognition and current management of the systemic mycoses, Med. Clin. N. Amer. **51**:519-527, 1967.
8. Weed, L. A.: Use of a selective medium for isolation of *Brucella* from contaminated surgical specimens, Amer. J. Clin. Path. **27**:482-485, 1957.

9. Weed, L. A.: Technics for the isolation of fungi from tissues obtained at operation and necropsy, Amer. J. Clin. Path. **29:**496-502, 1958.

10. Weed, L. A., McDonald, J. R., and Needham, G. M.: The isolation of "saprophytic" acid-fast bacilli from lesions of caseous granulomas, Proc. Staff Meet. Mayo Clin. **31:**246-259, 1956.

11. Wood, W. H., Oldtsone, M., and Schultz, R. B.: A re-evaluation of blood culture as an autopsy procedure, Amer. J. Clin. Path. **43:**241-247, 1965.

PART FOUR
Methods for identification of pathogenic bacteria

16 The micrococci, including aerobic, facultative, and anaerobic species

Micrococci are ubiquitous and exist as free-living saprophytes, parasites, and pathogenic forms. The great majority of pathogenic micrococci probably fall within the genus *Staphylococcus,* but nonstaphylococci are being isolated from clinical sites with increasing frequency. Collectively, the organisms of the group are spherical, gram-positive cocci occurring singly, in pairs, tetrads, packets, and irregular clusters. They are strongly **catalase positive,** usually nonmotile and occasionally pigmented. Nitrates are usually reduced to nitrites. With the exception of the peptococci, which are obligate anaerobes, the micrococci are aerobic or facultative anerobes.

AEROBIC AND FACULTATIVE MICROCOCCI
Genus Staphylococcus

Staphylococci are frequently found on the skin, in the nasal and other mucous membranes of man, and in various food products. Natural reservoirs are the human nose and nasopharynx. Two species are recognized—*Staph. aureus* and *Staph. epidermidis.*

Staphylococcus aureus

Staphylococcus aureus is a gram-positive, nonmotile coccus, occurring singly, in pairs, in short chains, or in irregular clusters. The last arrangement is probably the most characteristic (Fig. 16-1). The Greek word "staphyle," meaning a bunch of grapes, is used descriptively as the stem of the generic term. In older cultures the cells tend to lose their ability to retain the crystal violet and may appear gram variable or even gram negative. Typically, on initial isolation the organism produces a golden yellow pigment, which is soluble in alcohol and ether and is classed as a **lipochrome.** This characteristic, however, is variable; white or pale colonies may arise after laboratory cultivation and are not infrequently isolated from clinical sources. For such reasons we tend to agree with the discussion in the seventh edition of *Bergey's Manual of Determinative Bacteriology,*[2] wherein *albus* and *citreus* are considered varieties of the one species, *Staph. aureus.*

Colonies are usually opaque, circular, smooth, and entire, with a butyrous consistency. The organism grows well on trypticase soy agar or nutrient agar but develops larger colonies on blood agar. The hemolytic activity is variable. Both surface and subsurface colonies of some strains are hemolytic. The surface colonies of the majority of virulent strains are hemolytic, but hemolysis also occurs around subsurface colonies. The hemolytic property may be lost in stock cultures or after a number of transfers.

Most strains of *Staph. aureus* **ferment mannitol** and can tolerate relatively high concentrations of salt (7.5% to 10%).

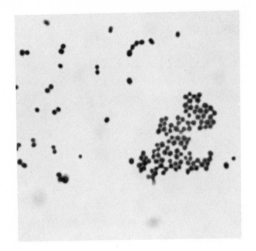

Fig. 16-1. *Staphylococcus aureus,* showing typical grapelike clusters. (4000×.)

These characteristics help promote their isolation from material such as feces, in which a large and varied bacterial flora exists. **Mannitol salt agar** as a selective medium is recommended for the isolation of staphylococci when contamination by other organisms is high. Colonies of salt-tolerant staphylococci appear on this medium, surrounded by a readily visible yellow halo, which indicates mannitol fermentation, after 24 to 48 hours. Mannitol fermentation, among other tests, distinguishes *Staph. aureus* from *Staph. epidermidis.*

The best single test for identifying pathogenic staphylococci is the **coagulase tube test,** which demonstrates "free" coagulase. Citrated rabbit plasma,* 0.5 ml. of 1:4 dilution, in a small tube is inoculated heavily with a 24-hour culture of the organism and incubated at 36° C. in a water bath. Complete or partial coagulation in 1 to 4 hours is interpreted as **positive.** A variation of this test may be performed by using a single colony in a 1:5 dilution of plasma; the test is observed for a period up to 24 hours.[9] Typical strains of *Staph. epidermidis* are coagulase negative,

whereas pathogenic strains of *Staph. aureus* are coagulase positive.

A word of caution should be offered to those who have adopted the practice of using plasma from outdated human bank blood for this test. Inhibitory factors are found in some human blood, and control tests with known strongly and weakly positive and negative strains should always be included.*

The **coagulase slide** test, which demonstrates "bound" coagulase, or clumping factor, is used by many laboratories and gives results comparable to the tube test. It is performed by emulsifying growth from a typical colony in a drop of water on a slide and adding a loopful of fresh **human** plasma. This is mixed thoroughly for 5 seconds. If the reaction is **positive,** easily visible white clumps will immediately appear; if no such clumping appears, the reaction is **negative** and **should be checked by the tube test,** since some strains of *Staph. aureus* may be negative.

Some investigators report that there is good correlation between **phosphatase** production and pathogenicity; yet some coagulase-negative strains are more strongly phosphatase-positive than some coagulase-positive strains. The amount or degree of phosphatase activity is also known to vary with the phage types of the staphylococci.

Strains of *Staph. aureus* isolated from human sources can elaborate a variety of metabolites. Some of these are toxic and of pathological significance, whereas others are either nontoxic or of low toxicity but have some diagnostic significance. There are recorded differences in the exotoxins produced by human and animal pathogenic strains. According to the report of Elek and Levy,[3] human pathogens normally produce alpha and delta lysins, whereas the animal pathogens produce alpha, beta, and delta lysins. Other

---

*Dehydrated sterile rabbit plasma is available commercially.

*Citrated plasma may also be coagulated by organisms other than *Staph. aureus,* such as some citrate-utilizing enterococci.[1]

exotoxins produced by the pathogenic staphylococci are **leukocidin,** which is probably the same as the delta lysin, a **dermonecrotic toxin,** a **lethal toxin,** and an **enterotoxin.** The last named is the most common cause of food poisoning. There are at least three types of enterotoxin, all of which are heat stable and resistant to the enzymes pepsin and trypsin.

Alpha hemolysin does not lyse human erythrocytes but does lyse rabbit and sheep red cells. Beta lysin lyses sheep red cells, and delta lysin lyses horse erythrocytes. Alpha hemolysin serves as a convenient index of virulence because the more virulent the strain the more alpha hemolysin it produces. Filtrates of virulent broth cultures usually yield a high titer. The procedure for carrying out the alpha hemolysin test may be found in Chapter 41.

Tests for the dermonecrotic toxin may be carried out by intradermal injection of a rabbit's back and tests for lethal toxin by intravenous injection. Tests for the enterotoxin are carried out in human volunteers, rhesus monkeys, and kittens. Much controversy still continues concerning the value of the kitten and the frog tests for enterotoxigenicity.

Among the nontoxic metabolites produced by the staphylococci are phosphatase and coagulase, to which reference was made previously, **hyaluronidase, deoxyribonuclease, staphylokinase,** or **fibrinolysin,** and in addition **lipase, gelatinase,** and **protease.** Most if not all of these are antigenic.

A close correlation also between deoxyribonuclease activity and coagulase activity has been demonstrated by Weckman and Catlin[11] and confirmed by other investigators. Deoxyribonuclease activity may be tested by streaking the surface of the commercially available deoxyribonuclease test medium* with the staphylococcus strain. Enzyme activity may be checked by flooding the surface with **normal hydro-**chloric acid** or 0.1% toluidine blue. The appearance of a **clear zone** around the streak constitutes a positive test. Plates flooded with toluidine blue show a bright **rose pink** zone around colonies of deoxyribonuclease producers. The formula for the medium is given in Chapter 39.

Although basically *Staph. aureus* is susceptible to most of the chemotherapeutic agents that are active on gram-positive bacteria, the organism can develop resistance to the widely used antibiotics, as well as to new ones being brought into use, with surprising facility. It would appear also that the phage groups of staphylococci show differences in antibiotic resistance. It is estimated that at least 80% of the hospital strains of staphylococci are resistant to penicillin. There seems to be a correlation between virulence and penicillinase production; the more resistant such strains are to penicillin the more virulent they appear to be. Laboratory personnel are thus advised to consider each isolate or strain individually and to test it for its susceptibility to a number of antibiotics by one of the accepted methods of assay. Resistance to two or more antibiotics is relatively rare in phage group II.* The reader is referred to Chapter 34 for information on the methods of antibiotic susceptibility testing of microorganisms.

Staphylococcus epidermidis

Staphylococcus epidermidis is a gram-positive, nonmotile coccus occurring singly, in pairs, and in irregular clusters, resembling the morphology of *Staph. aureus.* The colonies are circular, smooth, and usually a pale translucent white.

The organism is very salt tolerant, as is *Staph. aureus,* but differs from that species in being **coagulase negative** and **mannitol negative.** In other biochemical reactions it resembles the pathogenic species. The organism is parasitic rather than pathogenic, but it appears to be unquestionably involved in certain clinical syndromes. It

---

*Baltimore Biological Laboratory, Cockeysville, Md.; Difco Laboratories, Detroit.

*Parker, M. T.: Personal communication.

may be found normally on the skin and mucous membranes of man and other animals.

Phage typing of staphylococci

Since the recognition of prevalent hospital phage types is of interest in outbreaks, some laboratories are involved in the phage typing of staphylococci. The value of this procedure in epidemiological studies cannot be questioned, but because of the rather involved and precise procedures that are necessary, it is recommended that smaller laboratories not attempt to establish a routine typing service.

Phage typing is carried out by the Laboratory Centre for Disease Control, Ottawa, by The Cross Infection Laboratory, C. P. H. Laboratory, London, by the State Public Health laboratories, and at the Center for Disease Control in Atlanta. Table 16-1 lists the lytic groups of the international set of staphylococcal basic typing phages.

Diseases caused by staphylococci

The most common disorders caused by the pathogenic staphylococci are pimples, boils, carbuncles, impetigo, acne, postoperative and wound infections, pyelitis, and

Table 16-1. Lytic groups of the international staphylococcal typing phages*

GROUP	PHAGE TYPES
I	29, 52, 52A, 79, 80
II	3A, 3B, 3C, 55, 71
III	6, 7, 42E, 47, 53, 54, 75, 77, 83A
IV	42D
Not allotted	81, 187

*Compiled from data provided by M. T. Parker, Cross Infection Laboratory, Central Public Health Laboratory, London, N.W. 9, England.

Phages 81 and 187 cannot be allotted to any of the lytic groups. Phage 81 lyses many strains that otherwise have Group I patterns, but it also forms part of some Group III patterns. Phage 187, however, lyses strains that are sensitive to this phage only.[4] Strains that are lysed only by phage 81 are placed in phage Group I.

Additional information may be gained from the report in Internat. Bull. Syst. Bact., **21:**165, 167, 171, 1971.

cystitis. The most common type of food poisoning is staphylococcal food poisoning, caused by the enterotoxin produced in food. It has been reported that the production of this toxic metabolite is restricted mainly to phage Groups III and IV.[12]

Among the most serious staphylococcal infections are septicemia, endocarditis, meningitis, puerperal sepsis, pneumonia, and osteomyelitis. Staphylococcal enteritis may develop with great rapidity after prolonged oral antibiotic therapy if resistant strains develop. There is suggestive evidence that certain syndromes may be associated with specific phage types of staphylococcus. In phage Group II, strains reacting only with phage 71 appear to be specifically associated with vesicular skin lesions, such as impetigo and pemphigus of the newborn.

Staph. epidermidis can cause lesions such as stitch abscesses, but it is also involved in bacterial endocarditis after cardiac surgery. As the etiological agent of this disease, it produces a prolonged illness, but the onset is usually milder than that of the progressive acute form caused by *Staph. aureus.* Additionally, it is often cited as the cause of a persistent bacteremia following ventriculovenous shunts for the control of hydrocephalus.[8]

A blood culture is required to establish a diagnosis of staphylococcal endocarditis, and it is invariably positive. If clinical observations suggest this syndrome, four to six blood cultures should be made at convenient intervals before treatment, but in critical cases needing immediate treatment, three cultures should be made within a period of several hours.

Other micrococci

Gaffkya tetragena is an opportunistic parasite found on the skin and in the upper respiratory tract. It has been isolated from suppurative lesions, from empyemas following pneumonia, and from septicemic infections. The organism usually appears in tetrads on direct smear or when grown on blood media, but it is often found in pairs or small clusters from non-supplemented agar media. *G. tetragena* is coagulase negative; it ferments glucose anaerobically and does not reduce nitrates.

Sarcina species are parasitic in nature and occur in cubical packets of eight or more cells. Aerobic species are difficult to distinguish from species of *Micrococcus*. The latter are saprophytes and only rarely are found in human infections.

ANAEROBIC MICROCOCCI
Genus Peptococcus

The species of this genus are obligate anaerobes, and although they are found normally in the respiratory and genital tracts of humans, they have been isolated from breast abscesses, a case of prostatitis[7] and in mixed culture from lung and genital infections, and from miscellaneous abscesses.[10] They occur in irregular masses.

Peptococci are incriminated in anaerobic infections less frequently than peptostreptococci, and they may be distinguished from the latter by the fact that peptococci are **catalase positive** and do not produce a sharp, pungent odor. Recently, gas chromatography has provided a more sophisticated means of distinguishing between the anaerobic cocci. In this procedure, the characteristic patterns of these volatile fatty acids were observed by Pien and co-workers,[6] who reported that the peptostreptococci showed a single sharp, acetic acid peak, while the peptococci showed either a two-peak pattern of butyric and acetic acids or a small peak of acetic acid alone. Low budget laboratories

may not be able to employ gas chromatographic methods, unless laboratory supervisors or their personnel can adapt existing equipment from the clinical chemistry laboratories.

REFERENCES

1. Bayliss, B. G., and Hall, E. R.: Plasma coagulation by organisms other than *Staphylococcus aureus,* J. Bact. **89**:101-105, 1965.
2. Breed, R. S., Murray, E. G. D., and Smith, N. R.: Bergey's manual of determinative bacteriology, ed. 7, Baltimore, 1957, The Williams & Wilkins Co.
3. Elek, S. D.: *Staphylococcus pyogenes,* London, 1959, E. & S. Livingstone, Ltd.
4. Parker, M. T.: Phage typing and the epidemiology of *Staphylococcus aureus* infection, J. Appl. Bact. **25**:3, 1962.
5. Parker, M. T., and Williams, R. E. O.: Further observations on the bacteriology of impetigo and pemphigus neonatorum, Acta Paediat. Upps. **50**:101-112, 1961.
6. Pien, F. D., et al.: Clinical and bacteriologic studies of anaerobic gram-positive cocci, Mayo Clin. Proc. **47**:251-257, 1972.
7. Potoir, A., and Moriw, J. E.: Thirty-five cases of human anaerobic infections, Can. J. Pub. H. **48**:317-322, 1957.
8. Quinn, E. L., Cox, F., and Fisher, M.: The problem of associating coagulase negative staphylococci with disease, Ann. N. Y. Acad. Sci. **128**:428-442, 1965.
9. Recommendations, Subcommittee on Taxonomy of Staphylococci and Micrococci, Int. Bull. Bact. Nomenclat. Taxon. **15**:109-110, 1965.
10. Thomas, C. G. A., and Hare, R.: The classification of anaerobic cocci and their isolation in normal human beings and pathological processes, J. Clin. Path. **7**:300-304, 1954.
11. Weckman, B. G., and Catlin, B. W.: Desoxyribonuclease activity of micrococci from clinical sources, J. Bact. **72**:747, 1957.
12. Williams, R. E. O., Rippon, J. E., and Dowsett, L. M.: Bacteriophage typing of strains of *Staphylococcus aureus* from various sources, Lancet **1**:510-514, 1953.

17 The streptococci, including the enterococci and anaerobic streptococci

The streptococci are members of the genus *Streptococcus.* They are widely distributed in nature and may be found in milk and dairy products, in water, in dust, in vegetation, and in the normal respiratory tract and intestinal tract of various animals, including man. The majority are probably saprophytic and nonpathogenic, but certain species are human and animal pathogens.

The single streptococcal cell is characteristically spherical but may appear elliptical on occasion. The cells normally occur in chains of varying lengths. Cell size varies from 0.5 to 1μ in diameter, depending on growth conditions and age of culture. Large cells are seen occasionally with normal-sized cells in a chain in an aged culture, whereas undersized cells may appear with normal cells in anaerobic culture.

Liquid cultures normally yield longer chains (Fig. 17-1) than cultures grown on agar. Beta hemolytic streptococci frequently exhibit long chains in human infections and in milk from a diseased cow.

Streptococci are characteristically **gram positive** but may become gram negative as the cells age. Although a few motile forms have been reported, the cells are normally **nonmotile.** Virulent forms are usually encapsulated and contain an abundance of hyaluronic acid in the capsule. Streptococcal colonies are small, translucent to slightly opaque, circular, generally less than 1 mm. in size, convex, and appear as minute beads of moisture on a moist agar surface. On drier surfaces, colonies are less moist and almost opaque. By comparison, pneumococcal colonies are flatter and translucent.

Colony variation occurs quite commonly, showing the **mucoid, smooth** or **glossy,** and **matt** or **rough** forms. The mucoid and matt forms contain relatively large amounts of M protein and are virulent, whereas the smooth or glossy forms contain very little of this substance and are usually avirulent.

The organisms grow well on most enriched media and are facultative in relation to their oxygen requirements. They produce quite large amounts of lactic acid without gas in fermentable carbohydrates. **Inulin** is usually not fermented, and the organisms are **not soluble** in bile salts. The latter criterion, especially, will distinguish streptococci from pneumococci. They differ also from the staphylococci in being **catalase negative** (described on p. 397).

CLASSIFICATION OF STREPTOCOCCI

Various classifications for the streptococci have been used over the years, of which the one used in *Bergey's Manual of Determinative Bacteriology*[3] is an example, wherein the **pyogenic** group, the **viridans** group, the **enterococcus** group, and the **lactic** group make up the system. The Brown[4] classification, based on the reactions in blood agar and to which reference is made later, is another example, but the Lancefield[9] system, based on the antigenic characteristics of the **group-specific C** substance is probably the most reliable. The C substance is a polysaccharide, and this permits the arrangement of the streptococci into a number of antigenic groups identified as Lancefield groups A, B, C, D, and so forth.

HEMOLYTIC REACTIONS ON BLOOD AGAR

The most useful method for the preliminary differentiation of human streptococcal strains is their appearance (hemolytic activity) on blood agar. As originally postulated by Brown in 1919,[4] several types, including a recently described subtype, can be observed, particularly in subsurface colonies in pour plates[8]:

Alpha—an indistinct zone of partially lysed (fixed) red cells surrounds the colony, frequently accompanied by a greenish or brownish discoloration, which is best seen around subsurface colonies.

Beta—a clear, colorless zone surrounds the colony, indicating complete lysis of the red blood cells, which is best seen in deep colonies in a pour plate (along with the spindle shape of the colony). Surface colonies, on the other hand, may appear as alpha or nonhemolytic, due to inactivation of one of the hemolysins—streptolysin O—which is oxygen labile; streptolysin S, an oxygen-stabile hemolysin, may be present in only small amounts in these poor "surface hemolyzer" strains.

Gamma—colonies show no apparent hemolysis or discoloration in either surface or subsurface colonies.

Alpha-prime, or wide-zone alpha—a small zone of alpha hemolysis surrounds the colony, with a zone of complete or beta hemolysis extending beyond the zone into the medium. This can be confused with a beta hemolytic colony when only surface growth is observed.

Hemolysis of mammalian erythrocytes by the streptococci is a complex system of many variables, including the influence of the basal medium, the production of streptolysins O and S by a given strain, the effect of various types of blood (sheep, horse, rabbit, human), aerobic or anaerobic incubation, and so forth. To begin with, a good basal medium, such as soybean-casein digest agar* **without** dextrose (the acid produced by carbohydrate fermentation inactivates streptolysin S) at pH 7.3 to 7.4, to which 5% defibrinated sheep blood has been added, is recommended. This medium will support the growth of fastidious strains and permit good differentiation of types of hemolysis.

Although much has been written about the species of red blood cells and their effect on hemolytic patterns, it appears that this is restricted to the enterococci, in which about 90% show alpha hemolysis on sheep blood agar but are beta hemolytic on other mammalian blood agar media.[15] Sheep blood is recommended especially for its inhibitory action on the growth of *Haemophilus hemolyticus,* a normal throat commensal whose beta hemolytic colonies

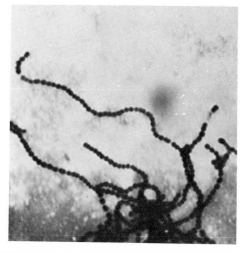

Fig. 17-1. *Streptococcus pyogenes,* showing chain arrangement. (4000×.)

---

*BBL trypticase soy agar, Difco tryptic soy agar.

may be confused with those of beta hemolytic streptococci. When in doubt, the examination of a Gram stain of the suspected colony will rapidly differentiate gram-positive streptococci from gram-negative bacillary forms of *H. hemolyticus.* Outdated human blood bank blood is **not** recommended for preparation of blood agar, because of its content of possible inhibitory factors, such as antibacterial substances, antibiotics, or an excess of citrate ion.

Although incubation under increased CO_2 tension (candle jar or CO_2 incubator) for 18 to 24 hours is recommended for the isolation and recognition of beta hemolytic streptococci on sheep blood agar, it should be noted that anaerobic incubation, or prolonged aerobic incubation and overnight refrigeration may increase hemolytic activity of these organisms.

PREPARATION OF POUR PLATES

The ideal method of identification of hemolytic activity is by the microscopic examination of subsurface colonies in a blood agar **pour plate.** Although not always practical, it is not a difficult procedure, and may be carried out as follows:

1. Melt a tube of sterile soy-casein digest agar (15 to 20 ml.) and cool to approximately 45° C.
2. Aseptically add 0.6 to 0.8 ml. of sterile defibrinated sheep blood.
3. Inoculate this with one **small** loopful (drained against inner wall of tube) of a suspension prepared from the original throat swab suspended in 1 ml. of sterile broth.
4. Mix the inoculated medium by rotating the tube between the palms of the hands, and pour into a sterile Petri plate. Allow to harden; incubate overnight at 36° C. If desired, the surface of the plate also may be streaked with a loopful of the suspension.

When colonies are examined microscopically, the illumination is reduced, and the low-power objective is focused on both the colony and the layer of blood cells in approximately the same plane. A colony near the bottom of the plate should be selected for examination, in order to minimize the effect of the intervening agar in focusing on the inverted plate. Surface colonies are examined with the lid of the dish removed.

BETA HEMOLYTIC STREPTOCOCCI

As previously indicated, beta hemolytic streptococci produce surface or subsurface colonies surrounded by a clear, cell-free zone of hemolysis. Although seen best in pour plates, beta hemolysis may also be demonstrated by making several stabs with the inoculating loop into the agar at the time of plate streaking. This permits subsurface growth and participation by both streptolysins O and S, if present.

Although beta hemolytic streptococci of Groups A, B, C, D, E, F, G, H, K, L, and others are found in Lancefield's classification,* Group A strains *(S. pyogenes)* are most often associated with communicable disease in humans and are the etiological agents in streptococcal pharyngitis, scarlet fever, epidemic wound infections, and so forth. They are subdivided serologically into 44 M types and 26 T agglutination types, depending on type-specific surface antigens.[7] According to Moody and coworkers,[12] more than 90% of Group A strains can be typed by a combination of M precipitin and T agglutination procedures.

Human pathogens may also occur among other serological groups, including Groups B and D. Group B strains, such as *S. agalactiae,* are normally present in the human vagina and may be associated with a maternal septicemia and neonatal meningitis;[2] animal strains are involved in bovine mastitis. Group D strains, including enterococci, are discussed in a subsequent section.

Occasionally, strains of Group C, G, H, and K streptococci have been reported in

---

*Antisera are available for these and other groups from Difco Laboratories and Burroughs Wellcome Co.

Table 17-1. Identification of beta hemolytic streptococci*

CHARACTERISTICS	GROUP A	GROUP B	GROUP C HUMAN	GROUP D (ENTEROCOCCI)‡
On blood agar				
Surface colonies	White to gray, opaque, hard, dry; 2 mm. zones of hemolysis	Gray, translucent, soft; narrow hemolytic zone; a few RBC may be observed microscopically under the colonies	Similar to Group A	Gray, translucent, soft; zone of hemolysis wider than the colony
Subsurface colonies	2- to 2.5-mm. zones of hemolysis, with sharply defined edges; spindle-shaped	0.5-mm. zone of hemolysis after 24 hours; 1-mm. zone of hemolysis after 48 hours; refrigeration produces double zones of hemolysis	Similar to Group A	3- to 4-mm. zones of hemolysis
Bacitracin susceptibility	Susceptible	Resistant	Resistant	Resistant
Sodium hippurate	Not hydrolyzed	Hydrolyzed	Not hydrolyzed	Not hydrolyzed
Growth at 10° C.	−	−	−	+
Growth at 45° C.	−	−	−	+
Bile esculin medium†	−	−	−	+
SF medium†	No growth	No growth	No growth	Growth, acid reaction (usually)
6.5% NaCl broth	No growth	No growth	No growth	Growth
Source	Throat, blood, wounds, rarely spinal fluid	Urine, peritoneum, rarely blood, occasionally human throat, genital tract of females	Throat, nose, vagina, intestinal tract	Urine, peritoneum, feces; milk and milk products
Pathogenicity	Septicemia, tonsillitis, scarlet fever, puerperal sepsis, pneumonia, erysipelas	Rare cases of endocarditis and meningitis; female genital tract infections; maternal septicemia; neonatal meningitis	Erysipelas, puerperal sepsis, throat infections; opportunist pathogen	Subacute bacterial endocarditis; urinary tract infections; bacteremia

*Note: Groups A, B, and C will not grow at pH 9.6 or in 0.1% methylene blue milk—two criteria that will distinguish them from Group D (enterococci).
†Baltimore Biological Laboratory, Cockeysville, Md.; Difco Laboratories, Detroit.
‡See Table 17-3, p. 123.

human infections, including bacteremias, and in respiratory and genitourinary tract infections.[5] These strains generally produce either alpha or beta hemolysis on blood agar.

Table 17-1 shows some useful tests for differentiating among several groups of beta hemolytic streptococci.

Benzyl penicillin (penicillin G) and other penicillins, including ampicillin, are the antibiotic drugs of choice in the treatment of infections due to streptococci of Groups A, B, C, G, and others. Sulfonamides or erythromycin is sometimes used prophylactically against Group A streptococcal infections in patients with rheumatic heart disease who are allergic to penicillin.

THE BACITRACIN DISC TEST FOR GROUP A STREPTOCOCCI

A useful **presumptive** test for differentiating Group A from other groups of beta hemolytic streptococci is the bacitracin disc test, first introduced by Maxted[11] and modified by Levinson and Frank.[10] The test depends on the selective inhibition of Group A streptococci on a blood agar plate by a paper disc containing 0.04 units of bacitracin.* (**Caution:** Use the differential disc containing 0.04 units, not the sensitivity disc containing 10 units.) A high degree of correlation between bacitracin and serological tests with Group A streptococci can be expected, especially if the following conditions are fulfilled[8]:

1. A **pure culture** of a beta colony is to be used—do not use on a primary plate of a mixed culture.
2. A fresh, moist blood agar plate must be used—an old, dried-out plate will reduce diffusion of the bacitracin, giving a false negative reading.
3. The inoculum size must be such to ensure **confluent growth**—a light growth of other streptococci may show inhibition zones.

*Available from Baltimore Biological Laboratory (Taxo A); Difco Laboratories (Bacto Differentiation Disc Bacitracin).

4. The bacitracin discs must be stored in the refrigerator with a desiccant and should be checked periodically for performance with known Group A and non–Group A strains.
5. The test should be used only to differentiate **beta** hemolytic gram-positive streptococci—some alpha strains may show moderate zones of inhibition (8 to 10 min.).
6. Group A strains characteristically exhibit zones of **10 to 18 mm.** In reporting a positive test, the result should be worded: "Beta hemolytic streptococcus (probably Group A)." Further identification procedures include the use of immunofluorescence technics and the Lancefield serological procedures, described in Chapters 35 and 37.

ALPHA HEMOLYTIC STREPTOCOCCI

Streptococci that do not possess group antigens and that generally produce alpha hemolysis on blood agar are known as **alpha hemolytic** (viridans) streptococci. They are constantly present in the human oropharynx and include such species as *S. salivarius, S. mitis,* and *S. sanguis.* These organisms are the commonest cause of subacute bacterial endocarditis, an insidious and fatal infection (if untreated) that usually follows dental or surgical procedures or instrumentation in patients with a previously damaged heart valve or other lesions of the endocardium. *Streptococcus MG,* a rarely isolated species, has been used in an agglutination test for the serological diagnosis of *Mycoplasma pneumoniae* infections (primary atypical pneumonia).

Colonies of alpha hemolytic streptococci must be distinguished from those of pneumococci or enterococci, both of which also may produce alpha hemolysis on blood agar. Table 17-2 shows some tests that are useful in identifying alpha hemolytic streptococci; other differential tests are indicated in subsequent sections.

Penicillin G appears to be the antibiotic

Table 17-2. Identification of alpha hemolytic streptococci

CHARACTERISTICS	STREPTOCOCCUS FAECALIS*	STREPTOCOCCUS SALIVARIUS	STREPTOCOCCUS MG†	STREPTOCOCCUS SANGUIS	STREPTOCOCCUS MITIS GROUP
On blood agar					
Surface colonies	Relatively large, gray, shiny, translucent; no greening after 24 hours, slight greening at 48 to 72 hours	Small, raised, convex, opaque; narrow zone of hemolysis with or without greening, depending on blood used	Nonhemolytic after 24 hours but may give alpha appearance after 48 hours; variations due to type of blood	Alpha hemolysis in 24 to 48 hours	Alpha hemolysis in 24 to 48 hours
Subsurface colonies	Large, non-hemolytic after 24 hours; definite greening after 48 to 72 hours; hemolysis evident after 24 hours of refrigeration	Typical alpha appearance— fixation of cells, greening, and hemolysis after 24 to 48 hours	Nonhemolytic after 24 hours, changing to alpha at 48 hours; refrigeration for 24 hours produces typical appearance	Alpha hemolysis in 24 to 48 hours	Alpha hemolysis in 24 to 48 hours
Colonies on 5% sucrose agar	Small, compact	Large, raised, mucoid	Small, compact; fluorescent in ultraviolet light	Small, compact	Small, compact
Growth in 5% sucrose broth	No change	No change	No change	Gelling of medium	No change
Bile esculin medium	Growth	No growth	No growth	No growth	No growth
SF medium	Growth	No growth	No growth	No growth	No growth
Methylene blue milk	Reduction	No reduction	No reduction	No reduction	No reduction
Mannitol	+	−	−	−	−
Inulin	−	+	−	+	−
Source	Intestines and genitourinary tract, blood	Respiratory tract, blood	Respiratory tract	Blood	Respiratory tract, blood
Pathogenicity	Urinary tract infections, subacute bacterial endocarditis; opportunist pathogen	Respiratory infections, subacute bacterial endocarditis	Possibly associated with primary atypical pneumonia	Subacute bacterial endocarditis	Respiratory infections, subacute bacterial endocarditis

*Streptococcus faecalis var. liquefaciens shows the same cultural characteristics as Streptococcus faecalis, but the former liquefies gelatin and the latter does not.
†Streptococcus MG can be identified specifically by the quellung reaction, using Streptococcus MG antiserum.

drug of choice in the treatment of subacute bacterial endocarditis; ampicillin with streptomycin has also been used successfully. It is essential that the organism be isolated promptly from blood cultures (see Chapter 6) and its antibiotic susceptibility determined.

These antibiotic agents are also recommended as prophylactic therapy for patients with rheumatic or congenital heart disease who may undergo dental or surgical procedures. In penicillin-allergic patients, erythromycin or lincomycin is a satisfactory substitute.

GROUP D STREPTOCOCCI, INCLUDING ENTEROCOCCI

The streptococci that react serologically with Lancefield Group D antisera comprise two different categories. The **first** contains the enterococci *S. faecalis* and *S. faecium,* of human intestinal origin, and *S. avium,* a member of Lancefield's Group Q. The former are important agents in human infections; the latter is apparently not pathogenic for humans. The **second** category includes the nonenterococci of Group D, *S. bovis* and *S. equinus,* which are of animal origin but occasionally cause infection in man. Although most human Group D isolates are *S. faecalis, S. bovis* is being recovered frequently from patients with subacute bacterial endocarditis. The incidence of other members of the group has not been accurately determined.

Generally, Group D streptococci appear as alpha or nonhemolytic colonies on sheep blood agar; occasional varieties of *S. faecalis* (var. *zymogenes*) will produce wide zones of beta hemolysis. Some Group D strains have a distinct "buttery" odor on the medium. Traditionally, Group D strains have been differentiated from other streptococci by their ability to grow at 45° C. and by being thermostable at 60° C. for 30 minutes; however, streptococci of other groups may demonstrate these characteristics, especially when standardized test conditions have not been met. It has become apparent from reports by Facklam

and Moody[6] at CDC, that the most accurate presumptive test for recognizing Group D streptococci is with **bile esculin medium** (BEM). By incorporating bile (oxgall) into an agar or broth medium containing esculin,* it becomes selective for the growth of enterococci (also *Listeria monocytogenes* and Enterobacteriaceae) that are capable of hydrolyzing esculin to 6,7 dehydroxycoumarin, which reacts with an iron salt in the medium to form a **dark brown or black** compound. The agar slant or broth is inoculated with a pure culture and examined after 48 hours' incubation at 36° C. for the production of a brownish black color. Plus-minus reactions are read as negative.

Two other useful biochemical tests for enterococci are: (1) growth and fermentation of dextrose after 48 hours (indicated by a color change of the brom cresol purple indicator) in **SF broth†** and (2) growth in heart infusion broth containing **6.5% sodium chloride** in 18 to 24 hours (Table 17-3).

The results of these tests may be interpreted as follows[6]:
1. BEM positive, SF or salt tolerance positive = enterococcus
2. BEM positive, SF or salt tolerance negative = Group D streptococcus, not enterococcus
3. BEM negative, SF or salt tolerance negative = non Group D streptococcus (serological grouping suggested)

The antibiotic drug of choice in the treatment of serious enterococcal infections is ampicillin; if penicillin allergy exists, erythromycin and streptomycin, or vancomycin and streptomycin in combination may be used. Most strains of *S. bovis* are susceptible to penicillin or clindamycin.

---

*Several bile esculin media are available: Difco BE agar contains 4% oxgall, Pfizer PSE agar and BBL enterococcus agar contain 1% oxgall plus sodium azide. (Some viridans strains may grow and hydrolyze esculin in the latter media.) Broth media also are available.

†There may be some lot-to-lot variation in this medium, resulting in discrepant reactions. (See Facklan, R. R.: Appl. Microbiol. **26:**138-145, 1973.)

Table 17-3. Differentiation of Group D streptococci, including enterococci*

CHARACTERISTICS	STREPTO- COCCUS FAECALIS	STREPTOCOCCUS FAECALIS VAR. LIQUIFACIENS	STREPTOCOCCUS FAECALIS VAR. ZYMOGENES	STREPTO- COCCUS FAECIUM	STREPTO- COCCUS DURANS	STREPTO- COCCUS BOVIS
Hemolysis on sheep blood agar	Alpha to gamma	Alpha to gamma	Beta	Gamma	Alpha or gamma	Gamma
Bile esculin medium	Growth	Growth	Growth	Growth	Growth	Growth
6.5% NaCl broth	Growth	Growth	Growth	Growth	Growth	No Growth
Acid in SF medium	+	+	+	+	+	−
Growth at 10° C	+	+	+	+	+	−
Growth at 45° C	+	+	+	+	+	+ (−)
Gelatin hydrolysis	− (+)	+	+ (−)	−	−	−
Acid in litmus milk	+	+	+	+	+	+
Acid from						
Glycerol	V	+	+	−	−	−
Mannitol	+	+	+	+	−	+ (−)
Sorbitol	+	+	+	− (+)	−	−
Sucrose	+ (−)	+	+	+	V	+

*Enterococci will survive a temperature of 60°C. for 30 minutes and grow in pH 9.6 broth and 0.1% methylene blue milk. Other streptococcal groups may also react positively with these tests.[6] + = positive reaction; ( ) = occasional; V = variable.

GAMMA (NONHEMOLYTIC) STREPTOCOCCI

Colonies of these streptococci, sometimes called indifferent streptococci, produce no change in surface or subsurface colonies in blood agar. They are rarely isolated from clinical material and have been poorly defined taxonomically.

ANAEROBIC AND MICROAEROPHILIC STREPTOCOCCI

These streptococci, some aerotolerant and some strictly anaerobic, are involved most frequently in mixed infections with other anaerobes or facultative organisms. They are found in pleuropulmonary infections and are secondary to infections in other parts of the body, such as abscesses of the liver, brain, and so forth. The anaerobic forms (peptostreptococci) also have been incriminated in puerperal sepsis, bacteremia, subacute endocarditis, and suppurating deep wound infections.

Material suspected of harboring anaerobic streptococci, particularly in **foul-smelling** pus, should be plated without delay on fresh blood agar and enriched thioglycollate medium and placed immediately into an **anaerobic** atmosphere (see Chapter 13 for anaerobic methods) and examined after 48 hours' incubation. Concurrently, a second set of blood agar plates also should be inoculated with the material; these are incubated **aerobically** and compared with the anaerobic plates for the presence of colonies of peptostreptococci, which appear as minute, shiny, smooth nonhemolytic colonies. Some strains, however, may show alpha or beta hemolysis. In the thioglycollate medium, gas and a foul odor are generally apparent; subcultures incubated aerobically and anaerobically should reveal only anaerobic growth of a catalase-negative, gram-positive streptococcus; this may sometimes exhibit a sharp pungent odor.[14]

A Gram stain of a **direct smear** of the purulent material also should be examined; a positive report to the clinician of the presence of tiny gram-positive cocci in chains may alert him to the possibility of a peptostreptococcal infection.

The taxonomy of the microaerophilic and anaerobic streptococci has not been well defined; the use of standardized media and various biochemical tests, along with the analysis of metabolic end products by gas chromatography, will aid in differentiating them.[13]

The antibiotic drug of choice in the treatment of infections due to anaerobic streptococci appears to be benzyl penicillin (penicillin G); clindamycin has also proved effective, especially in mixed anaerobic infections.[1] There is some evidence that the peptostreptococci may act synergistically with *Bacteroides* or *S. aureus* to produce tissue necrosis in mixed infections.[13]

REFERENCES

1. Bartlett, J. G., Sutter, V. L., and Finegold, S. M.: Treatment of anaerobic infections with lincomycin and clindamycin, New Eng. J. Med. **287:**1006-1010, 1972.
2. Braunskein, H., Tucker, E. B., and Gibson, B. C.: Identification and significance of *Streptococcus agalactiae,* Amer. J. Clin. Path. **51:**207-213, 1969.
3. Breed, R. S., Murray, E. G. D., and Smith, N. R.: Bergey's manual of determinative bacteriology, ed. 7, Baltimore, 1957, The Williams & Wilkins Co.
4. Brown, J. H.: Monograph no. 9, New York, 1919, Rockefeller Institute for Medical Research.
5. Duma, R. J., Weinbert, R. T., Medrek, T. F., and Kunz, L. J.: Streptococcal infections, Medicine **48:**87-127, 1969.
6. Facklam, R. R., and Moody, M. D.: Presumptive identification of Group D streptococci; the bile-esculin test, Appl. Microbiol. **20:**245-250, 1970.
7. Griffith, F.: The serological classification of *Streptococcus pyogenes,* J. Hyg. **34:**542, 1934.
8. Hall, C. T., In: Summary analysis of results for the proficiency testing survey in bacteriology (Jan. 9, 1970), National Communicable Disease Center, May 7, 1970.
9. Lancefield, R. C.: A serological differentiation of human and other groups of hemolytic streptococci, J. Exp. Med. **57:**571-595, 1933.
10. Levinson, M. L., and Frank, P. F.: Differentiation of group A from other beta hemolytic streptococci with bacitracin, J. Bact. **69:**284-287, 1955.
11. Maxted, W. R.: The use of bacitracin for identifying group A hemolytic streptococci, J. Clin. Path. **6:**224-226, 1953.
12. Moody, M. D., Padula, J., Lizana, D., and Hall, C. T.: Epidemiologic characterization of group A streptococci by T agglutination and M precipitation tests in the public health laboratory, Health Lab. Sci. **2:**149-162, 1965.
13. Pien, F. D., Thompson, R. L., and Martin, W. J.: Clinical and bacteriologic studies of anaerobic gram-positive cocci, Mayo Clin. Proc. **47:**251-257, 1972.
14. Rogosa, M.: Peptococcaceae, a new family to include the gram-positive, anaerobic cocci of the genera *Peptococcus, Peptostreptococcus,* and *Ruminococcus,* Inter. J. Syst. Bact. **21:**234, 1971.
15. Updyke, E. L.: Laboratory problems in the diagnosis of streptococcal infections, Publ. Health Lab. **15:**78-80, 1957.

18 Pneumococci

STREPTOCOCCUS PNEUMONIAE

The pneumococci are represented by a single species, *Streptococcus (Diplococcus) pneumoniae,* commonly known as the **pneumococcus.** These organisms consist of gram-positive lanceolate cocci, characteristically appearing as diplococci, but occasionally appearing as short, tight chains or as single cocci. In fresh specimens of sputum, spinal fluid, or other exudates, they are frequently surrounded by a capsule. Based on a specific capsular polysaccharide, 82 capsule types have been recognized.[9]

The normal habitat of the pneumococcus is the upper respiratory tract of humans; from there it may invade the lungs and the systemic circulation. Consequently, the pneumococcus is the commonest cause of lobar pneumonia (80% to 90%) in adults. Bacteremia occurs in about one fourth of these patients, usually early in the course of the disease; the organisms also may disseminate to the endocardium and pericardium, the meninges, joints, and so forth, with ensuing complications. Pneumococci have also been implicated in infections of the middle ear, mastoid, or eye; they are occasionally isolated from peritoneal fluid, urine, vaginal secretions, wound exudate, and other clinical specimens. Carrier rates in the respiratory tracts of healthy adults may vary from 5% to 60%, depending on the season of the year.

Pneumococci require an enriched medium for their primary isolation; trypticase or brain-heart infusion agar enriched with 5% defibrinated sheep, horse, or rabbit blood is recommended. Since a small percentage of strains require incubation in an increased CO_2 tension to grow on primary plate culture, it is necessary to incubate such media in a candle jar or CO_2 incubator.

CLINICAL SPECIMENS

The specimens usually submitted for culture include sputum (from the lower respiratory tract—not saliva), blood, cerebrospinal fluid, throat or, preferably, nasopharyngeal swabs (particularly in the pediatric patient), purulent exudates, serous effusions, and so forth. Sputum, if tenacious, may be homogenized by repeated mixing with a small volume of broth in a sterile syringe. Otherwise, a carefully selected portion of bloody or purulent material is teased out with the split halves of a sterile wooden applicator stick and one portion inoculated to blood agar. The other portion may be used for preparing a direct smear for gram staining. If quantitative studies of the bacterial flora of sputum are desired, a mucolytic agent, such as dithiothreitol,* may be used.

---

*Available commercially as Sputolysin, Calif. Biochem., La Jolla, Calif.

CULTURAL CHARACTERISTICS

After overnight incubation, typical pneumococcal colonies on blood agar are round and glistening with entire edges, **transparent,** mucoid, and about 1 mm. in diameter. They are surrounded by an approximate 2 mm.-zone of alpha hemolysis (beta hemolysis when incubated anaerobically[8]), due to the activity of hyaluronidase. Young colonies are usually dome shaped but on aging become flattened with a raised margin and central portion, giving the appearance of a **checker** or nail head. This is best seen with a dissecting microscope and oblique surface illumination. Alpha hemolytic streptococci, by contrast, produce small, raised, **opaque** colonies. Colonies of type 3 pneumococcus (occasionally, other types) are larger, more mucoid and confluent than those already described, and resemble droplets of oil on the agar surface. These eventually flatten on drying, and exhibit characteristics similar to those of other types.

The pneumococcus undergoes variation of the classical type. Using the terminology of Dubos[5] and Dawson and colleagues,[4] the encapsulated form is designated **M** or mucoid, the nonencapsulated form **S,** and the true rough form **R.** The M form is virulent, whereas the S and R forms are avirulent. The capsular substance is a polysaccharide, known as the specific soluble substance (SSS). This determines both virulence and capsular type.

An additional variant, known as the phantom colony, or the P-C form, may be found on blood agar plates when virulent forms, especially those isolated from the lungs, undergo rapid autolysis. Such forms show up after about 72 hours of incubation at 36° C. In such instances, one can observe only the impression left by the colony and its zone of green discoloration. Autolysis may be prevented by incubating the plates in a 10% carbon dioxide atmosphere.[6]

IDENTIFICATION TESTS FOR PNEUMOCOCCI

In addition to the colony characteristics by which pneumococci may be tentatively identified, there are certain tests that provide a reliable means of differentiating these organisms from the alpha streptococci or other cocci. Among them are the tests for (1) **bile solubility,** (2) **inulin fermentation,** (3) **optochin susceptibility,** (4) the **Neufeld (quellung) reaction,** and (5) **mouse virulence.**

Bile solubility test. Surface-active agents, such as bile, bile salts (sodium deoxycholate or taurocholate), or sodium dodecyl sulfate ("Dreft"), will act upon the cell wall of pneumococci and bring about lysis of the cell. The following modification of the bile solubility test used at the Mayo Clinic* utilizes a tube of veal infusion broth (5 ml.) containing 0.1 ml. horse serum. After inoculation, the tube is incubated at 36° C. until visibly turbid, and several drops of 4% "Dreft" are added. If positive, that is, **bile soluble,** the tube clears rapidly.

It should be noted that some strains of pneumococci are insoluble in bile, whereas some strains of alpha hemolytic streptococci are soluble; the test, therefore, is not infallible.

Inulin fermentation test. Most strains of pneumococci will ferment this carbohydrate. However, because some strains of *Streptococcus sanguis* and *Streptococcus salivarius* will also ferment it, the test should not be used alone, but in conjunction with other confirmatory tests, such as bile solubility or optochin sensitivity, as an identification procedure.

Optochin growth inhibition test. This is currently the most widely used test for differentiating pneumococci from other alpha hemolytic streptococci. It was first described by Bowers and Jefferies[3] in 1954 and modified by Bowen and co-workers[2]

---

*Courtesy of Dr. John Washington II, Director, Section of Clinical Microbiology, Mayo Clinic, Rochester, Minn.

in 1957. The test is carried out by placing an absorbent paper disc containing 5 μg. of ethylhydrocupreine HCl (Optochin)* on a blood agar plate heavily inoculated with a pure culture of the suspected strain. After overnight incubation at 36° C., pneumococci will usually exhibit a zone of inhibition **greater than 15 mm.** in diameter.

Experts consider the test over 90% reliable in differentiating pneumococci from other streptococci; however, they note that occasional strains are not inhibited. The Center for Disease Control (CDC) has received cultures of "penicillin-resistant pneumococci" that proved to be alpha streptococci, exhibiting a moderate inhibition zone (10 to 12 mm.) to Optochin, especially when light inocula were used.[7]

Note: Each new lot of Optochin discs should be checked with known strains of pneumococci and alpha hemolytic streptococci before use.

Mouse virulence test. The white mouse is particularly susceptible to infection by small inocula of pneumococci and may be used to advantage in certain unusual situations, especially with mixed cultures. For example, if 4 to 6 hours after intraperitoneal injection of 1 ml. of emulsified sputum the mouse's abdomen is entered with a sharp-tipped capillary pipet, the peritoneal fluid will contain organisms in pure culture, which may be used for capsular swelling tests, culture, and other procedures. It should be noted that some pneumococcus types, including type 14, may be avirulent for mice.

Avery's "artificial mouse" also may be used with mixed cultures. This culture medium—1% glucose and 5% defibrinated rabbit blood in meat infusion broth at pH 7.8—promotes rapid growth and encapsulation of pneumococci after overnight incubation at 36° C. and provides a good source for capsular swelling tests.

*Taxo P discs, Baltimore Biological Laboratory, Cockeysville, Md.; Optochin discs, Difco Laboratories, Detroit; and others.

Neufeld quellung reaction. The **quellung** reaction is the most accurate, reliable, and specific test for the identification of pneumococcal types, but it is infrequently used today because the treatment of pneumococcal infections is no longer dependent upon identification of the specific capsular type. The procedure, nevertheless, has proved of value in epidemiological investigations. The consensus is that the capsular swelling reaction, as it is popularly referred to, is not an actual swelling. Fig. 18-1 illustrates capsules seen in the quellung reaction. The phenomenon is probably caused by a change of the refractive properties brought about by the action of the specific antiserum, which makes the outline of the capsule more readily visible. Capsular antisera for the pneumococci are not readily available commercially because of the lack of demand.* An omnivalent serum, which reacts with all 82 pneumococcal types, has been prepared by Lund and Rasmussen.[10] This "omniserum" has proved very useful as a reliable reagent for

*Six pools of 33 types are available from Difco Laboratories, Detroit; Omniserum, 9 pools and 46 monotypic sera are available from the State Serum Institute, Copenhagen.

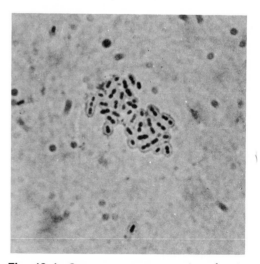

Fig. 18-1. *Streptococcus pneumoniae,* showing capsular swelling. (4000×.)

the rapid identification of pneumococci in clinical material, such as sputum,* cerebrospinal, pleural, or synovial fluid, or in positive blood culture bottles.

Directions for performance of the capsular swelling test are given in Chapter 35.

SIGNIFICANCE OF RECOVERY OF PNEUMOCOCCI

Because of the relatively high seasonal carrier rate of pneumococci in the respiratory tract, their recovery from this site may be difficult to interpret. However, isolation of the recognized virulent strains,[1] such as types 1, 3, 4, 7, 8, or 12 from blood-tinged sputum of a patient with lobar pneumonia is of clinical significance. Further, isolation of pneumococci from blood culture not only affords definite evidence of a pneumococcal infection; but if the bacteremia persists, it may even suggest a guarded prognosis or more intensive therapeutic measures.

Pneumococci are highly susceptible to penicillin, which remains the drug of choice in treating infections by this organism; lincomycin and erythromycin appear to be second-choice agents in the penicillin-allergic patient.

Since the advent of antibiotics, especially penicillin, the overall mortality of adults with pneumococcal pneumonia has been reduced from 84% to 17%[1] (type 3 infections have a higher mortality); results in

the pediatric age group appear to be even better. Pneumococcus types 14, 3, 6, 18, 19, and 23 are isolated most frequently from infections of infants and children.

It should be noted that pneumococci are usually not recoverable from sputum culture after 6 to 12 hours of adequate penicillin therapy.[1] It therefore behooves the clinician to secure material for culture **before** initiation of any antimicrobial therapy.

REFERENCES

1. Austrian, R.: The current status of pneumococcal pneumonia and prospects for prophylaxis, Ninth Annual Infectious Disease Symposium, Delaware Academy of Medicine, May 5, 1972.
2. Bowen, M. K., Thiele, L. C., Stearman, B. D., and Schaub, I. G.: The optochin sensitivity test; a reliable method for identification of pneumococci, J. Lab. Clin. Med. **49:**641-642, 1957.
3. Bowers, E. F., and Jeffries, L. R.: Optochin in the identification of *Str. pneumoniae,* J. Clin. Path. **8:**58, 1955.
4. Dawson, M. H., et al.: Variations in the hemolytic streptococci, J. Infect. Dis. **62:**138-168, 1938.
5. Dubos, R. J.: The bacterial cell, Cambridge, Mass., 1946, Harvard University Press.
6. Eaton, M. C.: Variants characterized by rapid lysis and absence of normal growth under the routine method of cultivation, J. Bact. **27:**271-291, 1934.
7. Hall, C. T.: Summary analysis of results for the proficiency testing survey in bacteriology (Jan. 9, 1970), National Communicable Disease Center, May 7, 1970.
8. Lorian, V., and Popoola, B.: Pneumococci producing beta hemolysis on agar, Appl. Microbiol. **24:**44-47, 1972.
9. Lund, E.: Laboratory diagnosis of *Pneumococcus* infections, Bull. WHO **23:**5-13, 1960.
10. Lund, E., and Rasmussen, P.: Omniserum—a diagnostic pneumococcus serum reacting with the 82 known types of *Pneumococcus,* Acta Path. Microbiol. Scand. **68:**458-460, 1966.

---

*It has been shown recently that the direct quellung (omni) test on sputum smears was more accurate than Gram stains when correlated with culture results. (Merrill, C. W., et al., New Eng. J. Med. **288:**510, 1973.)

19 Neisseriae and related forms

The members of the family Neisseriaceae are gram-negative cocci occurring in pairs or in masses. They may range from aerobic through facultatively anaerobic to anaerobic, depending on the genus. They may be found in the mouth, the intestines, and the genitourinary tract of man and other animals.

There are ten recognized species in the genus *Neisseria,* two of which are well-known pathogens of man, whereas the remainder are of doubtful pathogenicity, although certain species have been associated with respiratory disorders and meningeal infections. The species of *Neisseria* are either aerobic or facultatively anaerobic.

NEISSERIA GONORRHOEAE (GONOCOCCUS)

The **gonococcus** is a gram-negative diplococcus in which the paired cells have flattened adjacent walls. The organism is fastidious, requiring enriched media such as blood agar or chocolate agar for its cultivation. It is an obligate human parasite not found in other animals, and it causes a number of human infections, including urethritis, cervicitis, salpingitis, proctitis, and other complications in adults, vulvovaginitis of children, and ophthalmia of the newborn.

The disease is transmitted by sexual intercourse in most instances, and causes acute purulent urethritis in men and cervi-

citis (frequently asymptomatic) in women. Infection may also occur by extension to other sites, including the joints, rectal crypts, blood, and so forth. Less fequently, gonorrheal ophthalmia of the newborn may occur following passage through an infected birth canal; gonorrheal vulvovaginitis in the preadolescent female is usually the result of criminal assault or by contact with an infected fomite. Man is the only known host.

Isolation. Neisseria gonorrhoeae is a fastidious parasite, requiring an enriched culture medium and incubation under increased CO_2 (3% to 10%) for its recovery from clinical material. Luxuriant growth occurs on chocolate agar supplemented with yeast extract or a similar enrichment,* or on Thayer-Martin medium (TM), a modified chocolate agar made selective for the gonococcus and meningococcus by the addition of certain antibiotics.† Thayer-Martin is the preferred medium for isolation of *N. gonorrhoeae;* its use has been described in detail in Chapter 11.

On moist TM medium after 20 hours' incubation at 35° to 36° C‡ in a candle jar, typical colonies of *N. gonorrhoeae* appear

---

*Available as IsoVitaleX (BBL), Supplement B (Difco).

†Available from Baltimore Biological Laboratory, Cockeysville, Md.; Difco Laboratories, Detroit; and others.

‡Many strains of gonococci will not grow well at 37° C.

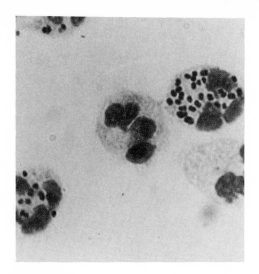

Fig. 19-1. *Neisseria gonorrhoeae.* Smear of exudate, showing intracellular gonococci. (Gram stain; 4000×.)

as small (1 to 2 mm.), transluscent, raised, moist grayish white colonies with entire to lobate margins. They are usually mucoid and tend to come off as whole colonies when fished from the agar surface. Colony size varies, depending upon age of the culture or crowding on the plate. Plates without growth should be returned to the incubator for 48 hours' incubation.

Presumptive identification. All members of the genus *Neisseria,* as well as some members of other genera (such as *Pseudomonas, Moraxella,* and *Aeromonas*) are **oxidase positive.** In the oxidase reaction, a bacterial enzyme called indophenol oxidase will oxidize a redox dye, which results in a color development of the bacterial colony. In the oxidase test (described in Chapter 41), a freshly prepared 1.0% solution of the redox dye tetramethyl-p-phenylenediamine dihydrochloride is applied with a platinum loop to a suspected colony on the TM plate; if **positive,** a color change in the colony will be observed—pink, progressing to maroon, to dark red, and finally to black. At this point (generally in 5 to 10 minutes) the organisms have been killed; however, they retain their gramstaining and FA-staining properties. Subcultures to a fresh TM plate should be

made at the pink color stage. If no characteristic colonies are observed, the surface of the TM plate may be flooded with the oxidase reagent, to detect inapparent colonies.

A thin smear of the oxidase-positive colony is then prepared, gram stained, and examined microscopically under oil immersion. Typical neisseriae will appear as gram-negative diplococci with their flattened sides adjacent; the microscopist should note, however, that certain gramnegative diplobacilli (for example, *Moraxella osloensis*) may resemble gonococci morphologically.

A nationwide program is presently under way to provide physicians with a rapid laboratory diagnosis of gonorrheal infection. The Center for Disease Control (CDC) has recommended that material from the genitourinary tract inoculated to TM (or Transgrow) medium showing growth of typical oxidase-positive colonies consisting of gram-negative diplococci provides sufficient criteria for a presumptive identification of *N. gonorrhoeae.*[1] This may be reported as: "Presumptive identification—*N. gonorrhoeae.*"* Isolation of neisseriae from other sites, such as a pharyngeal culture, or in special social or medicolegal situations, should be identified by the following, more definitive procedure.

Confirmatory identification. A presumptive identification of *N. gonorrhoeae* may be confirmed by carbohydrate fermentation reactions (preferred) or by direct FA staining. *N. gonorrhoeae* **ferments glucose only,** producing acid but no gas (Table 19-1). The recommended base medium—cystine trypticase agar (CTA), pH 7.6—readily supports the growth of fresh isolants of gonococci and meningococci, although occasional strains of the former may grow poorly or not at all. The carbohydrates used are added in 1% concentration to the sterilized medium (see Chapter 39 for preparation), which may be stored in the refrigerator.

---

*Approximately 98% of isolates from urogenital sites are confirmed by carbohydrate fermentation or FA.

Table 19-1. Differentiation of the neisseriae*

	CARBOHYDRATE FERMENTATION REACTIONS (CTA BASE, 1 TO 4 DAYS' INCUBATION AT 36° C.)						GROWTH	
ORGANISM	GLUCOSE	MALTOSE	SUCROSE	LACTOSE	FRUCTOSE	MANNITOL	ON TM	ON NUT. AGAR AT 22° C.
N. catarrhalis	−	−	−	−	−	−	−‖	+
N. gonorrhoeae	A†	−	−	−	−	−	+	−
N. meningitidis	A	A	−	−	−	−	+	−
N. sicca	A	A	A	−	A	−	−‖	+
N. lactamica	A	A	−	A(slow)	−	−	+	
N. hemolysans§	A	A	A	−	A	−		+
N. flava	A	A	−	−	A	−	−‖	+
N. perflava	A	A	A	−	A	A		+
N. subflava‡	A	A	−	−	−	−	−‖	+
N. flavescens‡	−	−	−	−	−	−	−‖	+
Mima polymorpha var. oxidans (Moraxella osloensis)	−	−	−				− or +	at 36° C.

*Chart courtesy of Dr. Charles T. Hall, Licensure and Proficiency Testing Branch, Center for Disease Control, Atlanta, Georgia.
†A, acid production (no gas).
‡Bacterial growth shows a yellowish pigmentation on Loeffler's serum medium.
§Beta hemolysis by second or third day.
‖A heavy inoculum can yield growth.

The CTA media are inoculated with a heavy suspension prepared from a purification plate (taken from a single colony on the TM plate) in about 0.5 ml. trypticase soy broth. A loopful of this suspension is transferred to the surface and an area immediately below the surface of the tubes of media. If preferred, 2 to 3 drops of the suspension may be deposited on the surface of the medium, using a sterile, cotton-plugged capillary pipet.

The screw caps of the tubes are then **tightened** and are incubated without added CO_2 at 35° to 36° C.* They are examined daily for evidence of growth and production of acid, indicated by turbidity and a yellow color of the upper layer of medium. After 48 to 72 hours' incubation, the tubes showing acidity are checked for purity by microscopic examination of a Gram stain. If only glucose is fermented, report: "*Neisseriae gonorrhoeae* isolated." An occasional strain of gonococcus may fail to ferment glucose on a first attempt; repeat-

*Some workers prefer incubation in a candle jar, with loose screw caps. A wet paper towel is included.

ing the inoculation of the CTA medium enriched with 10% ascitic fluid may encourage fermentation by these strains.

The direct FA staining procedure[8] is a less time-consuming and laborious procedure for identifying colonies of *N. gonorrhoeae* than are fermentation tests; but owing to some lack of sensitivity (due to cross reaction with Group B meningococci), the procedure appears less satisfactory for primary cultures. For the interested reader, however, the technique is described in Chapter 37.

NEISSERIA MENINGITIDIS (MENINGOCOCCUS)

The normal habitat of *Neisseria meningitidis* is in the human nasopharynx; many persons may carry the organism for indefinite periods without symptoms. In susceptible persons, however, the meningococcus gains access to the central nervous system via the hematogenous route (with a bacteremic precursor) or by extension from the sinuses through the lymphatics or bone. A suppurative infection of the meninges then occurs, producing the characteristic syndrome of a bacterial meningitis. The bloodstream invasion may result in an early petechial rash (smears of which may show meningococci) or take the form of a massive adrenal hemorrhage accompanied by disseminated intravascular coagulation, with a rapidly fatal outcome (Waterhouse-Friderichsen syndrome).

The meningococcus is similar to the gonococcus in morphology and staining reactions. The paired cocci have a miniature coffee bean appearance. Some of the cells may be swollen and appear larger than the norm in 48-hour cultures. The organism is fastidious in its growth requirements and may be cultivated on enriched media such as chocolate agar or blood agar, on which it develops relatively large, bluish gray, smooth, raised colonies. No hemolysis is produced on blood agar. The colonies tend to autolyze fairly rapidly. The colonies may be identified as *Neisseria* by the oxidase test, which shows the same color changes that are observed with gonococcal colonies.

The organism is **extremely sensitive** to temperature and dehydration; thus the same facilities provided for the culture of the gonococcus must be accorded the culturing of this species. Cultivation in a 3% to 10% carbon dioxide atmosphere greatly enhances growth.

N. meningitidis may be classified antigenically into Groups A, B, C, and D. Groups A and C are encapsulated, whereas Group B strains are usually nonencapsulated. The former give rise to larger and more mucoid colonies than one observes with strains of Group B, in which the colonies are smaller, rougher, and yellowish. Group A strains are generally involved in epidemics of meningitis, whereas Group B strains are isolated in the majority of sporadic cases between outbreaks.

The majority of strains of *N. meningitidis* isolated from patients with meningococcal infections or from healthy carriers fall into serological groups B and C; Group A and D strains are rarely isolated in the United States. In 1961, Slaterus[9] in Holland reported the isolation of meningococci that did not react with antisera prepared against Group A, B, C, or D strains, and provisionally labeled these new types as X, Y, and Z. Vedros and co-workers[10] also described a new strain, Group E; presently, meningococci of Group Z[1] (CDC Group 29E), Boshard (Slaterus Y), and 135 are recognized.

Isolation. The organism may be isolated from cerebrospinal fluid, from the nasopharynx, from joint fluids, from the blood, from petechiae of the skin, and from miscellaneous sites, such as the eye. Material from such sources is streaked on blood or chocolate agar plates and incubated in a candle jar at 36° C. A positive oxidase test on the colonies would be presumptive evidence of the presence of the organism. Confirmation by the Gram stain will indicate the typical morphology. Recognized colonies of the organism may then be transferred to blood agar slants to obtain pure cultures for biochemical tests.

Identification. Inoculation of the recommended carbohydrate media—glucose, maltose, and sucrose—in either the CTA or the serum or ascitic fluid semisolid agar (Chapter 39) will help to identify the meningococcus, which ferments only **glucose** and **maltose.**

The serological group may be determined by the capsular swelling reaction, although it should be noted that only Groups A and C possess capsular antigen; Group B strains are identified by agglutination reaction. Only cells from actively growing cultures, or cells directly from clinical material (for example, sediment from cerebrospinal fluid) showing a sufficient number of gram-negative diplococci, should be used for capsular swelling tests. The current method of choice in most diagnostic laboratories, however, is the **slide agglutination test,** using first polyvalent and then monovalent antisera. The reader is cautioned about the limitations of this technique, especially in reference to Group B meningococci, since cross relationships exist between this organism and other neisseriae. There also appears to be some antigenic cross relationship between Groups A and C.

The direct FA staining procedure also may be used for the identification of meningococci in cerebrospinal fluid; fluorescein-labeled antimeningococcal conjugates are available commercially.*

Techniques for capsular swelling and slide agglutination tests are described in Chapter 35; the immunofluorescent procedure is described in Chapter 37.

ANTIBIOTIC SUSCEPTIBILITY OF PATHOGENIC NEISSERIAE

In a comparative study of the susceptibility of gonococci to penicillin in the United States, Martin and co-workers[6] reported that in a 10-year period, from 1955 to 1965, there was an increase from 0.6% to 42% in cultures requiring more than 0.05 unit per milliliter of penicillin to inhibit growth. Procaine penicillin G (or ampicillin), however, remains the antibiotic of choice in most patients; those allergic to the penicillin drugs may be treated effectively with spectinomycin or tetracycline.

Present knowledge indicates that penicillin G in large dosage is the drug of choice in the treatment of meningococcal disease but is unsatisfactory for the prophylactic treatment of carriers.[3] It also appears that most of these strains could be eliminated by treatment with rifampin[2] or minocycline, since many are resistant to sulfadiazine.

OTHER NEISSERIA SPECIES

Since other *Neisseria* species may be isolated from sputum, the throat, the nasopharynx, and occasionally from cerebrospinal fluid, they should be recognized and differentiated from the gonococcus and the meningococcus. For these reasons they are included in Table 19-1. *Neisseria*

catarrhalis grows as a grayish white friable colony, granular and difficult to emulsify; *N. flava, N. perflava, N. subflava,* and *N. sicca* generally produce small yellowish to greenish colonies that may be smooth, or hard and wrinkled, and are considered to be nonpathogenic. *N. flavescens* and *N. lactamica* produce colonies similar to *N. meningitidis,* but may have a yellowish pigment on primary isolation; they have been recovered from some pathological processes, including meningitis.

Although previous authors had reported the isolation of **lactose-utilizing** strains of neisseriae,[5,7] little attention was paid until Hollis and co-workers at CDC[4] published results of a study of organisms referred for confirmation as *N. meningitidis,* but were found to be lactose-positive.* Since then, these workers have proposed a new species, *N. lactamica,* and described its characteristics. The majority of strains have been recovered from pharyngeal and nasopharyngeal specimens, but occasionally strains have been recovered from sputum, tracheal aspirations, amniotic and cerebrospinal fluids, lung tissue, and so forth. Reports, although scarce, indicate that *N. lactamica* is of little clinical significance. It has been most frequently misidentified as *N. meningitidis* or *N. subflava* (Table 19-1).

GENUS VEILLONELLA

The organisms in the genus *Veillonella* occur in pairs, short chains, and irregular clumps. They are gram negative and strictly **anaerobic.** The cells are smaller than neisseriae and are less fastidious in their nutritional requirements. All forms are parasitic, occuring in the respiratory tract, the intestinal tract, and the genitourinary tract of man and other animals. Two species have been associated with disease in man, *V. alcalescens* and *V. parvula;* but normally *Veillonella* is not considered pathogenic.

---

*Polyvalent antiserum (Groups A through D) from Sylvana Co., Millburn, N. J.; Difco Laboratories, Detroit.

*Probably because most laboratories were not using lactose (CTA) in their routine biochemical workup for neisseriae.[4]

REFERENCES

1. Balows, A., and Printz, D. W.: CDC program for diagnosis of gonorrhoeae, Letter, J.A.M.A. **222:**1557, 1972.
2. Deal, W. G., and Sanders, E.: Efficacy of rifampin in the treatment of meningococcal carriers, New Eng. J. Med. **281:**641-649, 1969.
3. Dowd, J. M., Blink, D., Miller, C. H., Frank, P. F., and Pierce, W. E.: Antibiotic prophylaxis of carriers of sulfadiazine resistant meningococci, J. Infect. Dis. **116:**473-480, 1966.
4. Hollis, D. G., Wiggins, G. T., and Weaver, R. E.: *Neisseria lactamicus* sp. n.; a lactose-fermenting species resembling *Neisseria meningitidis,* Appl. Microbiol. **17:**71-77, 1969.
5. Jensen, J.: Studien uber gramnegative kokken, Zentr. Bakteriol. Parasitenk. Abt. I. Orig. **133:**75-88, 1934.
6. Martin, J. E., Jr., Lester, A., Price, E. V., and Schmale, J. D.: (Note) Comparative study of gonococcal susceptibility to penicillin in the United States, 1955-1969, J. Infect. Dis. **122:**459-461, 1970.
7. Mitchell, M. S., Rhoden, D. L., and King, E. O.: Lactose-fermenting organisms resembling *Neisseria meningitidis,* J. Bact. **90:**560, 1965.
8. Peacock, W. L., Welch, B. G., Martin, J. E. Jr., and Thayer, J. D.: Fluorescent antibody technique for identification of presumptovely positive gonococcal cultures, Public Health Rep. **83:**337-339, 1968.
9. Slaterus, K. W.: Serological typing of meningococci by means of micro-precipitation, Antonie van Leewenhoek J. Microbiol. Serol. **27:**304-315, 1961.
10. Vedros, N. A., and Culber, G.: A new serological group (E) of *Neisseria meningitidis,* J. Bact. **95:**1300-1304, 1968.

20 Gram-negative enteric bacteria

The classification of the family Entero-bacteriaceae has been revised since the third edition of this book was published and will appear in its revised form in the eighth edition of *Bergey's Manual of Determinative Bacteriology.* The reader will notice that the new taxonomic position of certain genera is recognized by the authors of this text and are included in this chapter on the gram-negative rods.

FAMILY ENTEROBACTERIACEAE

The members of the Enterobacteriaceae are gram-negative, straight rods, some of which are motile and some nonmotile. The motile species possess **peritrichous** flagella, differing from the members of the Pseudomonadaceae, which have polar flagella. Several strains of *Salmonella, Shigella, Escherichia, Klebsiella, Enterobacter,* and *Proteus* possess **fimbriae** or **pili.**[5] The latter are not organs of locomotion, are considerably smaller than flagella, and bear no antigenic relationship to them.[10] They are readily observed under the electron microscope.

All species ferment glucose. Aerogenic and anaerogenic forms are found. The absence of gas in the fermentation of carbohydrates is characteristic of some genera. Nitrates are usually reduced to nitrites. Indophenol-oxidase is not produced.

The family is composed of a large and unwieldy group of organisms varying in antigenic structure and biochemical properties. The genera within the family have been established mainly on the basis of biochemical characteristics, whereas original species—the names of many of which still remain—were established on both biochemical and ecological bases. The antigenic complexity of the groups has led to the development of antigenic schemas, patterned after the Kauffmann-White schema for *Salmonella,* in which one finds numerous serotypes listed. Many of these serotypes are biochemically similar and can be distinguished only by serological procedures.

The organisms are found in the intestines of man and other animals, in the soil, and on plants. Many are parasites; others are saprophytes. Many species are pathogenic for man, producing intestinal diseases and septicemic infections.

Culturally, the enteric gram-negative bacteria produce similar growth on blood agar, usually appearing as relatively large, shiny, gray colonies, which may or may not be hemolytic. Species that produce hydrogen sulfide show a definite greening around subsurface colonies in blood agar. On trypticase soy agar, nutrient agar, or meat infusion agar, the colonies may vary in size, depending on the genus or group, but they are usually grayish white, translucent, and slightly convex. Some colonies are large and mucoid, as in the case of *Klebsiella,* certain types of *Shigella,* and certain variants of *Salmonella,* especially *S.*

typhimurium. Colony variation does occur, giving rise to smooth and rough forms. Individual species or type colony characteristics will be described in a discussion of the genera to which they belong.

ISOLATION OF ENTEROBACTERIACEAE

Because this book is confined to diagnostic procedures for the **pathogenic** microorganisms, emphasis will be placed only on the pathogenic members of the family.

The general procedures described in Chapters 5 and 9 for the isolation of the gram-negative enteric bacteria should be followed in order to obtain good results. Since these organisms may be isolated from various clinical sources, their isolation from fecal material, blood, urine, and other body fluids will be presented again in this chapter. Laboratory personnel are reminded that clinical specimens may contain relatively few pathogens, and appropriate enrichment procedures are not only highly recommended but often necessary. One is further reminded that enrichment procedures satisfactory for *Salmonella* and *Shigella* may not be applicable for members of the coliform and *Proteus* groups, since the latter are usually inhibited by their use.

Isolation from stools. The number of pathogenic enterobacteria may decrease during the period between collection and handling in the laboratory. If there is any delay in their arrival at the laboratory, specimens should be preserved by one of the procedures described earlier (Chapter 9). A specimen collected from a suspected carrier may show very few pathogens, and in certain cases the appearance of these organisms may be only intermittent. Stuart's medium[47] is highly recommended as a transport medium for such specimens if they are to be shipped to a diagnostic laboratory. Ewing and co-workers[18] have reported excellent recovery of *Salmonella typhi* from stools in transit in this medium after 1 week.

Because of the wide and varied flora present in fecal material, **enrichment media**—inhibitory for the normal intestinal inhabitants—must be used. The enrichment media normally employed are **GN broth** (Hajna), the **selenite broth** of Liefson, and the **tetrathionate broth** of Mueller. Kauffmann's modification[28] of tetrathionate gives excellent recovery of salmonellae but inhibits many shigellae. To either medium, tubed in 10 ml. amounts, approximately 1 gm. of the undiluted fecal specimen should be added. If mucus is present, some of this should also be placed in the enrichment broth. If preserved diluted specimens are used (Chapter 5), at least 2 ml. should be placed in the enrichment broth. After 24 hours of incubation at 36° C., suitable plating media may be streaked with inoculum from these tubes. We have found that it can be beneficial to hold the enrichment media for an additional 24 hours and repeat the plating if the first plates are negative.

There are numerous **plating media** in use today. Some of these are selective, whereas others are inhibitory. Desoxycholate citrate agar, Salmonella-Shigella (SS) agar, Hektoen enteric agar (HEA), bismuth sulfite agar, brilliant green agar (BGA), eosin-methylene blue (EMB) agar, xylose lysine desoxycholate (XLD) agar, and MacConkey agar are among the most widely used.* Most laboratories prefer to employ one selective medium, such as SS or desoxycholate citrate agar, and one inhibitory medium, such as MacConkey or EMB agar. Two procedures are followed: (1) the **direct** procedure of plating the specimen on these media, and (2) the **indirect** procedure using enrichment first. The use of bismuth sulfite is highly recommended in cases where *Salmonella typhi* is suspected because it is still the most efficient medium for the isolation of this pathogen.

If one is attempting to isolate enteropathogenic *Escherichia coli, Klebsiella, Enterobacter,* or *Citrobacter* from fecal material, tetrathionate and selenite enrich-

*To inhibit the spreading of *Proteus* strains on MacConkey or EMB media, the agar concentration may be increased to 5%.

ment broths are **not** used, since both are inhibitory for most strains of these genera. In such instances, the less inhibitory media (either MacConkey or EMB agar) are used for primary isolation by the direct plating procedure in lieu of the more selective plating media that tend to inhibit most strains. Blood agar also is recommended by some investigators.

Some lactose-fermenting, gram-negative enteric bacteria can withstand the inhibitory substances present in the enrichment broths and the selective media, but these can be recognized readily by their appearance on the selective plates.

Lactose-negative forms such as *Salmonella* and *Shigella* give rise to small **colorless** colonies on desoxycholate citrate, MacConkey, EMB, XLD, and SS media. On HEA, salmonellae and shigellae appear bluish green. Colonies of *Proteus* may be confused with *Salmonella* and *Shigella,* especially on desoxycholate, and also on EMB and MacConkey media containing 5% agar, because of their lactose-negative characteristic. Colonies of **lactose-fermenting** organisms on desoxycholate citrate agar (if not inhibited), on MacConkey agar, and on SS medium appear red; on EMB agar they appear **dark purple** to **black** and often have a **metallic sheen.** On HEA, they appear **salmon** to **orange** in color.

On **bismuth sulfite agar,** if the colonies are well separated, *Salmonella typhi* gives rise to **black** colonies with a metallic sheen. The use of a poured plate is recommended in instances where only a few organisms are likely to be present. In this procedure a relatively large inoculum (3 to 5 ml.) of fluid stool or preserved stool specimen is placed in a Petri dish, and approximately 15 ml. of the melted medium is added and rotated thoroughly to mix. This technique promotes greater chances of recovery. Subsurface colonies usually have a typical appearance.* Some strains of salmonellae

(*S. paratyphi B* and *S. enteritidis*), as well as other members of Enterobacteriaceae, will give rise to black colonies on this medium. Thus, every black colony that appears should not be presumptively identified as *S. typhi.* Generally, salmonellae other than *S. typhi (S. paratyphi A, S. typhimurium,* and *S. choleraesuis)* develop as dark green, flat colonies or as colonies with black centers and green peripheries.

Isolation from blood. The specimen used in making a blood culture may have been collected from an individual with a febrile disease of unknown etiology, and therefore the media used for this will be dictated by the needs of the suspected organism. General methods have been discussed in Chapter 6, but if culture is desired for *Salmonella* or *Shigella* specifically (shigellae are only rarely found in the blood), approximately 10 ml. of blood should be placed in 90 to 100 ml. of bile broth and incubated at 36° C. If the culture is negative after 24 hours, incubation should be continued for 10 to 14 days before a negative result is reported.

Should blood serum be required for antibody detection tests when a blood specimen is received in coagulated (clotted) form, the blood clot may be centrifuged and the serum removed with a pipet and rubber bulb. The blood clot should be broken up and added to the bile broth in approximately the same proportion as previously specified.

If **typhoid fever** is suspected, a successful blood culture is usually attained during the first or second week after onset. In septicemias produced by other salmonellae, blood cultures* should be taken during the first week and, if negative, should be repeated during the second week and thereafter if considered necessary. Laboratory personnel are reminded that in typhoid fever a blood culture may be positive **before** stool cultures become positive. During the first week blood cultures are

---

*Subsurface colonies of *S. typhi,* if well separated, are circular, jet black, and well defined. Only those near the surface will exhibit the characteristic metallic sheen.

---

*Cultures of bone marrow also may be helpful in salmonelloses.

positive in about 90% of the cases, whereas stool cultures are positive in only about 10% of the cases during this period.

Plating of blood cultures on the appropriate type of medium—bismuth sulfite, SS, EMB, or MacConkey—is carried out as described for stool isolations. Many investigators find that a plate of either EMB or MacConkey agar will suffice at this phase of the procedure.

Isolation from urine. Salmonella typhi may be isolated from the urine in about 25% of the cases of typhoid fever. Additional members of the Enterobacteriaceae may also be isolated from this source in urinary tract infections, including other species of *Salmonella* and certain members of the *Escherichia, Klebsiella,* and *Proteus* genera. Direct plating of the specimen on selective and inhibitory media is recommended. Best results for isolating salmonellae from urine are usually obtained by centrifuging the specimen at 2,500 to 3,000 r.p.m. for 20 to 30 minutes to sediment the bacteria. Several loopfuls of the sediment may then be plated on selective and inhibitory media and the remainder added to enrichment broth. The enrichment procedure will aid in the recovery of *S. typhi.* Plating from the enrichment broth is carried out on the appropriate media after 24 or 48 hours. If centrifugation is not employed for specimens from suspected *Salmonella* infections, 2 to 3 ml. of the specimen should be added to the enrichment broth to compensate.

PRELIMINARY SCREENING OF CULTURES

Colonies of salmonellae and shigellae, the characteristics of which have been discussed earlier in this chapter, may be recognized by their **lactose-negative** appearance on isolation plates. Suspected colonies of these two genera should be fished carefully to the screening medium with an inoculating needle. If time permits, it is advisable to select two or three colonies from each plate, since mixed

infections are not infrequent. **It is poor technique to touch the agar surrounding a colony in order to test the temperature of the needle, as members of the inhibited flora may be still present and viable, although not visible. Only the center of the colony should be touched, using a cool needle.**

With the inoculating needle a part of the colony should be stabbed first into the butt of a slant of **triple sugar iron agar** (TSI), used as a screening medium, and then streaked in a zigzag fashion over the slanted surface. The tube is always closed with a cotton plug or a loose closure but never with a tightly fitting rubber stopper or cork. The latter procedure can lead to a misinterpretation of the results, as is explained in the following section. Table 20-2 shows the different reactions obtained and the possible organisms encountered. The formula for TSI is given in Chapter 39.

Examination and interpretation of reactions in TSI agar. TSI agar contains the three sugars, glucose, lactose, and sucrose; phenol red indicator to indicate fermentation; and ferrous sulfate to demonstrate hydrogen sulfide production (indicated by blackening in the butt). The glucose concentration is one tenth of the concentration of lactose and sucrose in order that the fermentation of this carbohydrate **alone** may be detected. The small amount of acid produced by fermentation of glucose is oxidized rapidly in the slant, which will remain or revert to alkaline; in contrast, under lower oxygen tension in the butt, the acid reaction is maintained. To promote the alkaline condition in the slant, **free exchange of air** must be permitted through the use of a loose closure, as stated earlier. If the tube is tightly closed with a stopper or screw cap, an acid reaction (caused solely by glucose fermentation) will also involve the slant. The reactions in the medium, which is **orange red** in color when uninoculated, are shown in Table 20-2. **It is to be emphasized that the reactions should be read ideally after 18 to 24 hours,**

Table 20-1. Reactions observed in TSI agar

REACTION	EXPLANATION
Acid butt (yellow), alkaline slant (red)	Glucose fermented
Acid throughout medium, butt and slant yellow	Lactose or sucrose or both fermented
Gas bubbles in butt, medium sometimes split	Aerogenic culture
Blackening in the butt	Hydrogen sulfide produced
Alkaline slant and butt (medium entirely red)	None of the three sugars fermented

Table 20-2. Interpretation of reactions on TSI agar

REACTION	CARBOHYDRATES FERMENTED	POSSIBLE ORGANISMS
Acid butt Acid slant Gas in butt No H_2S	Glucose with acid and gas Lactose and/or sucrose with acid and gas	*Escherichia*[*] *Klebsiella* or *Enterobacter* *Proteus* or *Providencia* Intermediate coliforms
Acid butt Alkaline slant Gas in butt H_2S produced	Glucose with acid and gas Lactose and sucrose not fermented	*Salmonella* *Proteus* *Arizona* (certain types) *Citrobacter* (certain types) *Edwardsiella*
Acid butt Alkaline slant No gas in butt No H_2S	Glucose with acid only Lactose and sucrose not fermented	*Salmonella*[†] *Shigella* *Proteus* *Providencia* *Serratia*
Acid butt Acid slant Gas in butt H_2S produced	Glucose with acid and gas Lactose and/or sucrose with acid and gas	*Arizona* *Citrobacter*
Alkaline or neutral butt Alkaline slant No H_2S	None	*Acaligenes*[‡] *Pseudomonas*[‡] *Herellea (Acinetobacter)*[‡]

[*]Rarely, a strain is H_2S positive.
[†]*Salmonella typhi* produces a small amount of H_2S but seldom gas.
[‡]Included here because colonies of these may be confused frequently with lactose-negative members of Enterobacteriaceae and may be selected from isolation plates.

and they cannot be properly interpreted if the slants are incubated for more than 48 hours.

IDENTIFICATION OF ENTEROBACTERIACEAE

As some colonies of *Proteus* species can be confused with salmonellae and other lactose-negative enterobacteria on initial isolation, all TSI cultures should be further screened on Christensen **urea agar slants,** or in Rustigian and Stuart tubed **urea broth** (weakly buffered). A heavy inoculum is prescribed, and this is spread over the surface of the agar slant or properly emulsified in the urea broth. Urease activity is observed by a change of color (red) in the indicator due to the production of ammonia. A positive test in 2 to 4 hours on urea agar slants at 36° C. indicates *Proteus*. A positive test after 24 hours might indicate a member of the *Klebsiella* or *Enterobacter* genera.

The reactions observed on TSI agar slants **together with the effect on urea** usually indicate the possible genus or

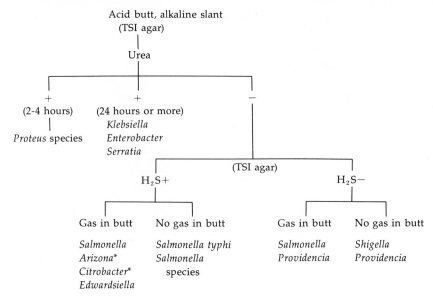

*Delayed lactose fermentation may not show up on TSI agar.

Diagram 1. Tentative differentiation of Enterobacteriaceae.

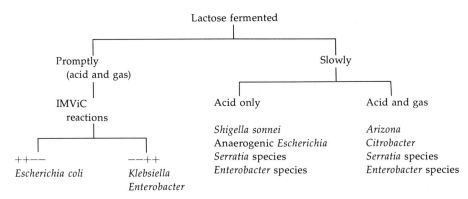

Diagram 2. Differentiation of lactose-fermenting Enterobacteriaceae.

group to which the isolate belongs (see reaction tree in Diagram 1). One should not draw hasty conclusions, however, because final identification depends on biochemical and serological confirmation. Many investigators use the growth from TSI agar slants for slide agglutination tests with polyvalent *Salmonella* or *Shigella* antisera. This is justified in cases of outbreaks or when rapid reporting is required. Cultures from TSI agar slants should be streaked on meat infusion agar, trypticase

soy agar, or MacConkey agar for purity. A single colony may then be transferred to a tube of broth for culture, followed by a complete series of tests.

Table 20-3 shows the tests on which group differentiation of the Enterobacteriaceae may be made. Tests to which reference is made in Table 20-3 are given in Chapters 39 and 41.

Lactose-fermenting members of Enterobacteriaceae. A large number of organisms within the family Enterobacteriaceae ferment the carbohydrate **lactose,** and those that do so may be tentatively identified by the reactions shown in Diagram 2. In recent years, lactose-fermenting strains of *Salmonella typhi* have been reported. Such strains are otherwise like *S. typhi* in possessing the Vi, O, and H antigens of this organism (antigenic structure is shown in Table 20-5). Some lactose-positive members of the family ferment the sugar promptly, whereas others exhibit a delayed reaction. In Diagram 2 the differentiation between these genera is indicated.

The IMViC reaction is used primarily to distinguish between the coliform bacteria, but it may be applied advantageously also to other organisms in the family. The letters stand for **indole, methyl red, Voges-Proskauer,** and **citrate** reactions. These may be carried out and interpreted as shown. (The "i" in IMViC is inserted for euphony.)

TEST FOR INDOLE. Inoculate tryptophan broth. Test after 48 hours by the addition of the Kovac or Ehrlich reagent. A **red** color indicates production of indole from the amino acid.

METHYL RED TEST. Inoculate MR-VP medium (Clark and Lubs dextrose broth medium). Test after 48 to 96 hours by adding 5 drops of methyl red indicator. A **red** color is read as positive. A yellow color is read as negative. Read the test immediately after adding the reagent.

VOGES-PROSKAUER TEST. Inoculate MR-VP medium (same as in previous test) and test for the production of acetylmethylcarbinol after 48 hours of incubation by adding to 1 ml. of culture 15 drops of 5% alpha-naphthol in absolute ethyl alcohol and 10 drops of 40% of potassium hydroxide. The necessary volume of culture may be pipetted to a small tube before the MR test above is performed. A positive test is the development of a red color in 15 to 30 minutes. An alternative test is given in Chapter 41.

CITRATE TEST. Inoculate Simmons citrate agar. A positive test is indicated by the development of a **Prussian blue color** in the medium, showing that the organism can utilize citrate as a sole source of carbon.

ANTIGENIC COMPLEXITY OF ENTEROBACTERIACEAE

The members of the Enterobacteriaceae exhibit a mosaic of antigens that fall into three main categories as follows:

1. The **K** (from German "Kapsel") or **envelope** antigens are those that, by concept, surround the cell. With certain exceptions these are heat labile. In the *Klebsiella* genus the subdivision into capsular types is based on the K antigens. K antigens mask the heat-stable somatic antigens of the cell and cause live cells to be inagglutinable in O antisera. Examples are the Vi antigen of *Salmonella typhi* and the B antigen found in certain types of *Escherichia coli.*

2. The **O** (from German "Ohne Hauch," nonspreading) or **somatic** antigens, which are heat stable, are located in the body of the cell, and are presumably near the surface. Chemically they are polysaccharide in nature. The O complex of antigens determines the somatic subgroup to which the organism belongs, in the *Salmonella, Arizona, Citrobacter, Escherichia, Providencia, Serratia,* and other schemae.

3. The **H** (from German "Hauch," spreading) antigens are the **flagellar** antigens. These are located in the flagella, are protein in nature, and are heat labile. The serotypes within the somatic groups in *Salmonella* and some other genera in the Enterobacteriaceae are determined by the H antigens.

IDENTIFICATION OF GENERA WITHIN THE FAMILY ENTEROBACTERIACEAE
Genus Salmonella

Salmonella cholerae-suis is the type species of the genus. Salmonellae are usually motile, but nonmotile forms do occur.[29] With the exception of *S. typhi* and *S. gallinarum,* they all produce **gas** in glucose. Suspected colonies of salmonellae on isolation media are inoculated to slants of TSI

Table 20-3. Differentiation of Enterobacteriaceae by biochemical tests*

	ESCHERICHIEAE		EDWARD-SIELLEAE	SALMONELLEAE			CITROBACTER		KLEBSIELLEAE		
---	---	---	---	---	---	---	---	---	KLEB-SIELLA	ENTEROBACTER	
	ESCHE-RICHIA	SHIGELLA	EDWARD-SIELLA	SALMO-NELLA	ARIZONA	FREUNDII	DIVERSUS		PNEU-MONIAE	CLOACAE	AEROGENES
Indol	+	− or +	+	−	−	−	+		−	−	−
Methyl red	+	+	+	+	+	+	+		− or +	−	−
Voges-Proskauer	−	−	−	−	−	−	−		+	+	+
Simmons' citrate	−	−	−	d	+	+	+		+	+	+
Hydrogen sulfide (TSI)	−	−	+	+	+	+ or −	−		−	−	−
Urease	−	−	−	−	−	dʷ	dʷ		+	+ or −	−
KCN	−	−	−	−	−	+	−		+	+	+
Motility	+ or −	−	+	+	+	+	+		−	+	+
Gelatin (22° C.)	−	−	−	−	(+)	−	−		−	(+) or −	− or (+)
Lysine decarboxylase	d	−	+	+	+	−	−		+	−	+
Arginine dihydrolase	d	d	−	+ or (+)	+ or (+)	d	+ or (+)		−	+	−
Ornithine decarboxylase	d	d⁽¹⁾	+	+	+	d	+		−	+	+
Phenylalanine deaminase	−	−	−	−	−	−	−		−	−	−
Malonate	−	−	−	−	+	− or +	− or +		+	+ or −	+ or −
Gas from glucose	+	−⁽¹⁾	+	+	+	+	+		+	+	+
Lactose	+	−⁽¹⁾	−	−	d	(+) or +	d		+	+ or (+)	+
Sucrose	d	−⁽¹⁾	−	−	−	d	− or +		+	+	+
Manntiol	+	+ or −	−	+	+	+	+		+	+	+
Dulcitol	d	d	−	d⁽²⁾	−	d	+ or −		− or +	− or +	−
Salicin	d	−	−	−	−	d	(+) or +		+	+ or (+)	+
Adonitol	−	−	−	−	−	−	−		+	+ or −	− or +
Inositol	−	−	−	d	−	−	−		+	d	+
Sorbitol	d	d	−	+	+	+	+		+	+	+
Arabinose	+	d	− or +	+⁽²⁾	+	+	+		+	+	+
Raffinose	d	d	−	−	−	d	−		+	+	+
Rhamnose	d	d	−	+	+	+	+		+	+	+

*Enteric Bacteriology Laboratories, DHEW-PHS-CDC, Atlanta, Ga. 30333, July, 1973.
(1) Certain biotypes of *S. flexneri* produce gas; cultures of *S. sonnei* ferment lactose and sucrose slowly and decarboxylate ornithine.
(2) *S. typhi, S. cholerae-suis, S. enteritidis* bioser. Paratyphi-A and Pullorum, and a few others ordinarily do not ferment dulcitol promptly.
(3) Gas volumes produced by cultures of *Serratia, Proteus,* and *Providencia* are small.
+, 90 percent or more positive in 1 or 2 days. −, 90 percent or more negative. d, different biochemical types [+, (+), −]. (+) delayed positive
NB. This chart is simply a guide. Users are urged to consult other publications, such as CDC publications entitled "Biochemical Reactions Given
percentage data, additional tests, and references.

| | KLEBSIELLEAE | | | | PROTEEAE | | | | | |
| ENTEROBACTER | | SERRATIA | | | PROTEUS | | | | PROVIDENCIA | |
HAFNIAE	AGGLOMERANS	MARCESCENS	LIQUEFACIENS	RUBIDAEA	VULGARIS	MIRABILIS	MORGANII	RETTGERI	ALCALI-FACIENS	STUARTII
−	− or +	−	−	−	+	−	+	+	+	+
− or +	− or +	− or +	+ or −	− or +	+	+	+	+	+	+
+ or −	+ or −	+	− or +	+	−	− or +	−	−	−	−
d	d	+	+	+ or (+)	d	+ or (+)	−	+	+	+
−	−	−	−	−	+	+	−	−	−	−
−	dw	dw	dw	dw	+	+	+	+	−	−
+	− or +	+	+	− or +	+	+	+	+	+	+
+	+ or −	+	+	+ or −	+	+	+ or −	+	+	+
−	d	+ or (+)	+	+ or (+)	+	+	−	−	−	−
+	−	+	+ or (+)	+ or (+)	−	−	−	−	−	−
d	−	−	−	−	−	−	−	−	−	−
+	−	+	+	−	−	+	+	−	−	−
−	− or +	−	−	−	+	+	+	+	+	+
+ or −	+ or −	−	−	+ or −	−	−	−	−	−	−
+	− or +	+ or −(3)	+ or −	d	+ or −	+	+ or −	− or +	+ or −	−
d	d	−	d	+	−	−	−	−	−	−
d	d	+	+	+	+	d	−	d	d	(+) or +
+	+	+	+	+	−	−	−	+ or −	−	d
−	− or +	−	−	−	−	−	−	−	−	−
d	d	+	+	+ or (+)	d	d	−	d	−	−
−	−	d	d	+ or (+)	−	−	−	d	+	− or +
−	d	d	+ or (+)	d	−	−	−	+	−	+
−	d	+	+	−	−	−	−	d	−	d
+	+	−	+	+	−	−	−	−	−	−
−	d	−	+	+	−	−	−	−	−	−
+	+ or (+)	−	d	−	−	−	−	+ or −	−	−

S. cholerae-suis does not ferment arabinose.

(decarboxylase reactions, 3 or 4 days). + or −, majority of cultures positive. − or +, majority negative. w, weakly positive reaction.
by Enterobacteriaceae in Commonly Used Tests" and Differentiation of Enterobacteriaceae by Biochemical Reactions" (W. H. Ewing,1973), for

agar. Isolates that produce acid, gas, and hydrogen sulfide in the butt and an alkaline slant in this medium and are **urease negative** should be tested with *Salmonella* polyvalent antisera. Pure cultures should be tested for motility and inoculated to the media shown in Table 20-4. Typical salmonellae will give the reactions shown.

Serological identification. The Kauffmann-White antigenic schema has a long list of *Salmonella* serotypes that are arranged in O or somatic subgroups. The H or flagellar antigens, as previously stated, determine the type.

Table 20-5 exemplifies a substantially reduced Kauffmann-White schema, but it will aid the student and laboratory worker in understanding the schematic arrangement of the serotypes. There are more alphabetized somatic groups in the schema than are shown. The tabulated types are not necessarily the most common.

Because approximately fifty different somatic antigens have been recognized,

Table 20-4. Biochemical reactions of genus *Salmonella*

TEST	REACTION	TEST	REACTION
Adonitol	−	Methyl red	+
Dulcitol	+	Voges-Proskauer	−
Glucose	+ with gas	Simmons citrate	+
Inositol	variable	KCN*	−
Lactose	−	Phenylalanine deaminase*	−
Mannitol	+	Sodium malonate*	−
Salicin	−	Lysine decarboxylase	+
Sucrose	−	Arginine dihydrolase*	+
Indole	−	Ornithine decarboxylase*	+

*Descriptions of these media and procedures for the tests performed may be found in Chapters 39 and 41. Although not used routinely in all laboratories, they aid in group differentiation.

Table 20-5. Some serotypes of the Kauffmann-White antigenic schema

TYPE	O ANTIGENS	H ANTIGENS PHASE 1	PHASE 2
	Group A		
S. paratyphi A	1, 2, 12	a	−
	Group B		
S. tinda	1, 4, 12, 27	a	e, n, z_{15}
S. paratyphi B	1, 4, 5, 12	b	1, 2
S. typhimurium	1, 4, 5, 12	i	1, 2
S. heidelberg	4, 5, 12	r	1, 2
	Group C_1		
S. paratyphi C	6, 7, Vi	c	1, 5
S. thompson	6, 7	k	1, 5
	Group C_2		
S. newport	6, 8	e, h	1, 2
	Group D		
S. typhi	9, 12, Vi	d	−
S. enteritidis	1, 9, 12	g, m	−
S. sendai	1, 9, 12	a	1, 5
	Group E_1		
S. oxford	3, 10	a	1, 7
S. london	3, 10	l, v	1, 6

serological screening with polyvalent O antisera is essential. These sera are obtainable commercially,* and they will agglutinate the majority of strains found in the United States and Canada. **Group** identification is determined by the O-grouping sera, and **type** identification is determined by H antisera. O-grouping sera for subgroups A, B, C_1, C_2, D, and E as well as H-typing sera for flagella antigens a, b, c, d, i, 1, 2, 3, 5, 6, and 7, also **Vi** antiserum, are available.*

The procedure usually employed in serological examination is the **slide agglutination test,** in which a concentrated suspension of the cells in saline is used. Details of the technique may be found in Chapter 35, but the reader's attention at this point is drawn to certain important considerations.

1. The culture to be tested must be **smooth** and not autoagglutinable in saline. A preliminary test with a 0.2% solution of acriflavine in 0.85% saline is an excellent indicator of smoothness. If a loopful of the test suspension as a control is mixed **gradually,** on a slide with a loopful of acriflavine and the cells remain in homogeneous suspension, the culture may be considered smooth.

2. If the culture suspension fails to agglutinate in the O diagnostic sera, heat it at 100° C. for 15 to 30 minutes, cool, and retest with the same sera. Some salmonellae possess K antigens, previously mentioned, and are inagglutinable in the live or unheated form in O antisera. Examples of envelope antigens are the Vi antigen of *S. typhi* and *S. paratyphi C,* the 5 antigen in somatic Group B, and the M antigen in some serotypes. Vi-containing types will agglutinate in the live or unheated form in Vi antiserum.

3. O-inagglutinable cultures should be referred to a reference center* for identification because it is better equipped to carry out complete serological analysis. **Cultures should be sent through the state or provincial laboratories.**

4. The majority of the salmonellae are **diphasic;** that is, the motile types may exhibit two antigenic forms referred to as phases in the sense of Andrews. These phases share the same O antigens but possess different H antigens, as seen in Table 20-5, and in order to identify the type, it is necessary to identify the H antigens in both phases. These may not always be in evidence, and phase-suppression procedures may become necessary in order to reveal the latent phase. This can be accomplished by inoculating the organism into a small Petri dish containing semisolid agar in which is incorporated antiserum against the antigen(s) of one identified phase. The homologous phase will be arrested at the site of inoculation, whereas the other phase will develop and may be identified. Such procedures can be carried out only by properly equipped laboratories. It should be noted that some flagellated salmonellae are nonmotile, and special procedures are required for their identification.[2]

5. To identify completely the O antigen and H antigen complexes, the use of absorbed **single factor** sera is required. These antisera are prepared by adding a concentrated suspension of cells, containing the appropriate antigens, to a suitable dilution of the multifactor serum; incubating in a water bath for 2 hours at 50° C. and refrigerating overnight. After centrifugation, the supernate will contain

---

*Lederle Laboratories, Pearl River, N. Y.; Difco Laboratories, Detroit; Baltimore Biological Laboratory, Cockeysville, Md.

*Enteric Bacteriology Unit, Center for Disease Control, Atlanta, Georgia; Central Public Health Laboratory, Colindale, London.

the unabsorbed and desired antibodies. Such a procedure is referred to as **agglutinin absorption.**

Example 1: To prepare single factor "4" O antiserum, use O antiserum "4, 12" (*S. typhimurium*) and absorb with live cells of *S. paratyphi A* containing "1, 2, 12" O antigens or *S. typhi* containing "9, 12" O antigens. Centrifuge to sediment the cells, and recover the supernatant fluid, which will contain factor "4" alone, since factor "12" has been bound and removed by absorption.

Example 2: To prepare factor "d" H antiserum, use *S. typhi* H antiserum containing 9, 12;d- antibodies, and absorb with a nonmotile strain of *S. typhi* (O901), containing 9, 12;– or with another Group D *Salmonella* type lacking the "d" antigen. The absorbed serum should be free of "9" and "12" antibodies and contain only "d" flagellar antibody.

Characteristics of Salmonella typhi. The typhoid organism, *Salmonella typhi,* exhibits its characteristic biochemical activity in the carbohydrates and produces small amounts of hydrogen sulfide in TSI agar. The organism is anaerogenic, a criterion

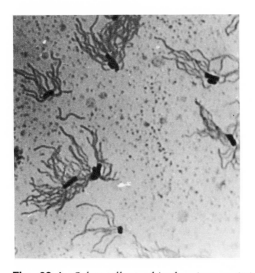

Fig. 20-1. *Salmonella typhi,* showing peritrichous flagellation using Gray's method. (4000×.)

that aids in its identification. A flagella stain of *S. typhi* is shown in Fig. 20-1.

It undergoes different types of variation, among which is **H-to-O** variation, involving the loss of flagella. The H form is motile, and the O form is nonmotile. There are two well-recognized variants of *S. typhi:* H901 (motile) and O901 (nonmotile), which are used widely in the preparation of H and O antigen suspensions, respectively, for the Widal test. The preparation of these antigens and the Widal test are discussed in Chapter 36.

Another type of variation exhibited by the typhoid organism is **V-to-W** variation, involving loss of the Vi antigen. In the **V** form the organism is virulent and inagglutinable in O antiserum. The **W** form readily agglutinates in O antiserum and is avirulent. V colonies on nutrient or starch agar medium appear orange-red by oblique light, whereas W colonies appear greenish blue. Only the V form is typable by the typhoid Vi phages. If the phage type of the organism is required for epidemiological purposes, a fresh culture in the V form should be sent to a reference laboratory.

The following criteria may serve to identify the organism: If a gram-negative isolate on TSI agar shows an acid butt, no gas, a small amount of hydrogen sulfide, and an alkaline slant and also agglutinates in Vi antiserum, it is usually *S. typhi.* Laboratory personnel should confirm this, however, with further biochemical studies and serology.

The reader is reminded of the possible, although rare, existence of lactose-positive variants of *S. typhi* (see earlier comments under lactose-fermenting members of Enterobacteriaceae).

Genus Arizona

Arizonae are gram-negative, short, motile rods that show a close relationship to the salmonellae. The type species of the genus is *Arizona hinshawii* (formerly *A. arizonae*). Lactose may be fermented with acid and gas in 24 hours, but most strains ferment the carbohydrate after 7 to 10 days. The reaction on TSI agar slants closely resembles that of the salmonellae, including hydrogen sulfide production. Gelatin is liquefied by these organisms in 7 to 30 days. They are sensitive to KCN (negative) and do not produce urease.

These organisms, as reported by Edwards and associates,[13] can be important in human infections. Many types cause disease in chickens, dogs, and cats. An antigenic schema has been established, and this contains 34 different O antigen groups with a total of about 300 serotypes.

The members of this group show the biochemical reactions listed in Table 20-6.

Because of the similarity they bear to the salmonellae, they are frequently mistaken for the latter in initial biochemical screening tests. Fermentation of lactose, although usually delayed, failure to ferment dulcitol, growth in sodium malonate, and slow gelatin liquefaction will distinguish arizonae from salmonellae.

Genus Citrobacter

Citrobacter includes what was formerly recognized as the *Escherichia freundii* group.[12] The type species is *Citrobacter freundii*. The members of this genus are gram-negative, motile rods that ferment lactose. Because of their biochemical reactions in preliminary screening, they are often confused with *Salmonella* and *Arizona*. They have been associated with cases of enteritis in man but are not considered truly pathogenic and are possibly opportunists. The organisms within this group exhibit the biochemical pattern shown in Table 20-7.

Certain *Citrobacter* strains possess the Vi antigen found in *Salmonella typhi*. The KCN test is positive for *Citrobacter* (the organisms grow in this medium), whereas *Salmonella* and *Arizona* are inhibited.

Considerable work has been done on the serological testing of members of this group, but a stable antigenic schema has not as yet evolved.

Table 20-6. Biochemical reactions of genus *Arizona*

TEST	REACTION	TEST	REACTION
Adonitol	−	Methyl red	+
Dulcitol	−	Voges-Proskauer	−
Glucose	+ with gas	Simmons citrate	+
Inositol	−	KCN	−
Lactose	+ or delayed	Phenylalanine deaminase	−
Mannitol	+	Sodium malonate	+
Salicin	−	Lysine decarboxylase	+
Sucrose	−	Arginine dihydrolase	+
Indole	−	Ornithine decarboxylase	+

Table 20-7. Biochemical reactions of genus *Citrobacter*

TEST	REACTION	TEST	REACTION
Adonitol	−	Voges-Proskauer	−
Dulcitol	+ or −	Simmons citrate	+
Glucose	+ with gas	Gelatin	−
Inositol	− or delayed	Urease	(+) or −
Lactose	+ or delayed	KCN	+
Mannitol	+	Phenylalanine deaminase	−
Salicin	variable	Sodium malonate	−
Sucrose	variable	Lysine decarboxylase	−
Indole	−	Arginine dihydrolase	+
Methyl red	+	Ornithine decarboxylase	variable

Genus Shigella

All of the *Shigella* organisms are **nonmotile;** they do not produce hydrogen sulfide; and with a few exceptions (biotypes of *Sh. flexneri* 6) they are anaerogenic. Although many shigellae produce catalase, for example, *Sh. flexneri,* this is an inconsistent property within the group. *Sh. dysenteriae* 1, the type species of the genus, is invariably negative.[6]

Colonies of shigellae are usually smaller than those of the salmonellae, but occasionally mucoid variants may be found in subgroup C. Lactose-negative colonies of suspected shigellae are picked from isolation plates to TSI agar slants. If these show an acid butt, no hydrogen sulfide, and an alkaline slant and prove to be urease negative, they may be tested with *Shigella* polyvalent sera. This procedure is recommended only for presumptive identification as cultures should be checked for purity and motility and then tested biochemically. The tests listed in Table 20-8 are recommended for the screening of *Shigella* cultures. The reaction tree, shown in Diagram 3, will serve as a guide in biochemical identification.

Serological identification. The *Shigella* schema is that which was proposed by Ewing in 1949 and modified and extended by the Shigella Commission of the Enterobacteriaceae Subcommittee. The schema is based in part on biochemical characteristics, on antigenic relationships, and on

tradition (the names of Shiga, Boyd, Flexner, and Sonne are obvious in the nomenclature). **Four** subgroups are presently identified: subgroup A—*Sh. dysenteriae,* types 1 to 10; subgroup B—*Sh. flexneri,* types 1 to 6; also X and Y variants; subgroup C—*Sh. boydii,* types 1 to 15; subgroup D—*Sh. sonnei.*

In **subgroup A** all types are mannitol nonfermenting, and each type exhibits a type-specific antigen not related to other shigellae.

In **subgroup B** the types are usually mannitol fermenting (mannitol-negative variants of *Sh. flexneri* 6 do exist) and in addition are interrelated through group or subsidiary antigens. Each type, however, possesses a type-specific antigen that differentiates it from other shigellae. Serotypes 1, 2, 3, and 4 have subtypes based on group antigen differences. The X and Y variants are forms that have lost the type-specific antigen.

In **subgroup C** the types are mannitol fermenting and possess individual type-specific antigens not related significantly to other shigellae.

In **subgroup D** there is only one type, *Sh. sonnei,* and this possesses a type-specific antigen that bears no significant relationship to other shigellae. There exist different "forms" of *Sh. sonnei* that differ antigenically.

The serological identification of the shigellae is a relatively simple procedure, but

Table 20-8. Biochemical reactions of genus *Shigella*

TEST	REACTION	TEST	REACTION
Adonitol	—	Methyl red	+
Dulcitol	variable	Voges-Proskauer	—
Glucose	+ no gas	Simmons citrate	—
Inositol	—	KCN	—
Lactose	variable	Phenylalanine deaminase	—
Mannitol	variable	Sodium malonate	—
Salicin	—	Lysine decarboxylase	—
Sucrose	—	Arginine dihydrolase*	—*
Indole	variable	Ornithine decarboxylase	—*

Sh. sonnei and *Sh. boydii* 13 are usually ornithine decarboxylase positive and some strains are arginine dihydrolase positive. The other shigellae are negative for all three amino acids.

in view of the number of types, preliminary serological screening is carried out with polyvalent antisera. Groups A, B, C, and D sera for slide agglutination tests are available commercially.* The technique is similar to that used for the salmonella O antigens. Since K antigens are found in some shigellae, live cells may fail to agglutinate in any of the pools, and one is advised to heat the test suspension in 0.5% NaCl at 100° C. for 30 to 60 minutes, cool, and retest with the antisera.

Type determination is carried out with monovalent antisera for the specific antigens. These are prepared from known types and are absorbed with suspensions of appropriate cultures to obtain type-specific diagnostic antisera. Agglutinin absorption is necessary for most types in

_____
*Lederle Laboratories, Pearl River, N. Y.; Difco Laboratories, Detroit; Baltimore Biological Laboratory, Cockeysville, Md.

Group B because of the group or minor antigens.

Colicin typing. In North America and in the United Kingdom, **Sh. sonnei** has become the most common etiological agent in bacillary dysentery or shigellosis. Colicin typing of strains of this type is being carried out routinely in some reference laboratories. Colicins, which are produced by different gram-negative enteric bacteria, are antibiotic-like substances that may have a lethal effect on other bacteria from the same habitat. This form of typing of the shigellae is based on colicinogeny rather than colicin susceptibility.

Fifteen colicin types of *Sh. sonnei* are recognized,[1] and these appear to be sufficiently stable to use in epidemiological investigations of cross infections in hospitals. Carpenter[6] reports that during 1964 to 1969 approximately one third of the strains isolated in England were noncolicinogenic (CTO), and of the typable strains one third

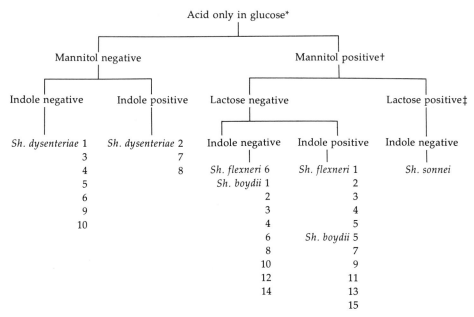

Sh. flexneri 6 varieties may be aerogenic (Newcastle and Manchester).
†Certain cultures of *Sh. flexneri* 4, *Sh. flexneri* 6, and *Sh. boydii* 6 may not produce acid from mannitol.
‡Lactose fermentation is delayed with *Sh. sonnei* (usually 4 to 7 days). Closure of the fermentation tube with a tightly fitting stopper will hasten the reaction.

Diagram 3. Biochemical differentiation of genus *Shigella*. (Adapted from Edwards, P. R., and Ewing, W. H.: Manual for enteric bacteriology, Atlanta, 1951, Center for Disease Control.)

are of type 7 and the remainder fall within the other fourteen types. The most prevalent of the latter in the United Kingdom are colicin types 1A, 2, 4, and 6.

Genus Escherichia

Within the genus *Escherichia* is included the former *Alkalescens-Dispar* group, which although anaerogenic, closely resembles the escherichiae biochemically and antigenically.

The type species of the genus is *Escherichia coli.*

Typical escherichiae are readily recognized and may be differentiated from other members of the Enterobacteriaceae by their rapid **fermentation of lactose** with acid and gas and the classical **IMViC** reaction (++ − −). There are some, however, that either fail to ferment lactose or do so slowly. Both motile and nonmotile forms occur.

Most strains produce lactose-fermenting colonies on MacConkey agar and EMB agar, and they are usually inhibited in enrichment broths and the more highly selective media, as stated earlier in this chapter. For the isolation of *Escherichia* types, the direct plating on the less inhibitory media, such as MacConkey agar or EMB agar, is recommended. Blood agar plates may also be used in isolation work because certain enteropathogenic strains may not develop on MacConkey agar[12] but will grow well on blood plates. Most strains are nonhemolytic.

Pure cultures of isolates of typical *E. coli* will give the reactions indicated in Table 20-9.

E. coli may be associated with various disease syndromes. Among these are (1) isolated but often severe and sometimes fatal infections, such as cystitis, pyelitis, pyelonephritis, appendicitis, peritonitis, gallbladder infections, septicemia, meningitis, and endocarditis; (2) epidemic diarrhea of the newborn; and (3) summer diarrhea.

The antigenic structure of the genus is just as complex as that of *Salmonella* and other genera within the Enterobacteriaceae. There are approximately 150 O antigen groups recognized. Many members of the group possess K antigens, which now number over 90, and they may be of the L, A, or B type and thus exhibit various physical and immunological properties.[12] *E. coli* serotypes have been established on the basis of the K and H antigens. By popular concept and for a relatively long period these K antigens have been considered to constitute a true or apparent sheath around the cell. This sheath masks or otherwise makes inaccessible the O complex. Certain reports have thrown some doubt on the sheath concept.[32] The O inagglutinability due to K antigens is analogous to the O and Vi relationship in *Salmonella typhi.*

Enteropathogenic Escherichia coli

Fourteen of the O antigen groups, to which prior reference is made, have been

Table 20-9. Biochemical reactions of genus *Escherichia*

TEST	REACTION	TEST	REACTION
Adonitol	−	Methyl red	+
Dulcitol	variable	Voges-Proskauer	−
Glucose	+ with gas	Simmons citrate	−
Inositol	−	KCN	−
Lactose*	+	Phenylalanine deaminase	−
Mannitol	+	Sodium malonate	−
Salicin	variable	Lysine decarboxylase	−
Sucrose	+	Arginine dihydrolase	−
Indole	variable	Ornithine decarboxylase	−

*Some are late or nonlactose fermenting.[12]

incriminated in epidemic diarrhea of the newborn. Serotypes from these groups are biochemically similar and therefore require serological identification.

Serological procedures. The *E. coli* types involved in infantile diarrhea are inagglutinable in O antisera and must be heated at 100° C. for 1 hour and retested. Live cells will agglutinate readily, however, in OB antiserum (antiserum prepared against unheated cells containing the B antigen). Each OB serogroup associated with this syndrome possesses a B type of K antigen that is serologically distinct from the other B antigens recognized in the other groups. The 16 enteropathogenic *E. coli* serogroups are listed as follows:

018:B20	086:B7	0124:B17
018:B21	0111:B4	0125:B15
020:B7	0112:B11	0126:B16
026:B6	0112:B13	0127:B8
028:B18	0119:B14	0128:B12
055:B5		

E. coli 0124 appears to be a definite pathogen for adults also.[22]

It should be recognized that type identification of the enteropathogenic forms is determined by the O, K, and H antigens.

Slide agglutination tests are normally used in serological identification, although tube tests are preferred by some investigators. Several suspected colonies from isolation plates may be tested first with OB antiserum pools, as a screening procedure, then with the individual OB antisera. A popular technique that is used during outbreaks of diarrhea in hospital nurseries is to collect a mass of cells by sweeping across a number of typical *E. coli* colonies on an isolation plate with an inoculating loop. This cell mass is then suspended in a small amount of saline (0.5 ml. of 0.5% NaCl) in a tube or in drops of saline on a slide. The suspension is subsequently tested with pooled or polyvalent OB sera.*

---

*These *E. coli* OB polyvalent and monovalent antisera are available from Lederle Laboratories, Pearl River, N.Y.; Difco Laboratories, Detroit; Baltimore Biological Laboratory, Cockeysville, Md.

If a strong positive reaction ensues, the test may be repeated with individual colonies, using monovalent sera.

Toxigenicity. It is duly recognized that certain types of enteropathogenic *E. coli* are toxic. Their toxigenicity may be tested by the injection of suspect broth cultures directly into ligated segments of rabbit gut. A positive test is readily recognized by dilatation of the ligated segment. The results obtained by De and co-workers[8] and by Taylor and co-workers[48] provide convincing proof of the enterotoxigenicity of some types.

Genus Edwardsiella

The type species of the genus *Edwardsiella* is *Edwardsiella tarda.* The members of this genus are motile, and their biochemical pattern conforms with that of the family Enterobacteriaceae. The first isolates were reported in 1959 and examined under biotype 1483-59 by various workers between 1962 and 1964. The generic term was suggested by Ewing and co-workers[17] in 1965.

These organisms have been isolated from human cases of diarrhea, from normal stools, from blood, from urine, and also from warm- and cold-blooded animals. The biochemical reactions are shown in Table 20-3.

Genus Klebsiella

The separation and establishment of *Klebsiella* and *Enterobacter* as genera was effected in 1962, following the recommendations of the Enterobacteriaceae subcommittee.[14,23,34] The tribe Klebsielleae currently includes the genera *Klebsiella, Enterobacter,* and *Serratia. Pectobacterium* was formerly included, but there now exists much doubt concerning its taxonomic position.

The members of the *Klebsiella* genus are gram-negative, **nonmotile,** encapsulated, short rods, which possess the characteristics shown in Table 20-10. The type species is **K. pneumoniae.** The klebsiellae can cause severe enteritis in children and pneumonia and upper respiratory tract infections, septicemia, meningitis, peritonitis, and an increasing number of hospital-acquired urinary tract infections in adults. On blood agar, EMB agar, MacConkey agar, trypticase soy agar, and other routine plating media, they generally give rise to large mucoid colonies that have a tendency to coalesce. Colonial growth usually strings out when touched with a needle. Growth in broth also can be very stringy and difficult to break when making transfers.

Seventy-two capsular types have been identified by Kauffmann, Brooke, Ed-

wards, and others. Cross-reactions between types are known to occur, and in the preparation of pooled antisera, related types are placed in the same pool. The use of these pools* for pathologically significant types may be practical in small laboratories. It should be emphasized that greater accuracy in identification of the types is obtained by the quellung reaction rather than by using agglutinating antisera.

*Difco Laboratories, Detroit; Baltimore Biological Laboratory, Cockeysville, Md.

Table 20-10. Biochemical reactions of genus *Klebsiella*

TEST	REACTION	TEST	REACTION
Adonitol	+ or −	Voges-Proskauer	+
Dulcitol	− or +	Simmons citrate	+
Glucose	+ with gas	KCN	+
Inositol	+ with gas	Urease	+ (slow)
Lactose	+	Gelatin liquefaction	−
Mannitol	+	Phenylalanine deaminase	−
Salicin	+	Sodium malonate	+
Sucrose	+	Lysine decarboxylase	+
Indole	−	Arginine dihydrolase	−
Methyl red	−	Ornithine decarboxylase	−

Table 20-11. Biochemical reactions of *Enterobacter* species

TEST	E. CLOACAE	E. AEROGENES	E. HAFNIAE 37° C.	E. HAFNIAE 22° C.	E. AGGLOMERANS
Adonitol	− or +	+	−	−	−
Dulcitol	− or +	−	−	−	− or +
Glucose (+ gas)	+	+	+	+	− or +
Inositol (+ gas)	d	+	−	−	d
Lactose	+	+	− or (+)	− or (+)	d
Mannitol	+	+	+	+	+
Salicin	+ or (+)	+	d	d	d
Sucrose	+	+	d	d	d
Indole	−	−	−	−	− or +
Methyl red	−	−	+ or −	−	− or +
Voges-Proskauer	+	+	+ or −	+	+ or −
Simmons citrate	+	+	+ or −	d	d
KCN	+	+	+	+	− or +
Urease	+ or −	−	−	−	d
Gelatin liquefaction	(+) or −	− or (+)		−	+
Phenylalanine deaminase	−	−	−	−	− or +
Sodium malonate	+ or −	+ or −	+ or −	+ or −	+ or −
Arginine dihydrolase	+	−	−	−	−
Lysine decarboxylase	−	+	+	+	−
Ornithine decarboxylase	−	+	+	+	−

Genus Enterobacter

According to the Ewing classification (Table 20-11), the genus *Enterobacter* contains four currently recognized species—*E. cloacae, E. aerogenes, E. hafniae* (the former *Hafnia* group), and *E. agglomerans,*[16] which groups the former Herbicola-Lathyri bacteria. The clinical significance of these as intestinal pathogens is presently undetermined. The relationship between *E. hafniae* and *E. cloacae* was shown originally by Stuart and co-workers.[45]

The species of *Enterobacter* are found in the soil, in water, in dairy products, and in the intestines of animals, including man. They are now being isolated quite frequently in urinary tract infections, from septicemias and from other clinical sites.

Genus Serratia

Members of the genus *Serratia* are gram-negative motile rods, of which only a small percentage of strains are **chromogenic**. Three species are now proposed—*S. marcescens, S. liquefaciens, and S. rubidaea.** Their biochemical reactions are shown in Table 20-3. Those that are chromogenic produce a red non-water-soluble

---

*Ewing, W. H., et al.: Biochemical characterization of *Serratia liquefaciens* (Grimes and Hennerty), Bascomb et al. (formerly *Enterobacter liquefaciens* and *Serratia rubidaea* (Stapp) comb. nov. and designation of type and neotype strains, Internat. J. Syst. Bacteriol. **23:** 217-225, 1973.

pigment at room temperature, but seldom at 36° C. or above. For many years these organisms were considered innocuous and were found in natural bodies of water and in soil, occurring as relatively free-living forms. They have been incriminated in pulmonary infections and septicemias.[39,52] It has been suggested that the sulfonamides and antibiotics have been responsible for these organisms becoming opportunist pathogens. They appear to be multiresistant to the broad-spectrum antibiotics in general use.

Fifteen somatic groups for *Serratia* have been previously identified serologically, and there is evidence to suggest that serotyping can be of value in studying nosocomial infections. The biochemical pattern exhibited by the genus is shown in Table 20-12, whereas the biochemical speciation is shown in Table 20-3.

Genus Proteus

If the Ewing classification[15] is followed, the tribe Proteeae consists of two genera, *Proteus* and *Providencia*. There is sufficient biochemical similarity between these to warrant their inclusion in the one tribe but sufficient dissimilarity to permit differentiation between the two genera.

The *Proteus* species are gram-negative pleomorphic motile rods. The type species is *Proteus vulgaris*. They are **lactose negative** and usually rapidly **urease positive.**[12] The species are actively motile at 25° C. but often weakly motile at 36° C. On moist

Table 20-12. Biochemical reactions of genus *Serratia*

TEST	REACTION	TEST	REACTION
Arabinose	−	Voges-Proskauer	+
Adonitol	variable	Simmons citrate	+
Dulcitol	−	Gelatin (22° C.)	+
Glucose	+ gas variable	KCN	+
Inositol	variable	Phenylalanine deaminase	−
Lactose	− or delayed	DNase	+
Mannitol	+	Sodium malonate	−
Salicin	+	Lysine decarboxylase	+
Sucrose	+	Arginine dihydrolase	−
Indole	−	Ornithine decarboxylase	+
Methyl red	− or +		

agar or agar of 1.5% concentration (usual strength for plating media) some *Proteus species* (*P. vulgaris* and *P. mirabilis*) tend to swarm, producing a bluish gray confluent surface growth at both 25° and 36° C. On isolation media such as deoxycholate agar, colonies of *Proteus* do not tend to swarm. Swarming may be inhibited on MacConkey and on EMB plates also if the agar concentration is increased to 5%.

The appearance of the spreading growth will vary from a rippled form of growth, which is easily recognized, to one that is smooth and almost transparent. This growth may sometimes go unobserved in examining plates containing mixed cultures, but a loop drawn over a seemingly colony-free area readily reveals its presence. Because of careless technique, *Proteus* is often the contaminant in cultures made from selected colonies.

Due to their lactose-negative characteristic, discrete *Proteus* colonies are sometimes selected from isolation plates as suspected salmonellae, shigellae, or other gram-negative organisms that show up on such media. Because of this, screening on TSI agar, followed by a rapid urease test, is highly recommended, as was previously stated in discussing the isolation of the Enterobacteriaceae.

Proteus organisms are frequently found in large numbers in the stools of individuals undergoing oral antibiotic therapy. They are also incriminated in cases of gastroenteritis—both of a sporadic and of an epidemic nature—and frequently are responsible for infections of the urinary tract.

The biochemical reactions of the *Proteus* species are given in Table 20-13. Many laboratory personnel today recommend species' identification of *Proteus* isolates, since *P. mirabilis* infections may respond to penicillin therapy.

Genus Providencia

In 1951 Kauffmann[30] proposed the name *Providence* for those microorganisms described by Stuart and co-workers[45,46] under the designation of 29911 paracolon bacteria. The generic name of *Providencia* is now internationally accepted, and the type species is *Prov. alcalifaciens.* There is another species, *Prov. stuartii.*

The species are gram-negative motile rods that are lactose-negative and grow well on MacConkey, EMB, SS, and bismuth sulfite agar media. Because they are hydrogen sulfide negative and may or may not produce gas in glucose, they often resemble shigellae on TSI agar. If sucrose

Table 20-13. Biochemical characteristics of *Proteus* species*

	P. MIRABILIS	P. VULGARIS	P. MORGANII	P. RETTGERI
Adonitol	−	−	−	+
Dulcitol	−	−	−	−
Glucose	+ (gas)	+ (gas)	+ (gas)	+
Inositol	−	−	−	+
Lactose	−	−	−	−
Mannitol	−	−	−	+
Salicin	variable	+	−	variable
Sucrose	+ (3-8 days)	+	−	+ (delayed)
Xylose	+	+	−	−
Gelatin	+	+	−	−
H₂S(TSI)	+	+	−	−
Indole	−	+	+	+
Methyl red	+	+	+	+
Voges-Proskauer	− or +	−	−	−
Simmons citrate	+ (variable)	− (variable)	−	+
Phenylalanine deaminase	+	+	+	variable

*All *Proteus* species are urease positive.

is fermented, the reaction is usually delayed, and thus fermentation of this carbohydrate will not be detected in 48 hours in the TSI slant. The species of *Providencia* may be distinguished from the shigellae, however, by their motility and utilization of citrate. Positive phenylalanine and negative urease reactions differentiate *Providencia* from *Proteus*. *Providencia* may be distinguished also from other genera with-

Table 20-14. Biochemical reactions of *Providencia* species

	PROV. ALCALIFACIENS	PROV. STUARTII		PROV. ALCALIFACIENS	PROV. STUARTII
Adonitol	+	−	H₂S(TSI)	−	−
Arabinose	−	−	Indole	+	+
Dulcitol	−	−	Methyl red	+	+
Glucose (gas)	+ or −	−	Voges-Proskauer	−	−
Inositol	−	+	Urease	−	−
Lactose	−	−	Simmons citrate	+	+
Sodium malonate	−	−	KCN	+	+
Mannitol	−	d	Phenylalanine deaminase	+	+
Salicin	−	−	Lysine decarboxylase	−	−
Sucrose	d	d	Arginine dihydrolase	−	−
Gelatin	−	−	Ornithine decarboxylase	−	−

d = different biochemical types; + or − = majority are positive.

Table 20-15. Distinguishing properties of *Yersinia* species

PROPERTY	YERS. PESTIS	YERS. ENTEROCOLITICA	YERS. PSEUDOTUBERCULOSIS
Colony forms	Two	Two	Two
Optimal growth temperature	25° to 30° C.	25° to 30° C.	25° to 30° C.
Motility at 25° C.	−	+	+
Serogroups	One	Two	Five
Catalase	+	+	+
Oxidase	−	−	−
Coagulase	+*		−
Fibrinolysin	+		−
Hemolysis on blood agar	−	alpha	−
Medium containing bile salts	+	+	+
H₂S	−	−	−
Indole	−	−	−
Methyl red	+	+	+
Voges-Proskauer	−	+	−
Urease	−	+	+
Nitrates	Variable	Reduced	Reduced
Cellobiose	−	+	−
Glucose	+	+	+
Glycerol	Variable	+	+
Lactose	−	−	−
Melibiose	−	−	+
Maltose	+	+	+
Mannitol	+	+	+
Rhamnose	−	−	+
Salicin	+	−	+
Sucrose	−	+	−

*Using rabbit plasma.

in the Enterobacteriaceae by their biochemical characteristics, indicated in Table 20-14. Biochemical differences between the species are also shown.

The organisms are not highly pathogenic, but they have been incriminated in sporadic cases of human diarrhea and in urinary tract infections.

An antigenic schema has been established for the genus, in which there are 62 O antigen groups and about 175 serotypes.

Genus Yersinia

The genus consists of the species *Yersinia pestis, Yers., enterocolitica,* and *Yers. pseudotuberculosis.* The distinguishing properties of these species are shown in Table 20-15.

Yersinia pestis

Yers. pestis is the causative agent of **plague** (a disease with a very high mortality) in man and rats and is infectious for mice, guinea pigs, and rabbits. Three clinical forms of plague are recognized in man: **bubonic, pneumonic,** and **septicemic.** The main vector for transmission is the rat flea. Plague bacilli are short, plump, nonmotile, gram-negative rods, sometimes elongated and pleomorphic, usually appearing singly or in pairs, and occasionally in short chains. Bipolar staining can be demonstrated with polychrome stains, such as the Giemsa or Wayson stains, but **not** by the Gram stain. Coccoid, round, filamentous, elongated, and other forms commonly occur, especially in old cultures. A capsule can be demonstrated in animal tissue and in young cultures. The latter will give a good fluorescent-antibody test, by which a presumptive diagnosis may be made.

Yers. pestis grows slowly on nutrient agar and fairly rapidly on blood agar, producing small, nonhemolytic, round, transparent, glistening, colorless colonies with an undulate margin. Older colonies enlarge, becoming opaque with yellowish centers and whitish edges, and develop-

ing a soft, mucoid consistency that is due to capsular material. A characteristic type of growth occurs in old broth cultures overlaid with sterile oil, in which a pellicle forms, with "stalactite" streamers. The biochemical properties are shown in Table 20-15.

Yers. pestis is pathogenic for rats and guinea pigs. After subcutaneous inoculation, the animals die within 2 to 5 days and show certain postmortem characteristics, including ulceration at the site of injection, regional adenopathy, congested spleen and liver, and pleural effusion. The bacilli may be demonstrated in splenic smears; they may also be recovered by culture. Cultures may be identified by specific bacteriophage typing or by agglutination with specific antisera.

Great care must be taken by the laboratory worker when handling suspect cultures or pathological materials. The worker should be masked and wearing rubber gloves. The creation of aerosols should be avoided, and all contaminated materials should be autoclaved.

Yersinia enterocolitica

Yers. enterocolitica appears to be closely related to *Yers. paratuberculosis* and has been involved in a number of cases of human mesenteric lymphadenitis in Europe, while there have been only sporadic cases of infection reported in the United States. As Braunstein and co-workers[4] have suggested, it appears that the establishment of *Yers. enterocolitica* as an etiological agent of this disease syndrome in the United States will become firmer in the years ahead.

The organism is a gram-negative coccobacillary form, occasionally showing bipolar staining. Its biochemical and other properties are shown in Table 20-15, where it may be compared with other species. Braunstein[4] has suggested that routine culture of enlarged lymph nodes during surgical exploration will result in additional case diagnoses of human infections.

Yersinia pseudotuberculosis

Yers. pseudotuberculosis causes disease primarily in rodents, particularly guinea pigs, but it also causes two recognized forms of disease in human beings. Of these the most serious is a **fulminating septicemia,** which is usually fatal, while the more common form is a **mesenteric lymphadenitis,** which resembles appendicitis, yet does not normally affect that organ. The organism has been responsible for many cases of this syndrome in Europe (Knapp and Mashoff[31]). In recent years, cases have been reported in the United States by Weber and co-workers.[53]

The bacterium is a small gram-negative coccobacillary form, exhibiting pleomorphism, and found singly, in short chains, and in small clusters. It grows well on blood agar and on media containing bile salts, such as MacConkey agar. Two colonial forms may be observed: a smooth grayish yellow translucent colony in 24 hours at 25° to 30° C., and a raised colony, with an opaque center and a lighter margin with serrations, that develops with continued incubation. A rough variant with an irregular outline may also develop. The characteristics of the species are shown in Table 20-15.

Genus Vibrio

Those species of *Vibrio* of clinical significance that are associated with intestinal disease are included in this chapter on the gram-negative enteric bacteria for convenience and practicality and because of the frequent appearance of these organisms in epidemic and even pandemic proportions. The genus has been assigned to the family *Vibrionaceae* in the eighth edition of Bergey's manual.

The species of medical importance are *V. cholerae* and *V. parahemolyticus*. *V. El Tor* is not considered a separate species by most investigators, but as a biotype of *V. cholerae*. This assignment was adopted by the International Subcommittee on Cholera. Both *V. El Tor* and *V. cholerae* produce **cholera;** but because of their biochemical

and other differences, it is of epidemiologic significance to distinguish between them. The other species, *V. parahemolyticus* is the most important member of the latter group.

Vibrio cholerae

This species* and its biotype *El Tor* are gram-negative, actively motile rods possessing a single polar flagellum. With careful preparation of stained smears, one may observe the slightly curved rods. Most observations reveal that the *El Tor* type is less susceptible to environmental changes than the parent type *V. cholerae,* and for this reason is more readily recovered from clinical specimens (usually stool) submitted to the laboratory.

As has been pointed out by Balows and co-workers,[3] the most effective and rapid bacteriological diagnosis of cholera is accomplished by proper communication between the clinician and the laboratory. Clinical specimens may be obtained in various ways: but as advocated by the above investigators, a liquid stool is preferably collected by rectal catheter, and formed stools should be collected in disinfectant-free containers. Rectal swabs are most effective when inserted beyond the anal sphincter. In all cases, specimens should be taken prior to any antibiotic therapy and inoculated to laboratory media immediately at the laboratory. The Center for Disease Control (CDC) recommends the use of **two plating** media and **one enrichment** medium. Thiosulfate-citrate-bile salt-sucrose (TCBS) agar at pH 8.6 and gelatin agar, the latter, a non-inhibitory medium, give good results. The selective medium may be inoculated quite heavily.

On TCBS at 36° C after 18 to 24 hours, *V. cholerae* appears as medium-sized, smooth, yellow colonies with opaque centers and transparent periphery. On the gelatin medium, the colonies are some-

---

*The noncholera vibrios are those similar biochemically to *V. cholerae* but which fail to agglutinate in *V. cholerae* antiserum.

what flattened and transparent, surrounded by a cloudy halo. Refrigeration tends to accentuate this characteristic, which demonstrates gelatin liquefaction.

The enrichment medium recommended by Balows and co-workers is alkaline peptone water, which, by virtue of its high alkalinity, tends to suppress other intestinal bacteria. Culturing in this medium should not be extended beyond 18 to 20 hours, because suppressed forms may begin to develop. The plating media may be streaked from the enrichment culture after 6 to 8 hours of incubation. Suspicious colonies are selected and inoculated to appropriate media or tested with O antisera by slide test. The young enrichment broth culture may be examined for the characteristic darting motility by dark-field microscopy. The motility test, if positive, can be extended by carrying out the immobilization test with *V. cholerae* pooled O antiserum. If this is positive, it may be used as presumptive evidence of identity.

There are several tests that will help to distinguish *V. cholerae* from the noncholera vibrios and also differentiate between *V. cholerae* and its biotype *El Tor.* Among these are the "string test" of Smith[41] and Neogy and Mukherji,[38] the hemagglutination test of Finkelstein and Mukerjee,[19] the polymyxin B susceptibility test,[20] the phage IV susceptibility test,[37] the VP test and the hemolysis test.[12,25]

The **string test** consists of testing for the viscid character of a cholera culture in 0.5% sodium desoxycholate on a slide. The "string" is detected by lifting the loop of mixture from the slide. The **hemagglutination** test is carried out by mixing a loopful of washed chicken RBC with a heavy suspension of a pure culture of the organism on a slide. Visible clumping of the RBC is a positive test. The **polymyxin B susceptibility** test determines the inhibitory effect of the antibiotic on the organism by the appearance of a zone of growth inhibition around discs of the substance on a seeded plate. The **phage IV susceptibility** test aids in differentiating *V. cholerae* from *El Tor.* The **hemolysis** test, performed with washed sheep RBC, is claimed by some investigators also to be effective in differentiating these types (see Table 20-16).

Vibrio parahemolyticus

This species is the most clinically significant of the remaining vibrios. It can be the cause of gastroenteritis or of food poisoning associated with the consumption of contaminated seafood. It is most prevalent in the Orient but has been reported in other countries and recently in the United States. The organism is halophilic and grows well in peptone water medium containing 7% to 8% NaCl, but it does not normally utilize citrate. For the characteristics of this species see Table 20-16.

Table 20-16. Distinguishing characteristics of *Vibrio* species

TEST	V. CHOLERAE	V. EL TOR	V. PARAHEMOLYTICUS
String test after 45 to 60 seconds	+	+	−
Hemagglutination test	−	+	−
Polymyxin B susceptibility test	+*	−	·
Phage IV susceptibility test	+*	−	·
VP	−	+	−
Hemolysis of sheep RBC	−	+†	−
Sucrose	+	+	−
Salt-free broth	+	+	−
Broth containing 7% to 10% NaCl	−	−	+
Cholera red test	+	+	−
Agglutination in O group serum	+	+	−

*+ indicates susceptibility.
†Considerable variability in hemolytic activity has been reported.

IDENTIFICATION OF ENTEROBACTERIACEAE BY RAPID METHODS

Rapid procedures designed to reduce the time period for identifying members of the Enterobacteriaceae have been developed commercially and have been compared with the conventional methods used in the identification of gram-negative bacterial isolates from clinical material. The results have been evaluated and reported by various investigators. There are currently in use at least five different systems, and a brief description of these is given in alphabetical order. In the performance of each, however, emphasis is placed on the competence and technical skill of the person or persons carrying out the tests, and it should be recognized that the results will only be as good as the laboratory expertise shown. All systems have been rated as good, with a percentage accuracy correlation ranging from 87% to 96% or higher, depending on the number and variety of cultures tested.

API System (Analytab Products Inc.*). The API 20 enteric system utilizes **22** biochemical tests that can produce results from a single bacterial colony in 18 to 24 hours. The colony is emulsified in about 5 ml. distilled water to supply the inoculum, and inoculation is carried out with a Pasteur pipet. Viable cells are introduced into the small plastic cupules and tubes arranged on a plastic covered strip that is incubated in a plastic tray, to which water has been added to provide humidity, for 18 hours. This system has been evaluated by Washington and co-workers,[50] by Smith and co-workers,[43] and by Guillermet and Desbresles.[21] Correlation with conventional tests has been reported as high as 96.4%. Complete information on use of the system may be gained from the manufacturers.

Auxotab Enteric 1 System. The Auxotab system, a development of Colab Laboratories, Inc., Glenwood, Ill., consists of a card with ten capillary units containing ten different reagents designed to differentiate between the genera of the Enterobacteriaceae on a reduced time schedule and to differentiate the species of *Enterobacter* and *Proteus.* The system has been evaluated by Washington and co-workers,[51] and by Rhoden and co-workers,[40] who have reported on the reliability of the procedure in regard to accuracy of results and laboratory safety.

Inoculum is prepared from a single colony selected from a primary isolation plate, and from this suspension the battery of tests is set up. Readings can be recorded in 7 hours at 35° C. Complete details may be obtained from the manufacturers.

Enterotube System (Roche Diagnostics).* The new improved tube permits simultaneous inoculation and performance of **eleven** biochemical tests from a single colony. A colony is usually selected from MacConkey, EMB, or Hektoen agar plates that have been seeded with a urine, stool, blood, sputum, or other clinical specimen. The multitest tube is inoculated by touching the needle to a single colony and drawing it through all media in the tube. It has been reported that the citrate reaction tends to present the greatest problem in the reading of the test. Correlation has run as high as 99% with some investigators, using all members of the Enterobacteriaceae and keying them to the genus level and to the species level for certain members. Further information and complete instructions may be gained from the manufacturers. Reports by Morton and Monaco,[36] by Douglas and Washington,[9] and by Smith[44] provide evidence of the practical use of this system.

PathoTec Rapid I-D System.† This system consists of a set of **ten** test strips impregnated with various biochemical reagents in carefully measured concentra-

---

*Carle Place, New York.

*Division of Hoffmann-LaRoche Inc., Nutley, N. J.
†General Diagnostics Division, Warner-Lambert Co., Morris Plains, N. J.

tions. The selected test strips will reportedly identify approximately 95% of the Enterobacteriaceae, isolated from the conventional clinical sites, in approximately 4 hours after the initial isolation. The strips are added to prepared test tubes (13 × 100 mm.). Two of the strips, cytochrome oxidase and esculin hydrolysis, are inoculated by rubbing cells from selected colonies on the designated areas, while the remaining strips are placed in tubes to which a measured amount of cell suspension has been added. The cytochrome oxidase test is read after 30 seconds, while the remaining tests are recorded after approximately 4 hours at 36° C.

Detailed information may be obtained from the manufacturers. Success in the performance of this test is largely determined by the expertise of the microbiologist.

R/B System. The basic system contains **eight** biochemicals in two tubes that permit readings on phenylalanine deaminase, hydrogen sulfide, indole, motility, lysine and ornithine decarboxylase, gas from glucose, and lactose. Two additional tubes have been introduced, namely, Cit/Rham and Soranase, permitting an expansion to fourteen biochemicals in four tubes. The addition of these tubes permits differentiation of *Enterobacter* species and a differentiation of other genera.

This system has been evaluated by different investigators, including Smith and co-workers,[45] Isenberg and Painter,[26] and McIlroy and co-workers.[33] On the basis of the reported findings, it seems appropriate to recommend to qualified personnel that the method be tried, because in this and other rapid procedures, the rapid reporting of accurate results will obviously assist the physician in making an early diagnosis.

Adequate literature and instructions are provided with the materials obtained from the manufacturers.*

---

*Diagnostic Research Inc., 25 Lumber Road, Roslyn, N. Y.

REFERENCES

1. Abbott, J. D., and Shannon, R.: A method for typing *Shigella sonnei* using colicin production as a marker, J. Clin. Path. **11:**71-75, 1958.
2. Bailey, W. R.: Studies on the transduction phenomenon I. Practical applications in the laboratory, Can. J. Microbiol. **2:**549-553, 1956.
3. Balows, A., et al.: The isolation and identification of *Vibrio cholerae*—a review, Atlanta, Ga., 1971, Center for Disease Control.
4. Braunstein, H., et al.: Mesenteric lymphadenitis due to *Yersinia enterocolitica,* Amer. J. Clin. Path. **55:**506-510, 1971.
5. Brinton, C. C.: Non-flagellar appendages of bacteria, Nature **183:**782-786, 1959.
6. Carpenter, K. P.: Personal communication, 1969.
7. Carpenter, K. P., and Lachowicz, K.: The catalase activity of *Sh. flexneri,* J. Path. Bact. **77:**645-648, 1959.
8. De, S. N., Bhattacharya, K., and Sarker, J. K.: A study of the pathogenicity of strains of *B. coli* from acute and chronic enteritis, J. Path. Bact. **71:**201-209, 1956.
9. Douglas, G. W., and Washington, J. A.: Identification of Enterobacteriaceae in the Clinical Laboratory, Atlanta, Ga., 1970, Center for Disease Control.
10. Duguid, J. P., et al.: Non-flagellar filamentous appendages ("Fimbriae") and haemagglutinating activity in *Bacterium coli,* J. Path. Bact. **70:**335, 1955.
11. Edwards, P. R., and Ewing, W. H.: Manual for enteric bacteriology, Atlanta, 1951, National Communicable Disease Center.
12. Edwards, P. R., and Ewing, W. H.: Identification of Enterobacteriaceae, ed. 2, Milwaukee, 1962, Burgess Publishing Co.
13. Edwards, P. R., Kauffmann, F., and van Oye, E.: A new diphasic *Arizona* type, Acta Path. Microbiol. Scand. **31:**5-9, 1952.
14. Enterobacteriaceae Subcommittee: Third report, Int. Bull. Bact. Nomenclat. Taxon. **8:**25-70, 1958.
15. Ewing, W. H.: The tribe Proteeae; its nomenclature and taxonomy, Int. Bull. Bact. Nomenclat. Taxon. **12:**93-102, 1962.
16. Ewing, W. H., and Fife, M. A.: Biochemical characterization of *Enterobacter agglomerans,* Dept. of Health, Education and Welfare Publ. No. (HSM) 73-8173, Aug. 1972.
17. Ewing, W. H., McWhorter, A. C., Escobar, M. R., and Lubin, A. M.: *Edwardsiella,* a new genus of *Enterobacteriaceae* based on a new species E. *tarda,* Int. Bull. Bact. Nomencl. Taxon. **15:**33-38, 1965.
18. Ewing, W. H., et al.: Transplant media in the detection of *Salmonella typhi* in carriers, J. Conf. State Prov. Public Health, Lab. Directors **24:**63-65, 1966.
19. Finkelstein, R. A., and Mukerjee, S.: Hemagglutination; a rapid method for differentiating *Vibrio*

cholerae and El Tor vibrios, Proc. Soc. Exptl. Biol. Med. **112**:355-359, 1963.

20. Gangarosa, E. S., et al.: Differentiation between *Vibrio cholerae* and *Vibrio* biotype *El Tor* by the polymyxin B disc test: comparative results with TCBS, Monsur's, Mueller-Hinton and nutrient agar media. Bull. W. H. O. **35**:987-990, 1967.

21. Guillermet, F. N., and Desbresles, A. M. B.: A Propos de l'Utilisation d'une Micromethode d'Identification des Enterbacties, Revue de l'Instit Pasteur de Lyon **4**:71-78, 1971.

22. Hobbs, B. C., Thomas, M. E. M., and Taylor, J.: School outbreak of gastro-enteritis associated with a pathogenic paracolon bacillus, Lancet **257**:530, 1949.

23. Hormaeche, E., and Edwards, P. R.: A proposed genus *Enterobacter,* Int. Bull. Bact. Nomenclat. Taxon. **10**:71-74, 1960.

24. Hormaeche, E., and Munilla, M.: Biochemical tests for the differentiation of Klebsiella and Cloaca, Int. Bull. Bact. Nomenclat. Taxon. **7**:1-20, 1957.

25. Hugh, R.: A comparison of *Vibrio cholerae,* Pacini and *Vibrio El Tor,* Int. Bull. Bact. Nomenclat. Taxon. **15**:61-68, 1965.

26. Isenberg, H. D., and Painter, B. G.: Comparison of conventional methods, the R/B System, and modified R/B System as guides to major divisions of Enterobacteriaceae, Appl. Bact. **22**(6):1126-1134, 1971.

27. Julianelle, L. A.: A biological classification of *Encapsulatus pneumoniae* (Friedlander's bacillus), J. Exp. Med. **44**:113-128, 1926.

28. Kauffmann, F.: Ein kombiniertes Anreicherungsverfahren fur Typhus und Paratyphusbazillen, Zbl. Bakt. **119**:148-152, 1930.

29. Kauffmann, F.: Die Bakteriologie der Salmonella-Gruppe, Copenhagen, 1941, Einar Munksgaard.

30. Kauffmann, F.: Enterobacteriaceae, Copenhagen, 1951, Einar Munskgaard.

31. Knapp, W., and Masshoff, W.: Zur Ätiologie der abszedierenden retikulozytären Lymphadenitis: einer praktisch wichtigen, vielfach unter dem Bilde einer aktuen Appendizitis verlaufenden Erkrankung, Deutsch Med. Wschr. **79**:1266-1271, 1954.

32. McDade, J. E., and Bailey, W. R.: Further studies on the antigens of *Escherichia coli,* Canad. J. Microb. **12**:249-254, 1966.

33. McIlroy, Gary T., et al.: Evaluation of modified R-B system for identification of members of the family Enterobacteriaceae, Appl. Bact. **24**(3):358-362, 1972.

34. Minutes of the Enterobacteriaceae subcommittee meeting, Montreal, 1962; report of the subcommittee on taxonomy of the Enterobacteriaceae, Int. Bull. Bact. Nomenclat. Taxon. **13**:69-93, 139, 1963.

35. Moeller, V.: Diagnostic use of the Braun KCN test within Enterobacteriaceae, Acta Path. Microbiol. Scand. **34**:115-126, 1954.

36. Morton, H. E., and Monaco, M. A. J.: Comparison of enterotubes and routine media for the identification of enteric bacteria, Amer. J. Clin. Path. **56**:64, 1971.

37. Mukerjee, S.: The bacteriophage susceptibility test in differentiating *Vibrio cholerae* and *Vibrio El Tor,* Bull. **28**:333-336, 1963.

38. Neogy, K. N., and Mukherji, A. C.: A study of the string test in *Vibrio* identification, Bull. W.H.O. **42**:638-641, 670.

39. Patterson, R. H., et al.: Chromobacterial infection in man, Arch. Intern. Med. **90**:79-86, 1952.

40. Rhoden, D. L., et al.: Auxotab—a device for identifying enteric bacteria, Appl. Microbiol. **25**:284-286, 1973.

41. Smith, H. L.: A presumptive test for vibrios: the "string" test, Bull. W. H. O. **42**:817-818, 1970.

42. Smith, P. B., et al.: Evaluation of the modified R/B system for identification of Enterobacteriaceae, Appl. Bact. **22**(5):928-929, 1971.

43. Smith, P. B., et al.: The API system—a multitube micromethod for identification of Enterobacteriaceae, Appl. Microbiol. **24**:449-452, 1972.

44. Smith, P. B.: Roundtable on *Enterobacteriaceae,* American Society for Microbiology Meeting, Miami Beach, May 1973.

45. Stuart, C. A., et al.: Biochemical and antigenic relationships of the paracolon bacteria, J. Bact. **45**:101-119, 1943.

46. Stuart, C. A., et al.: Further studies of one anaerogenic paracolon organism, type 29911, J. Bact. **52**:431-438, 1946.

47. Stuart, R. D.: Transport medium for specimens in public health bacteriology, Public Health Rep. **74**:431-438, 1959.

48. Taylor, J., Maltby, M. P., and Payne, J. M.: Factors influencing the response of ligated rabbit-gut segments to injected *Escherichia coli,* J. Path. Bact. **76**:491-499, 1958.

49. Waisbren, B. A., and Carr, C.: Penicillin and chloramphenicol in the treatment of infections due to *Proteus* organisms, Amer. J. Med. Sci. **223**:418, 1952.

50. Washington, J. A., et al.: Evaluation of accuracy of multitest micromethod system for identification of Enterobacteriaceae, Appl. Microbiol. **22**:267-269, 1971.

51. Washington, J. A., et al.: Evaluation of the Auxotab Enteric 1 System for identification of Enterobacteriaceae, Appl. Microbiol. **23**:298-300, 1972.

52. Wassermann, M. M., and Seligmann, E.: *Serratia marcescens* bacteriophages, J. Bact. **66**:119-120, 1953.

53. Weber, J., et al.: Mesenteric lymphadenitis and terminal ileitis due to *Yersinia pseudotuberculosis,* New Eng. J. Med. **283**:172-174, 1970.

21 Nonfermentative, gram-negative bacilli

Pseudomonas
Alcaligenes
Acinetobacter
Moraxella

The statement by Dr. Rudolph Hugh[11] —"It appears that almost any bacterial species can produce disease in man"—is well exemplified by this group of organisms. Predominately opportunistic in nature, they owe their invasiveness or infectivity to an altered or already debilitated host, who has been compromised by potent medications, varied instrumentation, or dramatic and prolonged surgical procedures not heretofore possible.

In the past few years major advances have been made in the characterization and taxonomy of these nonfermentative bacteria*; an important landmark was the introduction by Hugh and Leifson of their fermentation (O-F) medium,[12] enabling the laboratorian to determine whether an isolant was oxidative, fermentative, or inactive with respect to carbohydrate metabolism. These workers recognized that conventional fermentation media contained a high content of peptone (1%) that, when attacked, gave rise to alkaline amines capable of neutralizing any acidity formed in the fermentation. Although this was not observed when large amounts of acid were produced by active "fermenters," the oxidizative bacteria, which produced low levels of acidity, might be missed due to the accumulation of the alkaline amines.

By developing a carbohydrate medium low in peptone (0.2%), the oxidative, fermentative, or inactive properties could thus be recognized.

The base medium* with bromthymol blue indicator is sterilized by being autoclaved (see Chapter 39 for preparation); and after it is cooled, a filter-sterilized solution of the desired carbohydrate is added to give a final concentration of 1%. The recommended carbohydrates include glucose, lactose, sucrose, maltose, mannitol, and xylose. For each isolant, **two** tubes of glucose O-F medium are inoculated with a light stab from a young culture, and one of the tubes is overlaid with at least $1/4$ inch of sterile, stiff petrolatum or vaspar. They are then incubated at 36° C. for several days and examined daily. An acid reaction is indicated by a color change from the uninoculated blue-green color to a **yellow** color. The chart (p. 163) from Hugh and Leifson's paper,[12] indicates reactions of characteristic groups.

Generally, these nonfermentative bacteria are gram-negative, nonsporulating, aerobic bacilli that produce no change on TSI agar (occasionally an alkaline slant) and are indole and ornithine decarboxylase negative. For purposes of simplicity, they may be classified as several major groups (Gilardi[5]):

---

*See papers by Gilardi,[8] Hugh,[11] Pickett and Pedersen,[16] and others, for excellent descriptions of this group of organisms.

*Available from Baltimore Biological Laboratory, Cockeysville, Md.; Difco Laboratories, Detroit.

| ORGANISM | GLUCOSE | | GROUP |
	OPEN	COVERED	
Alcaligenes faecalis	–	–	I Nonoxidizers Nonfermenters
Pseudomonas aeruginosa	A	–	II Oxidizers Nonfermenters
Shigella dysenteriae	A	A	IIIa Fermenters (Anaerogenic)*
Salmonella enteriditis	A	AG	IIIb Fermenter (Aerogenic)*

*The criterion of gas production applies to vaspar-sealed tubes only.

1. Oxidase-positive motile rods—*Pseudomonas, Alcaligenes*
2. Oxidase-negative nonmotile diplobacilli—*Acinetobacter*
3. Oxidase-positive diplobacilli—*Moraxella*

Only those organisms of proved clinical significance will be discussed in detail; the interested reader is referred to authors already cited for further information.

PSEUDOMONAS SPECIES
Pseudomonas aeruginosa

Pseudomonas aeruginosa is the most frequently implicated member of the genus in human infections; it may colonize burn sites, wounds, tracheostomies, and the lower respiratory tract, particularly in patients whose defenses have been compromised, and may spread by dissemination to the blood, resulting in a serious and sometimes fatal septicemia. Although *Ps. aeruginosa* may be isolated from the skin and feces of normal humans (it has been responsible for severe outbreaks of infantile diarrhea), most of the infections are exogenous in origin.[18,20] Since the organism is part of the hospital environment—it can survive and even multiply in moist environments with minimal amounts of organic matter—it has been incriminated in from 5% to 15% of all hospital-acquired infections.[2]

Ps. aeruginosa is a polar **monotrichous,** gram-negative rod occurring singly, in pairs, or in short chains. On blood agar, the organism grows as a large, flat colony* with a ground-glass appearance, and produces a zone of hemolysis. The colonies tend to spread and give off a characteristic **grapelike odor.** Most strains excrete pyocyanin and fluorescin, giving the colony a characteristic **blue-green** color; approximately 4% are apyocyanogenic.

Ps. aeruginosa is oxidase positive by Kovacs' method[13] (described in Chapter 41) and utilizes glucose **oxidatively** in the O-F medium; gluconate is oxidized to ketogluconate† (but not by other pseudomonads). Most strains grow at 42° C. on trypticase agar slants; they also grow on agar media containing cetyltrimethylamine bromide (Cetrimide).‡ Other pseudomonads are generally inhibited on the latter.

Because *Ps. aeruginosa* grows on EMB or MacConkey agar as a nonlactose fermenting organism, it is frequently selected and streaked on TSI agar as a suspicious colony from stool cultures and may be incorrectly identified because of the alkaline slant or butt in the latter. One other pseudomonad—*Ps. putrifaciens*—also produces an appreciable amount of H_2S.[22]

A serological identification system that uses a standardized technique of bacteriophage typing has been recommended for tracing *Pseudomonas* strains during an epidemiological investigation.[19]

Ps. aeruginosa is susceptible to the aminoglycoside, gentamicin, the drug of choice in the treatment of serious *Pseudomonas* infections. Many strains are also sensitive to carbenicillin, a semisynthetic penicillin, which is recommended for therapy when renal impairment is a problem. These organisms are also inhibited by the

*Mucoid strains are frequently isolated from the sputum of patients with cystic fibrosis.
†Gluconate tablets, from Key Scientific Products Co., Los Angeles.
‡Available as Pseudosel agar, Baltimore Biological Laboratory, Cockeysville, Md.; Cetrimide agar, Difco Laboratories, Detroit.

polymyxin antibiotics polymyxin B and colistin, which may produce side effects of renal or neural toxicity.

Pseudomonas maltophilia

This pseudomonad is ubiquitous in nature; it is being increasingly isolated not only from the blood, cerebrospinal and other body fluids, sputum, urine, and abscesses of infected humans, but also from the environment, such as natural waters, sewage, spoiled foods, and so forth. This free-living organism is thus a true opportunist.

Ps. maltophilia is multitrichous, with tufts of two or more flagella per pole. It produces a yellow to tan pigment on trypticase soy agar and an ammoniacal odor. On glucose O-F medium, it produces an early (overnight) alkaline reaction, becoming weakly acid on further incubation; on **maltose** O-F medium, acidity is promptly produced oxidatively.

Ps. maltophilia is oxidase negative, ONPG and D'Nase positive; it does not reduce nitrate to nitrogen gas. Most strains are susceptible to polymyxin and colistin; varying results occur with other agents.

Other opportunistic pseudomonads

Ps. fluorescens and *Ps. putida* are only occasionally isolated from humans; sources have included urinary tract infections and contaminated blood-bank blood. They fail to grow at 42° C. (25° C. optional). *Ps. fluorescens* will grow at refrigerator temperatures. Both produce fluorescin, but not pyocyanin or pyorubrin; they are generally resistant to carbenicillin and sensitive to kanamycin.[14]

Ps. stutzeri is a polar monotrichous, nonfluorescent pseudomonad of questionable pathogenic significance that produces a yellow, wrinkled colony resembling that of *Ps. pseudomallei*. Oxidative in glucose O-F medium, it reduces nitrate to nitrogen gas and grows in 6.5% NaCl broth, which aids in differentiating it from other pseudomonads.

Ps. cepacia (synonyms: *Ps. multivorans, Ps. kingii,* EO-1) has occasionally been recovered from clinical material, including blood cultures[6] and contaminated detergent solutions in urinary catheter kits,[9] as well as hospital water supplies. Included in its characteristics are a variable oxidase reaction, lack of growth on Salmonella-Shigella agar, presence of lysine decarboxylase, and resistance to polymyxin and colistin.[6] Many strains produce a nonfluorescent yellow pigment that diffuses into the agar medium.

Ps. alcaligenes, Ps. acidovorans, and *Ps. putrefaciens* (produces H_2S on TSI agar) are examples of rarely isolated but potentially pathogenic pseudomonads. A description of these and related pseudomonads may be found in papers by the authors previously cited.[8,16,17]

Pseudomonas pseudomallei

Pseudomonas pseudomallei is the causative agent of **melioidosis** in man. This disease presents a varying clinical picture, ranging from unsuspected asymptomatic infection to acute, toxic pneumonia or overwhelming and highly fatal septicemia. The organism has been isolated frequently from moist soil, market fruits and vegetables, and from well and surface waters in Southeast Asia. Although apparently rare in natives (incidence is approximately 8%), the disease has been an important and frequently fatal infection in the U. S. Armed Forces in Vietnam. *Ps. pseudomallei* is multitrichous with a tuft of three or more flagella per pole and of similar morphology to *Ps. aeruginosa,* but the colonies frequently are wrinkled. An oxidative acidity is produced in glucose O-F medium; the oxidase reaction is positive, and growth occurs at 42° C., but pyocyanin or fluorescent pigment are not produced.

These pseudomonads may be cultivated on most laboratory media, growing well on trypticase soy agar, on blood agar, and on MacConkey agar, but not on Salmonella-Shigella (SS) agar or cetrimede agar. A selective medium for isolating *Ps. pseudomallei* from contaminated clinical material has been described.[3]

Ps. pseudomallei is generally susceptible

Table 21-1. Some characteristics of species of *Pseudomonas**

ORGANISM	PYOCYANIN CHCl₃ SOLUBLE	OXIDASE (KOVACS)	GROWTH AT 42°C.	NO₃ → NO₂	NO₃ → GAS	GROWTH ON SS	OXIDIZES MALTOSE
Ps. aeruginosa	+ (−)	+	+	+	+	+	−
Ps. fluorescens	−	+	−	−	− (+)	+	var
Ps. putida	−	+	−	−	−	+	var
Ps. cepacia	−	var	−	+/−	−	−	+
Ps. stutzeri	−	+	+	+	+	− (+)	+
Ps. maltophilia	−	−	+	− (+)	−	−	+
Ps. pseudomallei	−	+	+	+	+	−	+

*+ = positive reaction; − = negative reaction; ( ) = occasional strain; var = variable.

to tetracycline (the drug of choice) and kanamycin, but resistant to polymyxin, gentamicin, and carbenicillin.[7]

Thomason and associates[21] have reported that identification of *Ps. pseudomallei* in mixed culture on a slide may be made by the fluorescent antibody technique.

Some characteristics of *Pseudomonas* species are seen in Table 21-1.

Pseudomonas mallei

This pseudomonad, formerly classified as *Actinobacillus mallei,* is the causative agent of **glanders,** an infectious disease of horses that has occasionally been transmitted to man by direct contact through trauma or inhalation. *Ps. mallei* is a nonmotile, coccoid- to rod-shaped organism, sometimes occurring in filaments or with branching involution forms under special conditions. Colonies on infusion agar (especially when glycerol is added) appear in 48 hours as grayish white and translucent, later becoming yellowish and opaque. *Ps. mallei* oxidizes glucose, fails to grow at 42° C, and may be weakly oxidase positive. Male guinea pigs injected intraperitoneally with culture material develop a tender and swollen scrotum 2 to 4 days later (Straus test) if *Ps. mallei* is present.

ALCALIGENES FAECALIS

These bacilli are motile by means of **peritrichous flagella,** a useful aid in distinguishing them from pseudomonads. They are oxidase and phenylalanine deaminase positive, but **do not oxidize glucose** O-F medium. Growth occurs on MacConkey agar but is inhibited on SS agar. Most isolants are saprophytic, but *Alc. faecalis* has been recovered from a variety of clinical sites, including blood from patients after heart surgery. [23] Some strains have been incriminated in nosocomial infections. Chloramphenicol and sulfonamides appear to be effective therapeutic agents.[4]

ACINETOBACTER SPECIES

Acinetobacter species, formerly members of DeBord's tribe Mimeae, have undergone a major taxonomic revision, and the organisms known previously as *Herellea vaginicola* and *Mima polymorpha* are considered by medically oriented microbiologists as members of the genus *Acinetobacter.*[15] Although differentiation of species within the genus is far from clear, nearly all strains associated with human infections can be classified as either *Ac. anitratus* (*H. vaginicola*) or *Ac. lwoffii* (*M. polymorpha*).*

These organisms are oxidase negative, nonmotile diplococcoid to coccobacillary forms that **do not reduce nitrate** (some may show weak activity) and that grow well on MacConkey agar but poorly or not

*The International Committee on Nomenclature recently has recommended that the genus *Acinetobacter* comprise a single species, *Ac. calcoaceticus,* containing the organisms previously referred to as *H. vaginicola* and *M. polymorpha.* (Inter. J. Syst. Bact. **21**:213-214, 1971.)

at all on SS agar. *Ac. anitratus* produces acidity in the open tube of glucose O-F medium (oxidizer) as well as in other carbohydrate O-F media and on 10% lactose agar; *Ac. lwoffii* is **inactive** on these media. Both species produce strongly positive tests for catalase but fail to demonstrate decarboxylase, dihydrolase, or deaminase; neither do they grow on cetrimide agar. A characteristic reaction on Sellers medium (blue slant, yellow band, green butt) is produced by *Ac. anitratus*, which readily differentiates it from other nonfermentative bacilli.

Ac. anitratus is the most frequently isolated member of the genus, having been recovered from a wide variety of clinical sources, including the upper and lower respiratory tract, urinary tract, wounds, nosocomial infections, and so forth. Its role in these has been one of an opportunistic pathogen, generally occurring in mixed cultures from low-grade infections, although occasional cases of septicemia, emphysema, or pneumonia have been reported in debilitated hospital patients.[4]

Ac. lwoffii's role in human infections is more difficult to assay, since most isolants are of questionable clinical significance. It is part of the normal flora of the skin, external genitalia, and other sites, and may be recovered as a commensal from these areas.[14] It is probably best considered as an occasionally opportunistic agent, particularly in superficial infections.

Both organisms are susceptible in vitro to the aminoglycosides kanamycin and gentamicin, as well as to colistin and polymyxin B[7,14,23]; *Ac. lwoffii* appears to be susceptible to a wider variety of antimicrobial agents. Both are resistant to penicillin.

MORAXELLA SPECIES

Some of these organisms that have been variously classified with the genera *Achromobacter, Acinetobacter,* DeBord's Mimeae, and others, by recent studies and rules of nomenclature, are illegitimate epithets,[15] and are better included in the genus *Moraxella.* Moraxellae are nonmotile bacilli or coccobacilli that are **oxidase positive** and **penicillin sensitive;** most strains are biochemically inactive with respect to carbohydrate oxidation, denitrification and the production of deaminase, decarboxylase, and dihydrolase.[5] Some strains are fastidious, requiring enriched media or increased humidity for growth. Several species are generally recognized by medical microbiologists and include the following:

1. *M. lacunata*
2. *M. osloensis*
3. *M. kingii*

M. lacunata (Morax-Axenfeld bacillus) was originally described as the etiological agent in chronic conjunctivitis. It is infrequently isolated; characteristically it causes pitting, or lacunae, on the surface of Loeffler slants, with subsequent digestion of the medium. Serum is required for its growth, preferably in increased CO_2. It is oxidase positive.

M. osloensis[1] was previously classified as *Mima polymorpha* var. *oxidans* (DeBord); the latter was considered an illegitimate epithet by some taxonomists.[15] Although its involvement in human infections is comparatively rare, *M. osloensis* is easily confused with *N. gonorrhoeae* from genitourinary sources, since it is **oxidase positive** and appears as gram-negative coccobacillary forms resembling gonococci. *M. osloensis* **does not ferment** CTA glucose medium and grows on nutrient agar, whereas *N. gonorrhoeae* ferments the glucose medium and does not grow on nutrient agar.

M. kingii is a newly described species[10] named in honor of the late Elizabeth O. King. Unique among the moraxellae, it is catalase negative and oxidizes glucose O-F medium; it is oxidase positive and produces **beta hemolysis** on blood agar; poor growth is observed on nonenriched media. *M. kingii* has been isolated from blood and throat cultures of humans.[10]

Moraxella species are uniformly susceptible to penicillin, ampicillin, tetracycline, and erythromycin in vitro.[7]

REFERENCES

1. Bøvre, K., and Henriksen, S. D.: A new *Moraxella* species, *Moraxella osloensis,* and a revised description of *Moraxella nonliquifaciens,* Intern. J. Syst. Bact. **17:**127-135, 1967.
2. Eickhoff, T. C.: Hospital infections, Disease-a-Month, (September) Chicago, 1972, Year Book Medical Publishers, Inc.
3. Farkas-Himsley, H.: Selection and rapid identification of *Pseudomonas pseudomallei* from other gram-negative bacteria, Amer. J. Clin. Path. **49:**850-856, 1968.
4. Gardner, P., Griffin, W. B., Swartz, M. N., and Kunz, L. J.: Nonfermentative gram-negative bacilli of nosocomial interest, Amer. J. Med. **48:**735-749, 1970.
5. Gilardi, G. L.: Diagnostic criteria for differentiation of pseudomonads pathogenic for man, Appl. Microbiol. **16:**1497-1502, 1968.
6. Gilardi, G. L.: Characterization of EO-1 strains *(Pseudomonas kingii)* isolated from clinical specimens and the environment, Appl. Microbiol. **20:**521-522, 1970.
7. Gilardi, G. L.: Antimicrobial susceptibility as a diagnostic aid in the identification of nonfermenting gram-negative bacteria, Appl. Microbiol. **22:**821-823, 1971.
8. Gilardi, G. L.: Characterization of nonfermentative nonfastidious gram-negative bacteria encountered in medical bacteriology, J. Appl. Bact. **34:**623-644, 1971.
9. Hardy, P. C., Ederer, G. M., and Matsen, J. M.: Contamination of commercially packaged urinary catheter kits with the pseudomonad EO-1, New Eng. J. Med. **282:**33-35, 1970.
10. Henriksen, S. D., and Bøvre, K.: *Moraxella kingii* sp. nov., a hemolytic, saccharolytic species of the genus *Moraxella,* J. Gen. Microbiol. **51:**377-385, 1968.
11. Hugh, R.: A practical approach to the identification of certain nonfermentative gram-negative rods encountered in clinical specimens, J. Confer. Publ. Health Lab. Directors, **28:**168-187, 1970.
12. Hugh, R., and Leifson, E.: The taxonomic significance of fermentative versus oxidative metabolism of carbohydrates by various gram-negative bacteria, J. Bact. **66:**24-26, 1953.
13. Kovacs, N.: Identification of *Pseudomonas pyocyanea* by the oxidase reaction, Nature **178:**703, 1956.
14. Pedersen, M. M., Marso, M. A., and Pickett, M. J.: Nonfermentative bacilli associated with man. III. Pathogenicity and antibiotic susceptibility, Amer. J. Clin. Path. **54:**178-192, 1970.
15. Pickett, M. J., and Manclark, C. R.: Nonfermentative bacilli associated with man. I. Nomenclature, Amer. J. Clin. Path. **54:**155-163, 1970.
16. Pickett, M. J., and Pedersen, M. M.: Nonfermentative bacilli associated with man. II. Detection and identification, Amer. J. Clin. Path. **54:**164-177, 1970.
17. Riley, P. S., Tatum, H. W., and Weaver, R. E.: *Pseudomonas putrefaciens* isolates from clinical specimens, Appl. Microbiol. **24:**798-800, 1972.
18. Sutter, V. L., and Hurst, V.: Sources of *Pseudomonas aeruginosa* infection in burns; study of wound and rectal cultures with phage typing, Ann. Surg. **163:**596-603, 1966.
19. Sutter, V. L., Hurst, V., and Fennell, J.: A standardized system for phage typing *Pseudomonas aeurginosa,* Health Lab. Sci. **2:**7-16, 1965.
20. Sutter, V. L., Hurst, V., Grossman, M., and Calonje, R.: Source and significance of *Pseudomonas aeruginosa* in sputum, J.A.M.A. **197:**854-858, 1966.
21. Thomason, B. M., Moody, M. D., and Goldman, M.: Staining bacterial smears with antibody. II. Rapid detection of varying numbers of *Malleomyces pseudomallei* in contaminated materials and infected animals, J. Bact. **72:**362, 1956.
22. Von Grazvenitz, A., and Simon, G.: Potentially pathogenic, nonfermentative, H$_2$S-producing gram-negative rod (1 b), Appl. Microbiol. **19:**176, 1970.
23. Washington, J. W. II: Antimicrobial susceptibility of enterobacteriaceae and nonfermenting gram-negative bacilli, Mayo Clin. Proc. **44:**811-824, 1969.

22 Gram-negative, coccobacillary, aerobic bacteria

Pasteurella
Francisella
Bordetella
Brucella
Haemophilus
Actinobacillus
Calymmatobacterium
Noguchia

The members of this bacterial grouping are small gram-negative rods, occurring singly, in pairs, in short chains, and in other arrangements. Encapsulation may occur, and some forms show bipolar staining. Others exhibit pleomorphism. The organisms are aerobic to facultatively anaerobic. Carbon dioxide in excess of normal atmospheric concentration may favor the growth of some species, whereas serum or blood as culture medium enrichments will enhance the growth of others. X and V factors (p. 172) are required for the cultivation of certain fastidious species.

Some species can invade living tissue after gaining entrance through the mucous membranes of the skin. Zoonotic species may also be transmissible to man, producing diseases such as brucellosis, tularemia, respiratory illnesses, meningitis, and others. Many species are obligate animal parasites.

GENUS PASTEURELLA

Some of the former species of this genus, namely *Past. pestis* and *Past. tularensis,* have been given new generic status and are therefore appropriately described in their assigned taxonomic positions as *Yersinia pestis* and *Francisella tularensis.* The remaining species of clinical significance is *Past. multocida.*

Pasteurella multocida

Past. multocida is primarily an animal pathogen, causing a form of **hemorrhagic septicemia** in the lower animals and **cholera** in chickens. Man most frequently becomes infected from a bite or scratch of a cat or dog[4] or through contact with a diseased carcass, which may occur in abattoir workers and veterinarians. Latent respiratory tract infections also may occur, particularly in patients suffering from bronchiectasis. Clinical specimens include sputum, pus, blood, spinal fluid, and tissues.

Past. multocida is a small, coccoid, nonmotile, gram-negative rod often showing bipolar staining. It grows well at 35° C. on chocolate agar or blood agar, where it produces small, nonhemolytic, translucent colonies with a characteristic musty odor.* Four colony forms are recognized. Many strains isolated from the respiratory tract or from chronic infections produce mucoid (M), relatively avirulent colonies. This form is highly pathogenic for animals, however. Highly virulent strains for humans produce smooth (S), fluorescent colonies. Nonfluorescent, smooth, transitional forms are weakly virulent, and the R form, which is granular and dry, is

---

Past. multocida is inhibited on bile-containing media, such as SS, XLD, or Hektoen agar.

Table 22-1. Characteristics of *Pasteurella multocida*

PROPERTY	OBSERVATION	PROPERTY	OBSERVATION
Colony forms	four	Voges-Proskauer	−
Optimal growth temperature	35° to 37° C.	Urease	−
Motility at 25° C.	−	Nitrates	Reduced
Serogroups	Five	Cellobiose	·
Catalase	+	Glucose	+
Oxidase	+	Glycerol	−
Coagulase	−	Lactose	−
Fibrinolysin	−	Melibiose	−
Hemolysis on blood agar	−	Maltose	−
Medium containing bile salts	No growth	Mannitol	+
H_2S	+	Rhamnose	−
Indole	+	Salicin	−
Methyl red	−	Sucrose	+

avirulent. The biochemical properties and other characteristics are shown in Table 22-1.

Serologically, *Past. multocida* has been divided into five groups, A, B, C, D, and E. The human strains fall into Groups A and D.[2] Animal pathogenicity tests and serological tests with specific typing sera should be carried out if possible. Alternatively, a suspected culture should be sent to a reference laboratory. The organism is very **susceptible to penicillin** (2-unit disc) in vitro—an observation that frequently leads one to suspect its presence on routine culture plates streaked with sputum or bronchoscopic secretions.

GENUS FRANCISELLA (PASTEURELLA)
Francisella (Pasteurella) tularensis

Fran. tularensis causes **tularemia,** a disease of rodents (particularly rabbits) that is directly transmissible to man through the handling of infected animals or indirectly transmissible by blood-sucking insects. In culture, the organism is a minute, highly pleomorphic, nonmotile, gram-negative rod with capsules occurring in vivo. The organism reproduces by different methods, including budding,[3] binary fission, and the production of filaments. The organism exhibits a filterable phase, and in this respect it resembles members of the pleuropneumonia group. Clinical specimens include blood (first week), sputum, pleural fluid, and conjunctival scrapings.

Fran. tularensis requires special enriched media, such as the blood-cystine-dextrose agar of Francis (Chapter 39), and will not grow on plain agar. On Francis' medium, minute, transparent, droplike, mucoid, readily emulsifiable colonies are formed after 2 to 5 days of incubation at 36° C. The organism is an obligate aerobe and grows optimally at 36° C. Glucose, maltose, and mannose are fermented without gas; other carbohydrates are attacked irregularly.

Further identification procedures include the testing for susceptibility to specific bacteriophages, agglutination by specific antisera, and the demonstration of virulence by intraperitoneal inoculation of guinea pigs. The danger of handling infected animals and virulent cultures of *Fran. tularensis* cannot be overemphasized. **Many laboratory workers have become infected, and some have died of tularemia.**

GENUS BORDETELLA

The genus *Bordetella* consists of three species that are minute, gram-negative, motile or nonmotile coccobacilli. Some require complex media for primary isolation, and all are associated with whooping cough, or an infection clinically resembling it, in man.

Bordetella pertussis

Bord. pertussis, the type species, is the causative agent of **whooping cough,** or **pertussis.** It requires Bordet-Gengou agar (potato-blood-glycerol agar; see Chapter 39) for primary isolation. The addition of 0.25 to 0.5 unit of penicillin per milliliter is recommended for reducing overgrowth of gram-positive organisms. On this medium, small, smooth, convex colonies with a pearl-like luster (resembling **mercury droplets**) develop in 3 to 4 days. The colonies are mucoid and tenacious and are surrounded by a zone of hemolysis. The organism is **nonmotile** and may occur singly, in pairs, and occasionally in short chains. The cells tend to show bipolar staining, and may be encapsulated.

Indole is not produced by the organism, and citrate is not utilized. Nitrates are not reduced, nor is urea hydrolyzed. X and V factors (p. 172) are not required; catalase is produced. With the exception of glucose and lactose, the carbohydrates are not attacked.

When isolated from patients with pertussis, the organism gives rise to smooth, encapsulated phase I colonies on Bordet-Gengou medium. The other phases of the organism (II, III, and IV) may be determined by antigenic analysis. The identification of *Bord. pertussis* is further confirmed by a slide agglutination test with specific antiserum.* Fluorescent antibody staining may also be used as an identification aid.[5]

Bordetella parapertussis

Bord. parapertussis is occasionally isolated from patients with an acute respiratory tract infection resembling mild whooping cough. It is morphologically and colonially similar to *Bord. pertussis,* but it develops a large colony on Bordet-Gengou agar and produces a **brown** pigment in the underlying medium. It is **nonmotile** and does not require the X or V factors for growth. Indole is not produced, and carbohydrates are not fermented. The organism is normally urease and catalase positive, and utilizes citrate.

Although it is serologically homogeneous, *Bord. parapertussis* shares common somatic antigens with *Bord. pertussis* and *Bord. bronchiseptica* and may cross-agglutinate with these organisms. An absorbed high-titer antiserum is available for the serological identification of *Bord. parapertussis* (Chapter 35).

Bordetella bronchiseptica

Bord. bronchiseptica has been isolated occasionally from patients with a pertussis-like disease. It differs from *Bord. pertussis* in that it is **motile** and possesses peritrichous flagella. The organism grows readily on blood agar, producing smooth, raised, glistening colonies with hemolytic zones. Indole is not produced, and none of the carbohydrates is fermented. **Urea** is split quite rapidly (4 hours), catalase is formed, nitrates are often reduced, and citrate is utilized as a source of carbon. Cross-agglutination occurs with *Bord. pertussis* and *Bord. parapertussis. Bord. bronchiseptica* was once thought to cause canine distemper, and it is a common cause of bronchopneumonia in guinea pigs and rabbits; it may also occur in these latter animals as a normal inhabitant of the respiratory tract.

GENUS BRUCELLA

The genus *Brucella* consists of three main species that are nonmotile, gram-negative, rod- to coccoid-shaped cells; they are pathogenic for a variety of domestic animals and man. There is, however, a fourth species, *Br. neatomae,* isolated in 1957 by Stoenner and Lackman[9] from a desert rat. It is of no current importance to man.

The brucellae are **obligate parasites,** characterized by their intracellular existence, and are capable of invading all animal tissue, where they cause various bruceloses, including contagious abortion in goats, cows, and hogs, as well as undulant fever in man.

A definitive diagnosis of brucellosis is

---

*Difco Laboratories, Detroit.

established by the isolation and identification of the organism from clinical specimens. **Blood** is the material most frequently found to be positive on culture, particularly when drawn during the febrile period (first 3 weeks) of illness. Brucellae may be recovered occasionally from cultures of the bone marrow, from biopsied lymph nodes and other tissue, and also from urine and cerebrospinal fluid. The use of the modified Castañeda bottle (Chapter 39) is strongly recommended for culturing the blood from multiple specimens collected. Incubation of these cultures in a candle jar or CO_2 incubator (2% to 10% carbon dioxide) is imperative; cultures should be held for a minimum of 30 days before being discarded as negative. Blood cultures, as a rule, are negative after the acute symptoms have subsided, which generally coincides with the development of humoral antibodies in the patient.

The agglutination test (Chapter 36), which uses a standardized, heat-killed, smooth *Brucella* antigen, is the most reliable of the serological tests. The indirect FA test has been successfully used for detecting antibody in human sera.[1] The opsonocytophagic test is subject to great variations and is of doubtful value.

The three important species may be differentiated by their susceptibility to certain bacteriostatic dyes, by their reaction in carbohydrates, and by their requirements for additional carbon dioxide for primary isolation on laboratory media (Table 22-2).

Guinea pigs are susceptible to all species and will develop an infection within 30 days after injection of primary smooth (S) isolates.

Brucellae are aerobic and grow best at 35° C. *Br. abortus* requires an increased carbon dioxide tension (2% to 10%) for primary isolation, although many strains lose this requirement on subculture. The nutrition of these organisms is complex. The best growth may be obtained on enriched media, such as liver infusion, tryptose,* trypticase,† or brucella‡ agar at pH 7 to 7.2 (pH 7.5 to 7.8 in 5% carbon dioxide). After 24 to 48 hours' incubation, small, convex, smooth, translucent colonies appear, which become brownish with age. Brucellae may also be cultivated on synthetic media containing amino acids, vitamins, mineral salts, and glucose.

The three significant species reduce nitrates, and *Br. abortus* and *Br. suis* carry the reduction to nitrogen gas. **Urea** is rapidly hydrolyzed by *Br. suis* but slowly, if at all, by *Br. melitensis* and *Br. abortus*. All strains are catalase positive, with *Br. suis* being the most active. *Br. suis* strains isolated in the United States are active producers of hydrogen sulfide (lead acetate paper), whereas other species produce only smaller amounts. Brucellae do not liquefy gelatin or produce indole and are M.R. and V.P. negative.

The species of *Brucella* show a **differen-**

---

*Difco Laboratories, Detroit.
†Baltimore Biological Laboratory, Cockeysville, Md.
‡Albimi Laboratories, Flushing, N. Y.

Table 22-2. Differential characteristics of typical strains of the three species of genus *Brucella*

SPECIES	CO_2 REQUIREMENTS (5%)	H_2S PRODUCTION	GROWTH	
			THIONINE	BASIC FUCHSIN
Br. melitensis	−	− to +/− (throughout 4 days)	+	+
Br. abortus	+	+ (first 2 days only)	−	+
Br. suis	−	+ (throughout 4 days)	+	−

There is some strain variation; these reactions are the most typical.

tial sensitivity to a number of aniline dyes, such as thionine, basic fuchsin, crystal violet, pyronin, azure A, and so forth. These may be incorporated in the agar medium in concentrations of 1:25,000 to 1:100,000, depending on the dye content of each lot of dye and on the medium used. Inhibition of growth may also be determined by the use of dye tablets* similar to the antibiotic discs. An inhibition zone of 4 mm. or more in diameter around the discs is interpreted as susceptibility.

Br. melitensis can be differentiated from *Br. abortus* and *Br. suis* by the agglutination technique, using absorbed monospecific antisera. The latter two species, however, cannot be so differentiated, since they share an equal concentration of two identical antigens.

GENUS HAEMOPHILUS

Haemophilus species are small, nonmotile, gram-negative rods that require hemoglobin in the culture medium, or are stimulated by its presence. Whole blood contains the following two factors that are necessary for the growth of the type species of the genus, *H. influenzae:*

1. **X factor**—a heat-stable substance, hemin, associated with hemoglobin.
2. **V factor**—a heat-labile substance, which is coenzyme I, nicotinamide-adenine-dinucleotide (NAD), supplied by yeast, potato extract, and certain bacteria in addition to that found in blood. Table 22-3 shows the

——————
*Medical Research Specialties, Loma Linda, Calif.

characteristics of the pathogenic species.

Although all members of the genus are parasitic in nature and require growth factors, they may or may not be pathogenic for man. Some are members of the normal flora of the respiratory tract; others are important human pathogens, capable of causing severe respiratory tract disease, meningitis, pyogenic arthritis, or subacute bacterial endocarditis.

Haemophilus influenzae

H. influenzae, originally regarded by Pfeiffer and others as being the causative agent of influenza, is found in the respiratory tract of man. It plays an important etiological role in acute respiratory tract infections and conjunctivitis and may also cause a septicemia, a subacute bacterial endocarditis, and a purulent meningitis in children. This form of meningitis occurs only rarely in adults.[7] It also produces a characteristic obstructive laryngotracheal infection that may prove fatal in children 2 to 5 years of age.

H. influenzae is a fastidious organism, requiring an infusion medium containing X and V factors. Luxuriant growth occurs on **chocolate agar** (Chapter 39). Growth on this medium appears in 18 to 24 hours as small (1 to 2 mm.), colorless, transparent, moist colonies with a distinct "mousy" odor. On richer media, such as Levinthal's transparent agar, the colonies may be larger; on blood agar the organism grows poorly, if at all, producing only tiny colonies. Colonies are large and characteristic

Table 22-3. Hemolytic activity and the X and V requirements of members of genus *Haemophilus* pathogenic for man

ORGANISM	INFECTION SITE	X FACTOR	V FACTOR	HEMOLYSIS
H. influenzae	Respiratory tract, meninges, blood, and other areas	+	+	−
H. aegyptius	Conjunctiva	+	+	−
H. haemolyticus	Respiratory tract	+	+	+
H. parainfluenzae	Respiratory tract	−	+	−
H. parahaemolyticus	Respiratory tract	−	+	+
H. ducreyi	Genital region	+	−	+/−

on blood agar, however, when they are growing near colonies of staphylococci, neisseriae, pneumococci, and others capable of synthesizing an extra supply of V factor. This diffuses into the surrounding medium and stimulates growth of *H. influenzae* in the vicinity of such colonies. This phenomenon is known as **"satellitism."**

All strains of *H. influenzae* reduce nitrates to nitrites and are soluble in sodium desoxycholate*; indole is produced by the encapsulated organisms. Fermentation reactions are variable: Glucose and other carbohydrates are attacked by some strains but not by others. Six serological types (**a, b, c, d, e,** and **f**) are recognized by the quellung† method or by the precipitin reaction. Most meningeal infections are caused by **type b,** but the majority of respiratory strains are not type specific and are apparently less virulent. Virulent *H. influenzae* strains appear to be immunologically related to the pneumococcus; cross-reactions occur between the capsular substances of these two organisms. For example, type **b** cross-reacts with pneumococcus types 6 and 29.

Haemophilus aegyptius (Koch-Weeks bacillus)

H. aegyptius, which is associated with the highly communicable form of conjunctivitis known as **pink eye,** resembles *H. influenzae* morphologically. It requires both X and V factors for its growth, and it produces small, transparent, nonhemolytic colonies on blood agar. They exhibit satellitism with neighboring *Staphylococcus* colonies. On transparent agar the colonies show a bluish sheen with transmitted light. Indole is not produced, and the reaction in carbohydrates is inconsis-

tent. Nitrates are reduced to nitrites, and the organism is bile soluble. *H. aegyptius* is serologically related to, but not identical with, *H. influenzae.* It will agglutinate human red blood cells readily, giving a stronger reaction than the latter.

Haemophilus haemolyticus

H. haemolyticus is found normally in the upper respiratory tract of man but rarely causes infection. It requires both X and V factors for growth. Colonies on blood agar resemble *H. influenzae* but are surrounded by a wide zone of **beta hemolysis.**

In examining throat cultures on blood agar* plates, it is important that colonies of *H. haemolyticus* be differentiated from those of beta hemolytic streptococci, since both are small colonies surrounded by a zone of clear hemolysis. *H. haemolyticus* colonies are generally soft, pearly, and translucent in contrast to the firm, white, opaque colonies of Group A streptococci. A Gram stain of the colony in question will readily differentiate between the two.

Haemophilus parainfluenzae and Haemophilus parahaemolyticus

H. parainfluenzae and *H. parahaemolyticus* are found in the normal respiratory tract of man and are rarely associated with subacute bacterial endocarditis. Both require **only the V factor** for growth. *H. parainfluenzae* resembles *H. influenzae* both morphologically and colonially; it is not hemolytic, and it exhibits satellitism around staphylococcal colonies. *H. parahaemolyticus* resembles *H. haemolyticus,* although it produces somewhat larger colonies on blood agar. These are surrounded by a zone of beta hemolysis. Biochemically, both species are similar to other members of the genus.

Haemophilus ducreyi

H. ducreyi is the causative agent of an ulcerative venereal disease known as

---

*To 0.8 ml. of a young broth culture, add 0.2 ml. of a 10% solution of sodium desoxycholate. Incubate for 2 hours at 35° C. The tube should become clear except for a slight opalescence.
†Polyvalent and type-specific antisera are available from Hyland Laboratories, Los Angeles; Difco Laboratories, Detroit.

---

*Rabbit and horse blood do not contain inhibitory substances against hemoglobinophils; human and especially sheep blood greatly inhibit their growth.

chancroid (soft chancre) in man. (This is one of the classic five venereal diseases.) The small gram-negative rods occur in long strands in smears from the genital ulcer, where they are usually associated with other pyogenic bacteria. *H. ducreyi* is grown with considerable difficulty and grows best in fresh clotted rabbit, sheep, or human blood heated to 55° C. for 15 minutes. Smears made from this after 1 to 2 days of incubation will show the tangled chains of *H. ducreyi,* if present.

Patients infected with this microorganism develop a hypersensitivity to it, which can be detected by the intradermal injection of heat-killed cells. This is a useful diagnostic aid, since the test is positive 1 to 2 weeks after infection.

Haemophilus-like organisms

Mention has already been made of *H. (Corynebacterium) vaginalis,* a related organism isolated from the human urogenital tract (Chapter 11). Its growth requirements appear to be different from those of the *Haemophilus* species previously discussed, and its etiological significance still remains uncertain.

Another *Haemophilus*-like, gram-negative rod that has been isolated from patients with endocarditis[10] and meningitis is *H. aphrophilus.* It is a nonmotile pleomorphic rod requiring the X factor only (under CO_2) and grows as fluffy clumps on the inner walls of the tube. It appears to be similar, both biochemically and serologically, to *Actinobacillus actinomycetemcomitans,* and it may be confused with this organism when it occurs in actinomycotic lesions.[6]

Other genera of clinical significance in this group

Within the genera *Actinobacillus, Calymmatobacterium,* and *Noguchia* may be found certain human pathogens. Some are primarily animal pathogens but are transmissible to man.

The glanders bacillus, formerly named *Actinobacillus mallei,* has been improperly classified in this genus but rightly belongs in the genus *Pseudomonas* (see Chapter 21). Another species, *A. lignieresii,* is the cause of actino bacillosis or **wooden tongue** in cattle and is only rarely involved in human infections. *A. actinomycetemcomitans* is primarily associated with human diseases and has been isolated alone and in combination with *Actinomyces israelii* from human **actinomycosis.**

Calymmatobacterium granulomatis is the etiological agent in the venereal disease **granuloma venereum.** Oval- to bacillary-shaped bodies, referred to as **Donovan bodies,** that are surrounded by a dense capsule may be observed in large mononuclear cells in smears prepared from lesions. The initial lesion is a swelling or **bubo** in the groin area. Later the genitals, buttocks, and abdomen may become involved.

Noguchia granulosis may be found in cases of follicular conjunctivitis in man and other animals. The organism is motile, pleomorphic, and encapsulated, and grows best in the range of 15° to 30° C. on semisolid medium.

REFERENCES

1. Biegeleisen, J. Z., Jr., Bradshaw, B. R., and Moody, M. D.: Demonstration of *Brucella* antibodies in human serum, a comparison of the fluorescent antibody and agglutination techniques, J. Immun. **88:**109-112, 1962.
2. Carter, G. R., and Byrns, J. L.: A serological study of the hemorrhagic septicemia *Pasteurella,* Cornell Vet. **43:**223-230, 1953.
3. Hesselbrook, W., and Foshay, L.: The morphology of *Bacterium tularense,* J. Bact. **49:**209-231, 1945.
4. Holloway, W. J., Scott, E. G., and Adams, Y. B.: *Pasteurella multocida* infection in man, Amer. J. Clin. Path. **51:**705-708, 1969.
5. Kendrick, P. L., Eldering, G., and Eveland, W. C.: Application of fluorescent antibody techniques; methods for the identification of *Bordetella pertussis,* Amer. J. Dis. Child. **101:**149-154, 1961.
6. King, E. O., and Tatum, H. W.: *Actinobacillus actinomycetemcomitans* and *Hemophilus aphrophilus,* J. Infect. Dis. **111:**85-94, 1962.
7. Merselis, J. G., Sellers, T. F., Jr., Johnson, J. E., and Hook, E. W.: *Hemophilus influenzae*

meningitis in adults, Arch. Intern. Med. **110**:837-846, 1962.

8. Philip, C. B., and Owen, C. R.: Comments on the nomenclature of the causative agent of tularemia, Internat. Bull. Bact. Nomen. Taxon. **11**:67-72, 1961.

9. Stoenner, H. G., and Lackman, D. B.: A new species of *Brucella* isolated from the desert wood rat, *Neatoma lepida* Thomas, Amer. J. Vet. Res. **18**:947-951, 1957.

10. Witorsch, P., and Gorden, P.: *Hemophilus aphrophilus* meningitis, Ann. Intern. Med. **60**:957-961, 1964.

23 Anaerobic gram-negative, nonsporeforming bacilli

Bacteroides
Fusobacterium

The identification of the anaerobic species of *Bacteroides* and *Fusobacterium* is based upon a number of morphological, physiological, and genetic characteristics much too detailed to be considered satisfactorily in a text of this type. The interested reader is strongly advised to consult the excellent anaerobic bacteriology manuals currently available, for full descriptions of technical procedures and methods of identification, including gas-liquid chromatography.[3,4,6,9]

Certain preliminary **grouping procedures**, however, utilizing observations on colonial and cellular morphology, susceptibility to antibiotic discs, and the results of a few simple biochemical tests, may be carried out; these data may then provide early and useful information to the clinician, while awaiting a more definitive identification.

These and other guides will be presented in the following sections.

PRELIMINARY GROUPING PROCEDURES

Assuming that a pure culture of an anaerobe on blood agar has been successfully isolated, as described by the methodology in Chapter 13, certain observations and tests should be made and recorded, as outlined by Sutter and colleagues[9]:

1. Determine the colonial morphology, pigmentation, presence or absence of hemolysis, or pitting of the agar medium.

2. Observe fluorescence under ultraviolet light (see under *B. melaninogenicus*).

3. Note the results of a Gram stain of the colony, including staining reaction and the morphology and the cells' particular arrangement.

4. Perform the spot indole reaction[10] by smearing a loopful of the growth from a trypticase soy blood agar plate on a filter paper saturated with 1% para-dimethylaminocinnamaldehyde in 10% hydrochloric acid.* A blue color indicates a **positive** reaction; no change or a yellow color indicates a **negative** reaction color (check with tube test).

5. Prepare a culture in thioglycollate medium (BBL-135 C) enriched with either 5% Fildes extract plus 1 mg./ml. sodium bicarbonate, or 25% ascitic fluid, incubate 4 to 6 hours or until barely visible growth ensues (half of the density of McFarland Standard No. 1).

6. Inoculate 0.5 ml. of the above on each of two blood agar plates with a capillary pipet and spread evenly over the surface with a cotton swab.

7. Using 3 discs per plate, place discs of colistin 10 µg., erythromycin 60 µg., kanamycin 1,000 µg., penicil-

---

*Stable up to 9 months when stored in a brown bottle at 4° C.

lin 2 units, rifampin 15 μg., and vancomycin 5 μg. on the surface of the seeded plates. Only the colistin and penicillin discs are available commercially—the others may be prepared from stock antibiotics (see Sutter & Finegold's[11] paper).

8. Incubate the blood agar plates anaerobically in a GasPak jar for 48 hours, and measure the diameters of the inhibition zones in millimeters. (Zones ≥ 10 mm = sensitive, ≤ 10 mm = resistant.)

9. Determine inhibition in bile (peptone–yeast extract–glucose broth [PYG] + 2% dehydrated oxgall + 0.1% sodium deoxycholate) by inoculating the above plus a tube of PYG without bile and deoxycholate and by observing inhibition of the growth.

10. Determine catalase activity by placing a drop of 3% hydrogen peroxide on a colony growing on a nonselective solid medium that does **not** contain blood. Perform this after the

Table 23-1. Group identification of nonsporeforming anaerobic gram-negative bacilli*

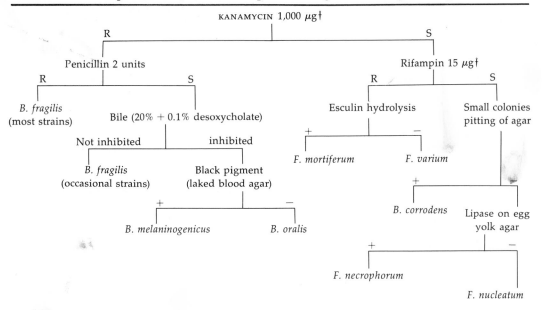

	B. CORRODENS	B. FRAGILIS	B. MELANIN-OGENICUS	B. ORALIS	F. NUCLEATUM F. NECRO-PHORUM	F. MORTI-FERUM	F. VARIUM
Confirmatory susceptibility tests							
Colistin 10 μg	S	R	V	V	S	S	S
Erythromycin 60 μg†	S	S	S	S	R	R	R
Vancomycin 5 μg	R	R	V	V	R	R	R
Other simple confirmatory tests							
Catalase	−	V	−	−	−	−	−
Indole production	−	V	V	−	+	−	+
Nitrate reduction	+	− (+)	−/(+)	−	−	−	−

*Permission for use granted by Sydney M. Finegold, M.D., and Vera L. Sutter, Ph.D. From *Anaerobic Bacteriology Manual*, published by Extension, University of California, Los Angeles, 1972. S = sensitive, zones ≥ 10 mm.; R = resistant, zones ≤ 10 mm.; + = positive; − = negative or absent; V = variable; (+) = some strains positive.

†These antibiotics may be obtained on written request from Baltimore Biological Laboratory, Cockeysville, Md. 21030. The others are available commercially from the same source.

plate has been exposed to the air for 30 minutes. The evolution of bubbles indicates a **positive** reaction.

From the results of the above tests, the group to which the isolant belongs is then determined (refer to Table 23-1). Additional biochemical tests and analysis of metabolic products by gas-liquid chromatography should be carried out for more precise speciation (see anaerobic manuals cited).

BACTEROIDES SPECIES
Bacteroides fragilis

B. fragilis is the anaerobe most frequently isolated from clinical infections; it is also the predominant organism of the normal human intestinal tract. The species has been further divided into numerous subspecies, including *B. fragilis* ss. *fragilis, distasonsis, ovatus, thetaiotaomicron, vulgatus,* and others. On blood agar, *B. fragilis* grows as a small (1 to 2 mm.), smooth, white to gray, nonhemolytic, translucent, glistening colony. A Gram stain shows a slender gram-negative bacillus with rounded ends; some are pleomorphic with filaments and bipolar staining. Bile is not inhibitory (it may be stimulatory); indole and catalase reactions are variable. *B. fragilis* is uniformly resistant to colistin, kanamycin, and vancomycin by the technique described in step 7, is usually resistant to penicillin, but is sensitive to erythromycin and rifampin. Subspeciation of *B. fragilis* requires a study of final pH in carbohydrates and metabolic products from growth in PYG by gas chromatography.[9]

Bacteroides melaninogenicus

This species of *Bacteroides* has been divided into three subspecies: *B. melaninogenicus* ss. *melaninogenicus, asaccharolyticus,* and *intermedius.* Normally part of the microflora of the oropharynx and upper respiratory tract, gastrointestinal tract, and genitourinary tract, it is less frequently isolated from human infections (an important pathogen in lung abscess) than is

B. fragilis.[8] Characteristically it produces a **brown to black colony** on blood agar after 5 to 7 days*; it grows more rapidly in media containing laked blood. Young colonies on blood agar enriched with hemin and vitamin K will show a **brick-red fluorescence** when examined under ultraviolet light.† By Gram stain, the organism appears coccobacillary and slightly pleomorphic in broth media. Bile is inhibitory to growth, catalase production is negative, and indole is variable; the organism is sensitive to erythromycin, penicillin, and rifampin, but is resistant to kanamycin and is variable to vancomycin by the technique described. Further biochemical tests and metabolic end product analysis are required for subspeciation.

Other Bacteroides species

B. oralis and *B. corrodens* (the obligately anaerobic variety)‡ are occasionally isolated from oral or pleuropulmonary infections[1] and are rarely isolated from blood cultures of bacteremic patients.[5] Colonies of *B. oralis* are similar to those of *B. fragilis,* while *B. corrodens* characteristically produces a pinpoint, translucent nonhemolytic colony that is depressed in the agar or **pits** the agar around the colony. Microscopically, *B. oralis* is similar to *B. fragilis* but is generally smaller; *B. corrodens* is bacillary with rounded ends, and rarely filamentous. Both species are inhibited by bile; indole and catalase are not produced. *B. oralis* is resistant to kanamycin but sensitive to erythromycin, penicillin, and rifampin; variable reactions occur with colistin and vancomycin. *B. corrodens* is sensitive to all of the above, except vancomycin. Other biochemical tests and analysis of metabolic products are required for definitive identification.

*Some gram-positive peptostreptococci also may produce black colonies—a Gram stain is required for differentiation.

†"Blak-Ray" ultraviolet lamp and viewbox from Ultra-Violet Products, San Gabriel, Calif.

‡Suggested new nomenclature for aerobic strains: *Eikenella corrodens* (Inter. J. Syst. Bact. **22**:73-77, 1972 and **23**:75-76, 1973).

FUSOBACTERIUM SPECIES

There are a number of species belonging to the genus *Fusobacterium;* included among those isolated from clinical specimens are *F. necrophorum* (formerly *Sphaerophorus necrophorus*), *F. nucleatum* (formerly *F. fusiforme*), and *F. mortiferum* (formerly *S. mortiferus*). Normally present in the upper respiratory tract, gastrointestinal tract, and genitourinary tract, the fusobacteria are increasingly responsible for serious pleuropulmonary, bloodstream, or metastatic suppurative infections (for example, abscesses of the brain, pelvic organs, lungs, and so forth).

Morphologically, members of this genus characteristically appear as long, slender, spear-shaped, gram-negative bacilli with **tapered ends**; some species, such as *F. mortiferum,* show a bizarre pleomorphism, with spheroid swellings along filaments and free round bodies. Colonies on blood agar are generally nonhemolytic (some are alpha or slightly beta) and vary from flat to convex, with opaque centers and translucent margins (for example, "fried egg" appearance of *F. mortiferum*). Convex, glistening alpha hemolytic colonies with a flecked internal structure are characteristic of *F. nucleatum.*

Bile is generally inhibitory and indole is produced (*F. mortiferum* is indole-negative and not affected by bile); *F. necrophorum* generally produces lipase on egg yolk agar. The fusobacteria are usually sensitive to colistin, kanamycin, and penicillin, but show variability with respect to erythromycin, rifampin, and vancomycin by Sutter's technique.[11] Other biochemical tests are required for precise identification.[9]

CLINICAL EFFECTIVENESS OF ANTIMICROBIAL AGENTS

The in vitro sensitivity testing of bacteroides and fusobacteria to various antimicrobial agents has not been uniformly standardized,* although reports of studies by various workers appear promising at this time.[7] Clinical studies, however, indicate that clindamycin (7-chlorolincomycin) is an effective agent* against a variety of these anaerobes.[2] Penicillin G remains the drug of choice for *B. melaninogenicus* infections, as well as for most of the fusobacteria, but is not indicated in infections due to *B. fragilis,* which is best treated with chloramphenicol. Metronidazole, a drug not yet available in the United States, may prove to be an excellent agent against serious *B. fragilis* infections, especially in bacterial endocarditis.[12] It should be noted that surgical drainage is an important part of the therapy of anaerobic infections.

---

*About one third of the strains of *F. varium* are clindamycin resistant (S. M. Finegold, personal communication, 1973).

REFERENCES

1. Bartlett, J. G., and Finegold, S. M.: Anaerobic pleuropulmonary infections, Medicine **51:**413-450, 1972.
2. Bartlett, J. G., Sutter, V. L., and Finegold, S. M.: Treatment of anaerobic infections with lincomycin and clindamycin, New Eng. J. Med. **287:**1006-1010, 1972.
3. Dowell, V. R., Jr., and Hawkins, T. M.: Laboratory methods in anaerobic bacteriology, Public Health Service Pub. No. 1803, Washington, D. C., 1968, U. S. Government Printing Office.
4. Dowell, Jr., V. R., and Thompson, F. S.: Identification of acid and alcohol products of anaerobic bacteria by gas liquid chromatography, Public Health Service, Center for Disease Control, May 1971.
5. Felner, J. M., and Dowell, V. R., Jr.: "Bacteroides" bacteremia, Amer. J. Med. **50:**787-796, 1971.
6. Holdeman, L. V., and Moore, W. E. C., editors: Anaerobe laboratory manual, Blacksburg, Va., 1972, Staff of the Anaerobe Laboratory, Virginia Polytechnic Institute and State University.
7. International Conference on Anaerobic Bacteria, Center for Disease Control, Atlanta, Ga., Nov. 27-29, 1972.
8. Moore, W. E. C., Cato, E. P., and Holdeman, L. V.: Anaerobic bacteria of the gastrointestinal flora and their occurrence in clinical infections, J. Infect. Dis. **119:**641-649, 1969.
9. Sutter, V. L., Attebery, H. R., Rosenblatt, J. E., Bricknell, K. S., and Finegold, S. M.: Anaerobic bacteriology manual, Los Angeles, 1972, Dept.

---

*For a further discussion of this topic, see Chapter 34.

of Continuing Education and Health Sciences, School of Medicine, UCLA.

10. Sutter, V. L., and Carter, W. T.: Evaluation of media and reagents for indole-spot tests in anaerobic bacteriology, Amer. J. Clin. Path. **58:**335-338, 1972.

11. Sutter, V. L., and Finegold, S. M.: Antibiotic disc susceptibility tests for rapid presumptive identification of gram-negative anaerobic bacilli, Appl. Microbiol. **21:**13-20, 1971.

12. Tally, F. P., Sutter, V. L., and Finegold, S. M.: Metronidazole versus anaerobes: in vitro data and initial clinical observations, Calif. Med. **117:**22-26, 1972.

24 Anaerobic gram-positive, nonsporeforming bacilli

Bifidobacterium
Propionibacterium
Eubacterium
Lactobacillus
Actinomyces

The identification and speciation of this group of anaerobes is somewhat difficult, since the taxonomy of its constituents has been confusing in the past. For example, some agents of actinomycosis or other human infections, such as *Actinomyces eriksonii* and *A. propionicus,* have been reclassified as *Bifidobacterium eriksonii* and *Arachnia propionica** as a result of chemical analyses of their cell walls and for other reasons. Currently, the criteria used in the classification of these anaerobes includes their microscopic morphology, results of biochemical tests, and analysis by gas-liquid chromatography of the fermentation products of glucose. The interested reader is referred to Holdeman and Moore's excellent manual on anaerobes for further details of these procedures.[4] Significant members of each genus are described below.

BIFIDOBACTERIUM ERIKSONII

This organism is part of the normal human intestinal microflora, and occurs generally in mixed infections. *B. eriksonii* grows as a white, convex, shiny colony with an irregular edge. In thioglycollate medium, growth is diffuse, and the Gram stain shows a diphtheroid to filamentous bacillus, branched or bifurcated. Catalase and indole are not produced, nitrate is not reduced; gelatin is not liquefied, but esculin is hydrolysed. Carbohydrate fermentation reactions and metabolic end products are characteristic.[4,6]

PROPIONIBACTERIUM ACNES (CORYNEBACTERIUM ACNES)

This anaerobe* produces propionic acid by fermentation of glucose and is part of the resident flora of normal skin; consequently, it is the most frequent contaminant of blood cultures. *P. acnes* characteristically grows as a small, white to pinkish, shiny to opaque colony with entire margins. A Gram stain of the colony shows slender, slightly curved rods, sometimes with false branching or a beaded appearance. Tests for catalase and indole production are positive; nitrates are reduced, gelatin is liquefied, and esculin is hydrolysed. Characteristic patterns of carbohydrate fermentation and organic acid end products are obtained.[4,6]

EUBACTERIUM SPECIES

Eubacterium species are infrequently isolated from wounds and are generally associated with other anaerobes or facultative bacteria; they are part of the normal fecal microflora. Growing as raised to convex, translucent to opaque colonies, they appear microscopically as pleomorphic bacillary to coccobacillary forms, occurring in

---

*This organism has recently been incriminated in human actinomycosis.[2]

---

*Some strains are microaerophilic.

pairs and short chains. *E. lentum* is relatively inactive biochemically; *E. limosum* hydrolyzes esculin and ferments several carbohydrates. Characteristic metabolic end products are observed.

LACTOBACILLUS SPECIES

These anaerobic members of the genus *Lactobacillus* produce primarily lactic acid from the fermentation of glucose and are only occasionally involved in human infections, usually of pleuropulmonary origin. One species, *L. catenaforme,* was formerly classified as a member of the genus *Catenabacterium;* however, some strains of the latter were shown to produce spores, and therefore belong to the clostridia.[4] *Lactobacillus catenaforme* grows as a convex, translucent colony, that on microscopic examination appears as pleomorphic gram-positive bacilli, sometimes in chains. Terminal swellings are also observed.

ACTINOMYCES SPECIES

The etiological agent of human **actinomycosis,** *A. israelii,* and the infrequent pathogen, *A. naeslundii,* are both found as a part of the normal microbiota of the mouth, and produce infection primarily as endogenous opportunists. While *A. israelii* grows best under anaerobic conditions, *A.*

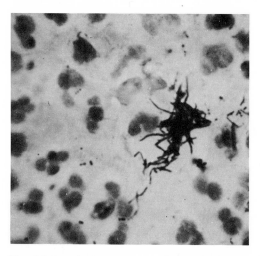

Fig. 24-1. *Actinomyces israelii,* crushed sulfur granule, showing hyphal segments and pus cells. (Gram strain; 4000×.)

naeslundii strains are either microaerophilic or facultative. *A. israelii* produces a characteristically heaped, rough, lobate colony resembling a **molar tooth;** microscopically it appears as long, filamentous, gram-positive bacilli, some of which may show branching. A Gram stain of "**sulfur granules**" from actinomycotic pus reveals a similar morphology, often with diphtheroid and coccal forms. *A. naeslundii* colonies generally are smooth, flat to convex, whitish, and transparent, with entire margins. Microscopically, they are gram-positive bacilli consisting of diphtheroidal forms and branched elements; some filaments may show clubbed ends.[3]

Both species of *Actinomyces* are catalase and indole negative, and may be identified by other biochemical tests; more recently, direct identification by immunofluorescent techniques have proved successful.[5]

Penicillin G remains the drug of choice in the treatment of infections caused by *Actinomyces* species, as well as in the treatment of those caused by *Eubacterium* species; tetracycline, lincomycin, and clindamycin also appear to demonstrate activity against these organisms.[1]

REFERENCES

1. Bartlett, J. G., and Finegold, S. M.: Anaerobic pleuropulmonary infections, Medicine **51:**413-450, 1972.
2. Brock, D. W., Georg, L. K., Brown, J. M., and Hicklin, M. D.: Actinomycosis caused by *Arachnia propionica;* report of 11 cases, Amer. J. Clin. Path. **59:**66-77, 1973.
3. Coleman, R. M., Georg, L. K., and Rozzell, A. R.: *Actinomyces naeslundii* as an agent of human actinomycosis, Appl. Microbiol. **18:**420-426, 1969.
4. Holdeman, L. V., and Moore, W. E. C., editors: Staff of the Anaerobe Laboratory, Anaerobe laboratory manual, Blacksburg, Va., 1972, Virginia Polytechnic Institute and State University.
5. Lambert, Jr., R. F., Brown, J. M., and Georg, L. K.: Identification of *Actinomyces israelii* and *Actinomyces naeslundii* by fluorescent antibody and agar-gel diffusion techniques, J. Bacteriol.**94:**1287-1295, 1967.
6. Sutter, V. L., Attebery, H. R., Rosenblatt, J. E., Brickwell, K. S., and Finegold, S. M.: Anaerobic bacteriology manual, Los Angeles, 1972, Dept. of Continuing Education and Health Sciences, School of Medicine UCLA.

25 Aerobic gram-positive, spore-forming bacilli

In this chapter, discussion of the clinical significance of the gram-positive spore-forming bacilli will be limited to certain species of *Bacillus.* The genus contains a large number of species that are aerobic, usually gram-positive, and spore forming. They are widely distributed in nature and are therefore frequent contaminants in laboratory cultures from clinical specimens.

It should be noted that certain species of *Bacillus,* other than the well-recognized pathogen *B. anthracis,* can be involved, even though infrequently, in human disease processes. *B. cereus, B. circulans, B. pumilus, B. sphaericus,* and *B. subtilis* have been variously incriminated in cases of meningitis, pneumonia, and septicemia as reported by different authors.[1,3,5,7] One fatal case of pneumonia with a bacteremia due to *B. cereus* in a patient with subacute lymphocytic leukemia has been reported by Coonrod and co-workers.[4] These authors have reported on an interesting study of the antibiotic susceptibility of *Bacillus* species in which they revealed that all species tested were susceptible to tetracycline, kanamycin, gentamicin, and chloramphenicol. Susceptibility to penicillin G, ampicillin, methicillin, and cephalothin was species related, but high for *B. subtilis,* intermediate for *B. pumilus,* and low for *B. cereus.* Such findings may indicate caution in the form of therapy used for patients with a low resistance.

B. cereus colonies on laboratory media will vary from the small, shiny, compact colonies to the large, feathery, spreading type found in *B. cereus* var. *mycoides. B. subtilis* colonies are normally large, flat, and dull with a ground-glass appearance.

BACILLUS ANTHRACIS

B. anthracis is the chief human pathogen in the genus *Bacillus,* but it is seldom encountered in the average hospital or public health laboratory. Nevertheless, because of its importance in some areas, its identification and pathogenicity should be discussed.

The organism is an aerobic, large, nonmotile rod, with an ellipsoidal to cylindrical spore. The sporangium is usually not swollen. The cells frequently occur in long chains, giving a **bamboo** appearance, especially on primary isolation from infected tissue or fluid discharge. The chains of virulent forms are usually surrounded by a **capsule.** Encapsulation will occur also in enriched media. Avirulent forms are usually nonencapsulated. Sporulation occurs in the soil and on inanimate media, but not in living tissue.

The colonies of *B. anthracis* are normally large, opaque, raised, and irregular, with a **curled** margin. On sheep blood agar, the colonies are invariably **nonhemolytic.** Smooth and rough colony forms may be observed, and both of these may be virulent.

The optimal temperature for growth is 36° C., and when it is cultivated at 42° to 43° C., the organism becomes **attenuated** or avirulent. This was shown by Louis Pasteur years ago. The loss of virulence is attributed to loss of the capsule. In broth, the bacillus produces a heavy pellicle with little if any subsurface growth. Biochemically the organism is characterized as follows:

> Carbohydrate fermentation—Dextrose, fructose, maltose, sucrose, and trehalose fermented with acid only; arabinose, galactose, lactose, mannose, raffinose, rhamnose, adonitol, dulcitol, inositol, inulin, mannitol, and sorbitol not fermented
> Gelatin—Inverted pine-tree growth; liquefaction
> Nitrates—Reduced to nitrites
> Starch—Hydrolyzed
> Voges-Proskauer—Positive

Pathogenicity of Bacillus anthracis. The organism is the cause of **anthrax,*** which in man may be manifested in three forms:

1. **Cutaneous anthrax** (the most common form in the United States). Infection is initiated by the entrance of bacilli through an abrasion. A malignant pustule usually appears on the hands and forearms. The bacilli are readily recognized in the serosanguineous discharge.
2. **Pulmonary anthrax,** or woolsorters' disease. The bacilli may be found in large numbers in the sputum. Spores are inhaled during shearing or sorting. This form can readily progress to fatal septicemia.
3. **Violent enteritis** (the most severe and rare form). The bacilli or spores are swallowed, thus initiating an intestinal infection. The organisms may be isolated from the stools. This form is also usually fatal.

The pathogenicity of the organism is determined most efficiently by injecting either 1 ml. of a centrifuged and washed 24-hour broth culture intraperitoneally into a guinea pig or 0.25 to 0.5 ml. into each of several white mice. Animals succumb in 2 to 4 days after inoculation, but they may remain asymptomatic until a few hours before death when labored breathing starts and respiratory collapse follows. Death is caused by a **septicemia,** and the organism is readily recovered from the heart, blood, spleen, liver, and lungs of the animal.

Ascoli test. The Ascoli test is a **precipitin test** used to diagnose anthrax in dead animals and to determine whether or not hides for industrial use have been removed from anthracized animals. The **antigen** is prepared by boiling a small piece of **spleen** or hide in 5 to 10 ml. of physiological saline for 15 minutes. This is cooled and filtered to clarify. One milliliter of the filtered antigen is then layered carefully over an equal volume of **antianthrax** serum, which has been prepared in rabbits against the encapsulated anthrax organism, using an agglutinating or capillary tube. A positive reaction is recognized by the formation of a **ring** of precipitate at the interface of the two reacting substances. The antigenic material is probably the high molecular weight **polypeptide,** contained in the capsule, which has diffused through the tissues.

REFERENCES

1. Allen, T. B., and Wilkinson, H. A.: A case of meningitis and generalized Schwartzman reaction caused by *Bacillus sphaericus,* Johns Hopkins Med. J. **125:**8-13, 1969.
2. Ascoli, A.: Der Ausbau meiner Präzipitinreaktion zur Milzbranddiagnose, Z. Immunitaetsforsch, Exper. Ther. **11:**103, 1911.
3. Curtis, J. R., et al.: *Bacillus cereus* bacteremia; a complication of intermittent haemodialysis, Lancet **1:**136-138, 1967.
4. Coonrod, J. D., et al.: Antibiotic susceptibility of *Bacillus* species, J. Infect. Dis. **123:**102-105, 1971.
5. Farar, W. E., Jr.: Serious infections due to "nonpathogenic" organisms of the genus *Bacillus;* review of their status as pathogens, Amer. J. Med. **34:**134-141, 1963.
6. Hospital of the University of Pennsylvania: A symposium on anthrax in man, Philadelphia, 1954, University of Pennsylvania Press.
7. Stopler, T. V., et al.: Bronchopneumonia with lethal evolution determined by a microorganism of the genus *Bacillus (B. cereus),* Romanian Med. Rev. **19:**7-9, 1969.

---

*Much valuable information on the anthrax syndrome may be gained from a 1954 symposium.[6]

26 Anaerobic gram-positive, spore-forming bacilli

The genus *Clostridium* contains the anaerobic, spore-forming true bacteria. There is a large number of species involved, and the majority are **obligate anaerobes.** Some species are said to be aerotolerant, showing sparse growth on enriched media in the presence of small amounts of oxygen. The sporangia are often characteristically swollen (Fig. 26-1), showing spindle, drumstick, and "tennis-racket" forms and containing central, subterminal, and terminal spores. With rare exceptions in the aerotolerant forms, the enzymes catalase, cytochrome-oxidase, and peroxidase are not produced. Many attack carbohydrates, and some are proteolytic.

True **exotoxins** are produced by the pathogenic clostridia, several species of which are important in medical science. The organisms are widely distributed in soil, dust, and water and are common inhabitants of the intestinal tract of animals, including man. They are often found in unclean wounds and wound infections. The spores of some pathogenic species may appear in improperly home-canned produce in which they can develop vegetatively under normal domestic conditions, in inefficiently sterilized surgical dressings and bandages, in plaster of Paris for casts, and on the clothing and skin of man.

GENERAL METHODS OF ISOLATION AND CULTIVATION OF PATHOGENIC CLOSTRIDIA

Many of the procedures for anaerobic culture that have been discussed in Chapter 13 are applicable to the clostridia. It should be reemphasized here that any material for anaerobic cultivation should be inoculated **immediately,** using prereduced media, without awaiting the results of aerobic culture. If this is impractical, the material may be held at room temperature in a tube of cooked meat with a petrolatum seal, and from this, plates may subsequently be inoculated. The use of thioglycollate medium is highly recommended for primary isolation. **Freshly prepared** blood agar plates may be streaked directly with the specimen, depending on its nature, or they may be streaked from the thioglycollate culture. Regardless, the plates are incubated anaerobically.

In most instances, the clostridia occur in mixed culture with gram-negative bacteria such as coliforms, *Proteus,* or *Pseudomonas.* The anaerobic plates, therefore, may be overgrown with such organisms, and isolation of the clostridia becomes difficult or impossible. To avoid the use of selective plating media, either of the two following methods may be used:

1. If the spore-forming anaerobe will sporulate in thioglycollate medium, the original culture, shown to contain gram-positive sporulating rods, may be **heat shocked** at 80° C. for 30 minutes and then streaked on blood agar plates for anaerobic cultivation. Alternatively, a fresh tube of thioglycollate medium may be inoculated from the original culture, heated immediately, and incubated for 24 to 48

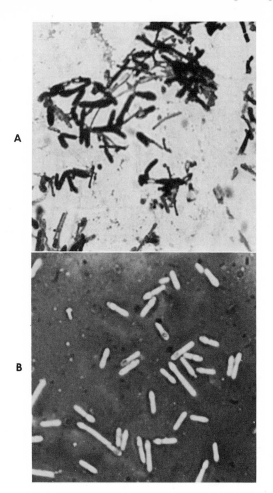

Fig. 26-1. A, *Clostridium botulinum,* showing the rarely occurring swollen sporangium with subterminal spores. **B,** *Clostridium sporogenes,* showing swollen sporangium with subterminal spores. (Both 4000×.)

hours. From this, the plates may be streaked.

2. Another method may be applied when *Clostridium perfringens* is suspected, since this organism seldom sporulates in thioglycollate medium. For the isolation of this species from mixed culture, the material may be placed in thioglycollate medium containing 0.6% glucose. After 48 to 72 hours at 36° C., the number of gram-negative bacteria is reduced considerably by the acid produced and, by plating this culture on blood

agar, recovery of the clostridia is enhanced. Discrete colonies should be transferred to thioglycollate medium for pure culture and final identification by biochemical methods.

Cooked-meat medium also may be used for cultivation of the spore-forming anaerobes and is preferred by some workers. Tubes of this medium require a petrolatum seal if obligate anaerobes are being cultivated. Proteolytic activity, recognized by digestion of the solid material of the medium, may be detected quite readily by this method, and the medium, therefore, offers certain advantages. Egg yolk or lecithovitellin medium is also a useful identification medium.[2]

The **Nagler** reaction[3] is recommended for the rapid identification of *Cl. perfringens.* The medium for this reaction, 10% egg yolk in blood agar base, is placed in a Petri dish, and one half the surface is smeared with a few drops of *Cl. perfringens* type A antitoxin* (antilecithinase). The culture is then streaked in a single line diagonally across the plate. The toxin lecithinase produces a precipitate (opalescence) in the absence of antitoxin but it is inhibited by its presence.†

IDENTIFICATION OF THE PATHOGENIC CLOSTRIDIA

By virtue of the disease syndromes they initiate, the pathogenic species may be placed in three categories or groups:

Group I, the **gas gangrene** group includes *Cl. perfringens* (type A), *Cl. novyi, Cl. septicum, Cl. sporogenes, Cl. bifermentans (sordellii), Cl. histolyticum,* and others. Of these, the three species first named are most important.

Group II, the *Cl. tetani* group.

Group III, the *Cl. botulinum* group.

In order to differentiate between these

---

*Available from Wellcome Reagents Division, Burroughs Wellcome Co., Research Triangle Park, N. C.
†*Cl. bifermentans, Cl. sordelli,* and *Cl. barati* also produce alpha toxin and will give a positive Nagler reaction.

species, special media are required. For carbohydrate fermentation, sugar-free thioglycollate medium or trypticase agar base may be used as the base to which carbohydrates may be added. To study liquefaction in gelatin and the reaction in litmus milk media, 0.05% sodium thioglycollate may be added as a reducing agent to allow growth of the organisms.

On the basis of their reactions in dextrose and lactose the clostridia may be placed in three fermentative groups, as shown in Diagram 4.

The media* recommended for species differentiation are listed on p. 188.

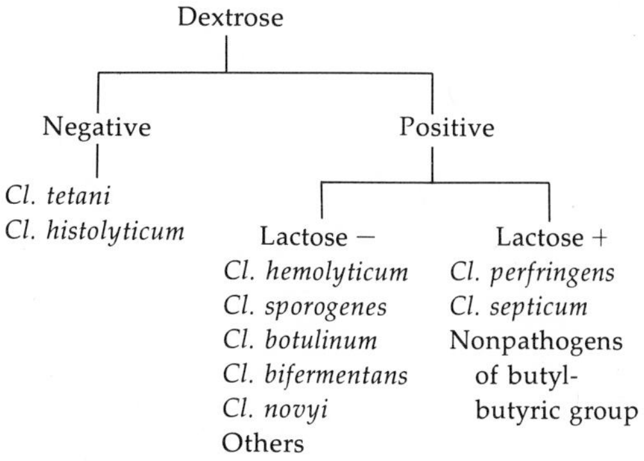

```
                    Dextrose
                       |
          ┌────────────┴────────────┐
      Negative                   Positive
                                     |
   Cl. tetani              ┌─────────┴─────────┐
   Cl. histolyticum    Lactose −            Lactose +
                       Cl. hemolyticum      Cl. perfringens
                       Cl. sporogenes       Cl. septicum
                       Cl. botulinum        Nonpathogens
                       Cl. bifermentans     of butyl-
                       Cl. novyi            butyric group
                       Others
```

Diagram 4. Fermentative groups of *Clostridium* species.

*Baltimore Biological Laboratory, Cockeysville, Md.

Table 26-1. Distinguishing characteristics of clostridia*

	CL. PER-FRINGENS	CL. SEPTICUM	CL. NOVYI	CL. BIFER-MENTANS	CL. SPORO-GENES	CL. BOTULI-NUM A, B	CL. HISTO-LYTICUM	CL. TETANI
Colonies on blood agar	Round, smooth, opaque	Medium, rhizoid	Medium, rhizoid	Crenated to ameboid	Large, rhizoid	Granular translucent, fimbriate	Smooth, transparent	Spreading, translucent, filamentous
Hemolytic zones	Double	Single	Single	Single	Single	Single	Single	Single
Spores	Rare, ovoid, eccentric†	Ovoid, eccentric	Ovoid, eccentric	Ovoid, eccentric	Ovoid, eccentric	Ovoid, eccentric	Ovoid, subterminal	Round, terminal, drumstick
Sporangia	Not swollen	Slightly swollen	Slightly swollen	Not swollen	Swollen	Swollen	Swollen	Swollen
Motility	−	+	+	+	+	+	+	+
Cooked-meat medium	Gas; no digestion	Gas; no digestion	Gas; no digestion	Gas; blackening	Gas; blackening, digestion	Gas; blackening, digestion	Gas; blackening	Slight gas; slow blackening
Dextrose	+	+	+	+	+	+	−	−
Lactose	+	+	−	−	−	−	−	−
Sucrose	+	−	−	−	−	−	−	−
Salicin	Usually −	+	−	+(−)	−	−	−	−
Indole	−	−	−	+	−	−	−	−
Nitrate reduction	+	+	−	−	−	−	−	−
Gelatin liquefaction	+	+	+	+	+	+	+	+

*Abstracted from a chart prepared by Harriette D. Vera, Baltimore Biological Laboratory, Cockeysville, Md.
†Most strains do not form spores readily; alkaline high-protein media are required.

1. Trypticase agar base—for motility, spore formation, and stock culture
2. Trypticase dextrose, lactose, sucrose, salicin agar—for fermentation
3. Thioglycollate plus gelatin*—for gelatin liquefaction
4. Trypticase blood agar—for hemolysis
5. Trypticase nitrate broth—for indole production and nitrate reduction
6. Cooked-meat medium under petrolatum—for proteolysis

Tubes should be deep filled with the medium and heated and cooled without agitation just prior to use. Discs containing the desired carbohydrates† are added aseptically, and the test culture is inoculated by the stab technique. All carbohydrate media should be examined periodically during the first 24 hours, since prolonged incubation can lead to color changes in the indicator and give erroneous results. Anaerobic jars are not needed with the foregoing media if handled as directed.

The cultural, morphological, and biochemical characteristics are summarized in Table 26-1. Although *Cl. perfringens, Cl. novyi, Cl. tetani,* and *Cl. botulinum* are the most important species, *Cl. sporogenes, Cl. sordellii, Cl. histolyticum,* and *Cl. septicum* are also given in the table, since they are often associated with the more highly pathogenic species in gangrenous infections.

Most of the human pathogenic clostridia are pathogenic also for guinea pigs, mice, rabbits, and pigeons. To establish the toxigenicity of isolated strains, white mice and guinea pigs are normally used in the laboratory. Susceptibility and rapidity of death may vary according to the virulence of the strain.

DETERMINATION OF TOXIGENICITY IN CLOSTRIDIA
Clostridium botulinum

Thioglycollate medium, cooked-meat medium, or chopped meat containing glucose and starch is inoculated with *Cl. botulinum* and incubated for 48 to 72 hours at 28° to 30° C. (optimal temperature for toxin production). The culture should be centrifuged; the supernatant fluid is decanted, and a few drops are given orally to a mouse or guinea pig with a syringe or plastic pipet. The animal is observed over a period of a few days. The usual symptoms of intoxication are the development of flaccid paralysis and death. First symptoms may appear in 24 to 48 hours, depending on the dosage. Type identification may be carried out by mixing aliquots of a culture with known antitoxin-containing sera prior to injection. Controls must be included.

Clostridium perfringens

Cl. perfringens is also cultured in thioglycollate or chopped meat medium containing glucose, but it is incubated at 36° C. for 18 to 24 hours. Young cultures are more toxic than older ones. The injection of 0.5 ml. of 10% lactic acid or 10% calcium chloride intramuscularly into the thigh or groin of a guinea pig is recommended for producing a mild tissue necrosis. With a syringe 0.5 ml. of the fluid culture is removed and injected at the same site into the animal. The animal is observed over a period of 1 week. At necropsy the skin at the site of injection is tight and reddish purple in color. The subcutaneous tissue is usually broken down and crepitant due to the presence of gas bubbles. An incision will release a thin red fluid and a foul-smelling gas. The toxin that produces the syndrome is a collagenase, called **kappa toxin,** that acts on the collagen of certain tissues.

The typing[4] of *Cl. perfringens* strains may be performed by protecting mice with known antitoxins, A, B, C, D, E, and F, respectively, prior to injection of a challenge culture.

Clostridium tetani

Cl. tetani may be cultivated in the same way as the other species and incubated at

---

*Thiogel, Baltimore Biological Laboratory, Cockeysville, Md.

†Taxo carbohydrate discs, Baltimore Biological Laboratory, Cockeysville, Md.

36° C. for 48 to 72 hours. A mouse is inoculated intramuscularly with 0.1 to 0.2 ml. of the liquid after an injection of the same volume of 10% lactic acid or 10% calcium chloride. The animal is observed over a period of 1 week, although symptoms usually appear in 2 to 4 days. Rigidity of the tail and the inoculated limb precedes death. Some investigators prefer to use the intraperitoneal route in lieu of the intramuscular route.

All of the clostridial toxins, except those of *Cl. tetani* and *Cl. botulinum,* will kill mice rapidly; a neurotoxin may require a number of days to cause death.

For a further study of the clostridia, the reader is referred to Sterne and van Heyningen's excellent discussion in Dubos and Hirsch's text[5] and also to Dowell and Hawkins' recommendations.[1]

REFERENCES

1. Dowell, V. R., and Hawkins, T. M.: Detection of clostridial toxins, toxin neutralization tests, and pathogenicity tests, Atlanta, 1968, Center for Disease Control.
2. McClung, L. S., and Toabe, R.: The egg yolk plate reaction for the presumptive diagnosis of *Clostridium sporogenes* and certain species of gangrene and botulinum, J. Bact. **53:**139-147, 1947.
3. Nagler, F. P. O.: Observations on a reaction between the lethal toxin of *Cl. welchii* (type A) and human serum, Brit. J. Exp. Path. **20:**473, 1939.
4. Oakley, C. L., and Warrack, G. H.: Routine typing of *Clostridium welchii,* J. Hyg. **51:**102-107, 1953.
5. Sterne, M., and van Heyningen, W. E.: The clostridia. In Dubos, R. J., and Hirsch, J. G.: Bacterial and mycotic infections of man, ed. 4, Philadelphia, 1965, J. B. Lippincott Co.

27 Gram-positive, nonsporeforming bacilli

Corynebacterium
Listeria
Erysipelothrix

Although there is some degree of pleomorphism in the three genera, this characteristic is perhaps most prominent in *Corynebacterium*.

GENUS CORYNEBACTERIUM

The corynebacteria (Gk. "koryne," a club) are gram-positive, nonsporulating, nonmotile (with occasional exceptions) rods. They are often club shaped and frequently banded or beaded with irregularly staining granules (Fig. 27-1). They frequently exhibit characteristic arrangements, resembling Chinese letters and palisades. The corynebacteria are generally aerobic, but microaerophilic and anaerobic species do occur. They have a wide distribution in nature. Some species are parasites or pathogens of plants and domestic animals; others are part of the normal human respiratory flora. The type species, *Corynebacterium diphtheriae*, produces a powerful exotoxin that causes **diphtheria** in man.

Corynebacterium diphtheriae

C. diphtheriae (Klebs-Loeffler bacillus) is an aerobic, nonmotile, slender, gram-positive rod that is highly pleomorphic. In addition to straight or slightly curved bacilli, club-shaped and branching forms are seen. The rods generally do not stain uniformly with methylene blue, but show alternate bands of stained and unstained material (septa) as well as deeply stained granules (metachromatic granules) that give the organism a beaded appearance. Individual cells tend to lie parallel or at acute angles to each other, resulting in V, L, or Y shapes. Although the appearance of *C. diphtheriae* in stained smears is highly characteristic, it **should not be identified by morphology alone;** many diphtheroids and actinomycetes stain in the same irregular fashion and are also pleomorphic.

On primary isolation, *C. diphtheriae* should be cultivated on enriched media, such as infusion agar with added blood; the most characteristic morphological forms, however, are found in smears from a 12- to 18-hour culture on Loeffler serum medium (Chapter 39), which seems to enhance pleomorphism. The addition of potassium tellurite to blood or chocolate agar provides a differential and selective medium. Contaminants are inhibited, and after one day *C. diphtheriae* appears as characteristic gray or black colonies. This characteristic aids in distinguishing the organism in mixed cultures and helps to differentiate the three types: *gravis, mitis,* and *intermedius*. Their characteristics are summarized in Table 27-1. The diphtheria bacilli on tellurite media are shorter, staining is more uniform, and the granules are less readily seen than when grown on Loeffler medium. For this reason most laboratories use **both** media.

In nature, *C. diphtheriae* occurs only in

the respiratory tract (rarely on the skin or in wounds) of infected persons or healthy carriers. The organisms are spread from these areas to susceptible individuals by droplets or direct contact. The initial focus

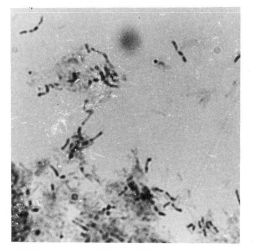

Fig. 27-1. *Corynebacterium diphtheriae,* granule stain. (4000×.)

of growth is on the mucous membrane of the respiratory tract where the exotoxin is produced. This metabolite is then absorbed by the body which, leads to the pathological syndrome that characterizes the disease.

Virulence test. All toxigenic strains of *C. diphtheriae* elaborate the same disease-producing exotoxin, and the one reliable criterion for identifying a diphtheria bacillus is **its ability to produce diphtheria toxin.** There are two methods for determining the toxigenicity of a suspect strain of *C. diphtheriae:* one in vivo and the other in vitro.

The **in vivo** test that yields satisfactory results for most investigators is the one prescribed by Fraser and Weld.[3] This is outlined below, and it may be applied to either a guinea pig or a white rabbit. The latter will permit a greater number of tests.

1. Inoculate a Loeffler slant and a tube of sugar-free infusion broth from a suspect colony of each culture to be

Table 27-1. Characteristics of *Corynebacterium diphtheriae* types and other corynebacteria

ORGANISM	MORPHOLOGY	COLONIES ON BLOOD-TELLURITE AGAR	HEMO-LYSIS	STARCH AND GLY-COGEN	GLUCOSE	SUCROSE	TOXI-GENIC
C. diphtheriae type *gravis**	Short, evenly staining	Large, dark gray, matt, striated ("daisy-head"), irregular	−	+	+	−	+
C. diphtheriae type *mitis*	Long, with many metachromatic granules	Small, black, shiny, convex, entire	+	−	+	−	+
C. diphtheriae type *intermedius*	Long, barred forms with clubbed ends	Small, flat, dry, gray with black, raised center	−	−	+	−	+
C. pseudodiph-theriticum†	Short, thick, with 1 or 2 unstained septa; granules absent	Resembles *mitis* type; not as dark	−	−	−	−	−
C. xerosis†	Rods showing polar staining; occasionally clubbed; resembles *C. diphtheriae*	Resembles *mitis* type; not as dark	−	−	+	+	−

*There is apparently no correlation between type and severity of disease in the United States.
†Diphtheroids, normal inhabitants of human beings; not associated with disease.

tested. Incubate both at 36° C. Check the slant culture for purity by staining after 24 hours. Use the broth culture after 48 hours of incubation for the test.

2. Prepare the test animal by clipping the back and sides closely. Disinfect the clipped area with alcohol or some suitable disinfectant. Make a line down the backbone with an indelible pencil and mark off a series of 2-cm.-square areas on either side of the line.

3. Using a 2-ml. syringe, graduated in tenths of a milliliter, fitted with a 24-gauge needle, take up 1 to 2 ml. of each 48-hour culture and inject 0.2 ml. intracutaneously into the marked squares respectively, leaving in each case the square immediately below for the control. Refrigerate the syringe and contents for later use. **Inject a similar amount of a known toxigenic strain into one square as a positive control.**

4. After 5 hours, inject 500 units of diphtheria antitoxin intravenously into the rabbit's ear and wait 30 minutes. If a guinea pig is used, inject the antitoxin intraperitoneally.

5. Using the refrigerated syringe cultures, inject 0.2 ml. of each culture into the corresponding square immediately **below** the original test site. This provides the **postantitoxin control.**

6. Read at 24 and 48 hours because the lesions are usually fully developed after the longer interval. A **toxigenic strain** will produce a **central necrotic area,** about 5 to 10 mm. in diameter, surrounded by an erythematous zone somewhat larger. The corresponding **control** should show only a **pinkish swollen area** of 5 to 10 mm. in diameter, without showing any necrosis.

7. If both test sites show necrosis, toxigenicity of the culture on test cannot be confirmed. This may be due to the fact that the culture was not toxigenic or was not pure, that the antitoxin

was either insufficient or mislabeled, or that the test culture was some other species of *Corynebacterium,* such as *C. ulcerans.*

The **in vitro** test, first reported by Elek,[2] has been modified by a number of workers. The procedure recommended here is the modification of Hermann and colleagues.[5] To an autoclaved and cooled basal medium (proteose peptone agar*) is added a nonserous enrichment (Tween 80, glycerol, and casamino acids*) and potassium tellurite in a Petri dish. This is then thoroughly mixed. Before the agar hardens, a 1 by 8 cm. sterile paper strip saturated with diphtheria antitoxin* is placed on the agar surface and pushed below the surface with sterile forceps. The plate is dried at 36° C. to ensure a moisture-free surface. A loopful of the culture of suspected *C. diphtheriae* is then streaked across the plate perpendicular to the paper strip in a single line, as shown in Fig. 27-2. Four to five cultures may be tested on a single plate. The plate is incubated at 36°

---

*Difco Laboratories, Detroit.

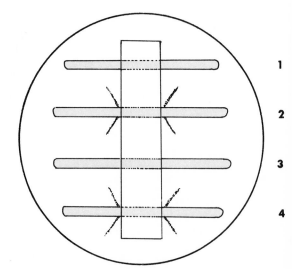

Fig. 27-2. Elek method for in vitro demonstration of toxigenicity of *Corynebacterium diphtheriae.* Filter paper strip impregnated with antitoxin: *1,* known nontoxigenic strain (negative control); *2,* known toxigenic strain (positive control); *3,* unknown culture that is nontoxigenic; *4,* unknown culture that is toxigenic.

C. for 1 to 3 days and examined. The antitoxin from the paper strip and any toxin produced by the growing diphtheria cultures will diffuse through the medium and react in optimal proportions to produce thin lines of precipitate, as shown. These lines arise at angles of approximately 45 degrees from those strains that are toxigenic. The test should include known **positive** and **negative** controls.

Brief mention should be made of the application of the FA test for the detection of *C. diphtheriae* in suspected cases of diphtheria. The test appears useful in the rapid **presumptive** diagnosis of diphtheria, but should be used in conjunction with conventional cultural procedures.[1]

DIPHTHEROID BACILLI

As their collective name implies, diphtheroid bacilli are morphologically similar and sometimes indistinguishable from the diphtheria bacilli (Table 27-1). Three species are observed variously in man: *Corynebacterium pseudodiphtheriticum,* which occurs in the normal throat and is not pathogenic for human beings or laboratory animals; *C. xerosis,* isolated from normal and diseased conjunctivae (its etiological role is questionable); and *C. acnes,** a microaerophile found in acne pustules but also occurring in other types of lesions as a saprophyte. None of these corynebacteria produces a soluble toxin.

GENUS LISTERIA

The genus *Listeria* is also a member of the family Corynebacteriaceae, and it contains the clinically significant species, *L. monocytogenes.* It is a small, motile, nonsporulating, nonencapsulated, gram-positive rod, 0.5 μ by 1 to 2 μ in size, isolated originally from rabbits with a disease characterized by a great increase in circulating mononuclear leukocytes. It has been suggested also as a cause of mononucleosis in man. The syndrome with this etiology in man, however, is not accompanied by het-

erophile antibodies. The organism has been isolated from organ lesions, meconium, blood, and cerebrospinal fluid of man and of wild and domestic animals, in all of which disease has occurred. In human subjects, the bacterium is responsible for an acute, highly fatal form of meningoencephalitis in infants and adults. It is implicated also in repeated abortions and stillbirths.

L. monocytogenes is a facultative organism that grows well on infusion media, on tryptose agar with or without added glucose and blood, or on modified McBride medium (Chapter 39). It is not always isolated from tissue or other clinical material on the first attempt. The bacterium grows well at 25° and 36° C. but poorly at 42° C. and slowly (4 to 7 days) at 4° C. On sheep blood agar incubated at 36° C. for 24 hours, it produces small (1 to 2 mm.), translucent, gray to white, **beta hemolytic** colonies; hemolysis is slow to appear on human and rabbit blood agar. When cultivated on a clear medium, such as McBride agar, for 18 to 24 hours and examined by unfiltered, oblique illumination with a scanning microscope,* colonies appear characteristically **blue green** in color. These are 0.2 to 0.8 mm. in diameter, translucent, round, slightly raised, and watery in consistency.

Indole and hydrogen sulfide are not produced by this organism; nitrate is not reduced; citrate is not utilized; urease is not formed; gelatin and coagulated serum are not liquefied. It is catalase positive and oxidase negative. The methyl red test is positive; the Voges-Proskauer reaction is negative with the O'Meara test but positive with Coblentz and other methods using alpha-naphthol, potassium hydroxide, and creatine.

The **motility** of *L. monocytogenes* is best demonstrated by the stab inoculation of two tubes containing 0.1% dextrose semisolid agar (Chapter 39), incubating one at room temperature (20° to 25° C.) and the other at 36° C. Motility is more pronounced

---

*Now classified as *Propionibacterium acnes* (see Chapter 24).

*The plate may also be placed on a laboratory tripod and examined with a hand lens, with 45-degree oblique illumination.

at room temperature than at the higher temperature, and a motile culture appears as a disc of growth about 2 mm. below the surface of the medium, spreading downward as a filmy cloud.

In meat extract broth with carbohydrates and bromcresol purple indicator (Chapter 39), cultures of *Listeria* grown at 36° C. for 1 week will produce acid in glucose, levulose, rhamnose, maltose, and salicin; will show irregular acidity in lactose, sucrose, xylose, arabinose, and dextrin; and will rarely, if ever, produce acid in mannitol, inositol, sorbitol, dulcitol, or raffinose.

Test for pathogenicity. The **ocular** test (Anton) is reliable for pathogenicity. It is carried out by introducing 2 to 3 drops of an overnight tryptose agar slant culture, suspended in 5 ml. of distilled water, into the conjunctival sac of a rabbit. A subsequent purulent conjunctivitis will develop in several days, which eventually heals completely.

Constantly mounting evidence indicates the important role *L. monocytogenes* plays in both human and animal infections. The medical bacteriologist should be fully aware that all small, gram-positive, diphtheroid-like organisms isolated from cerebrospinal fluid, blood, vaginal swabs, and so forth are not always contaminants. Some may prove to be *L. monocytogenes.* If facilities are not available for carrying out this identification, it is strongly recommended that a subculture be sent to a reference laboratory. For further information on listeriosis, the reader is referred to the excellent review by Gray and Killinger[4] and to Killinger's work in the *Manual of Clinical Microbiology.*[6]

GENUS ERYSIPELOTHRIX

Erysipelothrix rhusiopathiae (synonym, *E. insidiosa*) is the single member of the genus *Erysipelothrix.* It is a nonsporulating, nonencapsulated, nonmotile, gram-positive rod with a tendency to form long filaments. Isolated from cases of swine erysipelas and mouse septicemia, it usually causes a self-limiting infection of the fingers or hand of man. An erythematous, edematous lesion develops at the point of entrance of the organism, and the disease generally can be traced to contact with animals or animal products. On blood agar medium, smooth (S) colonies are clear and very small (0.1 mm.) and contain small slender rods. Rough (R) colonies are 0.2 to 0.4 mm. in diameter and contain long filamentous forms, showing beading and swelling. The latter colonies can spread out to simulate miniature anthrax-type colonies. Both forms stain evenly and may show deeply stained granules.

The organism is microaerophilic and grows best at 30° to 35° C. Hemolysis on blood agar occurs when 10% horse blood is used. A "test tube brush" type of growth (lateral, radiating projections) characteristically occurs in gelatin stab cultures at room temperature after 48 hours. Carbohydrate reactions are variable; most strains produce acid from glucose and lactose but not from sucrose and others. Hydrogen sulfide is produced; nitrate is not reduced; indole is not formed; and catalase is negative.

The diagnosis rests on the isolation of the organism from a skin biopsy; the skin fragment is incubated in thioglycollate broth overnight and then subcultured to blood agar. Inoculation of white mice and pigeons may be necessary to confirm the identification.

REFERENCES

1. Cherry, W. B., and Moody, M. D.: Fluorescent-antibody techniques in diagnostic bacteriology, Bact. Rev. **29:**222-250, 1965.
2. Elek, S. D.: The plate virulence test for diphtheria, J. Clin. Path. **2:**250-258, 1949.
3. Fraser, D. T., and Weld, C. B.: The intracutaneous virulence test for *C. diphtheriae,* Trans. Roy. Soc. Can. Sect. **20:**343-345, 1926.
4. Gray, M. L., and Killinger, A. H.: *Listeria monocytogenes* and *Listeria* infections, Bact. Rev. **30:**309-382, 1966.
5. Hermann, G. J., Moore, M. S., and Parsons, E. I.: A substitute for serum in the diphtheria in vitro test, Amer. J. Clin. Path. **29:**181-183, 1958.
6. Killinger, A. H.: Listeria monocytogenes. In: Manual of clinical microbiology, Baltimore, Md., 1970, American Society for Microbiology, The Williams & Wilkins Co., pp. 95-100.

28 Mycobacteria

The mycobacteria may vary from the coccobacillary form to long, narrow, rod-shaped cells ranging from 0.8 to 5 μ in length and about 0.2 to 0.6 μ in thickness. They do not stain readily but, once stained, will resist decolorization with acid-alcohol and are therefore called **acid-fast bacilli.** Occurring as single bacilli or in small irregular clumps, mycobacteria are sometimes beaded, banded, or pleomorphic in stained smears. In addition to saprophytic species, the group includes numerous organisms pathogenic for man, the most important of which is *Mycobacterium tuberculosis.*

LABORATORY DIAGNOSIS OF TUBERCULOSIS AND RELATED MYCOBACTERIOSES
General considerations

Tuberculosis is an infectious disease of a persistent and chronic nature that is usually caused by *Mycobacterium tuberculosis,* but occasionally by other species, such as *M. bovis, M. kansasii.* Although capable of infecting almost any organ of the body, tuberculosis is most commonly associated with the lungs, from which it spreads from man to man through coughing, sneezing, or expectoration. Tuberculosis is generally considered to be the most socioeconomically important specific communicable disease in the world today. Although it no longer ranks as the commonest cause of death in nations with a high standard of living, it still remains the leading killer in the world today.

Only within the last decade has it become generally accepted that mycobacteria other than *M. tuberculosis* can be the cause of human infections. Although *M. tuberculosis* is still the most commonly isolated mycobacterium from clinical specimens, *M. kansasii* and the Battey bacilli ("*M. avium–M. intracellulare*") are also well-established human pathogens, and their recognition becomes equally important. Mycobacteria other than *M. tuberculosis* may represent as much as 10% of all human mycobacterial infections.[25]

The definitive proof of a tuberculous infection is provided only by the demonstration of *M. tuberculosis* (or *M. bovis*) in clinical specimens from the suspected patient. The laboratory procedures used include (1) examination of a stained smear, (2) isolation by cultural procedures, and (3) recovery and virulence testing by animal inoculation. The best results are probably obtained by the combined use of all three methods.

These methods utilize characteristics of the tubercle bacillus that differentiate it from other microorganisms. They are as follows:

1. The resistance of stained tubercule bacilli to decolorization with strong decolorizing agents. Acid-alcohol (3% hydrochloric acid in 95% ethanol) is the usual agent.

2. The resistance of acid-fast bacilli to digesting agents, such as strong acids and alkalis. The use of these agents on material containing contaminating bacteria will destroy most contaminants without decreasing greatly the viability of any tubercle bacilli that may be present.

3. The ability of *M. tuberculosis* to produce progressive disease when small numbers of these organisms are injected into guinea pigs.

Collection of clinical specimens

In suspected mycobacterial infection, as in all other diseases of microbial origin, the **diagnostic procedure begins not in the laboratory but at the bedside of the patient.** In the collection of clinical specimens there should be the same careful attention to detail on the part of the attending physician, nurse, and ward personnel as is required of the bacteriologist in carrying out the cultural procedures.

Secretions from the **lung** may be obtained by methods such as the following: spontaneous or induced expectoration of sputum, aspiration of secretions during bronchoscopy, aspiration of gastric contents that contain swallowed sputum, or swabbing of the larynx.

Sputum

Since it may not be practical to provide constant supervision during the collection of a sputum specimen, it is necessary to elicit the intelligent cooperation of the patient by giving him detailed instructions for the collection and explaining the purpose and importance of the test. The importance of sputum expectoration "from deep down in the lungs" should be emphasized as opposed to the expectoration of saliva, nasopharyngeal secretions, or other extraneous materials. The specimen should not be collected immediately after the patient has used a mouth wash. Sputum submitted in tissue or other nonhygienic containers should be summarily discarded as unsatisfactory.

An appropriate container must be provided for the collection; the Falcon sputum collection kit* appears to be ideally suited for this purpose. It consists of a plastic disposable 50-ml. graduated conical centrifuge tube with an aerosol-free screw cap, which is fitted inside a funnel-like disposable plastic outer container in such a manner that the risk of accidental contamination by handling is almost eliminated. Not more than 5 to 10 ml. of sputum is collected in this container, which is then labeled with the patient's name, date, and other data.

If the collection kit just recommended is not available, a 4- to 6-ounce, clean, sterile wide-mouth glass jar with a screw cap and rubber or Teflon liner, or a disposable sterile plastic cup with tightly fitted lid may be used. However, the specimen must be transferred to an appropriate aerosol-free centrifuge tube in the laboratory—a procedure both unpleasant and hazardous to the laboratory worker.

For optimal recovery of mycobacteria, a series of 3 to 5 single **early morning** sputum specimens, not exceeding one fifth (10 ml.) of the volume of the centrifuge tube, should be collected on successive days from productive patients. If the amount of this specimen is insufficient, sputum may be collected over a 24-hour period. Extending the collection of a single specimen over a longer period or the pooling of several days' accumulation is **not** recommended, since a reduced number of positive isolations may result, along with an increased rate of contaminated cultures.[22]

Nebulized and heated hypertonic saline may be used to induce sputum production in patients unable to raise a satisfactory coughed specimen. The specimen is obtained (preferably by a member of the respiratory therapy service) 10 to 15 minutes after inhalation of 10% sodium chloride solution at 45° C.; it appears to give a

---

*Falcon Plastics, No. 9002, Division of Becton, Dickinson, and Co. Cockeysville, Md., and Los Angeles, Calif.

higher yield of positive cultures than those obtained by gastric lavage[25] and is generally more acceptable to both patients and personnel.

Gastric contents

A specimen of gastric contents is necessary if a patient is unable to raise a sufficient amount of sputum, if he is uncooperative, or if he cannot expectorate because he is comatose. Gastric lavage is frequently required in young children from whom it is difficult to obtain a sputum sample. A minimal number of three specimens should be submitted.

The gastric lavage should be performed before the patient gets out of bed in the morning, after he has fasted for at least 8 hours prior to the collection. A disposable plastic gastric tube* is used: it is first moistened with sterile water and then inserted into the tip of a nostril or into the mouth. As the tube is gently advanced, the patient is instructed to swallow small sips of sterile water to assist him in swallowing the tube. After the gastric tube is properly in the stomach, gastric contents may be aspirated with a sterile 50 ml. syringe and placed in a sterile flask. The patient should then be given 20 to 30 ml. of sterile water (commercially distilled water for parenteral administration), either by mouth or by injection through the gastric tube.† The gastric washings are again aspirated and added to the first collection. The specimen is then delivered **immediately** to the laboratory, where prompt processing (within 4 hours) is carried out to neutralize the adverse effects of gastric acids on the tubercle bacillus. If this is not practical, some means of neutralizing the gastric acidity must be carried out, such as the addition of 10% sodium carbonate to a pH of 7, using phenol red indicator.

Urine

A minimum of three early morning voided ("clean catch") or catheterized urine specimens is recommended; the entire volume of each voiding is collected in a sterile container. A 24-hour pooled specimen is sometimes used, although it is less satisfactory, since it is more likely to be contaminated and to contain fewer viable tubercle bacilli than a first-voided specimen.[11]

Other materials

Since tuberculosis may occur in almost any anatomical site of the human body, types of clinical material other than those previously mentioned may occasionally be examined. Specimens of cerebrospinal fluid, pleural and pericardial fluid, pus, joint fluid, bronchial secretions, feces, and resected lung tissue* and autopsy material may be submitted for study. Collection of these specimens does not generally require supervision by the bacteriologist; an adequate supply of sterile specimen containers must be available, however.

Processing of clinical specimens
Examination of stained smears

Although the demonstration of acid-fast bacilli in stained smears of sputum or other clinical material is only **presumptive** evidence of tuberculosis infection, the speed and ease of performance make the **stained smear** an important diagnostic aid, since it may be the first indication of a mycobacterial infection.

Because mycobacteria stain poorly by the Gram method, the conventional Ziehl-Neelsen carbolfuchsin stain or the fluorochrome staining technique, using auramine and rhodamine, is a required procedure. In the latter method, smears may be screened at 100× magnification, permitting a larger area of the slide to be examined in the same time as compared to 1,000× magnification using the conven-

---

*Falcon Plastics, Los Angeles.
†Water should not be injected until one is certain that the tube is in the stomach and not the trachea or esophagus.

---

*Tissue specimens may be frozen when a delay in processing is necessary.

tional technique. By use of the Truant[32] technique (described in Chapter 40) and a properly adjusted fluorescence optical system, the mycobacteria and acid-fast nocardia are readily discerned as bright, **yellow-orange fluorescent bacilli** against a dark background. It should be noted that an antigen-antibody reaction is not involved in the fluorochrome procedure. It should not be considered, therefore, as an FA or immunofluorescence test.

Preparation of the smear

Sputum specimen—direct smear

1. Transfer a portion of the sputum to a disposable Petri dish, and with a wooden applicator stick broken in half, tease out a small portion of purulent, bloody, or caseous material and transfer it to a clean **new** slide.
2. Press another slide on top of it, squeeze the slides together, then pull them apart. This should result in two preparations of the proper thinness for staining and microscopy.
3. Label both slides, air dry, and flame immediately two or three times.
4. Stain according to the Ziehl-Neelsen, or Kinyoun, acid-fast or Truant fluorochrome[32] methods (Chapter 40). Do not use staining dishes, as they permit transfer of mycobacteria between slides.
5. Dry in air or under an infrared lamp; do not blot.
6. Examine under oil immersion lens.

Sputum specimen—concentration. In the concentration method[26] the specimen is digested by sodium hypochlorite, and any mycobacteria present are concentrated in the sediment by centrifugation. This procedure will generally increase the number of positive smears, and it is preferably carried out on 24-hour specimens. Since sodium hypochlorite is actively tuberculocidal as well as an excellent digestant, the resulting sediments, although clean and compact, will be useful **only for stained smears** and **not** for subsequent cultural procedures or animal inoculation.

1. Mix an equal volume (5 to 10 ml.) of sputum and 5% sodium hypochlorite (Clorox household bleach is recommended) in a collection container of ample size; stir with wooden applicator stick to ensure complete mixing.
2. Shake for 2 to 3 minutes and keep at room temperature for 10 minutes or until complete digestion has taken place.
3. Transfer a portion to a 15-ml. centrifuge tube and centrifuge for 10 minutes at 3,000 r.p.m.
4. Decant the supernatant fluid and allow the centrifuge tube to stand in an inverted position on a paper towel in order to drain the sediment; no neutralization is necessary.
5. Transfer the creamy white sediment to a slide with a cotton-tipped applicator stick. Air dry (no fixing is required) and stain by usual method.

Reporting and interpreting the microscopic examination

Sputum smears stained by conventional methods should be examined by carefully scanning the long axis of the smear three times—right-left-right—before reporting it as negative.[33] Typical acid-fast bacilli are **red stained,** slender, slightly curved, long or short rods (2 to 8 μ), sometimes beaded or granular. Atypical forms are unusually thick or diphtheroid, very long, and sometimes coccoid. The presence of bacilli in sputum that are longer, broader, and more conspicuously banded than tubercle bacilli should lead to the suspicion that *Mycobacterium kansasii* is present.*

Results of microscopic examination should be reported simply **"positive for acid-fast bacilli"** or **"no acid-fast bacilli seen"**; a positive finding should be based only on typical forms, but atypical rods should also be noted. When large numbers of typical acid-fast bacilli are found, it is

---

*It is important to wipe the objective thoroughly with lens paper after every **positive** smear in order to prevent the transfer of acid-fast organisms to the next slide by way of oil remaining on the objective.

quite probable that they are *M. tuberculosis;* when atypical rods are seen, they may be other pathogenic mycobacteria, including members of the unclassified group.

It is preferable to report the number of acid-fast bacilli seen; the following code recommended by the National Tuberculosis Association may be used[25]:

Number of organisms seen	Report to read
1 to 2 in entire smear	(Report number found and request another specimen.)
3 to 9 per slide	Rare (+)
10 or more per slide	Few (++)
1 or more per oil immersion field	Numerous (+++)

When only one or two acid-fast bacilli are seen in the entire smear, it is recommended[18] that they should not be reported until confirmation is obtained by examining other smears from the same or another specimen.

The demonstration of mycobacteria in sputum or other clinical material should be considered only as **presumptive** evidence of tuberculosis, since it does not specifically identify *M. tuberculosis.* The examination of stained smears is the least sensitive of the diagnostic methods for tuberculosis; cultures should be carried out on all specimens examined microscopically.* Because of its simplicity and speed of handling, however, the stained smear is an important and useful test, since smear-positive patients ("infectious reservoirs") are the greatest risk to their environment.

Cultural methods

Culture of sputum and bronchial secretions by N-acetyl-L-cysteine-sodium hydroxide method[33]

It has long been recognized that the conventional chemical methods of diges-

tion and decontamination of sputum for the cultivation of tubercle bacilli result in the destruction of a large percentage of these organisms. For this reason, a milder decontamination and digestion procedure, using the mucolytic agent N-acetyl-L-cysteine (NALC), was advocated by Kubica and associates[19] at the National Communicable Disease Center and the National Jewish Hospital. (The preparation of the required reagents is described in Chapter 41.) The procedure is as follows:

1. Collect the sputum specimen as previously described, and transfer appproximately 10 ml. to a 50-ml. sterile, disposable, aerosol-free plastic centrifuge tube with a screw cap (Falcon Plastics, No. 2070).

2. Add an equal volume of NALC-sodium hydroxide solution.

3. Tighten screw caps, mix well in a Vortex mixer* for 5 to 20 seconds or until digested. Violent agitation is not recommended, because denaturation of the NALC may take place.

4. Allow to stand at room temperature for **15 minutes** to effect decontamination.†

5. Fill the tube within $1/2$ inch of the top with either sterile M/15 phosphate buffer, pH 6.8 (preferred), or sterile distilled water. This dilution acts to minimize the action of the sodium hydroxide as well as to reduce the specific gravity, thereby facilitating centrifugation.

6. Centrifuge at or near 3,000 r.p.m. (1,800 to 2,400 *g*) for **15 minutes** and carefully decant the supernatant fluid into a splash-proof can containing a phenolic disinfectant. Wipe the lip of the tube with a cotton ball soaked with 5% phenol. Retain the sediment.

7. If direct drug susceptibility tests are to be carried out (see step 12), prepare a smear of the sediment by using a sterile

---

*Authorities estimate that it requires from 10^4 to 10^5 organisms per milliliter of sputum to yield a positive direct smear, of which 10 to 100 organisms per milliliter are recoverable by culture.

*Vortex-type, Junior model, available from most laboratory supply houses.

†If contamination rate is high, the concentration of NaOH may be increased to 4%, but **not** the exposure time.

applicator stick or 3 mm. diameter wire loop to spread one drop of the material over an area approximately 1 by 2 cm. Stain by the Ziehl-Neelsen or fluorochrome method and determine the approximate number of AFB (acid-fast bacilli) per oil-immersion field.

8. Add 1 ml. of 0.2% of bovine albumin* to the sediment, using a sterile pipet. Shake gently by hand to mix. No neutralization is required, since the dilution of the sodium hydroxide by the phosphate buffer wash in step 5 and the strong buffering capacity of the bovine albumin makes this step unnecessary. Refrigerate these albumin-suspended sediments overnight if inoculation to media is not practical at this time. It is recommended that bovine albumin be added **before** smears are made if the volume of sediment is small.

9. Make a 1:10 dilution of this sediment by adding 10 drops of it to 4.5 ml. of sterile water.

10. Inoculate the following media, using 0.1 ml. of the sediment for each tube or plate:
   a. **Lowenstein-Jensen (L-J) slants**
      One tube with undiluted sediment
      One tube with 1:10 dilution
   b. **7H 10-Oleate-albumin-dextrose-catalase (OADC) bi-plates**
      One half plate with undiluted sediment
      One-half plate with 1:10 dilution
Spread with a glass spreader, making the 1:10 dilution first.

11. If the specimen showed a **positive** smear (step 7), dilute the concentrated sediment as follows:

AFB per oil-immersion field (avg. 20 fields)	Dilution of inoculum
Less than 1	Undiluted, 1:100
1 to 10	1:10, 1:1,000
More than 10	1:100, 1:10,000

Dilution is necessary in order to ensure an inoculum size that will yield at least forty to fifty colonies on the control plate when performing drug-susceptibility tests, but not large enough to permit the overgrowth of drug-resistant mutants, which occur spontaneously in drug-susceptible populations.

12. **Direct** drug susceptibility testing* is recommended on all specimens with **positive** smears. This may be carried out by inoculating the following drug media in Felsen quadrant 7H 10 plates† (see Chapter 39 for preparation), using sterile, disposable capillary pipets and 3 drops (0.15 ml.) of the two dilutions selected in step 11:

PRIMARY DRUG PLATE

Quadrant I—no drug (control)
Quadrant II—isoniazid (INH), 0.2 μg./ml.
Quadrant III—streptomycin, 2.0 μg./ml.
Quadrant IV—para-aminosalicylic acid (PAS), 2.0 μg./ml.

For the preparation of media containing the secondary antituberculous drugs, including kanamycin, viomycin, ethionamide, ethambutol, cycloserine, rifampin, and others, the reader is referred to the excellent Center for Disease Control (CDC) mycobacteriology manual by Vestal.[33] Many of these drug-containing media are available commercially as slants, some using 7H 10 agar and some with L-J medium.‡ These media are best stored in the dark at room temperature and have a shelf life of at least 2 weeks.

An alternate method is the use of paper discs impregnated with antituberculous drugs.§ These are placed in sectors of quadrant plates that are then filled with 7H 10 agar and incubated overnight to allow diffusion of the drug. They are then

---

*Bovine albumin fraction V, adjusted to pH 6.8; from Pentex, Inc., Kankakee, Ill.

*These tests should be performed on all suspected isolates of *Mycobacterium tuberculosis;* if not available, the culture should be referred to a state laboratory or other specialized facility for susceptibility tests and/or speciation.
†Falcon Plastics, Los Angeles, Calif.
‡Available from Difco Laboratories, Detroit; Baltimore Biological Laboratory, Cockeysville, Md.; Pfizer Diagnostics, Flushing, N.Y.; and others.
§Antimicrobial drugs for use in culture media, Baltimore Biological Laboratory, Cokeysville, Md.

inoculated with the test strain as previously described.

13. Incubate L-J slants (steps 10 and 11) in a horizontal position for 1 to 2 days at 35° to 36° C. in the dark, examining weekly for 6 to 8 weeks. It appears that an atmosphere of **carbon dioxide** is extremely beneficial to the growth of mycobacteria.[42] Lowenstein-Jensen slants (loosen screw caps) should be incubated in a carbon dioxide incubator for the first 2 weeks. Incubation in a candle jar is **not** recommended, since an increase in contamination and a decrease in the amount of growth may occur, apparently because of oxygen depletion within the jar.

14. Incubate the 7H 10 plates (steps 10 and 12) right side up in permeable polyethylene bags,* sealed by stapling, **in the dark** in a carbon dioxide incubator (5% to 10% CO_2) at 35° C. for 3 weeks.

15. Examine the 7H 10 plates for growth both macroscopically and microscopically (a dissecting microscope is recommended) after 5 to 7 days' incubation and weekly thereafter.† The transparent 7H 10 medium permits differentiation between corded and noncorded colonies under low-power magnification. **Group II scotochromogens** (see next section) are obvious by the presence of yellow, **noncorded** colonies on removal from the dark incubator; the development of **pigment** in colorless, noncorded colonies after exposure to light is characteristic of **Group I photochromogens.**

16. **Positive** cultures are reported as soon as noted, and the final report is made after 6 to 8 weeks' incubation of 7H 10 plates and L-J slants. Drug susceptibility tests are read as soon as possible, with the reservation of possible changes in the final reading because of the slow growth of some initially susceptible strains

---

*Falcon Plastics, Los Angeles, or Baggies, Colgate-Palmolive Co., New York.

†Runyon has written an excellent discussion of the identification of mycobacteria by microscopic examination of colonies.[29]

in the presence of certain drugs, particularly streptomycin.

In reporting the results of the drug susceptibility tests, the report should include:

1. Type of test—direct or indirect
2. Number of colonies on control sector
3. Number of colonies on the drug sector
4. Concentration of the drug in each sector

From these data, a rough approximation of the percentage of organisms resistant to the drug may be calculated as follows:

$$\frac{\text{No. of colonies on drug sector}}{\text{No. of colonies on control sector}} \times 100 = \text{\% resistance at that drug concentration}$$

Refer to manual previously cited[33] for examples and photographs of these drug susceptibility tests.

If strains resistant to isoniazid, streptomycin, and *p*-aminosalicylic acid (PAS) are isolated, the following drugs may be used[18]: kanamycin, ethionamide, viomycin, cycloserine, ethambutol, and capreomycin. In order of increasing resistance to the commonly used antituberculous drugs are the following mycobacteria: *M. tuberculosis, M. kansasii, M. intracellulare, and M. fortuitum.*

Culture of sputum by other methods

In addition to the NALC-sodium hydroxide method for digestion and decontamination previously described, there are two methods that, if properly performed, are acceptable substitutes.

Trisodium phosphate-benzalkonium chloride method.[44] The trisodium phosphate-benzalkonium chloride method requires digestion with trisodium phosphate for only 1 hour, as opposed to a 12- to 24-hour exposure in the original method, which is no longer an acceptable procedure, due to the low survival rate of mycobacteria after their lengthy exposure to the digestant. The preparation of the required reagents is described in Chapter 41. The technique is as follows:

1. Mix equal volumes of sputum and trisodium phosphate-benzalkonium chloride* in a 50-ml. disposable, leak-proof centrifuge tube†; shake in a shaking machine‡ for 30 minutes.
2. Allow the mixture to stand at room temperature for 20 to 30 minutes.
3. Centrifuge at 3,000 r.p.m. for 20 minutes.
4. Decant the supernatant fluid into a disinfectant, observing necessary precautions.
5. Resuspend the sediment in 10 to 20 ml. of M/15 sterile phosphate buffer, pH 6.6; recentrifuge for 20 minutes.
6. Again decant the supernatant fluid and inoculate the sediment to egg media. If nonegg media such as 7H 10 agar are used, residual benzalkonium chloride may inhibit the growth of mycobacteria. This inhibitory effect may be neutralized by adding 10 mg./100 ml. of lecithin to the sterile buffer[18] or by employing an additional wash with sterile buffer. This is not necessary when using egg-containing media, because neutralizing phospholipid compounds are already present.

Sodium hydroxide method[27]
1. Mix equal volumes of sputum and 3% to 4% sodium hydroxide containing 0.004% phenol red in a disposable leak-proof centrifuge tube; homogenize in a shaking machine for 10 minutes.
2. Centrifuge at 3,000 r.p.m. for 20 minutes.
3. Decant the supernatant fluid into a disinfectant, observing bacteriological precautions.
4. Using a sterile capillary pipet, add 2

N hydrochloric acid by drops until a definite yellow endpoint is obtained.
5. Back titrate with 4% of sodium hydroxide to a faint pink endpoint (neutrality).
6. Inoculate the sediment to L-J media and 7H 10 agar.*

When sputum specimens are consistently contaminated with *Pseudomonas* species and other organisms, the method of Corper and Uyei,[4] is recommended. It makes use of the decontaminating effect of oxalic acid as follows:

1. Mix equal volumes of sputum and 5% oxalic acid in a leak-proof centrifuge tube; homogenize in a Vortex mixer and allow to stand at room temperature for 30 minutes, shaking occasionally.
2. Add sterile physiological saline to within 1 inch of the top; this lowers the specific gravity, permitting better concentration of bacilli in the sediment.
3. Centrifuge at 3,000 r.p.m. for 15 minutes.
4. Decant the supernatant fluid into a disinfectant, using bacteriological precautions.
5. Neutralize the sediment with 4% sodium hydroxide containing phenol red indicator; inoculate the desired media.

Culture of gastric specimens

1. Add a pinch of NALC powder to about 25 ml. of specimen in a 50-ml. centrifuge tube.
2. Mix in Vortex as with sputum.
3. Centrifuge at 3,000 r.p.m. for 30 minutes.
4. Aseptically decant the supernatant fluid and resuspend sediment in 2 to 5 ml. sterile distilled water.
5. Add an equal volume of NALC-

---

*Zephiran R., Winthrop Chemical Co., New York, N.Y.
†Falcon Plastics, No. 2070, Los Angeles.
‡Paint conditioner, laboratory model No. 34, Red Devil Tools, Union, N. J.

---

*Recently, the use of L-J medium containing penicillin and nalidixic acid (Gruft medium[6]), along with digestion with 2% NaOH, has been found effective in the recovery of mycobacteria from heavily contaminated specimens received by mail.

NaOH reagent and proceed as with sputum.

6. If the gastric specimen is quite fluid, centrifuge directly and proceed with step 5.

Culture of urine

Urine specimens (clean-voided, early morning specimens are preferred) are handled as follows:

1. Pour approximately 50-ml. portions into one or more centrifuge tubes; centrifuge at 3,000 r.p.m. for 30 minutes.
2. Aseptically decant supernatant fluid, resuspend sediment in 2 to 5 ml. sterile water, and handle as described under gastric specimens.

An alternate method described in the CDC manual[33] is as follows:

1. Centrifuge and combine sediments as described above.
2. Add to sediment an equal volume of 4% H_2SO_4.
3. Mix in Vortex; let stand for 15 minutes.
4. Fill tube to nearly the 50-ml. mark with sterile distilled water, mix, and centrifuge as before.
5. After decanting supernatant fluid, add 0.2% bovine albumin to sediment and inoculate media as in sputum cultures.

Note: Decontaminate urines as soon as possible upon receipt in the laboratory, since prolonged exposure to urine may be toxic to mycobacteria.

Culture of cerebrospinal fluid and other body fluids

The isolation of *M. tuberculosis* from cerebrospinal fluid and other body fluids is generally more difficult than from sputum and gastric secretions. Since **very few** organisms may be present, success will often depend on the amount of specimen available for culture—the larger the volume the greater will be the chance of recovering the organism.

At least 10 ml. of fluid should be sub-mitted. If only a few milliliters are received, the entire specimen should be used for guinea pig inoculation; cultures and smears should not be attempted. A number of specimens may be pooled until at least 10 ml. are available.

The fluid specimen should be centrifuged for 30 minutes at 3,000 r.p.m. and the supernatant fluid discarded. (Cerebrospinal fluid supernate may be saved for other examinations, such as serological or chemical tests.) Half of the sediment is used for inoculating media and preparing a smear; the other half is used for guinea pig injection. Because these specimens will generally contain no other bacteria, the decontaminating procedure is neither necessary nor desirable. It could result in the loss of the few organisms present. If the fluid specimen contains a clot, this is cut into small pieces with sterile scissors and subsequently accorded the same treatment as a sputum specimen. Purulent material is also handled in the manner used for sputum.

Kubica and Dye[18] suggest that specimens that have been obtained aseptically, as cerebrospinal fluid, synovial fluid, pleural fluid, and small bits of biopsied tissue, be inoculated directly to a fluid medium such as Middlebrook 7H 9, Tween-albumin medium, or Proskauer and Beck medium, in a volume ratio of 1:5. These media are incubated at 36° C. in a carbon dioxide incubator, and acid-fast stained smears are prepared and examined weekly. If mycobacteria are seen, the medium is inoculated to egg slants or 7H 10 agar and incubated as previously described. If smears are negative after 4 weeks, the media are again subcultured at weekly intervals to L-J slants or 7H 10 agar for 4 more weeks before discarding.

Culture of feces

The examination of fecal material for *M. tuberculosis* is not very rewarding and its use should be limited to special cases only. The presence of tubercle bacilli in the feces does not necessarily suggest intestinal tuberculosis—it more likely indicates sputum swallowed by a patient with pulmonary disease.

In the culture of fecal material, suspend about 5 gm. in 20 to 30 ml. of distilled water and mix well. Add sufficient sodium chloride to make a saturated solution, and allow this to stand 30 minutes. Skim off the film of bacteria that forms on the surface with a sterile spoon and transfer to a sterile centrifuge tube. Add an equal volume of 4% sodium hydroxide, homogenize with vigorous shaking, incubate at 36° C. for 3 hours, shaking at intervals, neutralize with 2 N hydrochloric acid, and centrifuge for 20 minutes at 3,000 r.p.m. Inoculate five tubes of media with the **top layer** of the supernatant fluid. Save a portion of the supernatant fluid, discarding the rest, and combine with a portion of the sediment, to be later inoculated into a guinea pig if desired. The remaining sediment is inoculated to five tubes of media (**ten** tubes in all). This inoculation of an increased number of tubes is advised because of the increased likelihood of subsequent loss from contamination in culturing fecal material.

Culture of tissue removed at operation or necropsy

Tissue removed surgically (lymph nodes, resected lung, and so forth) generally is not contaminated and does not require sodium hydroxide treatment. The specimen is finely minced with sterile scissors and transferred to a sterile tissue grinder,* where it is ground to a pasty consistency with sterile alundum and a small amount of sterile saline or 0.2% bovine albumin V. After settling, 0.2-ml. portions of the supernatant fluid are removed and inoculated directly to egg media and 7H 10 agar (liquid 7H 9 medium also may be used), and a smear is made.

Tissue that is obviously contaminated (such as tonsils and autopsy tissue) is handled as just described, except that the supernatant fluid obtained after grinding is treated as is sputum (decontamination with NALC–sodium hydroxide solution, homogenization, centrifugation, neutralization, and so forth) before inoculation to culture media.

ANIMAL INOCULATION TESTS

The inoculation of a laboratory animal (generally the guinea pig) has been accepted in the past as the final criterion of the pathogenicity of an acid-fast bacillus. This is no longer true for human isolates, since some organisms—notably the mycobacteria other than tubercle bacilli and some isoniazid-resistant strains of *Mycobacterium tuberculosis*—do not produce progressive disease in the injected animal.[18]

Furthermore, the excellence of cultural procedures presently available and the more frequent use of **multiple specimens** from a suspected case of tuberculosis, makes it generally unnecessary for the hospital laboratory to inoculate laboratory animals.*

The procedure, however, is a useful diagnostic tool in the detection of **small numbers** of tubercle bacilli, as in cerebrospinal fluid, or in specimens that are consistently contaminated on culture. For these reasons, therefore, the technique has been included.

1. If clinical specimens are to be injected, decontaminate and concentrate them as described in previous discussions of culture methods.
2. After a portion of the concentrated and neutralized sediment has been inoculated onto appropriate media, suspend the remainder in 1 ml. of sterile physiological saline and inject either subcutaneously in the groin or intraperitoneally into a guinea pig.
3. If one is using pure cultures of mycobacteria, a dosage of approximately

---

*Ten Broeck tissue grinder, small size, heavy-walled Pyrex glass, Bellco Glass, Inc., Vineland, N. J.

*Specimens for animal inoculation are best handled in a mycobacteriology reference laboratory.

0.1 mg. (moist weight) is the usual inoculum. This is suspended in 1 ml. of physiological saline and injected subcutaneously into the right groin of each of **two** guinea pigs, the animals having been pretested and found negative to 0.1 ml. of 5% old tuberculin (O.T.) or a satisfactory substitute injected intracutaneously. A subculture is also made of the suspension, for control purposes.

4. Animals inoculated either with clinical material or with cultures should be examined weekly to detect the presence of enlarged lymph nodes draining the site of inoculation. Tubercle bacilli of human or bovine origin usually give rise to progressive disease in the guinea pig, with development of caseous nodes and involvement of the spleen and usually the liver and lungs. **Actual invasion of deep tissue** must be demonstrated—the microbiologist must not rely on the simple production of a local lesion alone.

5. If two guinea pigs have been injected, give 0.5 ml. of O.T. to one at the end of 4 weeks. In most instances an animal eventually proved tuberculous is so sensitized to tuberculin by this time that it will die within 1 to 2 days of O.T. injection. If the tuberculin reaction is negative, the animals are held another 4 weeks before killing and autopsying.

6. Examine tissue showing gross pathology, using the acid-fast stain.

7. The involvement of only the regional nodes is considered a **doubtful** test, and the test should be repeated.

8. If only extensive lung tuberculosis is observed, **spontaneous** infection should be suspected, and the test repeated; this also applies to animals that die in less than 4 weeks and that show negative autopsy findings.

9. If the aforementioned facilities are not at hand, refer the specimen to a specialized laboratory equipped for this procedure.

CULTURAL CHARACTERISTICS OF THE MYCOBACTERIA

It should be emphasized at this point that the **acid-fast** nature of all colonies developing in culture media must be confirmed first before they can be identified as mycobacteria. Cultures for the isolation of *M. tuberculosis* on L-J media should be incubated at 35° to 36° C. for a total of 6 to 8 weeks and examined at weekly intervals; 7H 10 media are held for 6 weeks. **Positive** cultures should be reported as soon as identification has been completed.

Mycobacterium tuberculosis

Colonies of human tubercle bacilli generally appear on egg media after 2 to 3 weeks at 36° C.; no growth occurs at 25° C. or 45° C. Growth first appears as small (1 to 3 mm.), dry, friable colonies that are rough, warty, granular, and buff in color. After several weeks these increase in size (up to 5 to 8 mm.); typical colonies have a flat irregular margin and a "cauliflower" center. Because of their luxuriant growth these mycobacteria are termed **eugonic.** Colonies are easily detached from the medium's surface but are difficult to emulsify. After some experience one can recognize typical colonies of human-type tubercle bacilli without great difficulty. If there is any doubt, however, final confirmation by biochemical tests or animal inoculation must be carried out.

Cultures of *M. tuberculosis* injected subcutaneously into guinea pigs will give rise to a progressive and usually fatal disease, unless an INH-resistant strain is encountered. Rabbits and mice are less affected, and chickens are unaffected. Virulent strains tend to orient themselves in tight, **serpentine cords,** best observed in smears from the condensation water or by direct observation of colonies on a direct cord medium.[23] Catalase is produced in moderate amounts but **not** after heating at 68° C. for 20 minutes in pH 7 phosphate buffer; human strains resistant to isoniazid (INH) are frequently catalase negative and have smooth colonies on egg media. The niacin test (p. 211) is useful in that most **niacin-positive** strains encountered in the diagnostic laboratory can be labeled as human varieties. Susceptibility to antituberculous drugs is characteristically high.

Mycobacterium bovis

Bovine tubercle bacilli are rarely isolated in the United States but remain a significant pathogen in other parts of the world. They require a longer incubation period—generally 3 to 6 weeks—and appear as tiny (less than 1 mm.), translucent, smooth, pyramidal colonies at 36° C. They adhere to the surface of the medium but are emulsified easily. Their growth is termed **dysgonic.** On 7H 10 the colonies are rough and resemble those of *M. tuberculosis.*

M. bovis will grow only at 36° C. It forms serpentine cords in smears from colonies on egg media; the niacin reaction and nitrate reduction tests are negative.* It is **virulent for rabbits** when injected intravenously; it is also virulent for guinea pigs and mice but not for chickens. Its susceptibility to the primary drugs is similar to that of *M. tuberculosis.*

Other Mycobacteria

Although it has long been recognized that acid-fast bacilli other than *M. tuberculosis* were occasionally associated with both pulmonary and extrapulmonary disease, it has only been within the last decade that these mycobacteria, variously called atypical, anonymous, or unclassified acid-fast bacilli, have been associated with pulmonary disease clinically diagnosed as tuberculosis.[2] Current evidence, obtained from skin testing, suggests that many persons in the United States may become naturally infected with these mycobacteria, even though most of them do not show clinical evidence of disease.[44]

In 1959, Runyon[28] proposed a schema for the separation of medically significant strains of the unclassified mycobacteria, dividing them into four large groups. This served as a rough classification until a more precise speciation could be established for members within each group. More recently, studies by individual workers[12,37] and cooperative groups[7] have demonstrated differences between clinically significant and insignificant strains of various mycobacteria by easily performed metabolic and biochemical tests (described in a following section).

Runyon's four groups, including well-defined species, are described next.

Group I photochromogens

Group I photochromogens possess the outstanding characteristic of **photocromogenicity**—the capacity to develop a pigment when exposed to light. A young, actively growing culture on L-J slants ex-

Table 28-1. Potential clinical significance of various mycobacteria*

ALWAYS OR SOMETIMES SIGNIFICANT	RUNYON GROUP	NEVER OR RARELY SIGNIFICANT
M. tuberculosis M. bovis		
M. kansasii	I	M. kansasii (low-catalase variety)
M. marinum (balnei)		
M. scrofulaceum	II	M. gordonae M. flavescens
M. avium† M. intracellulare (Battey)† M. xenope	III	M. gastri M. terrae complex M. triviale
M. fortuitum M. chelonei	IV	M. smegmatis M. phlei M. vaccae M. rhodochrous

*From Current Item No. 165 (updated), Laboratory Program, 1968, National Communicable Disease Center. Courtesy George Kubica, former Chief, Mycobacteriology Unit, NCDC, Atlanta.
†With the basic tests listed herein these two species cannot be separated and will be reported as *"M. avium–M. intracellulare."*

M. bovis is also susceptible to thiophen-2-carboxylic acid hydrazide (TCH), a useful test to distinguish it from other mycobacteria.[38]

posed to light for as little as 1 hour and reincubated in the dark will produce a bright lemon yellow pigment within 6 to 24 hours.

Two well-defined photochromogens comprise Runyon Group I: *M. kansasii* and *M. marinum*.

Mycobacterium kansasii. M. kansasii is responsible for a severe pulmonary disease in man, most frequently appearing in white, emphysematous males over the age of 45 years. The disease with its complications is indistinguishable from that caused by *M. tuberculosis*, but it follows a more chronic and indolent course. The disease is not highly communicable, in distinct contrast to that caused by the tubercle bacillus.

Many isolations of *M. kansasii* have been made in the midwestern part of the United States, but Hobby and co-workers[7] point out that this uncontrolled geographical distribution may reflect a biased interest on the part of the participating institutions.

Optimal growth of *M. kansasii* occurs after 2 to 3 weeks at 36° C. (slower at 25° C.); growth does not occur at 45° C. Colonies are generally smooth, although there is a tendency to develop roughness. They are cream colored when grown in the dark and become a bright **lemon yellow** if exposed to light.

A practical procedure for photochromogenicity testing has been suggested by Kubica[17]:

1. Two slants of L-J medium are inoculated with a barely turbid suspension of the culture; one tube is shielded from the light by wrapping with black x-ray paper or aluminum foil.
2. Both tubes are incubated at 36° C. (32° to 33° C. for suspected cultures of *M. marinum*) until visible growth occurs on the uncovered slant.
3. The shield is removed from the covered tube and any pigment is noted; if none is observed, recover only **one half** of the tube with the shield, while exposing the other half to ei-

ther a 60-watt bulb placed 8 to 10 inches from the tube, or to bright daylight, for 1 hour. Loosen cap during exposure.
4. Replace the shield, reincubate both culture tubes overnight with the cap still loose, and compare pigments in colonies grown (a) unshielded, (b) shielded, and (c) exposed to light for an hour. Three pigment classes can then be established:
   a. Scotochromogens—pigmented in dark
   b. Photochromogens—pigmented after exposure to light
   c. Nonphotochromogens—nonpigmented either in dark or light

If cultures are grown continuously in the light (2 to 3 weeks), bright orange crystals of beta carotene will form on the surface of colonies, especially where growth is heavy. Nonphotochromogenic variants of *M. kansasii* occur very rarely; these may require allergenicity studies in infected guinea pigs tested with different mycobacterial protein derivatives.[7]

Clinically significant strains of *M. kansasii* are strongly catalase positive,* even after heating at 68° C. at pH 7 (especially when using the semiquantitative test of Wayne,[36] described on p. 213). Most strains do not produce niacin, although aberrant strains have been noted; nitrates are reduced; mature colonies show loose cords.

Stained preparations of *M. kansasii* show characteristically long, banded, and beaded cells that are strongly acid fast. Such conspicuous cells in sputum smears should lead one to suspect *M. kansasii* infection. The organism is not pathogenic for guinea pigs or chickens but many produce lesions in mice. There is characteristically a slight to moderate susceptibility to INH and streptomycin, with mod-

---

*Low-catalase strains of *M. kansasii* have been described[21,36]; these were not associated with human pathogenicity.

erate resistance to para-aminosalicylic acid (PAS).

Mycobacterium marinum. M. marinum (as well as *M. ulcerans,* see below) is primarily associated with granulomatous lesions of the skin, particularly of the extremities, and usually follows exposure of the abraded skin to contaminated water ("swimming-pool granuloma").* It grows best at 25° to 32° C., with sparse to no growth at 36° C. (corresponding to the reduced skin temperatures of the extremities) and is never isolated from the sputum. Also as a **photochromogen,** it may be distinguished from *M. kansasii* by its source, its more rapid growth at 25° C., and its production of tail and foot lesions in injected mice.

Another mycobacterium, *M. ulcerans,* is also associated with skin lesions and is considered to be the causative agent of Buruli ulceration, a necrotizing ulcer found in African natives.[3] The organism is difficult to cultivate in vitro and requires several weeks' incubation at 32° C. Unlike *M. marinum,* this organism is **nonphotochromogenic.**

Group II scotochromogens

The Group II **scotochromogens** are pigmented in the dark (*scoto,* dark), usually a deep yellow to orange, which darkens to an orange or dark red when the cultures are exposed to continuous light for 2 weeks. This pigmentation in the dark occurs on nearly all types of media at all stages of growth—characteristics that clearly aid in their identification.[7]

The Group II scotochromogens are generally divided into two subgroups—the potential pathogen, *M. scrofulaceum,* and the so-called "tap water scotochromogens," isolated from laboratory water stills, faucets, soil, and natural waters. The latter organisms, formerly called *M. aquae,* are now classified as *M. gordoneae.*

Since the tap water scotochromogens

are not associated with human disease, it is important to differentiate them from the potentially pathogenic *M. scrofulaceum,* because the former may contaminate equipment used in specimen collection, as in gastric lavage.

M. scrofulaceum is a slow-growing organism, producing smooth, domed to spreading, yellow colonies in both the light and dark. When exposed to continuous light, the colonies may increase in pigment to an orange or brick red; the paper shield should remain in place (see under photochromogens) until visible growth occurs in the unshielded tube, then colonies in the shielded tube should be exposed to continuous light. Initial growth may be inhibited by too much light.[33] The hydrolysis of Tween-80 separates *M. scrofulaceum* from the tap water organisms in that the latter will hydrolyze it within 5 days, while *M. scrofulaceum* is negative up to 3 weeks.

Growth of *M. gordonae* appears late on L-J and 7H 10 media, usually after 2 weeks and frequently after 3 to 6 weeks, as scattered small yellow-orange colonies in both the light and dark. Hydrolysis of Tween-80 characteristically occurs. *M. flavescens* also produces a yellow-pigmented colony in both the light and dark, but is considerably more rapid in growth (usually within 1 week), and reduces nitrate as well as hydrolyzing Tween-80. Because of its rapid growth, some workers place *M. flavescens* in Runyon's Group IV (rapid growers). Human infection with this mycobacterium has not been reported.

Group III nonphotochromogens

Runyan Group III is made up of a heterogeneous variety of both pathogenic and nonpathogenic mycobacteria that **do not** develop pigment on exposure to light. The following species presently are recognized:

1. Battey-avian complex, or *"M. avium—M. intracellulare,"* (Battey bacilli are designated by some as *M. intracellulare*). Battey bacilli were first isolated at Battey

---

*A recent source of infection has been the tropical fish tank.[1]

State Hospital, Georgia, and are responsible for a pulmonary disease indistinguishable from that caused by *M. tuberculosis* and *M. kansasii,* although not occurring as frequently.

Except for their lack of pathogenicity for chickens and rabbits and their inability to grow at 45° C., Battey bacilli are similar to *M. avium,** a mycobacterium that causes tuberculosis in fowl, cattle, and swine. When avian bacilli are isolated from humans (rarely), they are virulent for chickens, and the human disease is frequently traceable to exposure to *M. avium* infections in poultry and other animals. On the other hand, Battey bacilli are usually nonpathogenic for chickens, and the patients give a history of exposure to poultry, swine, or other animals. We tend to agree with Kubica,[16] who sees "little need to separate avian from Battey bacilli in the clinical laboratory; both organisms respond poorly, if at all, to drugs, and little would be gained in identifying the exact species."

M. avium–M. intracellulare bacilli grow slowly at 36° C. (10 to 21 days) and 25° C. and produce characteristically thin, translucent, radially lobed to smooth, cream-colored colonies; rough variants† may show cording. These organisms are niacin negative, do not reduce nitrate, and produce only a small amount of catalase. Tween- 80 is not hydrolyzed in 10 days, but the organisms **reduce tellurite** within 3 days, a useful test to differentiate these potential pathogens from clinically insignificant members of Group III.

2. *M. xenopi* is of potential pathogenicity, having been isolated from the sputum of patients with pulmonary disease in Great Britain.[24] The organism resembles one that is isolated from a skin lesion of

the South African frog, *Xenopus laevis;* although it is recovered from a cold-blooded animal, the optimal temperature for growth is 42° C., and it fails to grow at 22° to 25° C. Four to five weeks' incubation is required to produce tiny dome-shaped colonies of a characteristic yellow color; branching filamentous extensions are seen around colonies on 7H 10 agar, resembling a miniature bird's nest. Tween 80 hydrolysis and tellurite reduction tests are negative; resistance is high to antituberculous drugs.

3. *M. terrae* complex ("radish" bacilli). A number of these mycobacteria have been isolated from soil and vegetables as well as from humans, where their pathogenicity is questionable. The clinical significance of these organisms, therefore, appears to be in excluding them from the Battey-avian complex. The slowly growing (36° C.) colonies may be circular or irregular in shape and smooth or granular in texture. They actively hydrolyze Tween 80 and reduce nitrate and are strong catalase producers, but they do not reduce tellurite in 3 days. *M. terrae* is resistant to INH and does not sensitize guinea pigs to Battey or avian protein derivatives.

4. *Mycobacterium gastri*[7] ("J" bacillus) has been described recently, occurring chiefly as single colony isolates from gastric washings, but it has not been associated with clinically significant diseases in humans. It bears a close similarity to low catalase-producing strains of *M. kansasii,* from which it may be readily differentiated by the photochromogenic ability of the latter. *M. gastri* may be differentiated from other members of Group III mycobacteria by its ability to hydrolyze Tween 80 rapidly and its loss of catalase activity at 68° C., and nonreduction of nitrate.

5. *Mycobacterium triviale* ("V" bacilli)[8] are rough colony mycobacteria occurring on egg media that may be confused with bacilli of *M. tuberculosis* or rough variants of *M. kansasii.* These bacilli have been recovered from patients with previous tuberculous infections but are considered to be

---

*Some workers suggest that Battey bacilli may eventually prove to be merely variants of *M. avium.*[7,37] There also may be a reservoir in the soil.

†Most strains of Battey bacilli show a small proportion of rough colonies, which may resemble colonies of tubercule bacilli.

unrelated to human infections. They are nonphotochromogenic, moderate to strong nitrate reducers, and maintain a high catalase activity at 68° C. They hydrolyze Tween 80 rapidly and do not reduce tellurite.

Group IV rapid growers

The Runyan Group IV mycobacteria are characterized by their ability **to grow out in 3 to 5 days** from a variety of culture media, incubated either at 25° or 36° C. Two members, *M. fortuitum* and *M. chelonei,* are associated with human pulmonary infections.*

M. smegmatis and *M. phlei* are considered to be **saprophytes** and are nonpathogenic. Both produce pigmented colonies and show filamentous extensions from colonies growing on cornmeal-glycerol agar. The ability of *M. phlei* ("hay bacillus") to produce large amounts of carbon dioxide is utilized to stimulate primary growth of *M. tuberculosis* on 7H 10 agar plates incubated in carbon dioxide–impermeable (Mylar) bags.

M. fortuitum has been incriminated in progressive pulmonary disease,[5] usually with a severe underlying complication, and has resulted in death. This potential pathogen is also a rapid grower (2 to 4 days), is generally nonchromogenic, and may be readily separated from the saprophytic rapid growers by giving a **positive 3-day arylsulfatase** reaction and by growing on MacConkey agar within 5 days, producing a change in the indicator. Both rough and smooth colonies are produced and show greening (dye absorption) on L-J medium. The organism is not virulent for guinea pigs or rabbits but produces multiple abscesses, particularly in the liver and spleen, of intravenously injected mice in 12 to 28 days. The niacin test is negative.

M. fortuitum is usually resistant to PAS and streptomycin, but it is susceptible to INH and the tetracyclines.

M. chelonei (formerly *M. borstelense*) comprises two distinct subspecies: *M. chelonei,* which fails to grow on 5% NaCl medium, and *M. chelonei* ssp. *abscessus,* which grows on the NaCl medium. This subspecies is considered a significant but rarely isolated pulmonary pathogen; *M. chelonei* (formerly *M. borstelense*) is considered a saprophyte.[33]

It should be noted that the majority of Group IV mycobacteria are not stained by the fluorochrome (auramine-rhodamine) stain.[10] All, however, are stained by the Ziehl-Neelsen technique.

Aids in identifying mycobacteria*

As soon as visible nonpigmented growth (5 to 6 days) is observed on any media, they should be exposed to light (previously described). After overnight incubation *M. kansasii* strains will become pigmented. All cultures of mycobacteria that are suspected of being members of Runyon's four groups, because of their rapid growth, pigmentation, colonial characteristics, microscopic appearance, and so forth, should be subjected to the following procedures, as recommended by Wolinsky[43]:

1. Subculture on three L-J slants in order to obtain isolated colonies.
2. Incubate one at room temperature (20° to 25° C.).
3. Incubate the second at 36° C. in light, either continuously exposed to fluorescent light or frequently exposed to bright light at hourly periods after growth appears.
4. Incubate the third at 36° C. in the dark, wrapped in aluminum foil or black paper and placed next to the second tube.
5. When slant No. 2 shows good

---

*M. fortuitum is also a common soil organism and may be frequently recovered from sputum without necessarily being implicated in a pathological process.

*The reader is referred to an excellent summary by Wayne and Doubek[38] of aids in the identification of most mycobacteria encountered in the clinical laboratory.

growth, remove foil or paper from slant No. 3 and compare them.

6. Next, loosen cap of slant No. 3 (grown in the dark) and expose to bright light and observe for development of yellow pigment the next day.

7. Perform a niacin test on one of the original slants if growth is sufficient; subculture and reincubate if growth is scanty.

8. Other tests, such as cording, catalase, nitrate reduction, Tween 80 hydrolysis, arylsulfatase activity, virulence for animals, ability to grow at 45° C., and so forth may be required. See following section for a description of these procedures.

Interpretation. M. tuberculosis and M. bovis will not grow at room temperature; rapid growers will produce full-grown colonies in several days, even at room temperature. *M. kansasii* will show yellow colonies in the light and white colonies when grown in the dark; the latter will turn yellow overnight after exposure to light. Group II scotochromogens will be yellow to orange in both the light and dark tubes, the pigment generally being deeper in the light tube. Battey-avian colonies will be cream colored at first, becoming a deeper yellow with aging. Light exposure has no effect on pigment production. *M. tuberculosis* only will give a positive niacin test; a negative test does not preclude the possibility of *M. tuberculosis,* and should be repeated at weekly intervals for a total of 6 weeks.

PROCEDURES USEFUL IN DIFFERENTIATING BETWEEN THE MYCOBACTERIA (INCLUDING CYTOCHEMICAL TESTS)

Although it is true that the identity of an acid-fast organism should be reported only after adequate cultural and drug susceptibility tests and animal pathogenicity tests have been carried out, it is also realized that many laboratories today are not equipped to perform all these studies. With the idea that some degree of identification

may be achieved by the use of in vitro biochemical and cytological tests, the following procedures, including the most reliable and useful, are presented. Descriptions of others may be found in the literature.*

It should be emphasized that if facilities for carrying out these special procedures are not at hand, the cultures should be **referred promptly** to a specialized mycobacteriological reference laboratory. An **interim report,** based on the available information (result of smear, growth on culture, niacin test and so forth) at the time, also **should be sent to the clinician promptly.**

Niacin test

Strong niacin production by an acid-fast bacillus from a clinical specimen is firm evidence of its identity as a **human** tubercle bacillus. Conversely, an accurately performed test resulting in a negative reaction generally indicates another species of *Mycobacterium. M. ulcerans* (causing skin ulcers in animals and man, growth at 30° to 33° C., not at 36° C.), *M. marinum* (causing fish tuberculosis), and particularly *M. microti* (causing vole tuberculosis) may produce smaller amounts of niacin and give doubtful or weakly positive results.

The niacin test devised by Konno[15] and modified by Runyon[30] depends on the formation of a complex color compound when a pyridine compound (niacin) from the organism reacts with cyanogen bromide (CNBr) and a primary or secondary amine.

If water of condensation is present on a culture slant of the organism to be tested (at least 3-week-old cultures on egg media should be used and have at least 50 to 100 colonies), it may be used for the niacin test. If it is not, the niacin may be extracted by adding a few drops of water or saline to the egg medium and placing the tube so that the liquid remains on or around the colonies for 15 minutes. This serves to

---

*Sommers and Russell's manual[31] is highly recommended.

extract niacin if it is present.* One or 2 drops of the extract are then transferred to a white porcelain spot plate, and 2 drops of each of the following reagents are added: (1) 4% aniline in 95% ethanol, which should be nearly colorless, and (2) 10% aqueous CNBr. When not in use, both reagents are stored in the refrigerator in brown dropper bottles and are made up fresh each month. A known positive strain of *M. tuberculosis* should always be included as a control.

Caution: The tests must be carried out under a chemical fume hood, since tear gas forms from CNBr. The production of an almost immediate **yellow color** indicates the presence of niacin. If a weak test is noted it should be repeated, using water in the place of CNBr, and the results compared. On completion of the tests, several drops of 4% sodium hydroxide or 10% ammonia are added to the spot plate to destroy the residual CNBr and arrest tear gas formation. The plate may then be reconditioned by placing in boiling water for several minutes. Commercial niacin paper test strips are available and are recommended.†

The only niacin-producing acid-fast organisms likely to be encountered in the clinical laboratory are those of *M. tuberculosis.* Most strains are niacin-positive on L-J medium in 3 to 4 weeks; others may take up to 6 weeks. It is well to keep in mind that a positive niacin test **does not indicate virulence.** Rather, it distinguishes human strains—regardless of virulence—from most other acid-fast bacilli.

Nitrate reduction test

The test for reduction of nitrate[34] (nitroreductase) is helpful in differentiating the slower-growing organisms, *M. tuberculosis* and *M. kansasii,* from members of Group II mycobacteria and clinically significant Battey-avian organisms. The

former are strongly positive nitrate reducers, whereas the latter are generally negative. Other Group III bacilli *(M. terrae, M. triviale),* which are not clinically significant, are also strongly positive.

Nitrate broth* tubed in 2-ml. amounts is heavily inoculated (spadeful) from a 4-week culture slant and incubated in a 36° C. water bath for 2 hours. The suspension is then acidified with 1 drop of a 1:2 dilution of hydrochloric acid, followed by 2 drops of sulfanilamide solution and 2 drops of the coupling reagent. (See Chapter 41 for preparation of these solutions.)

A **positive** test is indicated by the immediate formation of a **bright red** color (compared with the reagent control). A **negative** test should be confirmed by the addition of a pinch of zinc dust; a red color, due to the reduction of nitrate by the zinc, confirms the negative test. An uninoculated reagent control, and a negative and a positive control with a known strain of *M. tuberculosis* also should be included. Color standards (± to 5+) may be prepared from dilutions of sodium nitrate but are stable for only 10 to 15 minutes. (See Chapter 41 for preparation.) Positive and negative controls should be included.

Catalase activity

Acid-fast bacilli produce the enzyme **catalase,** which is detected by the breakdown of hydrogen peroxide and the active ebullition of gas (oxygen) bubbles. Detection of catalase activity, which should be tested routinely, has proved useful in several ways: (1) tubercle bacilli that become resistant to INH will lose or show reduced catalase activity; (2) the catalase activity of human or bovine strains may be selectively inhibited by heat[21]; and (3) a Group III mycobacteria *(M. gastri)* and isolates of clinically insignificant *M. kansasii* also lose their catalase activity at pH 7 and at 68° C.

---

*Puncturing the medium with the tip of the pipet aids in extracting niacin from the medium, especially if growth is confluent.
†Bacto TB Niacin Test Strips, Difco Laboratories, Detroit.

---

*Cater, J. C., and Kubica, G. P., unpublished data, have shown that Difco nitrate broth is a satisfactory substitute for the buffered nitrate substrate originally described.

Catalase activity may be determined in three ways:

1. At **room temperature** add 1 drop of a 1:1 mixture of 10% Tween 80* and 30% hydrogen peroxide† directly to a culture slant of modified L-J medium or the control quadrant of a 7H 10 agar plate. A **positive** catalase test is indicated by an active ebullition of grossly visible gas bubbles (nascent O_2 from breakdown of H_2O_2 by catalase) within 2 minutes.

2. To determine the effect of pH and temperature on catalase activity, scrape several spadefuls of growth from a culture slant and suspend in 0.5 ml. phosphate buffer pH 7 (M/15) in a screw-capped tube and place in a **68° C.** water bath for 20 minutes. After cooling to room temperature, add 0.5 ml. of the Tween 80–peroxide mixture, and observe the reaction for 15 or 20 minutes for gas bubbles.

3. A semiquantitative test, utilizing butt tubes (20 × 150 mm.) of L-J medium (available commercially‡) and a measurement of the height of the column of gas bubbles, has recently been introduced.[20,36] Inoculate the surface of the L-J medium with an actively growing egg slant or 7H 9 broth culture and incubate the tube, with loosened cap, at 36° C. for 2 weeks. Add 1 ml. of the Tween 80–peroxide mixture and after 5 minutes, measure the height of the gas bubbles; record as follows:

Negative = no bubbles
≤ 40 = less than 40 mm. bubbles (low)
≥ 50 = more than 50 mm. bubbles (high)

Interpretation. Except for *M. tuberculosis, M. gastri,* and low-catalase *M. kansasii* strains, all other mycobacteria commonly isolated from human sources retain their ability to produce catalase after heating at 68° C. at pH 7.

With the semiquantitative test, most Group II scotochromogens, Group IV rapid growers (including *M. fortuitum*), and *M. kansasii* strains, as well as some Group III nonphotochromogens *(M. terrae, M. triviale),* generally produce in excess of 50 mm. of bubbles.[18] *M. tuberculosis, M. bovis, M. avium—M. intracellulare,* as well as clinically insignificant strains of *M. kansasii* will show less than 40 mm. of foam; *M. marinum* and *M. xenopi* also produce less than 40 mm.

Tween 80 hydrolysis

Certain species of mycobacteria are capable of hydrolyzing Tween 80, a derivation of sorbitan monoleate, with the production of oleic acid. In the presence of a pH indicator, neutral red, this acid is indicated by a change in color of the solution from amber to pink.

The test is performed by the method of Wayne[40] (See Chapter 41 for reagents.) One 3-mm. loopful of an actively growing egg slant culture is suspended in a tube of the substrate, which is incubated at 36° C. A control tube of a known positive culture *(M. kansasii)* and an uninoculated (negative control) tube also are incubated. The tubes are examined after 5 days' and 10 days' incubation; a **positive** test is indicated by a color change of the fluid from amber to **pink** or **red.**

Generally most strains of *M. kansasii* are positive within 5 days, whereas many strains of *M. tuberculosis* are positive in 10 to 20 days. Clinically significant Group II and Group III cultures usually remain negative for 3 weeks, whereas the clinically insignificant strains are positive within 5 days.

Sodium chloride (NaCl) tolerance test[12]

This procedure is useful for the general separation of the rapid-growing mycobacteria (positive) from the slow growers (negative), also to aid in the identification of *M. triviale* (positive) and *M. flavescens* (sometimes positive).

---

*Atlas Chemicals Div., ICI America, Inc., Wilmington, Del.

†Superoxol, Merck & Co., Rahway, N. J.; must be refrigerated when not in use.

‡Lowenstein-Jensen Medium, Deeps, Difco Laboratories, Detroit.

The medium (not yet commercially available) is prepared by adding NaCl in a final concentration of 5% to either L-J or American Trudeau Society medium before inspissation. A light suspension (barely turbid) of growth from an L-J slant is prepared, and 0.1 ml. is inoculated to the surfaces of the NaCl medium and a control slant containing no NaCl. The tests are incubated at 36° C. and examined for growth at weekly intervals for 4 weeks; **growth on both media** is read as **positive.**

Arylsulfatase test

The enzyme arylsulfatase is present in varying concentration in many mycobacterial species, and its concentration in a controlled 3-day test is particularly useful in differentiating the potential pathogen, *M. fortuitum,* from other Group IV rapid growers.

This rapid test[35] is carried out by growing the suspected organism in Tween-albumin broth (available commercially) for 7 days and inoculating 0.1 ml. of this into a substrate containing 0.001 M tripotassium phenolphthalein disulfate,* along with known negative, weakly positive, and positive control cultures. These are incubated at 36° C for 3 days,† and 6 drops of 1 molar sodium carbonate are added to each tube. If appreciable amounts of arylsulfatase have been produced, phenolphthalein will be split from the sulfate and detected by alkalinizing with sodium carbonate. The resulting color is then compared with a set of standards (see Chapter 41 for preparation) immediately, since the color is not stable. A **positive** test shows a range from a faint pink (±) to a **light red** (3+); a negative test develops no color.

Growth on MacConkey agar

The procedure comprising growth on MacConkey agar, described by Jones and Kubica,[9] depends on the ability of *M. fortuitum* to grow on MacConkey agar in 5 days, thereby immediately differentiating it from other group IV rapid growers, which are inhibited.

A Petri dish containing 15 ml. of MacConkey agar* is inoculated on a turntable with a 3-mm. loopful of a 7-day Tween-albumin broth culture of the organism to be tested by rotating the plate and slowly moving the loop along the surface of the medium. The plate is incubated at 36° C. and observed at 5 days and 11 days for the presence of growth. A positive control of *M. fortuitum* and a negative control of *M. phlei* also should be included. Occasionally, Group IV organisms other than *M. fortuitum* will grow within 11 days.

Tellurite reduction

The tellurite reduction test[14] appears to be most valuable in the separation of the potentially pathogenic *M. arium–M. intracellulare* strains from the saprophytic nonphotochromogens. Middlebrook 7H 9 broth with ACD enrichment and Tween 80 is the base medium, to which is added a solution of tellurite. (See Chapter 39 for preparation.)

The medium is inoculated with the mycobacterium and incubated at 36° C. for 7 days. Two drops of a sterile 0.2% solution of potassium tellurite are then added to the medium, which is returned to the incubator and examined daily for reduction of the colorless tellurite salt to a black (or "dirty brown" in the case of highly pigmented strains) metallic tellurium.

The *M. avium-M. intracellulare* complex of Group III nonphotochromogens and the majority of Group IV rapid growers will reduce tellurite in 3 to 4 days; other mycobacteria, such as *M. terrae* and *M. xenopi,* react much more slowly.

Table 28-2 summarizes these tests.

---

*Nutritional Biochemical Corporation, Cleveland; L. Light & Co., Colinbrook, Bucks, England.
†A 2-week test, using a 0.003 M substrate, also may be used.

*Plates, rather than slant cultures should be used, since the latter medium will support growth of heavy inocula of *M. smegmatis.*

Table 28-2. Key differential features for clinically significant mycobacteria

SPECIES OF MYCOBACTERIA

TEST NO.	TUBERCULOSIS	AFRICANUM	BOVIS	KANSASII	MARINUM	SCROFULACEUM	GORDONAE	FLAVESCENS	XENOPI	AVIUM	INTRACELLULARE	GASTRI	TERRAE COMPLEX	TRIVIALE	FORTUITUM	CHELONEI*	VACCAE	SMEGMATIS	PHLEI
Group				GROUP I		GROUP II			GROUP III						GROUP IV				
(No. of strains)	239†	4	61	144	65	94	130	25	10	114	240	16	90	27	61	35	10	29	26
1. Speed of growth	Slow	Slow	Slow	Slow	Slow	Slow	Slow	Slow	Slow	Slow	Slow	Slow	Slow	Slow	Fast	Fast	Fast	Fast	Fast
2. Niacin	+(98)	+(75)	−(8)	−(2)	−	−	−	−	−	−(5)	−	−	−	−(4)	+(7)	−(3)	−	−	−
3. Nitrate reduction	+(98)	V(50)	−(2)	+(99)	−(3)	−(24)	−(4)	+	−	−(5)	−(8)	−	+(97)	+	+(98)	−	+(90)	+	+(96)
4. Catalase > 45 mm	−	−	−(2)	+(96)	−(22)	+(93)	+(92)	+(84)	+(90)	−(11)	−(3)	−	+	+(92)	+(98)	+(97)	+(90)	+(76)	+(96)
5. Catalase 68°/20 min	−(0.4)	−	−	+	V(41)	+	+(99)	+	+	+	+(94)	−	+(99)	+	+	+(94)	+	+(79)	+
6. Pigment, dark	−(0.4)	−	−	−(2)	−	+(95)	+(94)	+(96)	+‡	−(3)	−(2)	−	−(1)	−	−	−	V(50)	−	+
7. Pigment, photoactive	−	−	−	+(99)	+(91)	−§	−§	−	−	−‡(21)	−‡(1)	−	−(1)	−	−	−	+	−	−
8. Tween Hydrolysis — 5 day	−(20)	−	−	+(99)	+(97)	−	+(79)	+	−	−(2)	−(1)	+	+(94)	+(80)	−(26)	−(14)	+	+	+
8. Tween Hydrolysis — 10 day	−(39)	−	−	+(98)	+(98)	−	+(96)	+	−	−(2)	−(1)	+	+(99)	+	−(33)	−(20)	+	+	+
9. Tellurite reduction 3 days	−(2)	−	−	−	−	−(4)	−(1)	−(32)	−(10)	+(69)	+(83)	−	−	−	V(43)	V(51)	+(70)	+(90)	+(62)
10. NaCl tolerance	−(2)	−	−	−	−(3)	−	−	+(80)	−	−(2)	−(1)	−	−(10)	+	+	−*(80)	+(60)	+	+
11. Arylsulfatase 3 days	−	−	−	−(1)	−(6)	−	−	−	V(60)	−	−(2)	−(16)	−	−(26)	+(97)	+	−(20)	−(3)	−
12. MacConkey agar	−	−	−	−	−	−	−(1)	−	−	−(4)	V(45)	−	−(3)	−	+(97)	+(94)	−	−	−
Clinical significance	+	+	+	+	+	+	−	−	+	+	+	−	−	−	−	−	−	−	−

From Kubica, George: Amer. Rev. Resp. Dis. 107:9-21, 1973.

Definition of abbreviation and symbols: + or − indicates 100% of strains reacted as shown. Numbers in parentheses indicate percentage of strains positive in observed reaction, regardless of whether reaction was + or −. V indicates high degree of variability, with 40% to 60% of strains being reactive.

*Comprised of 2 distinct subspecies: M. chelonei (which contains the now illegitimate species M. borstelense) fails to grow on 5% NaCl, whereas M. chelonei ssp. abcessus does grow on NaCl.

†Figures indicate number of strains of each species.

‡Pigment increases with age and independent of light exposure.

§Pigment may intensity under influence of prolonged light exposure.

MYCOBACTERIUM LEPRAE (LEPROSY BACILLUS OR HANSEN'S BACILLUS)

Although the leprosy bacillus was discovered by Hansen in 1872 in the lepra cells (mononuclear epithelioid structures) of patients with leprosy, and although acid-fast bacilli have been isolated from these patients, the etiological agent has not yet been cultivated with certainty. The conditions concerning communicability of the disease are still unclear to us. Man is the only host; there is no animal or soil reservoir.

Typical cells of *Mycobacterium leprae* are found predominantly in smears and scrapings from the skin (not in the epidermis, but the corium) and mucous membrane (particularly the nasal septum) of patients with nodular leprosy. When stained by the Ziehl-Neelsen method, and decolorized and counterstained with 0.2% methylene blue in 4% sulfuric acid, large numbers of acid-fast bacilli are seen packed in lepra cells in parallel bundles, suggesting packets of cigars (globi). The bacilli are also found within the endothelial cells of blood vessels. Bacilli are rare or absent in tuberculous lesions and undifferentiated lesions.

The organism has not been grown successfully on artificial media or human tissue culture cells, nor has experimental infection been convincingly demonstrated in laboratory animals, although in 1960 Shephard was successful in producing infection in mouse foot pads with material from cases of human leprosy. Other myobacteria, such as *M. marinum* and *M. ulcerans,* also will grow in foot pads of mice, although the procedure is not practical as a diagnostic aid, due to the prolonged generation time of these organisms, especially *M. leprae.* Man is relatively resistant to infection. Geographical location is important.

There are no serological tests of value for diagnosis; serological tests for syphilis in lepers frequently yield biologically false positive results. Lepromin, a sterile extract of leprous tissue containing numerous *M. leprae* organisms, is of doubtful value as a diagnostic skin test.

Although they are bacteriostatic in their action, the sulfone drugs appear to be the current therapy of choice in the treatment of human leprosy. BCG vaccination also holds promise in prevention of the disease in contacts.

REFERENCES

1. Adams, R. M., Remington, J. S., Steinberg, J., and Seibert, J. S.: Tropical fish aquariums; a source of *Mycobacterium marinum* infections resembling sporotrichosis, J.A.M.A. **211**:457-461, 1970.
2. Chapman, J. S.: The anonymous mycobacteria in human diseases, Springfield, Ill., 1960, Charles C Thomas, Publisher.
3. Connor, D. H., and Lunn, H. F.: Buruli ulceration; a clinicopathologic study of 38 Ugandans with *Mycobacterium ulcerans* ulceration, Arch. Path. **81**:183-199, 1966.
4. Corper, H. J., and Uyei, N.: Oxalic acid as a reagent for isolating tubercle bacilli and a study of the growth of acid-fast nonpathogens on different mediums with their reaction to chemical reagents, J. Lab. Clin. Med. **15**:348-369, 1930.
5. Dross, I. C., Abbatiello, A. A., Jenney, F. S., and Cohen, A. C.: Pulmonary infections due to *M. fortuitum,* Amer. Rev. Resp. Dis. **89**:923-925, 1964.
6. Gruft, H.: Isolation of acid-fast bacilli from contaminated specimens, Health Lab. Sci. **8**:79-82, 1971.
7. Hobby, G. L., Redmond, W. B., Runyon, E. H., Schaefer, W. B., Wayne, L. G., and Wichelhausen, R. H.: A study on pulmonary disease associated with mycobacteria other than *M. tuberculosis;* identification and characterization of the mycobacteria. XVIII. A report of the Veterans Administration, Armed Forces Cooperative Study, Amer. Rev. Resp. Dis. **95**:954-971, 1967.
8. Jones, W. D., Abbott, V. D., Vestal, A. L., and Kubica, G. P.: A hitherto undescribed group of nonphotochromogenic mycobacteria, Amer. Rev. Resp. Dis. **94**:790-795, 1966.
9. Jones, W. D., and Kubica, G. P.: The use of MacConkey's agar for the differential typing of *M. fortuitium,* Amer. J. Med. Techn. **30**:187-190, 1964.
10. Joseph, S. W., Vaichulis, E. M. K., and Houk, V. N.: Lack of auramine-rhodamine fluorescence of Runyon Group IV mycobacteria, Amer. Rev. Resp. Dis. **95**:114-115, 1967.
11. Kenney, M., Loechel, A. B., and Lovelock, F. J.: Urine cultures in tuberculosis, Amer. Rev. Resp. Dis. **82**:564-567, 1959.
12. Kestle, D. G., Abbott, V. D., and Kubica, G. P.: Differential identification of mycobacteria. II. Subgroups of Groups II and III (Runyon) with different clinical significance, Amer. Rev. Resp. Dis. **95**:1041-1052, 1967.

13. Kestle, D. G., and Kubica, G. P.: Sputum collection for cultivation of mycobacteria; an early morning specimen or the 24- to 72-hour pool? Amer. J. Clin. Path. **48**:347-349, 1967.

14. Kilburn, J. O., Silcox, A., and Kubica, G. P.: Differential identification of mycobacteria. V. The tellurite reduction test, Amer. Rev. Resp. Dis. **99**:94-100, 1969.

15. Konno, K.: New chemical method to differentiate human-type tubercle bacilli from other mycobacteria, Science **124**:985, 1956.

16. Kubica, G. P.: Personal communication, 1968.

17. Kubica, G. P.: Differential identification of mycobacteria, Amer. Rev. Resp. Dis. **107**:9-21, 1973.

18. Kubica, G. P., and Dye, W. E.: Laboratory methods for clinical and public health mycobacteriology, Public Health Service Pub. No. 1547, Washington, D. C., 1967, U. S. Government Printing Office.

19. Kubica, G. P., Dye, W. E., Cohn, M. L., and Middlebrook, G.: Sputum digestion and decontamination with N-acetyl-L-cysteine-sodium hydroxide for culture of mycobacteria, Amer. Rev. Resp. Dis. **87**:775-779, 1963.

20. Kubica, G. P., Jones, W. D., Jr., Abbott, V. D., Beam, R. E., Kilburn, J. O., and Cater, J. C., Jr.: Differential identification of mycobacteria. I. Tests on catalase activity, Amer. Rev. Resp. Dis. **94**:400-405, 1966.

21. Kubica, G. P., and Pool, G. L.: Studies on the catalase activity of acid-fast bacilli. I. An attempt to subgroup these organisms on the basis of their catalase activities at different temperatures and pH, Amer. Rev. Resp. Dis. **81**:387-391, 1960.

22. Kubica, G. P., and Vestal, A. L.: Tuberculosis—laboratory methods in diagnosis, Atlanta, 1959, National Communicable Disease Center.

23. Lorian, V.: Direct cord reading medium for isolation of mycobacteria, Appl. Microbiol. **14**:603-607, 1966.

24. Marks, J., and Schwabacher, H.: Infection due to *Mycobacterium xenopei,* Brit. Med. J. **1**:32-33, 1965.

25. National Tuberculosis and Respiratory Disease Association: Diagnostic standards and classifications of tuberculosis, New York, 1969, the Association.

26. Oliver, J., and Reusser, T. R.: Rapid method for the concentration of tubercle bacilli, Amer. Rev. Tuberc. **45**:450-452, 1945.

27. Petroff, S. A.: Some cultural studies on the tubercle bacillus, Bull. Johns Hopkins Hosp. **26**:276, 1915.

28. Runyon, E. H.: Anonymous mycobacteria in pulmonary disease, Med. Clin. N. Amer. **43**:273-290, 1959.

29. Runyon, E. H.: Identification of mycobacterial pathogens utilizing colony characteristics, Amer. J. Clin. Path. **54**:578-586, 1970.

30. Runyon, E. H., Selin, M. J., and Harris, H. W.: Distinguishing mycobacteria by the niacin test—a modified procedure, Amer. Rev. Tuberc. **79**:663-665, 1959.

31. Sommers, H. M., and Russell, J. P.: The clinically significant mycobacteria; their recognition and identification, Chicago, 1967, Commission on Continuing Education, American Society of Clinical Pathologists.

32. Truant, J. P., Brett, W. A., and Thomas, W., Jr.: Fluorescence microscopy of tubercle bacilli stained with auramine and rhodamine, Henry Ford Hosp. Med. Bull. **10**:287-296, 1962.

33. Vestal, A. L.: Procedures for the isolation and identification of mycobacteria, Publ. Health Serv. Publication No. 1995 (available at the Center for Disease Control, Atlanta), 1969.

34. Virtanen, S.: A study of nitrate reduction by mycobacteria, Acta Tuberc. Scand. (supp.) **48**:1960.

35. Wayne, L. G.: Recognition of *Mycobacterium fortuitum* by means of a three-day phenolphthalein sulfatase test, Amer. J. Clin. Path. **36**:185-187, 1961.

36. Wayne, L. G.: Two varieties of *M. kansasii* with different clinical significance, Amer. Rev. Resp. Dis. **86**:651-656, 1962.

37. Wayne, L. G.: Classification and identification of mycobacteria. III. Species within Group III, Amer. Rev. Resp. Dis. **93**:919-928, 1966.

38. Wayne, L. G., and Doubek, J. R.: Diagnostic key to mycobacteria encountered in clinical laboratories, Appl. Microbiol. **16**:925-931, 1968.

39. Wayne, L. G., Doubek, J. R., and Diaz, G. A.: Classification and identification of mycobacteria. IV. Some important scotochromogens, Amer. Rev. Resp. Dis. **96**:88-95, 1967.

40. Wayne, L. G., Doubek, J. R., and Russell, R. L.: Classification and identification of mycobacteria; tests employing Tween 80 as substrate, Amer. Rev. Resp. Dis. **90**:588-597, 1964.

41. Wayne, L. G., Krasnow, I., and Kidd, G.: Finding the "hidden positive" in tuberculosis eradication programs; the role of the sensitive trisodium phosphate–benzalkonium chloride (Zephiran) culture technique, Amer. Rev. Resp. Dis. **86**:537-541, 1962.

42. Whitcomb, F. C., Foster, M. C., and Dukes, C. D.: Increased carbon dioxide tension and the primary isolation of mycobacteria, Amer. Resp. Dis. **86**:584-586, 1962.

43. Wolinsky, E.: Identification and classification of mycobacteria other than *Mycobacterium tuberculosis.* In Hobby, G. L., editor: Handbook of tuberculosis laboratory methods, Washington, D. C., 1962, Veterans Administration, U. S. Government Printing Office.

44. Youmans, G. P.: The pathogenic "atypical" mycobacteria, Ann. Rev. Microbiol. **17**:473-494, 1963.

29 Spirochetes

Spirillum
Borrelia
Treponema
Leptospira

The spiral bacteria, including members of the genera *Spirillum, Borrelia, Treponema,* and *Leptospira,* are unicellar organisms with a flexuous structure, varying in length from 1 to 500 μ. All exhibit at least one complete turn, and all are **motile,** but generally are not flagellated; movement is usually accomplished by spinning, turning, or flexing about their long axes. The spiral bacteria stain poorly by the ordinary bacterial stains; Wright or Giemsa stains usually yield the best results, if the organisms are stainable.

SPIRILLUM MINUS

This organism, already discussed in Chapter 6 on blood cultures, is one of the causes of rat bite fever (soduku), a disease characterized by an initially inflamed (occasionally ulcerating) wound associated with lymphadenopathy and an erythematous rash. The organism may be visualized by dark-field examination in material from the bite wound, affected lymph node, or in the blood. Blood films also should be stained by the Wright or Giemsa technique and examined microscopically.

When the microscopic examination is unrevealing, one must resort to animal inoculation, using at least several mice and a guinea pig. These are injected intraperitoneally with 1 to 2 ml. of the patient's blood, and the peritoneal fluid (mice) or defibrinated blood (guinea pig) is examined by dark-field microscopy in 3 to 4 weeks for the presence of short, thick, actively motile spiral forms of 2 to 6 spirals and bipolar tufts of flagella.

BORRELIA SPECIES

These spirochetes normally are parasitic for several species of arthropods, including body lice and ticks; man or other animals acquire infection (causing **relapsing fever**) by the bite of the infected vector. *Borrelia duttonii* (tick-borne), *B. novyi,* and *B. recurrentis* (both louse-borne) are the most frequently involved in human disease, but many other species or subspecies may be recovered[3].

Borreliae are best demonstrated in the blood of an infected individual early in the course of a febrile period by direct examination of stained blood films. When this is unrewarding, a similar specimen of blood is inoculated intraperitoneally into young rats, and the animals are examined daily for *Borrelia* organisms in films of tail blood. The organisms are 10 to 20 mm. long, with 5 to 7 open spirals, and are actively motile in a corkscrew fashion; they stain well by the Giemsa technique.[3]

TREPONEMA SPECIES

There are numerous species in the genus *Treponema,* and they comprise two major groups: those normally present in the mouth, urogenital tract, or gastrointestinal tract of man and other animals, and the several pathogenic species. In the former group are found *T. macrodentium* and *T. microdentium* from the oral cavity, *T. genitalis* from the genital area, and so forth. In the latter group are *T. pallidum,* the causative agent of **syphilis,** *T. pertenue,* from the tropical disease **yaws;** and *T. carateum,* which causes a chronic skin disease, endemic in Central and South America.

All the treponemes are actively motile and contain numerous tight, rigid coils. They are difficult to stain, but are best observed by **dark-field microscopy** (see Chapter 2). This is an important laboratory procedure, since a diagnosis of syphilis in the early stages can be immediately confirmed by demonstrating *T. pallidum* in material from suspected lesions (usually anogenital) or affected regional lymph nodes.

Dark-field examination for T. pallidum

The following procedure is recommended for obtaining material for a dark-field examination:

1. After donning rubber gloves, thoroughly cleanse the suspected lesion with a gauze sponge, removing crusts if present (after soaking in saline), and abrading it to produce fluid.
2. Pinch the lesion with gloved fingers so that a drop of fluid is expressed from its border or surface; if none is obtained (avoid blood), moisten with a drop of saline and after a minute or two, repeat the attempt to obtain fluid.
3. Touch a cover slip to the fluid, invert it over a microscope slide, and immediately examine for characteristic forms of motile *T. pallidum,* using dark-field microscopy and an oil-immersion objective.
4. Look for a thin, tightly wound, corkscrew-shaped organism of 8 to 14 uniform, rigid spirals, slightly longer than the size of an average red blood cell. Characteristic movement is slowly backward or forward, or a corkscrew rotation about the long

axis, sometimes with slight flexion.
5. Cautiously interpret a positive preparation from a lesion in the mouth—the commensal treponemes may be confused with *T. pallidum.*
6. A direct FA staining procedure is presently available for delayed examination of fluids to detect the presence of *T. pallidum* (see p. 79).

It is beyond the scope of this text to discuss the clinical manifestations of primary, secondary, or tertiary syphillis; nor is it possible in the space allotted to consider in detail the serological tests available for the laboratory diagnosis of syphilis. We shall delineate, therefore, the broad concepts of serological testing and refer the interested reader to standard texts on the subject.

Serological tests for syphilis

Since *T. pallidum* is not cultivable by the usual laboratory procedures, the serological test for syphilis (STS) becomes a valuable diagnostic aid, especially when preparations for dark-field examination of the primary lesion are no longer available. These tests are performed on blood (serum or plasma) and cerebrospinal fluid specimens, and are of two types[2,6]:

1. **Nontreponemal** antibody tests—utilize a cardiolipin-lecithin antigen and are used to detect an antibody-like agent called **reagin.** In theory, this substance results from the reaction of treponemes on body tissues, and the tests for its presence, such as the **VDRL** (Venereal Disease Research Laboratory) or **RPR** (rapid plasma reagin) **flocculation** tests, or the Kolmer **complement fixation** test, are not entirely specific for syphilis detection. Because of their sensitivity, ease of performance and low cost, the flocculation tests are now widely used as screening procedures and have largely replaced the more cumbersome and less sensitive complement fixation test.[2]
2. **Treponemal** antibody tests, such as

the **TPI** (*Treponema pallidum* immobilization) test and the **FTA—ABS** (fluorescent treponemal antibody-absorption) test employ as an antigen the causative agent of syphilis: *T. pallidum.* In the TPI test, live treponemes are combined with the patient's serum and complement; if the serum contains treponemal antibodies it causes immobilization of the treponemes. In the FTA-ABS test, the treponemes are fixed on a glass slide and combined with serum (previously absorbed to remove nonspecific treponemal antibodies). If the test serum contains syphilis antibodies, it will combine with the *T. pallidum* antigen and coat it; this reaction can be visualized by the addition of a fluorescein-labeled antihuman globulin that reacts with the serum-coated treponemes, causing them to fluoresce when examined with a fluorescent microscope. If no syphilis antibodies are present in the test serum, the treponemes will be invisible when examined by the fluorescence technique.

Infrequently, positive reactions with nontreponemal tests may result from a cause other than syphilis, such as atypical pneumonia, infectious mononucleosis, acute febrile illnesses (as in viral infections or smallpox vaccination), malaria, autoimmune or collagen disease, and so forth. These biological false positive **(BFP)** reactions are generally negative when repeated using a treponemal antigen, as in the FTA-ABS test. If the FTA-ABS test is reactive, however, a serological diagnosis of syphilis has been made, since the treponemal antibody tests have shown excellent specificity.[6]

GENUS LEPTOSPIRA

The principal sources of leptospirae infecting man are from natural waters, soils, vegetation, food stuffs, and so forth. The organisms enter through abrasions of the skin or via the mucosal surfaces of the body; occupational exposure is a prime factor in acquiring the infection. Clinical manifestations range from a mild illness to a severe icteric (jaundice) disease with liver and kidney complications.

The species of *Leptospira* are also thin, flexible, tightly coiled spirals of approximately the same size as the species of *Treponema.* One or both ends of the spirochete, being more flexible than the center portion, may be bent to form a hook. Spinning takes place on the long axis of the cell. If one end only is hooked, forward movement is in the direction of the straight end. With both ends hooked, a lashing effect may be noticed as the cell spins.

Leptospira icterohaemorrhagiae

L. icterohaemorrhagiae causes infectious leptospiral jaundice, or **Weil's disease.** The organism is aerobic and may be cultivated in a supplemented ascitic fluid medium, as can be spirochetes of *Borrelia,* or in the medium recommended by Fletcher (Chapter 39), consisting of salts, amino acids, and rabbit albumin.

The spirochetes can be observed in blood drawn from a patient during the initial stage of the disease (see Chapter 6 for discussion of blood culture of *Leptospira*) when the symptoms are chills, fever, headache, and intestinal discomfort. During the second stage, manifested by icterus and skin hemorrhages, the spirochetes may be readily observed in the urine by the dark-field technique.[1]

The **serological** detection of leptospiral antibodies in man and animals is an invaluable adjunct in the diagnosis of leptospirosis. Various techniques have been introduced, including the complement fixation test, the microscopic and the macroscopic agglutination test, the genus-specific hemolytic test, and the FA test. Stable, formalized suspensions of *Leptospira* serotypes are available* and can be used satisfactorily in a rapid macroscopic slide ag-

---

*Difco Laboratories, Detroit.

glutination test for serological diagnosis of the infection.

For further reading on leptospirosis, the reader is directed to an excellent chapter by Alexander and colleagues in the fifth edition of the American Public Health Association's *Diagnostic Procedures,* edited by Bodily and others, 1970.

Other Leptospira species

Other species of *Leptospira* will produce jaundice, but the disease is usually milder. *L. grippotyphosa, L. autumnalis, L. canicola, L. pomona,* and *L. hebdomadis* may be involved in human leptospirosis. The symptoms will vary in severity with the species.

REFERENCES

1. Alston, J. M., and Broom, J. C.: Leptospirosis in man and animals, London, 1958, E. & S. Livingston, Ltd.
2. Center for Disease Control: The laboratory aspects of syphilis, Atlanta, 1971, The Center.
3. Hardy, Jr., P. H.: Borrelia. In Blair, J. E., Lennette, E. H., and Truant, J. P., editors: Manual of clinical microbiology, Bethesda, Md., 1970, American Society for Microbiology.
4. Manual of tests for syphilis, Public Health Service Pub. No. 411, Washington, D. C., 1969, U. S. Government Printing Office.
5. Rhode, P., editor: BBL manual of product and laboratory procedures, Cockeysville, Md., 1969, Baltimore Biological Laboratory, Division of Bio-Quest, Becton, Dickinson and Co.
6. Wallace, A. L., and Norins, L. C.: Syphilis serology today, Prog. Clin. Path. **2:**198-215, 1969. (Reprinted by Center for Disease Control, Atlanta, 1971.)

30 Miscellaneous pathogens

Aeromonas
Streptobacillus
Mycoplasma

AEROMONAS HYDROPHILA

Members of the genus *Aeromonas* are gram-negative, motile rods with polar flagella that utilize carbohydrates **fermentatively,** with the production of acid or acid and gas, and are oxidase positive. Their normal habitat appears to be natural bodies of water, nonfecal sewage, marine life, and foods; they have also been isolated from fecal specimens of asymptomatic persons. *A. hydrophila* has been suggested as the type species of the genus.*

An increasing incidence in human infections caused by *A. hydrophila* has been reported,[6-9,13] and includes infected traumatic wounds (some with a history of exposure to soil or water), septicemia complicated by a hematological disorder, postoperative wound infections (usually in mixed cultures), and so forth.

A. hydrophila grows well on routine laboratory media, producing on blood agar small (1 to 3 mm.), smooth, convex **beta-hemolytic** colonies that become dark green after 3 to 5 days.[8] Good growth also occurs on the enteric isolation media, including EMB and MacConkey agar, SS agar, and others, both at 25° and 36° C. On TSI medium an alkaline slant over an acid and gas butt is observed; indole is produced from tryptophane, which sepa-

---

*The reader is referred to an excellent monograph on the *Aeromonas* group by Ewing and co-workers.[6]

rates aeromonads from pseudomonads, alcaligenes, flavobacteria, and others. Demonstration of a **positive oxidase reaction** (Kovacs' method) and **polar flagellation** are most useful in distinguishing these organisms from other facultative gram-negative bacilli of clinical importance.

Aeromonas hydrophila is highly susceptible to the aminoglycosides gentamicin, kanamycin, and streptomycin, and also to chloramphenicol and tetracycline; little activity is demonstrated with penicillin, carbenicillin, ampicillin, or cephalothin.[13]

GENUS STREPTOBACILLUS

Only one species, *Streptobacillus moniliformis,* is contained in the genus *Streptobacillus.* This is an **extremely pleomorphic, aerobic,** gram-negative organism that forms irregular chains of fairly uniform, small, slender rods 2 to 4 μ in length, to long, interwoven, looped or curved filaments up to 150 μ in length. These may be interspersed with fusiform enlargements and large, round, *Candida*-like swellings along the length of the filament (*moniliformis* = necklacelike), two to five times its width. These morphological forms depend largely on the medium used, the conditions of incubation, and the age of the culture; the regular, rod-shaped cells predominate in clinical specimens and under favorable artificial conditions.

Streptobacillus moniliformis is a normal inhabitant of the throat and nasopharynx of wild and laboratory rats and mice. It is extremely virulent for mice. Man usually acquires the infection by the **bite of a rat**, mouse, or other rodent, although ingestion of contaminated food, particularly milk, may also cause infection (Haverhill fever). The infection is characterized by fever, rash, and polyarthritis. The organism may be recovered from the blood, joint fluid, or pus.

Streptobacillus moniliformis requires the presence of natural body fluid, such as ascitic fluid, blood, or serum, for its growth on artificial media. The addition of 10% to 30% ascitic fluid to thioglycollate medium makes an excellent recovery medium. The streptobacillus grows in this in the form of "fluff balls" or "breadcrumbs" near the bottom of the tube or on the surface of the sedimented red cells if blood is cultured (Chapter 6). These colonies may be removed with a sterile Pasteur pipet and transferred to solid media or to slides for microscopic examination. (Giemsa or Wayson stains are preferable to the Gram stain.)

On slightly soft media, such as ascitic fluid agar or serum agar, incubated aerobically (some strains require increased carbon dioxide, as in a candle jar) and with an excess of moisture, small discrete colonies develop in 2 to 3 days. These are smooth and glistening, irregularly round with sharp edges, and colorless or grayish. Beneath or adjacent to the colonies of *Streptobacillus*, the microscopic L-form colony may develop; these measure 100 to 200 μ in diameter and show a dark center and lacy periphery ("fried egg" appearance). This L form is considered to be a variant of *Streptobacillus moniliformis*, arising spontaneously or by stimulation, and is characterized by a colony of special appearance. It is more resistant to penicillin than the bacterial-phase colonies of *Streptobacillus*; consequently penicillin can be used to separate the L component.

Conversely, L forms may persist when patients are treated with penicillin; treatment with another antibiotic may be required to eliminate the L organism. Morphologically, L-form colonies consist of tiny, bipolar-staining, coccoid or coccobacillary elements. These colonies may be maintained in pure culture and can be re-transformed to the *Streptobacillus* form by subculturing on fluid or semisolid media. L forms are nonpathogenic for mice, whereas *Streptobacillus moniliformis* is highly virulent; vaccines prepared from these L forms do not protect against infection by streptobacilli.

Agglutinins are formed in patients with *Streptobacillus* infections; titers of 1:80 or greater are considered indicative of infection.

For a further discussion of *Streptobacillus moniliformis*, the reader is referred to current texts.

MYCOPLASMA (PPLO) SPECIES

Dienes and Edsall in 1937 were the first to report pleuropneumonia-like organisms (PPLO) from a pathological process in man. Since that time, the role of these small microorganisms, which can be cultured on lifeless culture media, has been studied in infections of the urogenital and respiratory tracts, in wound infections, as tissue culture contaminants, and in animal hosts other than man. Mycoplasmas are generally considered as physiologically intermediate between the bacteria and rickettsiae, and may be either parasitic or saprophytic. One species, *Mycoplasma pneumoniae*, is a recognized human pathogen—the etiological agent of primary atypical pneumonia and bronchitis; other species are suspected as possible agents in nongonococcal urethritis, infertility, early abortion, rheumatoid arthritis, myringitis, erythema multiforme, and so forth.[3]

Presently there are six identified species of *Mycoplasma* indigenous to man. These include the following, showing their sites of recovery:

M. hominis	Oropharynx,[a] genitourinary tract,[a] blood,[b] pleural fluid,[b] anal tract[a]
M. fermentans	Genitourinary tract[a]
M. salivarium	Oropharynx[a]
M. orale types I,II,III	Oropharynx,[a] bone marrow[b]
M. pneumoniae	Oropharynx,[c] lung[b]
T-strains	Genitourinary tract[a]

---

[a]From both healthy and ill persons.
[b]Only from ill persons.
[c]Primarily from ill persons.

The parasitic mycoplasmas are best recognized by such characteristics as their growth requirements (enriched media with sterols, conditions of incubation, and so forth), colonial morphology, inhibition by specific antisera, and resistance to penicillin. Another clue to their identity is the anatomical site from which the specimen was obtained; for example, *M. pneumoniae* from lower respiratory tract secretions, T-strain mycoplasma from a genitourinary source, and so forth.

Growth of the parasitic mycoplasmas is best encouraged on an enriched medium containing heart infusion agar,* horse serum, yeast extract, and penicillin G (see Chapter 39 for preparation). Colonies are best observed microscopically under 45× magnification and appear as large (250 to 750 μ in diameter) to small (1 to 10 μ), raised, pitted, and lacy to coarsely pebbled colonies. The central portion of the colony grows into the agar medium,† appearing darker than the periphery when examined by transmitted light, thus giving it the characteristic "fried egg" appearance. Growth in enriched broth is not visible due to the small size of the organisms— subculture to solid media is required—and organisms are not demonstrable by the usual staining methods. Stained prepara-

tions are best studied by the Dienes method, in which an agar block bearing the selected colony is cut out with a sterile scalpel and placed upright on a glass microscope slide. A coverslip previously coated with the Dienes stain* and dried is carefully lowered, stain side down, upon the agar block, and the edges are sealed with valspar. The preparation may then be examined under oil immersion. Colonies of mycoplasma and bacteria (if selected in error) stain blue in a short time; after 15 minutes, however, the bacterial colonies will have lost their color due to the reduction of methylene blue by metabolizing organisms, while the mycoplasmas will retain their stain. The method of Clark and others[1] is also recommended for microscopic study of mycoplasmas.

Isolation from clinical material

Primary cultures of clinical specimens— usually swabs from the throat, genital tract, rectum, wounds, and so forth—are placed in a tube of a selective broth containing bacterial inhibitor,† incubated for several days at 36° C., then subcultured to freshly prepared solid media and incubated as before. *Mycoplasma pneumoniae* grows in air or 5% CO_2; *M. fermentans*, *M. orale*, and *M. salivarium* require an anaerobic atmosphere of 95% N_2 and 5% CO_2 in an evacuation-replacement jar. Plates are generally examined at 7, 14, and 21 days; those for *M. pneumoniae* should be held for 30 days before discarding. For primary cultivation of T-strain mycoplasmas, the use of a differential agar medium[10] containing urea and manganous sulfate is recommended.‡ The T-strain mycoplasmas have the unique property of hydrolysing urea and will appear on this medium as **dark, golden brown** colonies.

---

*Mycoplasma agar base, Baltimore Biological Laboratory; PPLO agar, Difco Laboratories, Detroit. Broth media are also available.
†Colonies cannot be transferred with conventional needles or loops—an agar block cut from the plate is used.

*Dienes stain contains methylene blue, agure II, and other ingredients; see Chapter 39 for preparation.
†See Chapter 39 for preparation.
‡Preparation of this medium is described in Chapter 39.

Identification by growth inhibition test

Early reports by various workers indicated that species–specific antisera prepared in rabbits were inhibitory to the growth of mycoplasmas and that this could be used in their identification.[2] Filter paper discs are saturated with individual growth-inhibiting antisera against prototype strains* and pressed onto an agar plate that has been seeded with a pure culture of the unknown strain. After incubation at 36° C. for 1 week (or until good growth occurs), the plate is examined microscopically (10 to 45 × magnification). A **clear zone** around any disc greater than the clear zone around a control disc that contains normal rabbit serum identifies the species of mycoplasma being tested. Since inhibition zones have been shown to be a function of the size of the inoculum and may vary from 0.1 to 17 mm. in diameter, the most frequent error of "no zones" around any disc indicates that too heavy an inoculum was used. This may be remedied by repeating the test with varying dilutions of the unknown strain or by cutting out an agar block of growth and pushing it across the surface of the test plate in **one direction only,** and placing two discs along the line of inoculation, 1 inch apart. Further details of the disc technique may be found in the late York Crawford's excellent chapter in the ASM manual.[3]

Other identification tests

Several other in vitro tests are available, directed primarily to the identification of M. pneumoniae; perhaps the most useful being the adsorption of erythrocytes onto colonies of M. pneumoniae. The test, as modified by Del Guidice,[5] is carried out by flooding a primary recovery plate with 2 ml. of a 1% suspension of sheep erythrocytes in phosphate buffer solution, allowing it to stand for 10 minutes at room tem-

perature, and then pouring off the red cell suspension. Next, the surface of the culture plate is washed over with a 1:100 dilution of Dienes stain, and examined microscopically under 10× and 45× magnification. The red cells will coat the colonies of M. pneumoniae, and the Dienes methylene blue stain will be taken up. Other mycoplasmas do not show this phenomenon.

M. pneumoniae colonies also will produce beta hemolysis when coated with a layer of blood agar prepared with sheep or guinea pig erythrocytes; other species are said to produce alpha, alpha-prime, or nonhemolytic reactions.[11]

M. pneumoniae colonies will reduce a tetrazolium salt incorporated in agar under aerobic conditions; the area around the colony becomes pink. Both M. pneumoniae and M. fermentans will utilize dextrose with the production of acid but no gas, a useful presumptive test.[3] Mycoplasma colonies stained with homologous FA conjugates also will show a characteristic yellow-green fluorescence when examined by incident light fluorescence microscopy.[4] These culture reactions apparently are not rewarding in the identification of other mycoplasmas; the recommended technique is the use of growth-inhibitions with specific antisera previously described.

Primary atypical pneumonia caused by M. pneumoniae has been most successfully treated with the tetracyclines and erythromycin; the penicillins, lincomycin, and ampicillin have not proved effective.

REFERENCES

1. Clark, H. W., Fowler, R. C., and Brown, T. M.: Preparation of pleuropneumonia-like organisms for microscopic study, J. Bacteriol. 81:500, 1961.
2. Clyde, W. A., Jr.: Mycoplasma species identification based upon growth inhibition by specific antisera, J. Immunol. 92:958-965, 1964.
3. Crawford, Y. E.: A laboratory guide to the mycoplasmas of human origin, Naval Medical Research Unit No. 4, 1972, Great Lakes, Ill. 60088.
4. Del Guidice, R. A., Robillard, N. F., and Carski, T. R.: Immunofluorescence identification of My-

*Hyperimmune rabbit antisera are available for the mycoplasma prototypes previously listed from Microbiological Associates, Inc., Bethesda, Md.

coplasma on agar by use of incident illumination, J. Bact. **93:**1205-1209, 1967.

5. Del Guidice, R. A., and Pavia, R.: Hemadsorption by *Mycoplasma pneumoniae* and its inhibition with sera from patients with primary atypical pneumonia, Paper M 143, p. 71, Bact. Proc., 1964.

6. Ewing, W. H., Hugh, R., and Johnson, J. G.: Studies on the Aeromonas group, Atlanta, 1961, National Communicable Disease Center.

7. Gifford, R. R. M., Lambe, D. W., Jr., McElreath, S. D., and Vogler, W. R.: Septicemia due to *Aeromonas hydrophila* and *Mima polymorpha* in a patient with acute myelogenous leukemia, Amer. J. Med. Sci. **263:**157-161, 1972.

8. Gilardi, G. L., Bottone, E., and Birnbaum, M.: Unusual fermentative, gram-negative bacilli isolated from clinical specimens. II. Characteristics of *Aeromonas* species, Appl. Microbiol. **20:**156-159, 1970.

9. Nygaard, G. S., Bissett, M. L., and Wood, R. M.: Laboratory identification of aeromonads from men and other animals, Appl. Microbiol. **19:**618-620, 1970.

10. Shepard, M. C., and Howard, D. R.: Identification of "T" mycoplasmas in primary agar cultures by means of a direct test for urease, Ann. N. Y. Acad. Sci. **174:**809-819, 1970.

11. Somerson, N. L., Taylor-Robinson, D., and Chanock, R. M.: Hemolysin production as an aid in the identification and quantitation of Eaton agent *(Mycoplasma pneumoniae)*, Amer. J. Hyg. **77:**122-128, 1963.

12. von Graevenitz, A., and Mensch, A. H.: The genus *Aeromonas* in human bacteriology, New Eng. J. Med. **278:**245-249, 1968.

13. Washington, J. A., Jr.; *Aeromonas hydrophila* in clinical bacteriological specimens, Ann. Intern. Med. **76:**611-614, 1972.

PART FIVE

Viruses and rickettsiae

31 Laboratory diagnosis of viral and rickettsial diseases

It is to be recognized that in a text of this type it would be virtually impossible to provide adequate detail on the diagnoses of rickettsial and viral diseases. Because of this, we have attempted to provide only some guidelines and instructions on the collection and handling of appropriate specimens. For information on virus identification the reader may refer to Lennette,[5] Lennette and colleagues,[6] Lennette and Schmidt,[7] Herrmann,[3] and Horsfall and Tamm.[4]

Most diagnostic microbiology laboratories are not capable of performing isolation procedures for viruses and rickettsiae due to lack of expertise and appropriate equipment. Serological methods, however, can be carried out on a limited basis in many laboratories.

Smaller laboratories should probably not become involved in viral diagnoses, but all laboratories can become knowledgeable about **collection, packaging,** and **transportation** of specimens to virus reference laboratories, the services of which are now more available.

GENERAL CONSIDERATIONS

The viruses are the smallest "living" things, and they exist as obligate intracellular parasites. As an order they can be divided into two groups, based on the type of nucleic acid they contain—ribonucleic acid (**RNA** viruses) or deoxyribonucleic acid (**DNA** viruses). On the basis of the host attacked, the viruses may be placed taxonomically in suborders—plant, insect, bacterial, or animal viruses. The animal viruses are generally destroyed by temperatures sufficient to kill most pathogenic bacteria, but they are preserved by freezing at low temperatures ($-20°$ C.). Some viruses can withstand lyophilization. Many are adequately preserved in 50% glycerol solutions for reasonable periods of time.

Because of their unique metabolic and reproductive mechanisms, viruses and rickettsiae have not yet been cultivated on inanimate media; they require living tissue for their propagation. With the advances in tissue and egg culture and the production of uniformly susceptible strains of mice, viral diagnoses can be made with greater facility and rapidity in reference laboratories.

Satisfactory antigens are available commercially,* which permit average-sized hospital laboratories to perform diagnostic serological tests almost routinely, especially those tests that involve complement fixation.

AVAILABLE TYPES OF LABORATORY EXAMINATIONS

Four types of laboratory examination are presently available for the diagnosis of viral and rickettsial infections:

---

*Microbiological Associates, Inc., Bethesda, Md.

1. Isolation and identification of the agent by inoculation of susceptible animals, chick embryo, or tissue culture
2. Detection and measurement of antibodies developing during the course of the disease (serological tests)
3. Histological examination of infected tissue, particularly material from fatal cases
4. Detection of viral antigens in lesions, using fluorescent antibody (FA) techniques

COLLECTION OF SPECIMENS FOR VIRAL AND RICKETTSIAL DIAGNOSIS

It is essential that appropriate and timely specimens be collected, packaged, and shipped, or stored with due consideration for the lability of rickettsial and viral organisms, if one is to gain any meaningful clinical information from subsequent analyses.[1] An ill-advised specimen collection, that is, one made randomly when a viral or rickettsial disease is being considered and shipped to a reference laboratory, misuses a viral diagnostic service and can only yield a negative result. A laboratory request for "viral studies" is meaningless —only a properly provided clinical history will suffice (see p. 233).

The following methods for the preparation of diagnostic specimens are those recommended by the Division of Laboratories, New Jersey State Department of Health (Martin Goldfield, M. D., Director).[2]

Stool specimens. Collect prune-sized (15 to 25 gm.) samples in dry, sterile, glass screw-capped jars; label with name and date collected, then store in a freezer until ready for shipment.

Throat washings. Obtain throat washings during the first 3 days of illness, while the patient is febrile. Using sterile tryptose phosphate or trypticase soy broth rather than saline, direct the patient to gargle repeatedly, cough to raise tracheal secretions, and then expectorate into a wide-mouthed, screw-capped jar provided for the purpose. This is sealed and refrigerated (not frozen). Collect specimens in the morning to enable delivery to the virus laboratory the same day in order to preserve viability and permit prompt inoculation.

Spinal fluid. Collect about 5 ml. of fluid in a dry sterile tube, label properly, and **freeze immediately.** Keep frozen until ready for shipment.

Postmortem tissues. Place aliquots of various specimens in dry, sterile, screw-capped jars, seal tightly, label, and store in a freezer. **Exception:** In cases of suspected viral encephalitis, remove the whole brain and refrigerate—**do not freeze.**

Conjunctival swabs. Use sterile cotton applicators and a test tube containing 2 ml. of sterile broth. Moisten applicator with broth; swab upper and lower palpebral conjunctivae; return swab to broth so that cotton tip is submerged; seal tightly, label, and store in freezer until ready for shipment.

Blood specimens. Whole blood specimens are collected in dry sterile tubes without anticoagulants. **Never freeze whole blood**—store in refrigerator for a period not longer than 3 days. It is preferable, however, to immediately separate the serum aseptically and store it in the freezer along with the refrigerated virus isolation specimens until ready for shipment.

Paired sera are required to establish a serological diagnosis. The first **(acute)** blood should be collected as soon as possible after onset of symptoms; the second **(convalescent)** blood should be collected 3 to 6 weeks later, except in suspect influenza cases, when a 2-week interval is advised.

The information in Table 31-1[2] will provide some guidelines for the diagnosis of the various types of viral diseases.

ISOLATION OF VIRUS FROM CLINICAL SPECIMENS

The material that follows will represent only generalizations relative to the isola-

Table 31-1. Isolation of viruses—specimens recommended

TYPE OF INFECTION		SPECIMEN RECOMMENDED	
Central nervous system			
Polioviruses	Lymphocytic choriomeningitis	Nonfatal case:	Fatal case:
ECHO viruses	Eastern, Western equine	3 stools	Brain and spinal cord
Coxsackieviruses	encephalitis	Spinal fluid	Segment of colon
Mumps	Rabies	Acute and convalescent	Postmortem blood
Herpes simplex	St. Louis, California, Japanese	bloods	
	B encephalitis		
Respiratory		Nonfatal case:	
Influenza		Throat washings	
Adenoviruses		Acute and convalescent bloods	
Parainfluenza viruses		Fatal case:	
Respiratory syncytial		Tracheobronchial swab	
Rhinovirus		Nasal and pharyngeal swabs	
Psittacosis		Throat washings, sputum	
Q fever		Blood; lung, spleen, and liver at postmortem	
Enteroviruses		3 stools, throat washings	
Morbilliform eruptions		Blood, throat swabs	
Measles		3 stools	
ECHO viruses		Throat washings	
Coxsackieviruses		Acute and convalescent bloods	
Typhus		Blood	
Rocky Mountain spotted fever		Blood (liver, spleen in fatal cases)	
Vesicular eruptions			
Herpes simplex		Throat washings, vesicle fluid	
Smallpox		Vesicle fluid or lesion scrapings	
Vaccinia		Acute and convalescent bloods, throat washings, vesicle fluid	
Rubella Varicella-zoster		Blood, throat and nasal secretions	
Rickettsialpox		Blood (liver, spleen in fatal cases)	
Enteroviruses		3 stools, throat washings	
Eye infections			
Adenoviruses		Conjunctival swab or corneal scrapings	
Herpes simplex		Acute and convalescent bloods	
Miscellaneous diseases			
Herpangina		3 stools, throat washings	
Pleurodynia		Acute and convalescent bloods, feces, throat washings	
Mumps		Acute and convalescent bloods, spinal fluid, throat washings	
Dengue		Blood	
Lymphogranuloma venereum		Buboes, lymph nodes	
Cytomegalovirus		Urine, throat swab, paired blood specimens, throat washings	
Gastroenteritis		3 stools	

tion and identification of viruses. It is recommended that the reader consult the references given at the end of the chapter.

The successful isolation of a viral agent depends upon proper collection of appropriate material from the patient, careful preservation while transporting it to the laboratory, and inoculation of the material—after eliminating viable bacteria—into susceptible animals, fertilized eggs, or suitable tissue cultures. The presence of virus is then demonstrated by the development of characteristic lesions in the animal or egg and by cellular changes (cytopathogenic effect, or CPE), observed microscopically in the tissue cultures. It is finally identified by immunological procedures, such as serum neutralization, followed by animal injection or hemagglutination inhibition tests.

Material for virus isolation must be kept frozen during transport; this requires the use of Dry Ice.* Since Dry Ice releases carbon dioxide, which is deleterious to most viruses, specimen containers must be tightly sealed. Specimens of cerebrospinal fluid, feces, sputum, throat swabs, vesicle fluid, tissue samples, and so forth in glass or plastic containers are closed with tightly fitting, rubber-sealed lids or stoppers, wrapped with tape, or sealed by dipping into melted paraffin. These specimens are packed well in Dry Ice, surrounded by insulating material such as paper or cotton, and shipped to the reference laboratory by the most rapid delivery service available (messenger or air express).

If Dry Ice is not available to keep the specimen frozen during transport, 50% neutral glycerol (reagent grade) may be used to preserve solid specimens, such as small pieces of tissue obtained at biopsy or autopsy, fecal material, or suspensions of mucus. The majority of bacteria die off in 3 to 5 days; many viral agents will remain alive for several days at moderate temperatures.

SEROLOGICAL DIAGNOSIS OF VIRAL AND RICKETTSIAL INFECTIONS

Serological tests are the most used and most economical of the diagnostic aids, and will provide information in the absence of a virus isolation. The diagnosis, however, may be in retrospect because a significant rise (fourfold or greater) in antibody titer must be demonstrated. Occasionally it may be necessary to carry out both isolation and serological procedures in order to evaluate the laboratory findings properly.

Past infection with a specific virus is indicated by a persistent antibody level, whereas a significant **rise in titer** during

*Dry Ice can be obtained from many ice cream plants.

convalescence offers further proof of the etiological role of the virus previously isolated from the patient. The serological tests in present use include the following: (1) complement fixation test, (2) neutralization test, (3) hemagglutination inhibition test, (4) nonspecific agglutination test, and (5) fluorescent antibody test.

Similar in principle to the bacterial complement fixation tests, the **viral complement fixation** tests are employed with increasing frequency for many viral diseases. In the past the difficulty with the reaction has been in obtaining concentrated antigens free of interfering substances. Recent refinements, including the development of purified commercially available antigens, and improvements in technique have made the complement fixation test a highly satisfactory procedure.

Generally, complement-fixing antibodies appear during the first to third week of illness, rise to a plateau, drop in titer for a few months, and disappear entirely within a year.

The principle underlying the **neutralization test** is that the specific virus-neutralizing antibodies can be measured by adding serum suspected of containing these antibodies to a suspension of virus and inoculating the mixture into a group of susceptible animals or tissue cultures. A control group that receives virus and serum **without** antibody should be set up. If the first group fails to develop the disease or produce CPE, while the control group becomes infected, the presence of neutralizing antibody has been established. In some infections, tissue cultures (in poliomyelitis) or embryonated eggs (in mumps) have been used instead of experimental animals.

A number of viruses have the capacity to agglutinate the red blood cells of chickens, guinea pigs, human Group O, and other species. This reaction may be specifically inhibited by immune or convalescent sera. This forms the basis of a test for detecting **hemagglutinating** viruses and their homologous antibodies and for measuring the amount of virus in a sample or the antibody content of a serum.

Specific virus agglutination tests are rarely used in the serological diagnosis of viral disease because of lack of the required amounts of pure viral antigen. This is due to difficulties in preparation. The **nonspecific agglutination** tests, however, are widely used in the diagnosis of rickettsial, viral, and other diseases. These include the Weil-Felix reaction used in the diagnosis of typhus and spotted fevers, the heterophile agglutination test in the diagnosis of infectious mononucleosis, and the cold agglutinin and *Streptococcus MG* agglutination tests in the diagnosis of primary atypical pneumonia. Since the reagents used in these tests are readily available in most laboratories, a description of some will be given in the section on miscellaneous serological procedures.

Fluorescent antibody tests may be effectively used for determining the concentration of serum antibody in a patient. Aliquots of infected cell suspension are smeared and dried on a slide. To these may be added several dilutions of the patient's serum, followed by the addition of an antihuman globulin-fluorescein conjugate. The end point is the highest dilution of serum bound to the cellular antigen as indicated by immunofluorescence.

Blood specimens for serological testing must be **paired;** that is, they must comprise both **acute** and **convalescent** phase sera, since only a rise in specific antibody titer can be considered indicative of current infection with the causative viral or rickettsial agent. The first specimen, 5 to 10 ml. of blood, is collected in a sterile tube within the first 3 days of onset of symptoms, and a second specimen is collected 2 to 6 weeks later. A third or fourth specimen may be required occasionally; these are collected biweekly.

CLINICAL HISTORY

A clinical history should always accompany any request for viral and rickettsial serology or isolation. It should indicate pertinent information such as the following: (1) date of onset of infection, (2) date specimen was collected, (3) clinical symptoms, essential laboratory data, (4) suspected clinical diagnosis or differential diagnoses, (5) vaccination history, history of exposure to animals, similar cases in family or vicinity, and so forth, and (6) antibiotic therapy—amount and dosage. Finally, the tests requested should be specified.

WHERE TO SEND SPECIMENS

If the hospital or other local laboratory is unable to carry out the desired viral or rickettsial studies, the following sources may be investigated:

1. Municipal, county, or state health department laboratories
2. The Center for Disease Control **(only through the local public health or state laboratory)**
3. Local medical or research centers
4. Military and Veterans Administration laboratories
5. Private laboratories engaged in virology

In suspected viral diseases of obscure etiology or in instances of major outbreaks, **prompt consultation with the virologist should be the rule rather than the exception.**

REFERENCES

1. Behbehani, A. M.: Human viral and rickettsial diseases, ed. 2, Kansas, 1968, University of Kansas Medical Center.
2. Goldfield, M.: Personal communication, 1966.
3. Hermann, E. C., Jr.: Viral diagnosis for medical practice, Mayo Clin. Proc. **42:**112-123, 1967.
4. Horsfall, F. L., and Tamm, I., editors: Viral and rickettsial diseases of man, ed. 4, Philadelphia, 1965, J. B. Lippincott Co.
5. Lennette, E. H.: Laboratory diagnosis of viral infections; general principles, Amer. J. Clin. Path. **57:**737-750, 1972.
6. Lennette, E. H., et al: Clinical virology; introduction to methods, Manual of clinical microbiology, Bethesda, Md., 1970, American Society for Microbiology.
7. Lennette, E. H., and Schmidt, N. J.: Diagnostic procedures for viral and rickettsial diseases, ed. 3, New York, 1964, American Public Health Association, Inc.

PART SIX

Fungi

32 Laboratory diagnosis of mycotic infections

Although the pathogenicity of certain fungi has been recognized since the first half of the nineteenth century, laboratory expertise in the handling of clinical specimens and in the subsequent isolation and identification of the causative agents of fungal disease has been developed only in comparatively recent years. It is recognized now that the **mycoses,** those diseases with fungal etiology, are more common than before, and there are suggestions that their incidence may have been increased through the widespread use of antibacterial agents, immunosuppressive drugs, and other patent modifications. It is incumbent on microbiologists, therefore, to become more knowlegeable about the pathogenic fungi and their identifying characteristics.

From basic microbiology, one is reminded that fungi are multicellular heterotrophic members of the plant kingdom that lack roots and stems and are referred to as **thallophytes.** In addition, they are devoid of chlorophyll and fundamentally consist of a basic, branching, intertwining structure called a **mycelium,** composed of tubular filaments known as **hyphae** (sing. **hypha**). The latter may possess cross walls or **septa,** in which case the mycelium is referred to as being **septate;** or in the absence of septa, where the filaments are continuous, the mycelium is said to be **aseptate.** The mycelial cottony mass constitutes the **colony.**

In septate hyphae, only one nucleus is found per segment or cell, whereas aseptate hyphae exhibit a multinucleated condition in the continuous filament. The mycelium also has two parts: the **vegetative** part that grows in or on the substrate, absorbing nutrients, and the **reproductive** or aerial part that projects above the substrate, producing fruiting bodies bearing characteristic spores. Mature spores, on dissemination when arriving on a suitable substrate, germinate by producing a **germ tube** that finally leads to a new mature microscopic plant.

Fungi may reproduce sexually or asexually, or by both means. **Sexual** reproduction is associated with the formation of specialized structures that facilitate fertilization and nuclear fusion, resulting in the production of specialized spores called **oospores, ascospores,** and **zygospores.** Fungi that exhibit a sexual phase are known as **perfect** fungi. **Imperfect** fungi are those in which no sexual phase has been demonstrated; the spores are produced directly by or from the mycelium. Most of the fungi of medical importance belong to the imperfect group, although the possibility of future recognition of a perfect phase must not be excluded.*

Since the form of sporulation and the type of spore are important criteria in the

---

*The perfect state of some dermatophytes has been described.[3,12]

identification of the various fungi, the following morphological data are provided to aid in the characterization of these microorganisms.

The simplest type of sporulation is the development of the spore directly from the vegetative mycelium, and the spore is known as a **thallospore.** Three types of thallospores are recognized: **blastospores** (Fig. 32-1, *A*), simple budding forms in which the daughter cell is abstricted from a single mother cell or forms laterally from a mycelium or pseudomycelium, as in *Candida* species; **chlamydospores** (Fig. 32-1, *B*), thick-walled, resistant, resting spores produced by the rounding up and enlargement of the terminal cells of the hyphae, as in *Candida albicans;* and **arthrospores** (Fig. 32-1, *C*), resulting from simple fragmentation of the mycelium into cylindrical or cask-shaped, thick-walled spores, as in *Geotrichum candidum.*

Conidia are asexual spores produced singly or in groups by specialized vegetative hyphal stalks called **conidiophores** (Fig. 32-2, *B* and *C*). The conidia are freed from the point of attachment by pinching off, or abstriction. Some conidiophores become swollen at the end, and over their swollen surfaces are formed numerous small flask-shaped stalks, from which conidia in chains **(catenate)** are pushed out. The swollen portion of the conidiophore is called a **vesicle;** the flask-shaped structures are **sterigmata** (Fig. 32-2 *C*). Many fungi produce conidia of two sizes: **Microconidia** (Fig. 32, *D* and *E*) are small, unicellular, round, elliptical or pear shaped (pyriform); **macroconidia** (Fig. 32-2, *F* to *H*) are large, usually septate, club shaped (clavate) or spindle shaped (fusiform). The microconidia may be borne directly on the hyphae (Fig. 32, *D*) and are said to be **sessile,** or they may develop directly from the end of a short conidiophore (Fig. 32-2, *E*) and are called **pedunculate.** If conidia have a rough or spiny surface, they are called **echinulate.**

Sporangiospores (Fig. 32-2, *A*) are asexual spores contained in **sporangia** produced terminally on **sporangiophores** (aseptate stalks). Sporulation takes place by a process called progressive cleavage. During maturation within the sporangium, the protospores divide into definitive uninucleate sporangiospores, which are released by irregular rupture of the sporangial wall (Fig. 32-2, *A*). This form of sporulation occurs in the Phycomycetes, which exhibit an aseptate mycelium.

Other types of spores, produced sexually by the perfect fungi, have been referred to earlier in this chapter. Additional information may be gained from texts on general mycology.[8,11]

Since the techniques commonly employed in medical bacteriology are not always practical in the identification of fungi, these microorganisms must be recognized primarily by their **gross** and **microscopic** characteristics. This brings to focus certain questions: Is the colony rapid growing (2 to 5 days) or slow growing (2 to 3 weeks)? Is it flat, heaped up, or regularly or irregularly folded? Is its texture creamy and yeastlike, or is it smooth and skinlike (glabrous)? Does it produce a mycelium that is powdery, granular, velvety, or cottony? Is a distinct

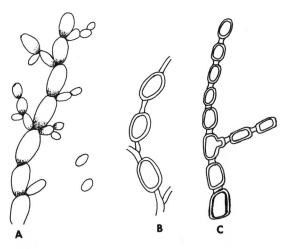

Fig. 32-1.   Asexual spores produced from hyphae. **A,** Blastospores. **B,** Chlamydospores. **C,** Arthrospores.

surface pigment observed, and is the pigment similar on the reverse side? These are the criteria that are important in the **gross** characterization of an unknown fungus.

A small portion of the colony and some of the medium may be removed with a 22-gauge Nichrome needle or loop, and the mycelium teased apart in a drop of lactophenol cotton blue mounting fluid on a glass slide. This is covered with a thin coverglass, heated *gently* over the pilot flame of a Bunsen burner, and examined **microscopically,** using both low- and

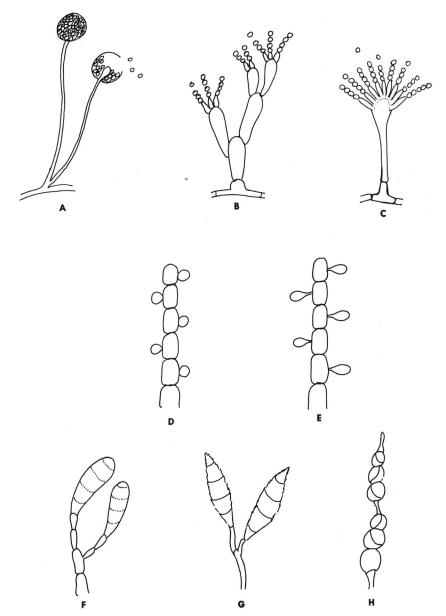

Fig. 32-2. Asexual spores produced on specialized hyphae. **A,** Sporangiospores in a sporangium. **B,** Conidia on branched conidiophore. **C,** Conidia on unbranched conidiophore. **D,** Sessile microconidia. **E,** Pedunculate microconidia. **F** to **H,** Various macroconidia.

high-power objectives. The size, shape, septation, and color of spores, if present, may be observed, and the morphology of the specialized structures bearing the spores is noted.

Frequently, it may be possible to identify a culture solely by this direct examination. Often, however, more sophisticated procedures may be required, and the preparation of a **slide culture** or hanging drop for the demonstration of the undisturbed relation of spores to specialized hyphae, such as conidiophores, is recommended. This technique will be described in a later section.

Special culture media designed to depress vegetative growth and stimulate **sporulation** of a fungus or to produce a special type of growth may also be required. Such media include potato dextrose agar, wort agar, brain-heart infusion agar, rice grain medium, chlamydospore agar, and others; their preparation and use is discussed in the section on culture media.

To identify a fungal culture properly, it may be necessary on occasion to carry out various **biochemical** studies, such as sugar fermentations, nitrate assimilation tests, carbon and nitrogen utilization studies, and specific vitamin requirements. When indicated, these will be described.

Finally, it may be necessary in certain instances to carry out **animal pathogenicity tests** in order to identify some species. These will be discussed under the specific diseases.

Saprophytic fungi are frequently found as **contaminants** of clinical specimens as well as in the laboratory, where they readily contaminate cultures. It is essential, therefore, that the medical mycologist become well versed in differentiating these forms from recognized pathogens. Furthermore, some of these common saprophytes are important **opportunistic invaders** in debilitated patients who may have been treated with the multiplicity of antimicrobial agents presently available to the clinician. Their recognition, therefore,

is becoming important. The laboratorian should become familiar with about twelve genera commonly considered as contaminants. Much help will be obtained from the excellent texts referred to at the end of this chapter.[12,14,17,24] Excellent training courses also are offered by the Mycology Unit of the Center for Disease Control (CDC).*

In familiarizing oneself with the characteristics of the saprophytic genera, including *Penicillium, Aspergillus, Paecilomyces, Fusarium, Scopulariopsis, Rhizopus, Mucor,* and others, not only does the student acquire experience in mycological techniques but he gains an opportunity to recognize fungal morphology and its relation to the taxonomy of the group.

GENERAL LABORATORY METHODS

Mycological examination of all clinical material should include a direct microscopic examination, culturing on appropriate media, and if indicated, inoculation of susceptible laboratory animals. Although it is true that mycelial fragments, spores, and spore structures can sometimes be demonstrated in unstained preparations, it is essential that **all specimens are cultured** in an attempt to isolate the causative agent.

Collection of specimens

A prime requisite to good medical mycology is a **properly collected and properly handled specimen.** Procedural details will be given under the specific diseases; therefore, the following are general instructions only.

To obtain specimens of **skin or nails,** the affected site is carefully washed with 70% isopropanol and, after drying, the lesion is scraped with a sterile scalpel and the material obtained is placed in a sterile Petri dish or on a piece of white paper carefully folded in a packet to prevent loss

---

*Contact the Office of Training Activities, Laboratory Branch, Center for Disease Control, Atlanta, Ga. 30333.

of the specimen. **Hairs** from infected areas are clipped or plucked and sent to the laboratory in a similar manner.

In the **subcutaneous mycoses** (see later in this chapter), a variety of materials may be submitted, including pus or exudate from draining lesions, material aspirated with syringe and needle from unopened abscesses or sinus tracts, or biopsied tissue.* These should be placed in sterile tubes or Petri dishes and submitted directly to the laboratory. If mailing of the specimen is necessary, the material first should be inoculated to a suitable culture medium. **Under no circumstances** should glass or plastic Petri dishes containing clinical specimens or fungus cultures be sent through the mails; they invariably break in transit. Neither should inoculated cotton swabs be mailed, as they are usually dried out on arrival. For best results, **only pure cultures** on agar slants should be mailed. Scrapings of skin, or nails and hairs can be mailed in appropriate containers.

Material from suspected cases of **systemic mycoses** includes such varied specimens as blood, cerebrospinal fluid, sputum, bronchial secretions and gastric washings, pus and exudates from abscesses and draining sinuses, sternal marrow, and tissue. These specimens should be placed in sterile tubes or bottles and submitted **promptly** to the laboratory.

In order to aid the laboratory in the proper examination of the specimen, some indication of the **suspected disease** should be noted on the laboratory request slip, along with the specimen source. This is necessary to guide laboratory personnel in the selection of the proper medium and method of incubation. For example, the recovery of *Actinomyces israelii* requires special enriched media and anaerobic incubation.

---

*Tissue for fungus culturing should **not** be ground in a tissue grinder—it fragments and may kill the larger fungal elements. Rather, the specimen should be minced with a sterile scalpel blade and pieces embedded directly into the culture medium.

Direct microscopic examination

The following clinical specimens are preferably examined in the **unstained state:** sputum and bronchoscopic secretions, gastric washings, pus and exudates, sediments of cerebrospinal fluid, pleural effusions, and urine. Several loopfuls of the material are placed on a clean glass slide, covered with a thin coverglass, and examined under both the low- and high-power objectives, using reduced light. If the material is opaque, 10% sodium hydroxide may be added and gentle heat applied. These preparations should be carefully searched for the following: broad mycelial fragments with septa (*Aspergillus* and *Penicillium* species) or without septa (*Mucor* species), arthrospores (*Geotrichum* species), budding cells (*Candida, Cryptococcus,* and *Blastomyces* species), or filamentous mycelial fragments (*Actinomyces* and *Norcardia* species).

Dried and fixed films may be stained by the Gram method (mycelium and spores are gram positive), Kinyoun's acid-fast stain for *Nocardia asteroides* and Wright or Giemsa stain to reveal the presence of *Histoplasma capsulatum* in the macrophages or phagocytes of blood or bone marrow. An India ink preparation of cerebrospinal fluid may reveal encapsulated forms of *Cryptococcus neoformans.* The sulfur granules present in actinomycotic pus, when washed and stained, may demonstrate gram-positive, nonacid-fast, narrow, branched filaments of *Actinomyces israelii.*

Cultural procedures

Although the basic principles of microbiological technique apply to the mycologist as well as the bacteriologist, certain differences should be noted. A 22-gauge Nichrome needle, flattened at the end to a spade shape, is used to transfer mycelial growth. When the needle contains infectious material, care should be exercised to prevent spattering when flaming it; the needle should be gradually heated in the less intense part of the flame. Two stiff, sharp-pointed, teasing needles in holders are useful in tearing apart the mycelial mat when immersed in mounting fluid on a glass slide.

Large (18 by 150 mm.) borosilicate test tubes without lips are to be recommended for solid culture media rather than Petri dishes to minimize the hazard of spore dissemination. Screw-capped test tubes used for cultures are **not** recommended; they promote anaerobiosis and retain moisture, both of which prevent maximum sporulation from occurring. The large cotton-plugged test tubes afford ease of storage and handling, are less easily broken, and allow for a **thick butt** of agar that will withstand drying during extended incubation. Pigment production is enhanced when there is free circulation of air, and a dry surface encourages the development of an aerial mycelium and spores.

Sabouraud dextrose agar (hereafter called Sabouraud agar) at pH 5.6, and brain-heart infusion agar, with or without added blood (hereafter called BHI agar), are the most useful media for primary isolation of most pathogens. (The preparation of these media is discussed in Chapter 39.) The addition of 0.5 mg. per milliliter of cycloheximide and 0.05 mg. per milliliter of chloramphenicol to these media will effectively inhibit the growth of contaminating bacteria and saprophytic molds, especially when material likely to contain these contaminants in large numbers is cultured, such as skin and nail scrapings, sputum, pus, or autopsy material. On BHI agar with added antibiotics, pathogenic fungi will develop their typical colonial morphology, color, and microscopic appearance and can generally be identified without further subculturing. It is well to note, however, that certain pathogens are partially or completely inhibited by these antibiotics. Included here are: *Cryptococcus neoformans, Candida* species (including *C. parapsilosis* and *C. krusei*), and *Trichosporon cutaneum.* The yeast phases of *Histoplasma capsulatum* and *Blastomyces dermatitidis* are susceptible to cycloheximide when incubated at 36° C. but not at 25° C.; *Allescheria boydii* and *Aspergillus fumigatus* are partially sensitive, but cycloheximide may inhibit sporulation. *Nocardia asteroides, N. brasiliensis* and other actinomycetes are mildly susceptible to chloramphenicol.

Some cultures neither sporulate nor produce pigment satisfactorily on Sabouraud agar. To induce these, special media, including potato-dextrose agar, potato-carrot agar, cornmeal agar, rice grain agar, or Sabouraud agar with added thiamine and inositol have proved useful and can be recommended.

All isolation media should be held for a minimum of 4 weeks before discarding.

Slide culture

Microscopic observation of fungi in the natural state is often necessary for identification. The following method (Fig. 32-3) has proved successful in our hands for the culture of fungi on a glass slide:

1. With forceps dip a clean slide in alcohol, flame, and burn off. Repeat the procedure. Place in a sterile Petri dish to cool.

2. Carefully apply aseptically quick-drying, tubed, plastic cement to the slide to form three sides of a square, equal in area to that of a coverglass, or slightly less. The cement walls should be 1 to 2 mm. high.

3. With a sterile dropping pipet, place sufficient melted Sabouraud agar at 45° C. in the square to fill the space to the top of the walls. Place the slide in the Petri dish and allow the agar to set firmly.

4. Apply a minimal amount of inoculum (spores) to the center of the agar surface and then cover squarely with a sterile coverglass (alcohol flamed and cooled).

5. Place a wide strip of filter paper in the Petri dish and moisten with a few drops of water. Cover with lid, which assures a moist atmosphere.

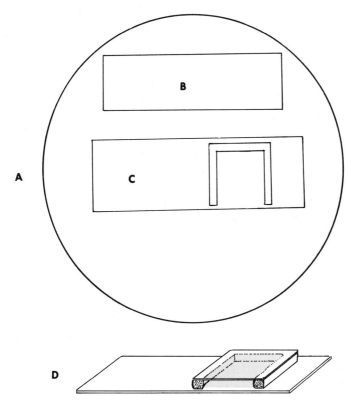

Fig. 32-3. Slide culture. **A,** Petri dish. **B,** Strip of filter paper. **C,** Slide with cement walls. **D,** Slide with agar medium and coverglass in place.

6. Incubate at the desired temperature and examine after 48 hours or until sporulation occurs. This is determined by quick microscopic examination under the low-power objective.

7. At the desired stage, remove the damp filter paper and add a dry strip. Add several drops of formalin to the strip and cover the dish.

8. Allow to remain for 30 minutes. Remove the slide, blot away excess moisture, and examine under high-dry objective, with appropriate light adjustment.

9. Make a permanent mount by applying nail polish or asphalt tar varnish to all four sides of the square.

Hanging drop culture

A hanging drop culture preparation is recommended also for examining fungi in the natural state. Introduce a minute inoculum of the culture into a drop of Sabouraud broth on a coverglass, which is inverted over a glass microslide ring* coated at both ends, with petroleum jelly. Place this assembly on a 3- by 1-inch glass slide (with several drops of water for moisture) and incubate at 25° C. for 2 to 3 days until sporulation occurs.

Do not make slide cultures of *Histoplasma, Blastomyces,* or *Coccidioides.* The procedure is too hazardous.

STUDY OF THE MYCOSES

Of the more than 50,000 valid species of fungi, only about fifty species are generally recognized as being pathogenic for man. These organisms normally live a saprophytic existence in soil enriched by

_____

*A. H. Thomas Co., Philadelphia.

decaying nitrogenous matter, where they are capable of maintaining a separate existence, with a parasitic cycle in humans or animals. The systemic mycoses are not communicable in the usual sense of man-to-man or animal-to-man transfer; man becomes an **accidental host** by the inhalation of spores or by their introduction into his tissues through trauma. Unusual circumstances, reflecting an **altered suceptibility** of the host, may also produce lesions by fungi that are normally considered saprophytes. Such conditions may occur in patients with debilitating diseases or diabetes or those suffering from impaired immunological mechanisms resulting from steroid or antimetabolite therapy. Prolonged administration of antibiotic agents may also upset the host's normal microbiota, resulting in a **superinfection** by one of these fungi. The neophyte mycologist is cautioned, therefore, not to discard cultures as "contaminants" without first checking the clinical history of the patient.

The **mycoses** may be conveniently placed in three groups, based on the tissues involved[6]:

1. The **dermatophytoses,** including the superficial and cutaneous mycoses.
2. The **subcutaneous** mycoses, which involve the subcutaneous tissues and muscles.
3. The **systemic** mycoses, which involve the deep tissues and organs. These are the most serious of the groups.

Dermatophytoses

As the name suggests, these fungal diseases include those infections that involve the superficial areas of the body, namely the **skin, hair,** and **nails.** The etiology for these diseases for the most part relates to the genera *Microsporum, Trichophyton,* and *Epidermophyton.* Such cutaneous mycoses are probably the most common fungal infections of man and are usually referred to as **tinea** (Latin word for gnawing worm or ringworm). The gross appear-ance is that of a ring that surrounds the infected area. They may be characterized or qualified by another Latin noun in the genitive form to designate the area involved, for example, tinea corporis (body), tinea cruris (groin), tinea capitis (scalp and hair), tinea barbae (beard), tinea unguium (nail), and others. These fungi break down and use keratin (keratinolytic) as a source of nitrogen, but are incapable of penetrating the subcutaneous layers.

Tinea versicolor (pityriasis), a disease of the skin characterized by superficial brownish scaly areas on the trunk, arms, shoulders, and face, is widely distributed throughout the world. It is caused by *Melassezia furfur,* of which the mycelial fragments and clusters of thick-walled, yeast-like spores may be observed microscopically in skin scrapings.

The reader is referred to some excellent texts for a further description of the less common dermatophytoses.[12,14,24]

Direct examination

The presence of the fungi can be readily demonstrated in direct slide preparations of digested skin scales, nail scrapings, or hair. The skin of the involved area is cleansed with 70% alcohol, and some epidermal scales at the active edge of the lesion are stripped off and placed on a microscope slide containing a drop of 10% sodium hydroxide. A coverglass is added, and the slide is gently warmed over a small flame to just short of boiling. The slide is then examined under low- and high-power magnification, using much-reduced light, for the presence of long branching threads of young hyphae or of older septate hyphae and barrel-shaped arthrospores. Occasionally, budding yeasts may be seen. Fungal elements must be differentiated from fibers of cotton, wool, and other fabrics as well as from a mosaic of cholesterol crystals and other artifacts.

Specimens of nail scrapings must be secured from the deeper layers of the infected nail; they are handled as previously described for skin scrapings.

Infected hairs are selected either by their characteristic appearance (broken-off hairs and twisted grayish stubs) or by their bright yellow-green fluorescence when examined under a Wood's lamp, using filtered ultraviolet light. Invasion of the inside of the hair (endothrix) or outside the hair shaft (ectothrix), as determined microscopically, can be hélpful in identifying the fungus involved.

Some workers find a stained preparation easier to examine than an unstained sodium hydroxide mount. The most convenient stain digestant is prepared from equal parts of Parker 51 blue-black ink and 10% sodium hydroxide and used as previously described. One part of ink plus 4 parts of hydroxide will give a lighter stain and is desired by some.

Cultural procedures

Specimens of skin and nail scrapings are obtained as described previously; hair stubs or scrapings of areas showing loss of hair (alopecia) are obtained without attempting to cleanse the scalp unless a fungicide has been applied, but a gentle wiping with 70% alcohol will result in less contamination.

The specimen may be either inoculated directly to media in the clinic or submitted to the laboratory in disposable Petri dishes or clean paper envelopes. The upper portion of the hairs should be clipped off with alcohol-flamed scissors; only the lower ends should be inoculated to media.

Primary isolation of the dermatophytes is readily accomplished by inoculating the hairs or scrapings on the surfaces of Sabouraud agar slants; the specimen should be partially embedded in the agar. It is advisable to inoculate duplicate sets of media, one of plain medium and one containing cycloheximide and chloramphenicol to inhibit the common bacterial and mold contaminants. These agents do not alter the cultural characteristics of dermatophytes but promote greater facility in isolation.

All media should be incubated at room temperature (not over 30° C.) and examined at 5-day intervals for at least a month before discarding. Sporulation of the dermatophytes generally occurs within 5 to 10 days of inoculation, and cultures should be examined during this period since characteristic colonial appearance and microscopic morphology are more easily recognized before the cultures age.

Common species

Species of *Microsporum* attack the hair and skin and include *M. audouinii, M. canis,* and *M. gypseum. Trichophyton* species are responsible for infection of the hair, skin, and nails and include principally *T. mentagrophytes, T. rubrum, T. tonsurans, T. schoenleinii, T. violaceum,* and *T. verrucosum. Epidermophyton* causes infection of the skin and nails and includes a single species, *E. floccosum.*

Since these dermatophytes generally present an identical appearance in microscopic examinations of infected skin or nails, final identification can only be made by culture.

Descriptions of the ten principal species of fungi involved in dermatophytoses in the United States follow, and other geographically limited species are described in the references cited.

Genus Microsporum

The genus *Microsporum* is immediately identified by the presence of large (8 to 15 μ by 35 to 150 μ), spindle-shaped, rough or spiny macroconidia with thick (up to 4 μ) walls and containing 4 to 15 septa (Fig. 32-4). The microconidia are small (3 to 7 μ), club shaped, and borne on the hyphae and are either sessile or on short sterigmata. Cultures of *Microsporum* develop slowly or rapidly and produce an aerial mycelium that may be velvety, powdery, glabrous, or cottony, varying in color from whitish, buff, and bright yellow to a deep cinnamon brown, with varying shades on the reverse side of the colony.

M. audouinii is the most important cause of epidemic tinea capitis among school

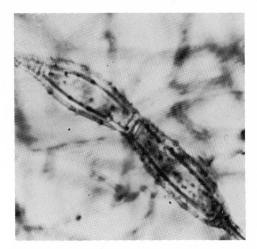

Fig. 32-4. *Microsporum audouinii,* showing macroconidium. (1800×.)

Fig. 32-5. *Microsporum gypseum,* showing several spindle-shaped, thick-walled, multi-celled macroconidia. (500×.)

children in the United States, but it rarely infects adults. The fungus is known as an **anthropophilic,** or "man-loving," fungus and is spread directly by means of infected hairs on headware, upholstery, combs, or barbers' clippers. The majority of infections are chronic in nature; some heal spontaneously, whereas others may persist for several years.

On Sabouraud agar, *M. audouinii* grows slowly, producing a flat gray to tan colony with short aerial hyphae and a radially folded surface. The reverse of the colony is generally reddish brown in color. *M. audouinii* sporulates poorly on Sabouraud agar, and the characteristic macroconidia may be lacking in some cultures. The addition of yeast extract will stimulate growth and production of both macroconidia and small club-shaped microconidia borne laterally along the hyphae. Abortive and bizarrely shaped macroconidia, hyphal cells with swollen ends (racquet hyphae), abortive branches (pectinate bodies), and chlamydospores are commonly observed.

M. canis is primarily a pathogen of animals **(zoophilic);** it is the most common cause of ringworm in dogs and cats in the United States. Children and adults acquire the disease through contact with infected animals, particularly puppies and kittens,

although human-to-human transfer has been reported. Hairs infected with *M. canis* **fluoresce** a bright yellow green under a Wood's lamp, which is a useful tool for screening animal pets as possible sources of human outbreaks. On direct examination in 10% sodium hydroxide, small spores (2 to 3 μ) are found outside the hair (ectothrix), although cultural procedures must be carried out for specific identification.

On Sabouraud agar, *M. canis* grows rapidly as a flat, disclike colony with a bright yellow periphery and possesses a short aerial mycelium. On aging (2 to 4 weeks), the mycelium becomes dense and cottony, a deeper brownish yellow or orange, and frequently shows an area of heavy growth in the center. The reverse side of the colony is **bright yellow,** becoming orange with age. Rarely, strains are isolated that show no reverse-side pigment. Microscopically, *M. canis* shows an abundance of large (15 μ by 60 to 125 μ), spindle-shaped, multicelled (4 to 8) macroconidia (Fig. 32-5) with knoblike ends. These are thick walled and bear warty (echinulate) projections on their surfaces. Microconidia, pectinate hyphae, racquet hyphae, and chlamydospores are rare.

M. gypseum is a free-living saprophyte

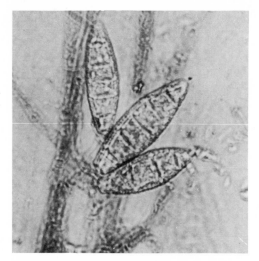

Fig. 32-6. *Microsporum canis,* showing ellipsoidal multicelled macroconidia. (750×.)

of the soil **(geophilic)** that only rarely causes human or animal infection. Infected hairs generally do not fluoresce under a Wood's lamp. However, microscopic examination of infected hairs shows them to be irregularly covered with clusters of spores (5 to 8 μ), some in chains. These arthrospores of the ectothrix type are considerably larger than those of other *Microsporum* species.

On Sabouraud agar *M. gypseum* grows rapidly as a flat, irregularly fringed colony with a coarse powdery surface with a fawn to buff or cinnamon brown color. The underside of the colony is a conspicuous orange to brownish color. Macroconidia are seen in large numbers and are characteristically large, ellipsoidal, and multicelled (3 to 9) with echinulate surfaces (Fig. 32-6). Although spindle shaped, these macroconidia are not as pointed at the distal end as are those of *M. canis.* Microconidia are rare.

Genus Trichophyton

Species of this genus are widely distributed and are the most important causes of **ringworm** of the feet and nails; they are occasionally responsible for tinea corporis, tinea capitas, and tinea barbae. They are most commonly seen in adult infections

and vary considerably in their clinical manifestations. Most cosompolitan species are anthropophilic; a few are zoophilic.

Generally, trichophyton-infected hairs **do not fluoresce** by Wood's lamp; the demonstration of fungal elements either inside the hair shaft, surrounding and penetrating the hair shaft, or within skin scrapings is needed in order to make a diagnosis of ringworm. Isolation and identification of the fungus is necessary for confirmation.

Microscopically, *Trichophyton* is characterized by club-shaped, thin-walled macroconidia with 8 to 10 septa ranging in size from 8 by 4 μ to 15 by 8 μ. The macrococonidia are borne singly at the terminal ends of hyphae or on short branches; the microconidia are usually spherical or clavate and 2 to 4 μ in size. Although a large number of *Trichophyton* species have been described, many have proved to be colonial variants. It now appears that there are distinct species; only the common species will be described.

T. mentagrophytes occurs in **two** distinct colonial forms: the so-called **"downy"** variety commonly isolated from human tinea pedis, and the **"granular"** variety isolated from suppurative ringworm acquired from animals (zoophilic). It is possible to convert the downy form to the granular form by animal passage; the reverse may occur spontaneously in laboratory cultures.

Growth of *T. mentagrophytes* is rapid and abundant on Sabouraud agar, appearing as white, cottony, or downy colonies to flat, cream-colored, or peach-colored colonies that are coarsely granular to powdery. The reverse side of the colony is rose brown, occasionally orange to deep red in color. The white downy colonies produce only a few clavate microconidia; the granular colony sporulates freely with numerous small, globose to club-shaped microconidia and thin-walled, slightly clavate, spindle- or pencil-shaped macroconidia measuring 6 by 20 μ to 8 by 50 μ in size, with 2 to 5 septa (Fig. 32-7). Spiral hyphae and nodular bodies may be

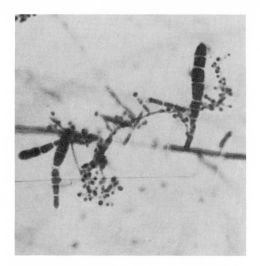

Fig. 32-7. *Trichophyton mentagrophytes,* showing numerous microconidia in grapelike clusters. Also shown are several thin-walled macroconidia. (500×.)

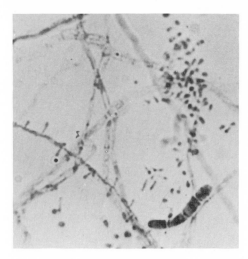

Fig. 32-8. *Trichophyton rubrum,* showing sausage-shaped macroconidia and numerous pyriform microconidia borne singly on hyphae. (750×.)

present. Macroconidia are demonstrated best in 5- to 10-day-old cultures.

T. rubrum is a slow-growing species, producing a flat or heaped-up colony with a white to reddish, cottony or velvety surface. This characteristic **cherry red** color is best observed on the reverse side of the colony, commencing at the margin or spreading concentrically, but it may disappear on subculture. Occasional strains may lack the deep red pigmentation on first isolation.* Microconidia are rare in most of the fluffy strains and more common in the velvety or granular strains, occurring as globose to clavate spores, 2 to 5 μ in size, and growing in clusters on the lateral sides of the mycelium. Macroconidia are rarely seen, although they are more common in the granular strains where they appear as thin-walled, sausagelike cells with blunt ends, containing 3 to 8 septa (Fig. 32-8).

T. tonsurans, along with *M. audouinii,*

---

*A useful method, utilizing in vitro hair cultures to differentiate aberrant forms of *T. mentagrophytes* from *T. rubrum,* has been described by Ajello and Georg.[5]

is responsible for an **epidemic** form of tinea capitis occuring most commonly in children but occasionally in adults. The fungus causes a low-grade superficial lesion of varying chronicity and produces circular, scaly patches of alopecia. The stubs of hair remain in the epidermis of the scalp after the brittle hairs have broken off and give the typical "black dot" ringworm appearance. Since the infected hairs do not fluoresce under Wood's light, a careful search for the embedded stub should be carried out in a bright light, using a magnifying head loop.

The direct microscopic examination of infected stubs mounted in 10% sodium hydroxide reveals the hair shaft to be filled with masses of large (4 to 7 μ) arthrospores in chains, an endothrix-type invasion. Cultures of *T. tonsurans* develop slowly on Sabouraud agar as flat, white, powdery colonies, later becoming velvety and varying in color from a gray through a sulfur yellow to tan. The colony surface shows radial folds, often developing a craterlike depression in the center with deep fissures. The reverse side of the colony is a yellowish to reddish brown color. Micro-

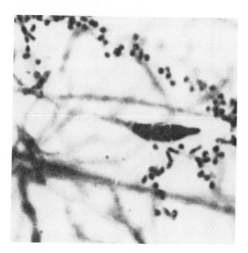

Fig. 32-9. *Trichophyton tonsurans,* showing numerous microconidia borne singly or in clusters. A single macroconidium (rare) is also present. (600×.)

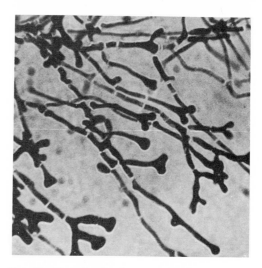

Fig. 32-10. *Trichophyton schoenleini,* showing swollen hyphal tips, resembling antlers, with lateral and terminal branching (favic chandeliers). Microconidia and macroconidia are absent. (500×.)

scopically, numerous microconidia are observed, borne laterally on undifferentiated hyphae or in clusters. These vary greatly in size, from 2 by 3 μ to 5 by 7 μ. Macroconidia, although rarely encountered, are clavate or irregular in shape with thin walls (Fig. 32-9). Chlamydospores are abundant in old cultures; swollen and fragmented hyphal cells resembling arthrospores are also seen. The addition of thiamine to the isolation medium will enhance the growth of *T. tonsurans.*

Favus is a severe type of ringworm of the scalp caused by *T. schoenleinii.* The infection is characterized by the formation of yellowish cup-shaped crusts, or scutula, resulting in considerable scarring of the scalp and sometimes permanent baldness. A distinctive invasion of the infected hair, the favic type, is demonstrated by the presence of large inverted cones of hyphae and arthrospores at the mouths of the hair follicles along with a branching mycelium throughout the length of the hair (endothrix). Longitudinal tunnels or empty spaces appear in the hair shaft where the hyphae have distintegrated, which, in sodium hydroxide preparations, readily fill

with fluid; air bubbles also can be seen in these tunnels.

T. schoenleinii grows slowly as a gray, glabrous, and waxy colony on Sabouraud agar, somewhat hemispherical at first, but later spreading to resemble a sponge placed on the medium. The irregular border consists mostly of submerged mycelium, which tends to crack the agar. The surface of the colony is yellow to tan, furrowed, and irregularly folded. Old atypical strains show a powdery or downy surface with short aerial hyphae. The reverse side of the colony is usually tan in color or nonpigmented.

Microscopically, one sees only a few microconidia, which vary greatly in size and shape. Macroconidia are not produced; the mycelium is highly irregular. The hyphae tend to become knobby and club shaped at the terminal ends (pin heads), with the production of many short lateral and terminal branches (favic chandeliers) (Fig. 32-10). Chlamydospores are generally numerous. All strains of *T. schoenleinii* may be cultivated in a vitamin-free medium and grow equally well at both room temperature and 36° C.

T. violaceum causes ringworm of the scalp and body, which occurs mainly in the Mediterranean region, the Middle and Far East, and occasionally in the United States. Hair invasion is of the endothrix type; clinically, the typical black dot ringworm is observed. Microscopically, direct examination of sodium hydroxide mounts of the short nonfluorescing hair stubs show dark thick hairs filled with masses of arthrospores arranged in chains, similar to the appearance in *T. tonsurans* infections. On Sabouraud agar, the fungus is very slow growing, beginning as a cream-colored, glabrous, cone-shaped colony, later becoming heaped up, verrucose (warty), and lavender to deep purple. The reverse side of the colony is purple or nonpigmented. Older cultures may develop a velvety aerial mycelium and sometimes lose their purple pigment. Microscopically, microconidia or macroconidia are generally absent; only sterile, thin, and irregular hyphae and chalmydospores are found. *T. violaceum* requires thiamine-enriched media to produce conidia.

T. verrucosum causes a variety of ringworm lesions in cattle (zoophilic) and in man; it is seen most often in farmers who are infected from cattle. The lesions are found chiefly on the beard, neck, wrist, and back of the hand; they appear as deep, boggy, suppurating formations with sinus tracts. On pressure, short stubs of hair can be recovered from the purulent lesion. On direct examination, the outside of the hair shaft reveals sheaths of isolated chains of large (5 to 10 μ) spores and mycelium within the hair (ectothrix and endothrix type). Masses of these spores are also seen in the pus and germinate to form long thin filaments.

T. verrucosum grows very slowly (10 to 14 days) and poorly on Sabouraud agar at room temperature but better at 36° C. Maximal growth is obtained on media enriched with thiamine and inositol or yeast extract; no growth occurs on vitamin-free media.[6] The colony on Sabouraud agar is small, heaped, and folded, occasionally flat and disc shaped. At first glabrous and waxy, the colony sometimes develops a short aerial mycelium on enriched media; colonies vary from a gray, waxlike color to a bright ocher. The reverse of the colony is usually yellow, but may be nonpigmented.

On Sabouraud agar a thin, irregular mycelium is produced, with many terminal and intercalary (between two hyphal seg-

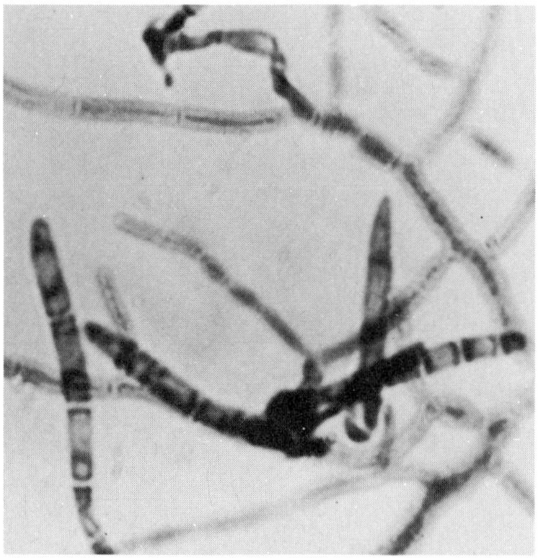

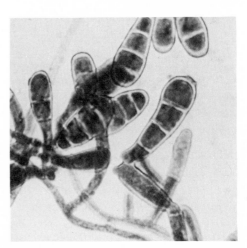

Fig. 32-11. *Trichophyton verrucosum,* showing multicelled, smooth, thin-walled macroconidia, which are rarely seen. (500×.)

Fig. 32-12. *Epidermophyton floccosum,* showing numerous smooth, multiseptate, thin-walled macroconidia with rounded ends (microconidia absent). (1000×.)

ments) chlamydospores; sometimes favic chandeliers are formed. Chlamydospores are more numerous on Sabouraud agar incubated at 36° C. On enriched media, *T. verrucosum* forms more regular mycelia and numerous small microconidia, borne singly along the hyphae. Macroconidia (Fig. 32-11) are rarely formed and vary considerably in size and shape.

Genus Epidermophyton

Epidermophytosis, although caused primarily by *Epidermophyton floccosum,* can be caused also by species of *Trichophyton* and *Microsporum.* The skin is usually attacked, but not the hair. *E. floccosum* does not attack the nails. In the direct examination of scrapings mounted in sodium hydroxide, the fungus is seen as fine branching filaments in young lesions that form chains of arthrospores or as round cells in older lesions.

E. floccosum grows slowly on Sabouraud agar; primary growth appears as yellowish white spots, developing into powdery or velvety colonies with centrally radiating furrows of a distinctive greenish yellow color. The reverse side of the colony is a yellowish tan color. After several weeks, the colony develops a white aerial mycelium (pleomorphic), which completely overgrows the colony.

Microscopically, numerous smooth, thin-walled, multiseptate (2 to 4) macroconidia are seen, rounded at the tip and borne singly or in groups of 2 or 3 on the hyphae (Fig. 32-12). Microconidia are absent, spiral hyphae are rare, and chlamydospores are usually numerous.

Subcutaneous mycoses

Subcutaneous mycoses are fungal infections that involve the skin and subcutaneous tissue, generally without dissemination to the internal organs of the body. The agents are found in several unrelated fungal genera, all of which probably exist as saprophytes in nature. Man and animals serve as **accidental hosts** through inoculation of the fungal spores into cutaneous

and subcutaneous tissue after trauma. Three subcutaneous mycoses will be considered here: **sporotrichosis, chromoblastomycosis,** and **maduromycosis.**

Sporotrichosis

Sporotrichosis is a chronic infection of worldwide distribution caused by the **dimorphic** fungus *Sporothrix schenckii,* whose natural habitat is in the soil and on living or dead vegetation. Man acquires the infection through an accidental wound (thorn, splinter) of the hand, arm, or leg. The infection is characterized by the development of a nodular lesion of the skin or subcutaneous tissue at the point of contact and later involves the lymphatic vessels and nodes draining the area, which break down to form an indolent ulcer that later becomes chronic. Only rarely is the disease disseminated. The infection is an occupational hazard for farmers, nurserymen, gardeners, florists, miners, and others.*

The **tissue forms** of *S. schenckii* appear as small, oval, budding, yeastlike cells, which are not usually demonstrable in unstained or stained smears of material from suspected lesions, except by immunofluorescence procedures (Chapter 37). However, the tissue form may be produced readily by inoculating mice or rats (see p. 252).

Pus from unopened subcutaneous nodules or from open draining lesions is inoculated to BHI agar incubated at 36° C. and on Sabouraud agar at room temperature. Chloramphenicol and cycloheximide should be added to the medium if contamination is suspected. *S. schenckii* is not inhibited by these agents.

The tissue (yeast) phase develops at 36° C., appearing in 3 to 5 days as smooth, tan, yeastlike colonies. Microscopically, such colonies show cigar-shaped (fusi-

---

*An outbreak of sportrichosis associated with sphagnum moss as the source of infections has been reported.[13] A report, in lighter vein, of an outbreak involving beer cans, bricks, and medical students is described in Arch. Intern. Med. **127:**482-483, 1971.

form) cells, measuring 1 to 4 μ by 1 μ or less, and round or oval budding cells 2 to 3 μ in diameter (Fig. 32-13, A). Occasionally, a few large, pyriform cells, 3 to 5 μ in size, may be produced.

On Sabouraud agar at room temperature, growth appears in 3 to 5 days as small, moist, white to cream-colored colonies. On further incubation these become membranous, wrinkled, and coarsely tufted, the color becoming irregularly dark **brown** or **black.** Microscopically, the mycelium is made up of delicate (2 μ thick),

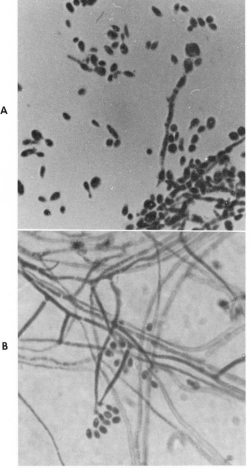

Fig. 32-13. *Sporotrichum schenckii.* **A,** Yeast phase, showing cigar-shaped and oval-budding cells. (500×.) **B,** Mycelial phase, showing pyriform to ovoid microconidia borne bouquetlike at the tip of the conidiophore. (750×.)

branching, septate hyphae that bear pyriform or ovoid to spherical microconidia 2 to 5 μ in diameter. These are borne, bouquetlike, in clusters from the tips of the conidiophores or directly on the sides of hyphae as dense sleeves of conidia (Fig. 32-13, B). Older cultures may produce larger, thick-walled chlamydospores.

Because of their morphology, saprophytic species of the genus may be confused with *S. schenckii,* and it is necessary to distingusih between them by in vitro and in vivo culture. For the former, moist slants of BHI agar are inoculated and incubated at 36° C. It may require several transfers before the characteristic tissue phase develops. Animal inoculation may be employed if laboratory culture is nonproductive; and to this end, white rats are injected intratesticularly with pus, yeast cells, or mycelial fragments, using approximately 0.2 ml. In 3 weeks, the animals are killed and examined for a purulent orchitis. Gram-stained pus will reveal grampositive cigar-shaped or oval budding forms of *S. schenckii.*

Chromomycosis

Chromomycosis is a chronic noncontagious skin disease characterized by the development of a papule at the site of infection that ulcerates and spreads to form warty or tumorlike lesions, later resembling cauliflower in appearance. The multiple pustules drain and may later ulcerate. The lesions are usually confined to the feet and legs, but may involve the head, face, neck, and other body surfaces.

The disease is widely distributed but most cases occur in tropical and subtropical areas. Occasional cases are reported from temperate zones, including the United States. The infection is seen most often in areas where barefoot workers suffer thorn or splinter puncture wounds through which the spores enter from the soil.

The etiological agents of chromoblastomycosis comprise a group of closely related fungi that produce a slow-growing, heaped-up, and slightly folded **dematia-**

ceous (dark-colored) colony with a grayish velvety mycelium. The reverse side of the colony is jet black. The different species are distinguished by the type of conidiophores they produce and include the following:

1. Cladosporium type *(Cladosporium carrionii).* Conidia in branched chains are produced by conidiophores of various lengths.

2. Phialophora type *(Phialophora verucosa).* Conidia are produced endogenously in flasklike conidiophores or phialides (Fig. 32-14).

3. Acrotheca type *(Fonsecaea [Hormodendrum] pedrosoi, F. compacta).* Conidia are formed along the sides of irregular club-shaped conidiophores (Fig. 32-15).

A laboratory diagnosis of chromoblastomycosis is essential and easily made. Scrapings or scales from encrusted areas mounted in 10% sodium hydroxide, xylol, or balsam show the presence of long, dark brown, thick-walled, branching septate hyphae 2 to 5 μ in width. In pus, tissue, or biopsy specimens thick-walled, rounded brown cells 4 to 12 μ in diameter may be observed. All of the fungi causing chromoblastomycosis have the same appearance.

Crusts, pus, and biopsy tissue are cultured on Sabouraud agar with antibiotics (Chapter 39) and incubated at room temperature. Identification of the dematiaceous isolates is based on the type of sporulation observed: *C. carrionii* exhibits only the cladosporium type of sporulation, and the conidial chains are quite long. *F. compacta* and *F. pedrosoi* may exhibit all three types of sporulation concurrently, although the cladosporium type predominates, with short chains of conidia. *P. verrucosa* exhibits only the phialophora type of sporulation. A *Cladosporium* species, considered to be a saprophyte, also produces a cladosporium type of sporulation but, unlike *C. carrionii,* will liquefy gelatin and hydrolyze a Loeffler serum slant.

Mycetoma (maduromycosis)*

Mycetoma is a chronic granulomatous infection that usually involves the lower extremities, but may occur on any part of the body. It was first described by Gill in 1842, while he was working in a dispensary near Madura, India. The term "Madura

---

*The interested reader is referred to Vanbreuseghem's excellent monograph on mycetoma.[23]

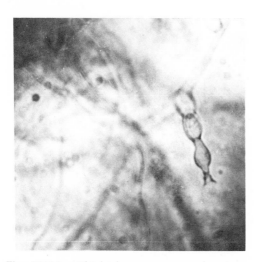

Fig. 32-14. *Phialophora verrucosa,* showing a single flasklike conidiophore. (1000×.)

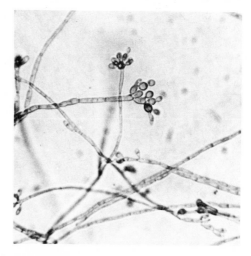

Fig. 32-15. *Fonsecaea (Hormodendrum) pedrosoi,* showing conidia produced terminally and in clusters on club-shaped conidiophores (cladosporium-type). (400×.)

foot" probably originated from the natives in describing the deformed foot seen in infected patients. The disease is characterized by swelling, purplish discoloration and tumorlike deformities of the subcutaneous tissue, resulting in the development of sinus tracts that drain pus containing yellow, red, or black granules. The infection gradually progresses to involve bone, muscle, or other contiguous tissue, ultimately requiring amputation.

Maduromycosis is common among the dark-skinned natives of the tropical and subtropical regions, whose outdoor occupations and shoeless habits often predispose them to trauma. These are significant factors in exposure to the fungus. More than 60 cases of mycetoma have been reported in the United States.[16]

There are two other types of mycetoma: actinomycotic mycetoma, or nocardiomycosis, caused by *Nocardia* and *Streptomyces* species, and fungal mycetoma, or maduromycosis, caused by a heterogeneous group of septate true fungi with broad hyphae. The most common cause of maduromycosis in the United States is the perfect fungus, *Allescheria boydii,* in the class Ascomycetes, since it produces sexual ascospores; its imperfect form is known as *Monosporium apiospermum.* The fungus is a common saprophyte in soil and sewage, and man acquires the infection after such injuries as a scratch, bruise, or penetrating wound, or by contaminating an open wound.

Allescheria boydii is a **hyaline** (glassy, transparent) organism, producing white or yellow granules in pus. These are composed of tightly meshed, wide, septate mycelia and numerous large, hyaline chlamydospores. On Sabouraud agar without antibiotics, *A. boydii* grows rapidly at room temperature as a white fluffy colony that changes in several weeks to a brownish gray mycelium. The reverse of the colony is gray black. Microscopic examination shows large, septate, hyaline hyphae and many conidia borne singly on short conidiophores. The conidia are pyriform to oval in shape, are unicellular and measure approximately 6 by 9 μ. Clusters of conid-

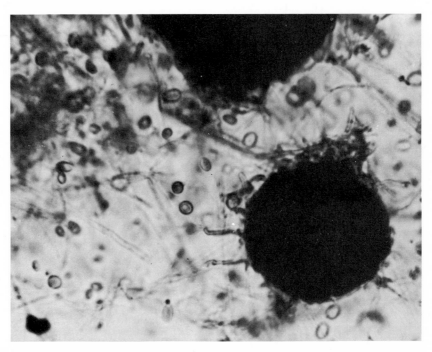

Fig. 32-16.   *Allescheria boydii,* showing perithecia and numerous ascospores. (750×.)

iophores (coremia) with conidia at the ends sometimes occur; these resemble ripened grains on sheaves of wheat.

Some strains of *A. boydii* produce **perithecia,** closed structures containing asci with ascopores. When the latter are fully developed, the large (50 to 200 μ), thin-walled perithecia rupture, liberating the ascia and spores (Fig. 32-16). The ascospores are yellow, oval shaped, and delicately pointed at each end and are somewhat smaller than the conidia.

It should be noted that media containing antibiotics are **not to be used alone** in culturing clinical specimens from mycetomas or draining sinuses, since some of the agents of maduromycosis may be inhibited in their growth. This is particularly true with species of *Nocardia, Actinomyces, Aspergillus,* and *Alloscheria boydii.* Since actinomycotic lesions may respond to specific therapy, whereas maduromycoses may not, this etiological differentiation must be made.

Nocardiosis

Nocardiosis is a mycotic infection of man and the lower animals caused by species of *Nocardia.* It may result in chronic suppuration and draining sinuses of the subcutaneous tissue (mycetoma) or in a primary pulmonary infection resembling tuberculosis, which may metastasize to other parts, especially the brain and meninges.*
The pulmonary infection is caused by *N. asteroides,* an **aerobic, partially acid-fast,** funguslike bacterium composed of branched mycelial filaments that fragment readily into bacillary and coccoid forms. The organism grows as a saprophyte in the soil, and systemic infection follows the inhalation of the fungus **(exogenous)** or

occurs through skin abrasions, expecially on the feet. *N. asteroides* may be demonstrated in direct smears of pus (opaque or pigmented granules are rarely present) or fresh morning sputum and in the sediment of centrifuged cerebrospinal fluid. Gram-stained smears reveal gram-positive, intertwining branching filaments 1 μ in diameter, with or without clubs, or coccoid and diphtheroid forms. An acid-fast stain* will show that some filaments retain the carbolfuchsin; young cultures are usually strongly acid fast, whereas older cultures or subcultures are less so.

All infected material in which delicate, branching, gram-positive filaments have been demonstrated should be inoculated heavily on two tubes each of infusion blood agar and Sabouraud agar **without antibiotics†;** both media should be incubated aerobically at room temperature and at 36° C.

It should be pointed out that *N. asteroides* frequently survives the sodium hydroxide or other decontamination procedures used in preparing sputum specimens for the isolation of *Mycobacterium tuberculosis* and will grow well on the usual isolation media.[7] *N. asteroides* appears as a moist glabrous colony on the tuberculosis media and grows out within 1 to 2 weeks. The colonies resemble those of the saprophytic or other mycobacterial species; slide cultures will reveal the branching acid-fast mycelium characteristic of *N. asteroides.*

On Sabouraud agar and blood agar, growth appears in 3 days as **small yellow colonies** resembling those of *M. tuberculosis.* After 5 to 10 days the colonies become waxy, cerebriform, irregularly folded, and yellow to deep orange in color. Microscopically, the colony is composed of delicate **branching,** mycelial filaments 1 μ in diameter that break up into bacillary

---

*Approximately 10% of patients with pulmonary alveolar proteinosis have either pulmonary or systemic nocardiosis (see Louria, D. B.: Deep-seated mycotic infections, allergy to fungi and mycotoxins, New Eng. J. Med. **277:**1065-1071, 1126-1134, 1967). The increased incidence appears to be associated with increasing use of steroids in patients with underlying disorders.

---

*The Kinyoun stain is recommended, followed by **light** decolorization with acid alcohol or, preferably, 0.5% to 1% aqueous sulfuric acid.
†*Nocardia asteroides* will not grow on media containing chloramphenicol.

forms. These are gram positive and acid fast.*

Because of the similarity between some species of *Nocardia* and species of *Streptomyces,* a variety of biochemical tests have been devised to differentiate them. The following tests are used by the Mycology Unit of the Center for Disease Control (CDC).

Demonstration of branched mycelium. Demonstration of the branched mycelium is best done in slide culture. *N. asteroides, N. brasiliensis,* and *Streptomyces* species form a branched mycelium; filamentous bacteria do not.

Acid fastness. As shown by the Kinyoun stain with 1% sulfuric acid decolorization, *N. asteroides* and *N. brasiliensis* are partially acid fast; spores of *Streptomyces* may be acid fast.

Hydrolysis of casein. See Chapter 41. *N. asteroides* does **not** hydrolyze casein, but casein is readily hydrolyzed by *N. brasiliensis* and *Streptomyces* species.

Growth in gelatin. See Chapter 41. *N. asteroides* either fails to grow or grows poorly, with a small, thin, flaky, white growth. *N. brasiliensis* grows well, forming discrete, compact, round colonies. *Streptomyces* species may grow well; growth is flaky or stringy.

Animal pathogenicity. Most isolates of *N. asteroides* are pathogenic for white mice and guinea pigs inoculated intraperitoneally; *N. brasiliensis* is usually not pathogenic. *Streptomyces* is not pathogenic, although some strains may cause a toxic death within 24 hours after inoculation.

As a test of pathogenicity for laboratory animals, a heavy suspension of the suspected culture should be prepared by grinding the growth from several slants of Sabouraud agar (incubated 1 to 2 weeks) with sterile saline in a sterile mortar and pestle. This is combined with an equal amount of 5% hog gastric mucin, and 1

*Pure cultures grown in litmus milk for several weeks will show a strong acid-fastness when stained as described above.

ml. is injected intraperitoneally into a small (200 gm.) guinea pig. The animal will generally die in 7 to 14 days, revealing a considerable amount of purulent peritoneal exudate at autopsy. Smears of this will reveal the characteristic gram-positive, acid-fast elements of *N. asteroides.* Since, however, there is considerable variation in strain virulence and guinea pig susceptibility in *N. asteroides,* animal injection is not reliable unless positive. *N. asteroides* is best identified by colonial and biochemical characteristics.

An early diagnosis of pulmonary nocardiosis, combined with vigorous treatment, is important to prevent metastasis to the brain; sulfonamides appear to be the drugs of choice. Recently, it has been suggested that a sulfonamide combined with ampicillin or trimethoprim may be more effective.[1]

Yeastlike fungi

This group of imperfect fungi resembles the true yeasts both morphologically and culturally. They produce yeastlike, creamy colonies on solid media and are generally unicellular, although some produce a pseudomycelium and true mycelium. The genera described here include *Cryptococcus, Candida,* and *Geotrichum.*

Cryptococcosis (torulosis)

Cryptococcosis is a subacute or chronic mycotic infection involving primarily the brain and meninges and the lungs; at times the skin or other parts of the body may be involved. It is caused by a single species of yeastlike organism, *Cryptococcus neoformans.* The organism was first isolated by Sanfelice in 1894 from peach juice. Subsequently, the source of infection in man and animals was erroneously assumed to be endogenous until Emmons, in 1950, reported the isolation of virulent strains of *C. neoformans* from barnyard soil. In 1955 he further reported a frequent association of virulent strains of *C. neoformans* with the excreta of pigeons and indicated that, until other sources are discovered, exposure to pigeon excreta was the most significant and important source of infection in man and animals. This hypothesis has been substantiated by numberous reports that pigeon habitats serve as reservoirs for human infection; the pigeon manure apparently serves as an enrichment for *C. neoformans* due to its chemical makeup.[3] The organism is apparently the only pathogenic yeast not found in the normal human flora.[1]

There is a strong association of cryptococcal infection with such debilitating diseases as leukemia, malignant lymphoma, tuberculosis, and so forth. The infection is probably more frequent than is commonly supposed—it is estimated that there are 2,000 undiagnosed cases for every proved case of infection due to *C. neoformans.*

All clinical material, especially the cerebrospinal fluid,* should be mixed with a drop of **India ink** (a cool loop must be used, since heat will precipitate the ink particles) on a slide and examined under a coverglass using the oil-immersion objective with reduced light. The India ink serves to delin-

---

*The demonstration of encapsulated forms in the urine may precede their presence in cerebrospinal fluid; urine may also be a good source for the isolation of *C. neoformans.*

eate the large capsule, since the ink particles cannot penetrate the capsular material. *C. neoformans* appears as an oval to spherical, single-budding, thick-walled yeastlike organism 5 to 15 μ in diameter, surrounded by a wide, **refractile, gelatinous capsule** (Fig. 32-17). This characteristic morphology occurs in India ink preparations of cerebrospinal fluid, sputum, pus, urine, infected tissue, or gelatinous exudates. Frequently these capsules are more than twice the width of the individual cells. In cerebrospinal fluid, *C. neoformans* may be mistaken for a lymphocyte and is often observed first in the spinal fluid counting chamber.

Dried, heat-fixed, or stained preparations are not generally recommended; distortion of the cryptococci may render them unrecognizable.

The infected material should be promptly cultured on infusion blood agar **without cycloheximide** (*C. neoformans* is inhibited) at 36° C. and on Sabouraud agar without cycloheximide at room temperature. In culturing cerebrospinal fluid, Utz[22] recommends that the **uncentrifuged** fluid should be inoculated in generous amounts (15 to 20 ml.) to a series of culture tubes, since the cryptococci may be present in very small numbers and the centrifugation may destroy the more fragile cells. After several days of incubation at either temperature, the organism produces a wrinkled, whitish colony, which on microscopic examination, may show only budding cells without capsules. On further incubation the typical slimy (*Klebsiella*-like), mucoid, cream- to brown-colored colony develops. This colony has no mycelium and flows down to the bottom of the slant. At this time, budding cells with large capsules can be readily demonstrated in India ink wet mounts, although some strains do not form large capsules (unless they are transferred several times), and generally produce a shiny, dry colony.

Of the yeastlike fungi, only members of the genus *Cryptococcus* (both saprophytic strains and *C. neoformans*) consistently

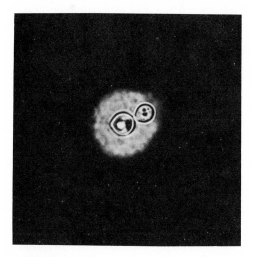

Fig. 32-17. *Cryptococcus neoformans,* in spinal fluid, showing a large, encapsulated, single-budding, yeastlike cell. (India ink; 1000×.)

produce **urease.** This can be detected by inoculating a urea agar slant (Christensen) with the suspected cryptococcus. If it is a *Cryptococcus* species, it will produce a positive reaction (red color) in the medium after 1 to 2 days of incubation at room temperature.* *C. neoformans* characteristically does **not** assimilate nitrate or lactose but can assimilate glucose, maltose, and sucrose as carbon sources.

The incorporation of cycloheximide and chloramphenicol in Sabouraud agar suffices in most instances for the isolation of pathogenic fungi from **heavily contaminated** material, except for *C. neoformans,* which is inhibited by cycloheximide. Staib[21] introduced a medium containing creatinine and an extract of *Guizotia abyssinica* (a canary seed constituent) as a color marker for the selective isolation of *C. neoformans,* the growth of which produces a **brown** color. This medium, however, was rapidly overgrown with saprophytic fungi when inoculated with material from pigeon nests. The addition of diphenyl and chloramphenicol apparently increased the efficiency of the original preparation.†

Most saprophytic strains of cryptococci will **not** grow at 36° C.‡ The pathogenicity of suspected strains of *C. neoformans,* especially from the sputum or skin, should be demonstrated in mice. This is carried out by injecting two to four white mice either with 1 ml. intraperitoneally or with 0.02 ml. intracerebrally (under light ether anesthesia) of a heavy saline suspension of a 4- or 5-day-old culture.

Mice injected **intraperitoneally** will develop lesions in the brain in about **3 weeks,** whereas mice injected **intracerebrally** will generally develop them in **less than 1 week.** Mice are killed at the end of 2 weeks if death has not occurred. At autopsy the animals will show gelatinous masses in the visceral cavity, lungs, and brain. India ink preparations will reveal the typical budding, encapsulated *C. neoformans,* and the fungus may be cultivated from these lesions.

Candidiasis

Candidiasis is an acute or subacute infection caused by members of the genus *Candida,* chiefly *C. albicans,* although all species may be pathogenic.* Although the fungus may be isolated from the stools, genitourinary tract, throat, and skin of normal persons **(endogenous),** it may produce lesions in the mouth, genitourinary tract, skin, nails, bronchi, or lungs in patients whose normal defense mechanisms may have been altered by underlying disease, overuse of antibiotics, or immunosuppressive agents, and so forth.[22] Bloodstream infection, endocarditis (in drug addicts), and meningitis caused by *Candida* species have also been reported.

The isolation of *Candida* species from clinical materials is difficult to evaluate, since positive cultures may be obtained from various anatomical sites of a large percentage of normal persons. The organisms must be recovered constantly and repeatedly in significant numbers from **fresh** specimens and to the exclusion of other known etioligical agents before a diagnosis of candidiasis can be entertained.

Skin and nail scrapings should be examined directly; they are mounted in 10% sodium or potassium hydroxide with a coverglass and heated gently. Sputum and exudate from the oropharynx or vagina or material from the intestinal tract should be crushed under a coverglass and examined fresh, either unstained or Gram stained. *Candida* appears as small (2 to 4 μ), oval

---

*Some strains of *Rhodotorula, Candida,* and *Trichosporon* occasionally hydrolyze urea.

†See Shields, A. B., and Ajello, L.: Medium for selective isolation of *C. neoformans,* Science **151:**208-209, 1966.

‡*C. laurentii, C. albidus* and *C. luteolus* sometimes grow at 36° C. and may show a mild degree of mouse virulence. Therefore, when reporting to the clinician, microbiologists must make certain that the cryptococcus isolated is not *C. neoformans.*

**C. tropicalis* and *C. parapsilosis,* although less commonly isolated from human infections, have been increasingly implicated in endocarditis and fungemia.

or budding, yeastlike cells, along with mycelial fragments of varying thickness and length **(tissue phase).** The yeastlike cells and mycelial elements are strongly gram positive. It is well to report the approximate number of such forms seen, since the presence of large numbers in a fresh specimen may have diagnostic significance.

Since saprophytic yeasts are similar microscopically to the pathogenic species, all infected material should be **cultured** on duplicate sets of Sabouraud agar with and without cycloheximide* and incubated at both room temperature and 36° C. Colonies of *Candida* species (and saprophytic yeasts) appear in 3 to 4 days as medium-sized, cream-colored, smooth, pasty colonies with a characteristic yeastlike appearance. Most strains grow well at either temperature. On microscopic examination a slide mount will show budding cells along with elongated unattached cells (pseudomycelia) with clusters of blastospores at the constrictions. Old dry colonies may show a fringe of mycelium in the agar (dimorphism; see explanation p. 263).

If the unknown culture is suspected of belonging to the genus *Candida,* subsequent procedures must be carried out, including the demonstration of chlamydospore production, germ tube production, or sugar fermentation and assimilation tests. Although other species of *Candida* may be encountered in candidiasis, *C. albicans* is the most frequently isolated species and is the usual etiological agent in oral or vaginal thrush, intertriginous or cutaneous monilial infection, paronychial infection, or bronchopulmonary candidiasis.

The demonstration of the production of germ tubes affords a rapid and reliable screening procedure for the identification of isolants of suspected *C. albicans.* A simple test recommended by Ahearn[2] is as follows:

1. Cells from a young (not more than 96 hours) colony are transferred by means of the tip of a plastic straw* into about 0.3 ml. pooled human serum contained in a clean 12 × 75 mm. test tube, leaving the straw immersed in the serum.
2. The tube is incubated for several hours at 36° C.; a drop of the suspension is placed on a glass slide, and a cover slip is applied, using the straw for transfer.
3. Microscopic examination of typical *C. albicans* reveals thin **germ tubes** 3 to 4 μ in diameter and up to 20 μ in length; unlike pseudohyphae, they are **not constricted** at their point of origin.
4. Arthrospores of *Geotrichum* or *Trichosporon* species may be mistaken for germ tubes by the inexperienced. For this reason, a known isolate of *C. albicans* and *C. tropicalis* should be included as controls.

Another method of identification, based on the morphology of *C. albicans* when grown on chlamydospore agar† containing 1% Tween 80, can be recommended, the procedure for which follows:

1. If a yeastlike colony has developed on Sabouraud agar or other solid medium, examine a wet-mount preparation for presence of yeast cells.
2. Melt and pour about 25 ml. of chlamydospore agar (Chapter 39) into a sterile Petri dish and allow to solidify.
3. Using a stiff (22-gauge Nichrome) inoculating needle, transfer a **small** portion of yeastlike colony to the surface of a sector of the plate and **cut it into the agar.** The inoculating needle should be held so that the cut

---

*A number of *Candida* species are inhibited by 0.5 mg./ml. of cycloheximide; these include *C. parapsilosis, C. krusei,* and strains of *C. tropicalis;* most strains of *C. albicans* are resistant.

*Commercial cocktail straws cut into approximately 100-mm. lengths, clean but not sterile.
†Although commercial media are generally satisfactory, better results may be expected from "homemade," yellow cornmeal agar plus 1% Tween 80.

is made at a 45-degree angle, across the plate, and down through the medium to the bottom of the dish. A single Petri dish may be used to identify four or five different cultures. Place an alcohol-flamed coverglass over each cutting.

4. Always inoculate each plate with a known culture of *C. albicans* as a **control** of the culture medium and incubation temperature.

5. Incubate plates at 25° C.—**not** at 36° C. (chlamydospores do not develop at this temperature)—for 1 to 3 days and examine the control strip for the presence of chlamydospores. This can be done by placing the dish on the microscope stage and examining under the low-power or high-power objective or both for the presence of pseudomycelia (elongated buds that fail to detach) with clusters of blastospores along the points of constriction and the thick-walled, round **chlamydospores** at the terminal ends of the pseudomycelium (Fig. 32-18). Only *C. albicans* produces those characteristic spores in abundance.*

6. When the control streak shows these chlamydospores, examine the test streaks.

7. If an unknown culture does not form chlamydospores but morphologically resembles *C. albicans,* hold the plates for 2 to 3 days longer and reexamine. If chlamydospores still fail to develop, cut from this streak to a **fresh plate** of medium and restudy as before.

8. If the unknown forms chlamydospores and has the morphology of *C. albicans,* report as *"C. albicans."*

9. If no mycelium has been formed, and the organism has been ruled out as *Cryptococcus neoformans,* report as a **"nonpathogenic yeast."**

10. If the morphology is that of a *Can-dida* species but not that of *C. albicans,* report as **"Candida species, not Candida albicans."** If further identification is desired, fermentation and assimilation tests may be carried out, as is described subsequently.

11. If only **arthrospores** are produced, the unknown is probably *Geotrichum candidum.*

12. If both blastospores and arthrospores are produced, the unknown may be a *Trichosporon* species.*

Animal virulence of *C. albicans* may be demonstrated by injecting 1 ml. of a 1% cell suspension in saline of a 24- to 48-hour culture intravenously into the ear of a rabbit. Death generally ensues within 3 to 4 days; smears made from the numerous small kidney abscesses will reveal the presence of gram-positive yeast cells. Intravenous injection of mice with strains of *C. albicans, C. tropicalis,* and *C. stellatoidea* frequently may aid in demonstrating pathogenicity.

Identification of Candida by fermentation tests

1. Obtain a pure culture by inoculating a tube of Sabouraud dextrose broth and incubating overnight at 36° C.

2. Shake the tube and inoculate a loopful to a blood agar plate and incubate overnight at 36° C.

3. Examine the plate and pick single colonies to Sabouraud agar slants; incubate overnight at 36° C.

4. Transfer to sugar-free beef extract agar slants for three successive transfers, incubating each transfer overnight at 36° C.

5. Inoculate growth from the third transfer to sugar media in the following manner:

   a. Make a suspension of the growth in 2 ml. of sterile saline.

Candida stellatoidea may produce similar structures.

Trichosporon species produce pseudomycelia, true mycelia, blastospores, and arthrospores. This rapidly growing yeast may be part of the normal skin flora; it has also been isolated from infected fingernails.

b. Pipet 0.2 ml. to each of five tubes containing 9.5 ml. of beef extract broth with 0.04% bromthymol blue indicator.

c. To each of these tubes add, respectively, 0.5 ml. of a filter-sterilized 20% stock solution (Millipore or Seitz) of the following: dextrose, maltose, sucrose, lactose, and galactose.

d. Overlay each tube with sterile melted petrolatum, or paraffin and petrolatum, to form a plug about 1 cm. thick.

e. Hold five uninoculated tubes containing the sugars as sterility controls.

f. Incubate all tubes at 36° C. for 10 days and record the presence of acid or acid and gas.

Refer to Table 32-1 for test results.

Identification of Candida by assimilation tests

1. Prepare a sterile solution of yeast-nitrogen base* by weighing out 6.7 gm. of the dehydrated medium. To this add 5.0 gm. of the appropriate carbohydrate* and dissolve in 100 ml. distilled water.

2. Sterilize by membrane filtration, and add to an agar solution (20 gm. in 900 ml. distilled water) that has been autoclaved and cooled to approximately 50° C.

3. Dispense into sterile, cotton-stoppered test tubes; solidify in the slanting position. The use of agar slants, rather than plates, facilitates handling and storage.

4. The inoculum is prepared from 24- to 36-hour cultures of the isolant in yeast-nitrogen broth plus 1 mg./liter of glucose, and 1 drop (approximately 0.01 ml.) is added to each carbohydrate slant. The tests are read after 96 hours' incubation, and evidence of growth is noted on each of the carbohydrate slants when compared against a control slant of the basal medium.

Refer to Table 32-1 for test results.

---

*Difco Laboratories, Detroit.

*Recommended: dextrose, maltose, sucrose, lactose, galactose, cellobiose, xylose, and raffinose.

Table 32-1. Fermentation and assimilation patterns of *Candida* species*

CANDIDA SPECIES	DEXTROSE	MALTOSE	SUCROSE	LACTOSE	GALACTOSE	CELLOBIOSE†	XYLOSE†	RAFFINOSE†	ADDITIONAL STUDIES	
									UREASE	GROW AT 37° C
C. albicans	F	F	+	0	F‡	0	+	0	0	+
C. stellatoidea	F	F	0	0	+	0	+	0	0	+
C. parapsilosis	F‡	+	+§	0	+§	0	+	0	0	+
C. tropicalis	F	F	F	0	F	+	+	0	0	+
C. pseudotropicalis	F	0	F	F	F‡	+	+	+	0	+
C. pseudotropicalis var. lactosa	F	0	F	F	F‡	+	0	+	0	+
C. krusei	F	0	0	0	0	0	0	0	+	+
C. guilliermondii	F	+§	F	0	+§	+	+	+	0	+
C. rugosa	+	0	0	0	0	0	0	0	0	0

*Chart courtesy of Charles T. Hall, Chief, Microbiology and Serology Unit, Proficiency Testing Section, Center for Disease Control, Atlanta, Ga. Abbreviations: F = fermentation and assimilation; + = assimilation only, or positive (under additional studies); 0 = no reaction.
†Used in assimilation tests only.
‡Fermentation sometimes weak.
§Occasionally a weak fermentation.

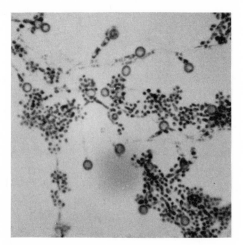

Fig. 32-18. *Candida albicans,* showing round, thick-walled chlamydospores, pseudomycelia, and numerous blastospores. (750×.)

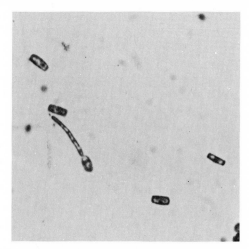

Fig. 32-19. *Geotrichum candidum,* showing barrel- to rectangular-shaped arthrospores, some with developing germ tubes. (500×.)

Geotrichosis

Geotrichosis is a rather rare infection caused by the yeastlike fungus, *Geotrichum candidum,* which reproduces by fragmentation of the hyphae into rectangular arthrospores. It may produce lesions in the mouth, bronchi, lungs, or intestinal tract. Since *Geotrichum* has been isolated from the mouths and intestinal tracts of normal persons, it must be recovered repeatedly and in large numbers from freshly obtained clinical specimens in order to be considered of etiological significance.

Sputum, pus, or flecks of bloody or purulent fecal material are pressed in a thin layer on a slide under a coverglass and examined directly. *Geotrichum* appears as rectangular (4 by 8 μ) or large spherical (4 to 10 μ) arthrospores that stain heavily gram-positive; no budding forms are seen.

Since these cells may be confused with those of the saprophytic *Oospora,* which frequently occurs as a contaminant, or with the filamentous *Coccidioides immitis,* the etiological agent of coccidioidomycosis, or with *Blastomyces dermatitidis,* the etiological agent of North American blastomycosis, all infected material should be **cultured.** The specimen is inoculated on duplicate sets of Sabouraud agar and infusion blood agar slants with and without chloramphenicol and cycloheximide; one set is incubated at room temperature and one at 36° C. for at least 3 weeks. At room temperature the fungus develops rather rapidly as a moist, creamy colony, or as a colony with a dry, mealy surface and radial furrows, or as one with a fluffy aerial mycelium. At 36° C. the slowly growing fungus develops only a small waxlike surface growth with a distinct zone of mycelium penetrating the subsurface of the medium. Microscopically, the septate, branching hyphae are fragmented into chains of rectangular, barrel-shaped, or spherical arthrospores that break apart readily. The rectangular cells frequently germinate by germ tubes from **one corner,** which are at first rounded and later elongated, a characteristic of *Geotrichum* (Fig. 32-19). Blastospores are **not** produced.

Animal inoculation or serological testing procedures are of no value in diagnosis.

Torulopsis glabrata

This organism, closely related to *Cryptococcus* and *Candida* species, was once considered a nonpathogenic saprophyte from the soil, being widely distributed in nature. However, recent reports have indicated its role as a potentially important opportunistic pathogen, particularly in the compromised or receptive host.[19]

On sheep blood agar, *T. glabrata* appears as **tiny,** white, raised, nonhemolytic colonies after 1 to 3 days' incubation at 36° C. Gram stain of these colonies reveals round to oval budding yeasts, 2 to 4 μ in diameter; no hyphae or capsules are seen. Cuttings into chlamydospore agar are negative for mycelia or pseudohyphae; the germ tube test is also negative. *Torulopsis glabrata* ferments glucose and trehalose only and does not assimilate carbohydrates.[20]

Systemic mycoses

Systemic mycotic infections may involve any of the internal organs of the body, including bone, subcutaneous tissue, and in some stages, the skin. **Asymptomatic** infections may go unrecognized clinically and may be detected only through skin sensitivity tests or serological procedures; in some cases x-ray examination may reveal healed lesions. **Symptomatic** infection may present signs of only a mild or self-limited disease, with positive supportive evidence by culture or immunologic findings. **Disseminated** or progressive infection may reveal severe symptomatology, with spread of the initial disease to all visceral organs as well as the bone, skin, and subcutaneous tissues. This type of disease is frequently fatal.

Collection of specimens

The obtaining of a proper specimen for the laboratory diagnosis of systemic mycoses is of prime importance: the success or failure of isolating the etiological agent may well depend on it.

The most satisfactory **sputum** specimen is a single, coughed specimen, after an early morning coughing spell, before eating, and after vigorous rinsing of the mouth with water after brushing the teeth. Twenty-four hour–specimens or those containing excessive amounts of postnasal discharge are **not** satisfactory and should be rejected. Sputum raised after a heated 5% saline aerosol (prepared fresh and sterilized) and material obtained by bronchial aspiration are also satisfactory for mycological examination. In all instances of suspected pulmonary infections, Georg[15] at CDC recommends that **at least six** sputum specimens be obtained, at 2- or 3-day intervals, on successive days. It should also be emphasized that all of these types of specimens be delivered promptly to the mycology laboratory and that they be **promptly cultured,** since *Histoplasma capsulatum* organisms die rapidly in specimens held at room temperature; furthermore, saprophytic fungi, *C. albicans,* and commensal bacteria may multiply rapidly and prevent the isolation of significant pathogens.

Biopsy specimens, such as scalene nodes, direct lung biopsies, and so forth are excellent specimens for the recovery of fungal pathogens; they should be submitted in sterile tubes or Petri dishes that are slightly moistened with sterile salt solution. Empyema fluid should be anticoagulated with heparin during aspiration to prevent clotting; gastric lavage specimens, cerebrospinal or synovial fluid, blood and bone marrow, urine, prostatic secretions (in blastomycosis), and lesions of skin and mucous membrane afford significant sources for recovery of systemic fungal pathogens.

Introduction to systemic mycoses

The systemic mycoses to be considered in this section include **histoplasmosis, coccidioidomycosis, blastomycosis,** and **paracoccidioidomycosis.** The fungi responsible for these infections, although unrelated generically and dissimilar morphologically and culturally, have one characteristic in common—that of **dimorphism.** The organisms involved exist in nature as the **saprophytic** form, sometimes called the **mycelial phase,** which is quite distinct from the **parasitic,** or tissue-invading form, sometimes called the **tissue phase.** The reader will note that reference has been made previously to this diphasic phenomenon in the fungal diseases candidiasis, sporotrichosis, and chromoblastomycosis, where distinct morphological differences may be observed both in vivo and in vitro. Temperature (36° C.), certain nutritional factors, and stimulation of growth in tissue not dependent on temperature have been among the factors considered necessary to effect the transformation of mycelial forms to the parasitic phase.

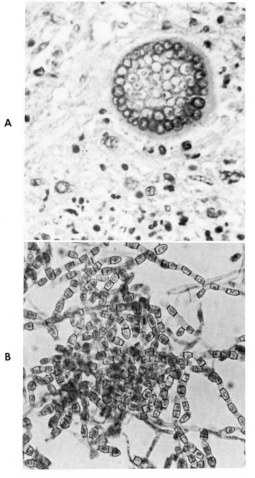

Fig. 32-20. *Coccidioides immitis.* **A,** Showing a spherule containing many spherical endospores. (1000×.) **B,** Showing thick-walled, rectangular- or barrel-shaped arthrospores in mycelial phase. (500×.)

Coccidioidomycosis

Coccidioidomycosis is an infectious disease caused by a single species of fungus, *Coccidioides immitis.* Generally an acute, benign, and self-limiting respiratory tract infection, the disease less frequently becomes disseminated, with extension to other visceral organs, the bone, skin, and subcutaneous tissue, where extensive burrowing abscesses may be formed.

Coccidioides immitis spores are found in the semiarid regions in the southwestern part of the United States and northern Mexico as well as in Central and South America; the disease, however, may be seen anywhere in the United States and can be traced to previous travel or residence in an endemic area.

Man acquires the infection by inhaling arthrospores from contaminated soil, particularly during the dry and dusty season. Less than 0.5% of persons who acquire the infection ever become seriously ill; dissemination does, however, occur most frequently in dark-skinned individuals or in those whose occupation results in continuous exposure to dust.

In direct microscopic examination (using wet, unstained preparations) of sputum, sediment from gastric washings or cerebrospinal fluid, exudates, or pus, *C. immitis* appears as a nonbudding, thick-walled (up to 2 μ) **spherule** or sporangium 20 to 200 μ in diameter, containing either granular material or numerous small (2 to 5 μ in diameter) **endospores** (Fig. 32-20, *A*). These endospores are freed by rupture of the cell wall, and empty and collapsed "ghost" spherules may be present. Small, immature spherules measuring 10 to 20 μ may be confused with nonbudding forms of *Blastomyces dermatitidis,* since they are thick walled, and endospores are not yet apparent. To check such structures, seal the edges of the coverglass with petrolatum and incubate overnight. If spherules are present, mycelial filaments will have developed from the endospores.

On some occasions the tissue phase cannot be demonstrated by direct microscopic examination. Consequently, **all material** from suspected cases should be cultured on **duplicate** sets of Sabouraud agar and infusion blood agar with and without chloramphenicol and cyclohexi-mide,* one set incubated at room temper-ature and one set at 36° C. Animal inocu-lation is also indicated on occasion.

Growth appears in 3 to 5 days at room temperature (more positive isolations at this temperature) as a moist, membranous colony growing close to the surface of the medium. This soon develops a white, cot-tony mycelium, turning from buff to brown with age. Frequently, the central area of the colony will remain moist and glabrous. The slant should be **wetted down** before removing any of the myce-lium (see following list of precautions). Microscopically, these cultures show a branching, septate mycelium, forming chains of thick-walled, rectangular or bar-rel-shaped **arthrospores,** 2 by 3 μ to 3 by 5 μ. In lactophenol cotton blue mounts, these chains show only **alternate** deeply stained arthrospores, with dried-out, transparent cell tags on either side. (Fig. 32-20, *B*). If such structures are observed, the identification should be confirmed by animal inoculation (see below).

At 36° C. only the saprophytic or my-celial phase develops, since spherule pro-duction generally cannot be induced on the usual artifical media. The tissue phase can be cultured, however, on embryonat-ed eggs or by injecting ground-up mycelia intratesticularly into male guinea pigs.† Mycelial suspension, 0.1 ml., is injected, and if orchitis does not develop (generally within 1 week), the animal is killed in 2

---

*The mycelial growth of *C. immitis* is not apprecia-bly affected by cycloheximide, whereas the white cottony growth of most saprophytes is inhibited.
†If guinea pigs are not available, white mice may be injected intraperitoneally with 1 ml. of the spore sus-pension. After about 1 week they will develop lesions or lymphatic exudates containing the mature spher-ules of *C. immitis*. These methods may also be used with sputum or gastric washings by adding 0.05 mg. per milliliter of chloramphenicol and incubating 1 hour with frequent shaking prior to injection.

to 4 weeks. In either case, the testicular exudate is examined for the presence of typical spherules, which verifies the iden-tification of *C. immitis*.

Old (more than 10 days) cultures in the arthrospore stage are in the **most danger-ous** phase of the fungus, and dissemi-nation of the highly infectious arthrospores in the air can lead to infec-tion of laboratory personnel. Therefore, the following precautions must be taken in handling such cultures:

1. Never use Petri dishes—always em-ploy cotton-plugged test tubes.
2. To prevent the escape of arthro-spores, **as soon as a cottony mold grows out** (usually within 3 days), **wet down the slant with sterile sa-line before introducing an inoculat-ing needle.**
3. Make mounts for microscopic exami-nation in lactophenol cotton blue, which kills the spores; subculture to Sabouraud slants if indicated.
4. Sterilize all contaminated equipment by autoclaving promptly.

In culture, *C. immitis* is differentiated from *Geotrichum* and *Oospora,* both of which produce arthrospores by mycelial fragmentation. The following features may be noted:

1. *Geotrichum* remains yeastlike on Sa-bouraud agar.
2. *Oospora* does not produce alternately stained arthrospores and is not viru-lent for animals.
3. *Coccidioides immitis* on animal injec-tion produces the characteristic en-dospore-filled spherules.

Histoplasmosis

Histoplasmosis is a mycotic infection of the reticuloendothelial system that may involve the lymphatic tissue, lungs, liver, spleen, kidneys, skin, central nervous sys-tem, and other organs. It is caused by the **dimorphic** (saprophytic and tissue forms) fungus *Histoplasma capsulatum,* which exists as a saprophyte in the soil. Man and animals acquire the infection by the inhalation of spores from the environment; the severity of the disease is generally related directly to the intensity of the exposure. The growth of *H. capsulatum* in nature appears to be associated with decaying or composted manure of chickens, birds (especially starlings), and bats ("cave disease"). A typical human case may result from the cleaning of a chicken house or silo that has not been disturbed for a long period, or from working in soil under trees that have served as roosting places for starlings, grackles, or other birds. Although histoplasmosis may infect dogs and cats, there is no evidence of contagion between animal and man or between humans. The domestic animals, as well as several species of wild animals, appear to be accidental hosts and play no role in distributing or encouraging growth of *H. capsulatum* in the soil.

Histoplasmosis, once considered a rare and generally fatal illness, is now recognized as a common and benign disease of endemic areas in the eastern and central United States, where it is estimated that 500,000 persons are infected annually. Further studies and proper utilization of laboratory facilities will probably reveal the disease to be global in distribution.

The most frequent site of **primary** infection in man is the respiratory tract, resulting in a mild or asymptomatic pulmonary infection with a cough, fever, and malaise. In some areas a positive histoplasmin skin test, indicating exposure to *H. capsulatum,* is elicited in 60% to 90% of the inhabitants, who give no history of an unusual respiratory illness at all.

A chronic **cavitary** form of histoplasmosis also occurs in humans, with a productive cough, low-grade fever, and an x-ray picture of pulmonary cavitation that strikingly resembles tuberculosis. Undoubtedly, many such cases have been misdiagnosed and treated and hospitalized for pulmonary tuberculosis.

In less than 1% of cases of histoplasmosis, a severe, **disseminated** form of the disease develops, with involvement of the reticuloendothelial system and any structure or organ and frequently ending fatally.

Since *H. capsulatum* is primarily a parasite of the reticuloendothelial system, it is rarely found extracellularly in tissue. Therefore, direct and stained smears of clinical material are generally inadequate to demonstrate the fungus. Films of buffy coats, sternal marrow, cut surfaces of lymph nodes, splenic and liver punch biopsies, sputum, and scrapings should be stained with the Giemsa or Wright stains and carefully examined with the **oil-immersion objective.** *H. capsulatum* occurs intracellularly as small, round or oval yeastlike cells, 2 by 3 μ to 3 by 4 μ in size, with a large vacuole and a crescent or half-moon–shaped mass of red-stained protoplasm at the larger end of the cell. These may be found within the cytoplasm of macrophages and occasionally in the polymorphonuclear leukocytes, or free in the tissue.

The following methods for the isolation of *H. capsulatum* from clinical material are those used by the Mycology Unit of the CDC[6] and are highly recommended. **Sputum** specimens should be requested in all cases in which pulmonary or disseminated disease is suspected. A series of **six** early-morning specimens should be collected in sterile bottles; 1 to 10 ml. quantities are adequate. If the specimen cannot be inoculated promptly to culture media **(immediate inoculation is recommended),** add 1 ml. of a stock solution of chloramphenicol* to 1 to 10 ml. of the specimen. It is advisable, however, to inoculate the specimen directly to duplicate sets of Sabouraud agar and infusion blood agar, with and without antibiotics; with antibiotics, pretreatment with chloramphenicol is unnecessary. **Never hold specimens at room temperature;** *Histoplasma* **will not survive.**

---

*To prepare a stock solution, suspend 20 mg. of chloramphenicol in 10 ml. of 95% ethanol and add 90 ml. of distilled water. Heat gently to dissolve. The solution is stable. The concentration is approximately 0.2. mg. per milliliter of sputum.

Gastric washings should be requested when sputum is unobtainable; a series of three to six specimens is adequate. These are centrifuged and the sediments inoculated to the media previously described. Cerebrospinal fluid is submitted only when cerebral or meningeal involvement is evident. It is also centrifuged, and the sediment is inoculated as described. Citrated blood and bone marrow are of value only in acute disseminated cases. The blood is centrifuged and the buffy coat used for inoculation of media or laboratory animals. Bone marrow is handled identically, without centrifugation.

Incubate the infusion blood agar without antibiotics at 36° C., and incubate the other media at room temperature. The yeast phase of *H. capsulatum* and other dimorphic fungi does not develop at 36° C. on media containing the antibiotics.

The use of **mouse inoculation** is often helpful in the isolation of *H. capsulatum* from clinical specimens. Sputum and gastric washings are liquefied by agitation with glass beads and an equal part of physiological saline; tissues are ground with alundum and physiological saline in a tissue grinder. Chloramphenicol is added (0.05 mg. per mililiter) for decontamination and the specimen incubated at 36° C. for 1 hour. This is not necessary for buffy coat of blood or cerebrospinal fluid. Two to four mice are inoculated intraperitoneally with 1-ml. aliquots of the material. They are killed in 4 weeks, and cultures are made of portions of liver and spleen on cycloheximide media at room temperature and on infusion agar without antibiotics at 36° C. These are examined at intervals for development of colonies of *H. capsulatum.*

On Sabouraud agar at **room temperature,** *H. capsulatum* grows slowly (10 to 14 days) as a raised, **white, fluffy mold,** becoming tan to brown with age.* Micro-

scopically, these cultures show septate, branching hyphae bearing delicate, round to pyriform, smooth **microconidia** (2 to 4 μ in diameter), either on short lateral branches or attached directly by the base (sessile). Although at this stage the culture may be mistaken for *Blastomyces dermatitidis,* further incubation will usually reveal the diagnostic **tuberculate macroconidia** (chlamydospores) (Fig. 32-21, *A*). These spores are large (7 to 25 μ in diameter), round, thick walled, and covered with knoblike projections (tuberculate) that are

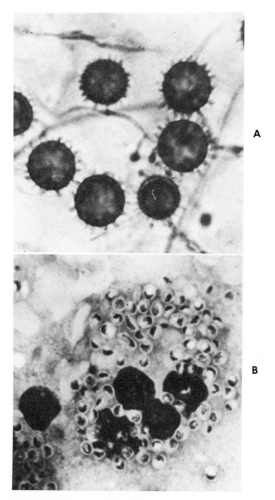

Fig. 32-21. *Histoplasma capsulatum.* **A,** Mycelial phase, showing characteristic tuberculate macroconidia. (1000×.) **B,** Blood smear, showing intracellular oval- to pear-shaped yeastlike cells, deeply stained. (2000×.)

---

*A white (albino) colony, showing only rare, smooth chlamydospores, may overgrow the typical colonies; the latter must be subcultured early in order to separate the two forms.

sometimes difficult to see when focusing in only one plane.

On **moist** infusion blood agar incubated at 36° C.,* *H. capsulatum* grows slowly as a white to brown, membranous, convoluted (cerebriform), yeastlike colony, resembling that of *Staphylococcus aureus,* or a very mucoid colony. A mycelial or mixed type of growth also may be produced. Although these colonies do not produce typical spores, a higher percentage of isolations results on this medium.† Transfer of these colonies to Sabouraud agar held at room temperature with ready access to air results in development of the typical macroconidia previously described, and this is recommended as a confirmatory procedure.

To convert the mycelial to the typical **yeast phase,** inoculate slants of moist infusion blood agar, seal with Parafilm, and incubate at 36° C. for several weeks. Growth appears as dull, white, yeastlike colonies, the contents of which appear microscopically as small (1 to 3 u), oval, **budding cells** similar to those seen in tissues. Strains in the mycelial phase that cannot readily be converted to the yeast phase culturally may be converted by animal inoculation.

Confirmation of the identification of *H. capsulatum* by **animal injection** is recommended. Inoculate several white mice intraperitoneally with a suspension of yeast phase cells harvested from several tubes of infusion blood agar in 5% hog gastric mucin‡ or a suspension of a 4- to 6-week mycelial growth ground in saline. One mouse is killed after 2 weeks and one at weekly intervals thereafter; impression smears of the involved organs (generally

liver and spleen) are made and stained with the Giemsa (**not** hematoxylin-eosin) stain. Microscopic demonstration of the typical intracellular organisms confirms the identification (Fig. 32-21, *B*). Yeast phase cultures also may be obtained from these tissues by culturing on enriched media incubated at 36° C.

Sepedonium, a saprophytic fungus found on mushrooms, may be confused with *H. capsulatum,* since it produces tuberculate chlamydospores. However, it will not form a yeast phase and it is not virulent for animals.

North American blastomycosis

North American blastomycosis is a chronic granulomatous and suppurative disease caused by the **dimorphic** fungus *Blastomyces dermatitidis.** The disease is limited to the continent of North America, extending southward from Canada to the Mississippi Valley, Mexico, and Central America. Some isolated cases also have been reported from Africa.[3] The largest number of cases occur in the Mississippi Valley region.

Blastomycosis, first described by Gilchrist in 1894, originates as a respiratory infection. Man probably acquires the infection through inhalation of the spores from the dust of his environment, although a saprophytic existence of *B. dermatitidis* has not yet been demonstrated. The infection may spread and involve the lungs, bone, and cutaneous tissue. It is not spread from man to man and generally occurs as a sporadic case. Small outbreaks appear to have been related to a common exposure; although blastomycosis is more common in the male, there is no apparent association with occupational exposure.

Material from cutaneous lesions is collected by scraping bits of tissue or taking swabs of pus from the edge of the lesion. Pus from unopened subcutaneous ab-

---

*Note: The yeast phase of the dimorphic fungi is suppressed by cycloheximide; media containing this agent must be incubated at room temperature.

†Incubation under an increased CO_2 tension appears to improve the growth of *H. capsulatum.*

‡Intravenous injection of 0.2 ml. of **chilled** (must be kept at 8° to 10° C. until the moment of injection) cell suspension (without mucin) into the tail vein of white mice frequently gives better results.[6]

---

*The perfect stage, *Ajellomyces dermatitidis,* has recently been described.

scesses should be aspirated with a sterile syringe and needle. Sputum, urine, and cerebrospinal fluid also should be examined in suspected systemic blastomycosis.

Material from such lesions is prepared for direct microscopic examination by placing it on a slide and pressing it into a thin layer with a coverglass. If the material is opaque, it may be cleared in 10% sodium hydroxide with gentle heating. The specimen is examined under high power, using subdued light. *B. dermatitidis* appears as a large, spherical, **thick-walled** cell, 8 to 20 μ in diameter, usually with a single bud that is connected to the mother cell by a **wide base.** Some walls may be sufficiently thick to give a double-contoured effect.

The infected material should be cultured on infusion blood agar incubated at 36° C. and on Sabouraud agar incubated at room temperature. Material likely to be contaminated with bacteria should be inoculated on the aforementioned media with chloramphenicol and cycloheximide and incubated at room temperature. Growth of the yeast phase of *B. dermatitidis* may be suppressed when grown on these media incubated at 36° C. In suspected pulmonary disease a fresh morning sputum specimen should be examined and cultured as previously described.

On Sabouraud agar at **room temperature,** *B. dermatitidis* generally forms a slowly growing, moist, grayish, mealy or prickly colony, which soon develops a **white cottony,** aerial mycelium, becoming tan (rarely, dark brown to black) with age. Microscopically, these filamentous colonies are made up of septate hyphae bearing small, oval (2 to 3 μ) or pyriform (4 μ to 5 μ) **conidia** (aleurospores) laterally, near the point of septation (Fig. 32-22, *A*). Older cultures develop conidia 7 to 15 μ in diameter with thickened outer walls that suggest the appearance of chlamydospores.

At 36° C. incubation on either Sabouraud agar or infusion blood agar, *B. dermatitidis* grows as a **yeastlike** organism. The fungus develops slowly (1 week)

as a creamy, wrinkled, waxy colony (similar to that of *Mycobacterium tuberculosis*) with a verrucose (warty) surface texture, cream to tan in color. Microscopic examination reveals thick-walled, budding, **yeastlike cells** 8 to 20 μ in diameter, resembling those seen in tissues or exudates (Fig. 32-22, *B*).

In order to identify an organism as *B. dermatitidis,* it is necessary to **convert** the mycelial phase at 25° C. to the tissue phase at 36° C. This is done by subculturing to **fresh** media (infusion blood

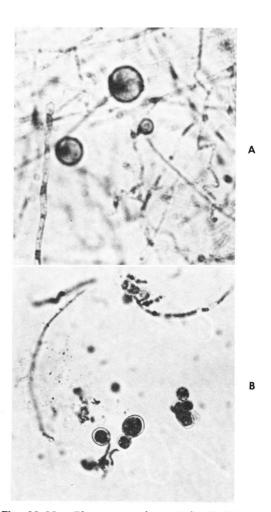

A

B

Fig. 32-22. *Blastomyces dermatitidis.* **A,** Mycelial phase, showing oval microconidia borne laterally on branching hyphae. (1000×.) **B,** Yeast phase, showing thick-walled, oval to round, single-budding, yeastlike cells. (500×.)

agar) and incubating at 36° C. Animal inoculation usually is unnecessary if the budding cells are demonstrated in direct smears of the infected material and the dimorphism is shown as described here.

If, however, the two phases are poorly defined, **animal injection** is required. A heavy suspension of either the mycelial or yeast phase is prepared, using physiological saline, and 1 ml. of this suspension is then injected intraperitoneally into several mice. These are killed in 3 weeks; microscopic examination of caseous material from the lesions or peritoneal fluid will show the thick-walled, budding, yeastlike tissue forms of *B. dermatitidis.* If these are not demonstrated, portions of the organs are cultured on infusion blood agar and incubated at 36° C. Yeast phase cultures are generally obtained.

Paracoccidioidomycosis (South American blastomycosis)

Paracoccidioidomycosis is a chronic progressive infection of the mucous membranes of the mouth (portal of entry) and nose, the lymph nodes of the neck, and metastatic lesions of the internal organs. It is caused by the **dimorphic** fungus *Paracoccidioides brasiliensis.* The disease is most common in Brazil, although it is seen

in many areas, including Mexico, Central America, and Africa.

Materials for **direct examination**** are secured and prepared as described for North American blastomycosis. *P. brasiliensis* appears as large, round to oval, budding cells 8 to 40 μ in diameter, thick walled, refractile, and with characteristic **multiple** buds (Fig. 32-23). Those cells with single buds are indistinguishable from those of *B. dermatitidis;* therefore, a search should be made for the diagnostic multiple budding cells. The daughter buds, 1 to 2 μ in diameter, usually are attached to the thick-walled mother cell by narrow connections, giving the whole a **steering wheel** appearance.

Infected material should be cultured on infusion blood agar at 36° C. and on Sabouraud agar at room temperature, as in the study of North American blastomycosis. At room temperature the fungus develops as a very **slowly growing** (2 to 3 weeks), heaped-up, folded colony with a short nap of white, velvety mycelium. Microscopically, small, delicate (3 to 4 μ), round or oval conidia may be seen sessile borne or on very short sterigmata on septate hyphae. Usually, however, only a fine septate mycelium and chlamydospores are seen.

At 36° C., *P. brasiliensis* grows slowly as a smooth, soft, yeastlike colony, cream to tan in color. These yeastlike colonies, which may appear either verrucose or smooth and shiny, are composed of single cells and multiple budding forms, identical with those seen in tissue and exudates. All cultures should be held at least 4 weeks before being discarded as negative.

As with *B. dermatitidis,* **conversion** of the mycelial to the yeast phase must be demonstrated by subculture and incubation at 36° C. If animal inoculation is required for further confirmation, guinea pigs injected intratesticularly (1 ml. of a heavy saline suspension of yeast phase organ-

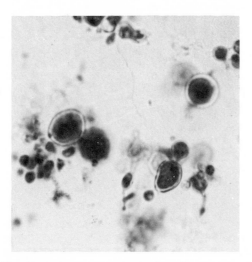

Fig. 32-23. *Paracoccidioides brasiliensis,* yeast phase, showing multiple budding. (400×.)

**Recommended as the most practical diagnostic method.[3]

isms) may be used. The guinea pigs are killed in 8 to 12 days and examined for the presence of the diagnostic multiple budding tissue forms of the fungus in pus from the draining sinuses.

Rare mycoses
Aspergillosis

Aspergilli are among the most common and troublesome contaminants in the laboratory; several are pathogenic and may produce either inflammatory or chronic granulomatous lesions in the bronchi or lungs, often with hematogenous spread to other organs. The external ear, cornea of the eye, nasal sinuses, and other tissues of man or animals are frequently infected. *Aspergillus fumigatus* is the species most frequently associated with pathological processes. This may include pulmonary aspergillosis of a severe and invasive type, which is being observed with increasing frequency in debilitated patients receiving antibiotic or steroid therapy and immunosuppressive or antimetabolite drugs.

Since aspergilli are found frequently in cultures of sputum, skin scrapings, and other specimens, it is essential that the fungus be **repeatedly demonstrated** in large numbers in direct smears of the fresh material and **repeatedly isolated** on culture to be considered etiologically significant. Direct microscopic examination of sputum or other infected material may reveal fragments of branched, septate mycelium (3 to 6 μ in width) and sometimes conidial heads.

Suspected material should be inoculated on Sabouraud agar without cycloheximide* and incubated at room temperature. *A. fumigatus* grows rapidly (2 to 5 days) and appears first as a flat, white filamentous growth, which rapidly becomes **blue green and powdery** as a result of the production of spores.

Microscopically, *A. fumigatus* is characterized by branching, septate hyphae,

some of which terminally bear a conidiophore that expands into a large, inverted, flask-shaped vesicle (sac) covered with small sterigmata (Fig. 32-24). These sterigmata occur only in a **single row** and around the **upper half** of the vesicle; from their tips are extruded parallel chains of small, rough-surfaced, green conidia, giving the whole structure a flaglike appearance.*

Animal inoculation is not necessary for the identification of *A. fumigatus*.

Phycomycosis (mucormycosis)

Phycomycosis is a rare but rapidly fatal disease caused by members of the class Phycomycetes, which produce aseptate mycelia and are ordinarily considered nonpathogenic laboratory contaminants. The genera involved include *Mucor, Rhizopus, Absidia,* and others. The fungus enters the nose of susceptible patients, particularly uncontrolled diabetics and patients receiving prolonged antibiotic, corticosteroid, or cytotoxic therapy, and pene-

---

*For further aid in identifying other species of aspergilli the reader is referred to Thom, C. T., and Raper, K. B.: The aspergilli, Baltimore, Md., 1920, The Williams & Wilkins Co.

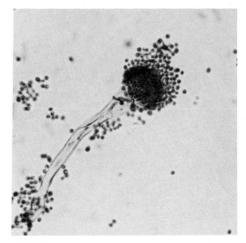

Fig. 32-24. *Aspergillus fumigatus,* showing conidia in chains, arising from a single row of sterigmata on the upper portion of the vesicle. (500×.)

Aspergillus fumigatus may be inhibited by cycloheximide.

trates the arteries to produce thrombosis. Later it invades the veins and lymphatics, producing infarcts. The disease assumes cerebral and pulmonary forms and, rarely, intestinal, ocular, and disseminated forms. It is usually fatal.

The diagnosis of phycomycosis is usually made by examination of tissue specimens taken at biopsy or autopsy, in sections of which can be demonstrated broad (4 to 200 μ thick), branching, predominantly **nonseptate hyphae.** The culture of sputum, cerebrospinal fluid, or exudate should be attempted in suspected cases.

On Sabouraud agar incubated at room temperature, *Rhizopus* species produce a rapidly growing (2 to 4 days), coarse, woolly colony, which soon fills the test tube with a loose, grayish mycelium dotted with brown or black sporangia. The fungus is characterized miscroscopically by a large, broad, **nonseptate,** hyaline mycelium that produces horizontal runners

(stolons), which attach at contact points (medium or glass) by rootlike structures called **rhizoids** (Fig. 32-25). From these contact points arise clusters of long stalks, known as **sporangiophores,** the ends of which are terminated in large, round, dark-walled **sporangia** (spore sacs). When mature, these sporangia are filled with spherical hyaline spores. Since *Rhizopus* species are common contaminants, the recovery of this organism in culture is not in itself diagnostic.

On Sabouraud agar at room temperature, *Mucor* species produce a rapidly growing colony that fills the test tube with a white fluffy mycelium, becoming gray to brown with age. Microscopically, the fungus is characterized by a nonseptate, colorless mycelium **without** rhizoids. The sporangiophores arise singly from stolons and branch profusely, with sporangia containing many spores arising from the apex of each branch. The columnellae (the per-

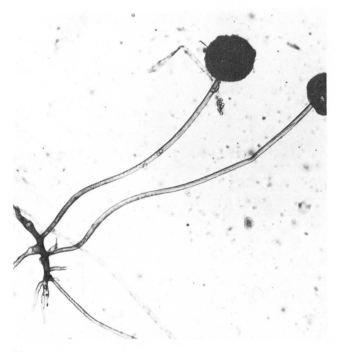

Fig. 32-25. *Rhizopus* species, showing sporangium on long sporangiophore arising from a nonseptate mycelium. Characteristic rhizoids are seen at the base of the sporangiophore. (250×.)

sistent dome-shaped apices of the sporangiophore) are usually well developed and of various shapes but **never hemispherical** as in *Rhizopus.* Empty sporangial sacs, still attached to the conidiophores after release of the spores, may be observed in both genera.

Organisms of the species *Absidia* are similar to the *Rhizopus,* except that the sporangiophores arise on stolons at a **point between two nodes** from which rhizoids are formed. The sporangia are pear shaped (10 to 70 μ in diameter) instead of being round.

Treatment of systemic mycoses

The available agents for treatment of systemic fungal infections include the polyene antibiotic amphotericin B, the synthetic drug 5-flucytosine, and iodides. Amphotericin B appears to be the drug of choice in blastomycosis, cryptococcal meningitis, systemic candidiasis, disseminated histoplasmosis, and coccidioidomycosis and is given parenterally. Flucytosine, given by mouth, has proved effective in candidemia and candidal urinary tract infections. Iodides by mouth is the specific drug for lymphocutaneous sporotrichosis.

The interested reader is advised to consult publications by Beneke[9] and by Bennett[10] for further information on therapy.

Use and interpretation of hypersensitivity tests and serological reactions in mycotic infections

Conclusive evidence of the presence of a fungal infection is best offered by the demonstration of higher fungi, either by direct examination or by cultural procedures, in the exudate or in diseased tissue. However, **indirect** evidence, by the demonstration of a hypersensitive state or an increasing titer of specific antibody, may prove exceedingly helpful.

Intradermal skin tests are widely used to aid in the diagnosis, treatment, and epidemiological survey of the systemic mycoses, particularly in blastomycosis, histoplasmosis, and coccidioidomycosis. In the hypersensitive patient, the injection of mycelial or whole cell extracts of these fungal agents (blastomycin, histoplasmin, and coccidioidin*) will elicit a **tuberculin-like response**—redness, swelling, and induration—within 24 to 48 hours. The significance of this positive reaction is simply that a mycotic infection has occurred at some time during the patient's life. Since cross-reactions occur among these three antigens, all three should be applied **simultaneously,** using a separate needle and syringe uncontaminated by previous use with other skin test antigens. Results should be interpreted with caution; when negative, they are valuable in ruling out the diagnosis of a systemic infection, particularly in histoplasmosis. Skin tests have been used to demonstrate hypersensitivity to *Paracoccidioides brasiliensis, Sporotrichum schenckii, Aspergillus fumigatus,* and other fungi, although standardized reagents are not yet commercially available.

Serum from patients infected with mycotic agents may contain **specific antibodies.** These may be demonstrated in the laboratory by utilizing complement fixation, precipitin, latex or colloidin particle agglutination, hemagglutination, agar gel immunodiffusion, or indirect immunofluorescence procedures.† These serological examinations are useful adjuncts in the diagnosis and prognosis of a mycotic infection. For example, in a case of a pulmonary infection in which no sputum can be obtained, or in a case where it is difficult or impossible to secure infected tissue or exudate, the diagnosis may well depend upon the application of serological tests and the interpretation of their results.

Precipitins and agglutinins appear early in mycotic diseases, and their presence may indicate subsequent cultural examinations. Complement fixation tests using an-

---

*Standardized skin test antigens may be obtained from Parke-Davis & Co., Eli Lilly & Co., Cutter Laboratories, and Michigan State Health Laboratories.
†The reader is referred to an excellent summary of the serology of the systemic mycoses by Kaufman.[18]

Table 32-2. Serological tests presently available*

DISEASE	TEST	ANTIGEN
Histoplasmosis	Complement fixation	⎰ Histoplasmin
	Agar gel	⎱ Yeast
Blastomycosis	Complement fixation	Yeast
	Agar gel	Yeast
Coccidioidomycosis	Complement fixation	Coccidioidin
	Agar gel	Coccidioidin
Paracoccidioidomycosis	Complement fixation	Yeast
	Agar gel	Yeast
Aspergillosis	Agar gel	*A. fumigatus*
Aspergillosis	Agar gel	*A. niger*
Candidiasis	Agar gel	*C. albicans*
Candidiasis	Latex agglutination	
Cryptococcosis	Indirect FA	
Cryptococcosis	Tube agglutination	
Cryptococcosis (antigen)	Latex agglutination	
Sporotrichosis	Tube agglutination	
Sporotrichosis	Latex agglutination	

*Data from Fungus Immunology Unit, Center for Disease Control, Atlanta, Ga.

tigens prepared either from the yeast or mycelial form of a dimorphic fungus are particularly useful in the diagnosis of histoplasmosis (a high titer is considered diagnostic) and generalized coccidioidomycosis; it is less useful in blastomycosis. As is true of all other serological examinations, a dependable diagnostic interpretation **cannot be made on a single serum specimen.** Blood samples taken early in the illness—one during the height of the disease, one during convalescence, and one after recovery—are to be hoped for, but rarely acquired, by the laboratory. Under such ideal serological conditions a **sharp rise** (fourfold or greater) and subsequent fall in titer usually corroborates a clinical diagnosis. Because of cross-reactions all three antigens should be used in the serological tests performed.

Prognostic interpretation of the immunological and serological tests is based upon the general observation that a patient reacting positively to an intradermal test and not showing a significant serological titer has a better prognosis than the patient with a negative skin test and a high titer of complement-fixing antibodies. This is particularly true in coccidioidomycosis. It is well to note that serum for histoplasmosis serological testing should be drawn **before,** or not later than 2 days after, an intradermal test is carried out because the latter may induce the production of demonstrable circulating antibodies.

REFERENCES

1. Adams, A. R., et al.: Nocardiosis, Med. J. Austral, **1:**669-674, 1971.
2. Ahearn, D. G.: Identification and ecology of yeasts of medical importance. In: Prier, J. E., and Friedman, H., editors: Opportunistic pathogens, Baltimore, Md., 1973, University Park Press.
3. Ajello, L.: Comparative ecology of respiratory mycotic disease agents, Bact. Rev. **31:**6-24, 1967.
4. Ajello, L.: A taxonomic review of the dermatophytes and related species, Sabouraudia **6:**147-159, 1968.
5. Ajello, L., and Georg, L. K.: In vitro hair cultures for differentiating between atypical isolates of *Trichophyton mentagrophytes* and *Trichophyton rubrum,* Mycopathologia **8:**3-17, 1957.
6. Ajello, L., Georg, L. K., Kaplan, W., and Kaufman, L.: CDC manual for medical mycology, Public Health Service Publication No. 994, Washington, D.C., 1963, U. S. Government Printing Office.
7. Ajello, L., Grant, V. Q., and Gutzke, M. A.: The effect of tubercule bacillus concentration procedures on fungi causing pulmonary mycoses, J. Lab. Clin. Med. **38:**486-491, 1951.
8. Alexopoulous, C. J.: Introductory mycology, ed. 2, New York, 1962, John Wiley & Sons, Inc.
9. Beneke, J. E.: In Thomas, B. A., editor: Scope

monograph on human mycoses, Kalamazoo, Mich., 1972, The Upjohn Co.

10. Bennett, J. E.: The treatment of systemic mycoses, Ration. Drug Ther. **7**(4), 1973.

11. Bessey, E. A.: Morphology and taxonomy of fungi, New York, 1950, Blakiston Division, McGraw-Hill Book Co.

12. Conant, N. F., Smith, D. T., Baker, R. D., and Callaway, J. L.: Manual of clinical mycology, ed. 3, Philadelphia, 1971, W. B. Saunders Co.

13. D'Alessio, D. J., Leavens, L. J., Strumpf, G. B., and Smith, C. D.: An outbreak of sporotrichosis in Vermont associated with sphagnum moss as the source of infection, New Eng. J. Med. **272**:1054-1058, 1965.

14. Emmons, C. W., Binford, C. H., and Utz, J. P.: Medical mycology, ed. 2, Philadelphia, 1970, Lea & Febiger.

15. Georg, L. K.: Personal communication (EGS), 1972.

16. Green, W. O., Jr., and Adams, J. E.: Mycetoma in the United States, Amer. J. Clin. Path. **42**:75-91, 1964.

17. Haley, L. D.: Diagnostic medical mycology, New York, 1964, Appleton-Century-Crofts.

18. Kaufman, L.: Serology of systemic fungus diseases, Public Health Rep. **81**:177-185, 1966.

19. Marks, M. I., Langston, C., and Eickhoff, T. C.: *Torulopsis glabrata*—an opportunistic pathogen in man, New Eng. J. Med. **283**:1131-1135, 1970.

20. Marks, M. I., and O'Toole, E.: Laboratory identification of *Torulopsis glabrata;* typical appearance on routine bacteriological media, Appl. Microbiol. **19**:184-185, 1970.

21. Staib, F.: Membranfiltration und Negersaat (Guizotia abyssinica)—Nährboden für den *Cryptococcus neoformans*—Nachweis (Braunfarbeffekt), Z. Hyg. Infektionskr. **149**:329-336, 1963.

22. Utz, J. P.: Recognition and current management of the systemic mycoses, Med. Clin. N. Amer. **51**:519-527, 1967.

23. Vanbreuseghem, R.: Early diagnosis, treatment and epidemiology of mycetoma, Rev. Med. Vet. Mycol. **6**:49-60, 1967.

24. Wilson, J. W., and Plunkett, D. A.: The fungous diseases of man, Berkeley, Calif., 1965, University of California Press.

PART SEVEN

Intestinal parasites
of clinical significance

33 Laboratory diagnosis of intestinal parasites

VICHAZELU IRALU, Ph.D., M.P.H.*

Approximately **100** species of animal parasites can infect the body[12]; about **70** of these are common and considered important. More than half can be detected by examination of fecal specimens because they reside in the gastrointestinal tract itself or are so located that they or their progeny find their way into the alimentary canal. A few parasites that are closely associated with this canal, such as those present in expectorated material, in sputum, and in urine, will also be mentioned briefly.

Of approximately 40 animal parasites that can be diagnosed by fecal examinations, more than a third of them are single-celled protozoa. The remainder are multicellular worms. While only half a dozen of the intestinal protozoa are clinically important, almost all of the worms are potentially pathogenic. In most cases, the **number** of worms harbored is the most important factor. Because of better sanitation and certain other favorable conditions, intestinal parasites are less frequently seen in the United States than in the less developed countries, especially those of the tropics. Notably rare are the trematode worms; almost all trematode infections seen in the continental United States are acquired from other countries.

Because of increased travel to other countries, especially to the tropical areas, and increased visits to the United States by people from these regions, laboratories should anticipate occasionally encountering even the more esoteric forms.

METHODS OF FECAL EXAMINATION

Two types of laboratories examine fecal specimens: those in hospitals and clinics, where **fresh** specimens are obtained from patients directly, and central and public health laboratories, where **preserved** specimens are submitted from a distance. The routine to be followed in processing fresh specimens will differ in some important details from that of handling preserved specimens. Each laboratory's regimen will vary in certain details.

Fresh fecal specimens should be examined macroscopically and microscopically. Direct wet mounts with or without saline, depending on the consistency of the stool, and an iodine stain should be prepared and examined microscopically for every specimen. When protozoa are present, especially trophozoites of the amebae, a Nair's stained smear should be examined. A permanent stain should be made for the confirmation of *Entamoeba histolytica*, for record and for the correct identification of other protozoa, some of which are hard to identify by other methods. Each fresh specimen should be examined by one of the concentration techniques. Preserved

*Assistant Professor of Pathology, School of Medicine, University of Pennsylvania and Assistant Microbiologist, William Pepper Laboratory, Philadelphia, Pa.

specimens in either Schaudinn or polyvinyl alcohol (PVA) fixative should be stained with trichrome or an equivalent permanent stain. Specimens preserved in formalin or other solutions should be examined in direct wet mounts as well as by concentration. A number of the more commonly used techniques and directions for obtaining suitable specimens are provided in this chapter.

COLLECTION AND HANDLING OF FECAL SPECIMENS

Fecal specimens should be examined immediately after their passage from the human body in order to detect the living trophozoites of protozoa, particularly the amebae, because many of these anaerobic organisms become nonmotile and die after 30 minutes. Degeneration of normal morphology follows rapidly. Cysts, eggs, and larval forms, however, retain their viability and forms for days. A specimen that cannot be examined right away should be preserved or at least refrigerated at about 5° C. Stools should not be kept at room temperature beyond a few hours, because this hastens degeneration of the trophozoites. Specimens that arrive too late for examination by competent laboratory day personnel, those that arrive in the evening or night, and those that are collected over the weekend should be placed in preservatives. Specimens that are to be mailed to a diagnostic center or a public health laboratory should also be suitably preserved. For techniques of fixing and preservation see the section on preservation of specimens.

In hospitals or clinics, specimens can be taken out of a clean, dry bedpan and examined directly, or the material can be transferred to a waxed, half-pint cardboard container. The patient can also defecate directly into the paper carton, or first on a newspaper placed over a toilet seat and the specimen then transferred into the carton. It should be free of urine and toilet bowl water. The specimen should consist of at least 5 to 10 gm., but the entire specimen may be submitted for a better evaluation. The carton should be closed and labeled with the patient's name, the date, and the time the specimen was taken. Since the actual collection is usually carried out by the patient, a lay person, careful instructions should be given. While diagnostic techniques are very important, obtaining proper specimens may determine the success of laboratory examinations.

Antidiarrheal compounds, antibiotics, antacids, oils, bismuth, barium, and various other drugs will render stool specimens unsuitable for parasitological examinations. Specimens should be obtained before these chemicals are taken or after their effects have practically disappeared. In the case of barium and bismuth administration, at least 7 to 10 days should be allowed to elapse before an attempt is made to obtain a specimen. It should be kept in mind that the tetracyclines are to some extent antiamebic in nature; hence the probability of detecting *E. histolytica* will be reduced if such an antibiotic has been used. If barium, bismuth, or other interferring drugs are present, the specimens should be rejected and more suitable ones requested. If the specimen submitted is of inadequate quantity, dehydrated, or aged, another should be requested.

The number of specimens to be examined to detect most parasitic infections varies with the severity of the infection, the care and thoroughness of examinations, and the quality of the specimens submitted. The greater the number of specimens examined, the greater the chances for detection. Even if a combination of techniques is used, fewer than 50% of *E. histolytica* infections will be detected from a single stool specimen, and at least six examinations are necessary to detect more than 90% of infections.[21] Most worm eggs are shed from day to day at a fairly uniform rate, but some of the protozoa, including *E. histolytica* and a few helminth eggs and segments, are shed irregularly. On some days very few are detected, while on other days many are released. In practice, three normally passed stools and one saline-purged stool are usually adequate. The specimens should be obtained on different days, preferably skipping a day or two in between. Sodium sulfate or buffered sodium biphosphate is recommended for purging.

GROSS INSPECTION OF SPECIMENS

The consistency of stools varies from firm to watery, and the color may reveal some underlying abnormality. Melena (a dark-colored stool) usually indicates bleeding high in the gastrointestinal tract, while the presence of frank blood in the stool suggests bleeding at a much lower level, perhaps as low as the rectum. In a few parasitic infections the presence of streaks of blood, with or without mucus, may be particularly significant. In amebic dysentery and in the rare disease of balantidiasis, considerable amounts of blood and mucus may be encountered. Oral intake of iron or bismuth may produce a black stool. Beets or sulfobromophthalein (Bromsulphalein) sodium may render it red. In the absence of barium, a very light color may be due to diminution or absence of bile pigments.

Certain adult intestinal helminths are so large that their presence in the stool is quite obvious. Parasites of this magnitude are adult *Ascaris* and segments of large tapeworms. Adult whipworms 3 to 5 cm. in length can easily be seen. The entire length of the dwarf tapeworm *(Hymenolepis nana)* is of this size. Hookworms and pinworms can be seen by keen observation with the naked eye. Some of these worms may be passed with a normal stool, but most are not seen in the stool except after medication. The intestinal parasites are diagnosed primarily by finding the eggs and larvae of the helminths and the cysts and trophozoites of the protozoa on microscopic examination. However, when adult worms or segments of tapeworms are present, they should not be missed. Information regarding the

presence of unusual characteristics enumerated above, including those that are not strictly parasitological, could be included in the report for the physician's evaluation, because he does not have the opportunity to observe them.

MICROSCOPIC EXAMINATION

A microscope, preferably binocular, equipped with low-power (10×), high dry (44×), and oil immersion (98×) objectives and 5× and 10× oculars, is suitable for parasitological work. If the ocular is fitted with a micrometer, accurate measurements can be made. The microscope should have a variable intensity bluish light source. Most of the helminth eggs and larvae can be recognized at 100× total magnification. At this level, the entire coverglass can first be scanned rapidly to spot important structures. The high dry objective may be used to identify several of the protozoa. For identification of minute details, the oil immersion objective is used. For the wet mounts described below, if oil is to be used, the coverslip preparations can be sealed with a heated paraffin-petrolatum mixture (approximately 1:1). This prevents the material under the glass from shifting. It also prevents desiccation of the specimen for as long as 2 to 3 days.

Three kinds of fecal preparations are used for microscopic examinations:
1. Direct wet mounts with saline and stains of fresh or preserved feces
2. Concentrated wet mounts of fresh or preserved feces
3. Permanently stained slides

For the first two, the preferred slide sizes are 1.5 × 3 inches or 2 × 3 inches (3.8 × 7.6 cm. or 5 × 7.6 cm.). Ordinary hematologic slides (1 × 3 inches) are used for permanent stains.

Direct wet mounts

Physiological saline mount. Routinely, two wet mounts, one with saline and one with iodine, placed a few millimeters apart on a slide, are examined first. To prepare the saline mount, a drop is placed on the left-hand side of a glass slide, and a few milligrams of selected fecal material are transferred into it with a wooden applicator. This material is emulsified, and an even spread is made, thin enough that newsprint could be read through it. A No. 1 coverglass (22 × 22 mm.) is placed over it. The natural characteristics of the living protozoan trophozoites, such as motility of the psuedopods, flagella, undulating membranes, cilia, and the large, actively lashing larvae of *Strongyloides,* can be seen in this preparation. Cysts and eggs are also detected in this mount. If the stool is soft, watery, or mucosanguineous, trophozoites can be expected. If present, cyst forms predominate in formed stools, while eggs and larvae are found in stools of all consistencies. If the stool is watery, saline dilution is unnecessary.

Iodine stain. The wet iodine stain preparation, placed on the right-hand side of the slide, is made in the same manner as the saline mount. Dobell and O'Connor's iodine solution (2 gm. of potassium iodide and 1 gm. of powdered iodine in 100 ml. water, the solution filtered or decanted) or diluted Lugol's solution (1:5) are generally used. These iodine solutions should be prepared fresh every 2 to 3 weeks because they lose strength. The nuclei of protozoan cysts are visible in the iodine stain; glycogen, if present, appears reddish brown; and the cytoplasm is yellow. Chromatoid bodies are not as conspicuous in iodine as they are in saline. The iodine stain is not suitable for trophozoites.

Nair's stain. In order to bring out details in the nuclei of trophozoites in fresh fecal preparations, Nair's[17] buffered methylene blue stain may be used. This stain has largely replaced the older Quensel stain. The properties of this stain result from the fact that at a lower pH this biological dye penetrates the cells more readily, thus staining the nuclei. A small amount of methylene blue (0.06%) in acetate buffer at pH 3.6 (made by mixing 46.3 ml. of 0.2

M acetic acid with 3.7 ml. of 0.2 M sodium acetate), for example, will give satisfactory results. Other acid pH values can be made by other combinations of these reagents. Commercially available buffer tablets giving a pH range of 3.8 to 6.0 can also be used. A wet mount of a fresh fecal specimen is made with Nair's stain in the same manner as the saline mount. In a correctly stained preparation, the nuclei of trophozoites are stained dark blue in 5 to 10 minutes, while the cytoplasm is light blue. The slide should be read within 30 minutes. Some flagellated trophozoites and *Dientamoeba fragilis* do not stain well. The stain does not penetrate living cysts.

Fecal specimens that have been preserved in formalin, Merthiolate-iodine–formaldehyde (MIF), or phenol-alcohol–formaldehyde (PAF) solutions can be examined as wet mounts, directly or after concentrations.

Concentrated wet mounts

The daily output of feces in an adult is 100 to 250 gm. Although parasites reproduce in vast numbers and leave the host body in large numbers, the sampling of a mere few milligrams of feces will frequently fail to reveal parasites. Helminth eggs, larvae, and protozoan cysts (but not trophozoites), not numerous enough to be detected easily by the direct mount examination, can be concentrated by flotation or by sedimentation. There are numerous procedures for concentration, but the two most useful techniques are zinc sulfate flotation and formalin-ether (FE) sedimentation. The former is not suitable for operculated worm eggs and schistosomes. There is some distortion of the small cysts, as in *Endolimax nana* and *Entamoeba hartmanni.* The FE technique overcomes these deficiencies, but it is not as effective as the former method in detecting cysts of *Giardia lamblia, Iodamoeba bütschlii,* and the eggs of *Hymenolepis nana.*

Zinc sulfate flotation. The original method developed by Faust and co-workers[11] included two more steps that are omitted in the modified method described below. In the original method, the fecal material is emulsified in a beaker, and the suspension is strained through gauze into a small tube.

The zinc sulfate solution should have a **specific gravity of 1.18.** It is prepared by adding 331 gm. of zinc sulfate (USP) to 1 liter of warm tap water. A hydrometer is used to check the specific gravity. If the reading is off, water or salt can be added as needed. Some workers prefer a specific gravity of 1.20. This heavier solution should be used when concentrating formalin-preserved specimen. The technique is as follows:

1. Using two applicators, comminute a fecal sample about the size of a small pea in a Wassermann tube (13 × 100 mm.) half filled with tap water. Particles are broken up, and an even suspension is formed.
2. Add additional water until the tube is two thirds full.
3. Centrifuge the preparation for 1 minute at approximately 2,500 r.p.m.
4. Decant the supernatant into a container holding an appropriate disinfectant, such as cresol or Amphyl.
5. Repeat this washing if the stool is extremely oily.
6. Add enough zinc sulfate solution to fill the tube half-way.
7. Using an applicator, thoroughly break up the packed sediment.
8. Add additional zinc sulfate solution to fill the tube to within 1 cm. of the top.
9. Centrifuge the suspension for 1 minute at 2,500 r.p.m. Proceed to either step 10 or 11.
10. a. Using a small wire loop bent at right angles to the stem, remove two or three loopfuls of material from the surface film to a 3 × 2 inch slide. A capillary pipet may also be used to transfer the surface material to the slide.
    b. Add a drop of iodine and cover with a 22 × 22 mm. No. 1 coverglass, and seal.

c. Examine under a microscope for cysts, eggs, and larvae.
11. a. Without shaking or spilling the solution, place the tube carefully in a rack.
    b. Using a pipet or dropper, slowly fill the tube to the brim with more zinc sulfate, with a minimum of disturbance to the surface.
    c. Place a clean, grease-free No. 1 coverglass (22 × 22 mm.) on top of the tube so that the undersurface touches the meniscus. Avoid air bubbles. Leave undisturbed for about 10 minutes.
    d. Remove the coverglass with a straight, upward motion. A drop containing eggs, larvae, and cysts (if present) will adhere to the underside of the coverglass.
    e. Lower this onto a drop of iodine stain placed on a clean 3 × 2 inch slide. Seal the preparation and examine.

Formalin-Ether (FE) Sedimentation. This method, introduced by Ritchie,[18] can be used to concentrate eggs, larvae, and cysts of parasites in stool specimens. It is especially effective when excessive amounts of fats and fatty acids are present in the stool. It also concentrates formalin-preserved specimens effectively.

For fresh specimens:

1. Comminute a portion of stool in sufficient saline or water so that upon centrifugation 10 ml. of emulsion will yield about 1 ml. of sediment. A portion about the size of a medium-sized walnut is adequate. The suspension may be prepared in the carton in which it is submitted, or in a beaker or flat-bottomed paper cup. If permanently-stained slides are to be made or if a portion of the specimen is to be preserved in PVA fixative, these should be done first.
2. Using a small glass funnel, strain about 10 ml. of the emulsion through one or two layers of wet gauze into a 15-ml. tapered centrifuge tube. Use

two layers of wide-mesh gauze or one layer of narrow-mesh gauze. To conserve glassware, a cone-shaped paper cup with the point cut off can be substituted for the glass funnel.

3. Centrifuge emulsion at 2,000 to 2,500 r.p.m. for 1 minute. Decant supernatant.

4. Resuspend the sediment in fresh saline, centrifuge, and decant as before. This step may be repeated if a cleaner sediment is desired.

5. Add about 10 ml. of 10% formalin to the sediment, mix thoroughly, and allow to stand for 5 minutes. The mixture can also be stoppered and held for later use.

6. Add 3 ml. of ether, stopper the tube, and shake vigorously in an inverted position for a full 30 seconds. Remove the stopper with care.

7. Centrifuge at 1,500 r.p.m. for about 1 minute. Four layers should result as follows: (1) ether at top, (2) plug of debris, (3) formalin solution, and (4) sediment.

8. Free the plug of debris from the sides of the tube by scraping with an applicator stick, and carefully decant the top three layers. Use a cotton swab to remove any debris adhering to the sides of the tube.

9. Mix the remaining sediment with the small amount of fluid that drains back from the sides of the tube (or, if necessary, add a small amount of formalin or saline), and prepare iodine and unstained mounts in the usual manner for miscroscopic examination.

For formalin-preserved specimens:

1. Stir the formalinized specimen thoroughly and, depending on the size and density of the specimen, strain a sufficient quantity through gauze into a 15-ml. tapered centrifuge tube to give the desired amount of sediment indicated below.

2. Add tap water, mix thoroughly, and centrifuge at 2,000 to 2,500 r.p.m. for 1 minute. The resulting sediment should be about 0.5 to 1 ml.

3. Decant supernatant fluid and, if desired, wash again with tap water.

4. Proceed with the addition of formalin and ether as described above for the concentration of fresh specimens, from steps 5 through 9.

Other concentration methods. In addition to the zinc sulfate flotation and FE sedimentation methods, several other concentration techniques may be mentioned. Concentration by gravity sedimentation or centrifugal sedimentation are useful for recovering parasites in the living condition and especially for recovering schistosome and operculated eggs of helminths. Another sedimentation technique particularly used for the recovery of helminth eggs is the acid-ether method, but the procedure is not suitable for protozoan parasites. Specimens preserved in MIF and PAF are also suitable for concentration. For details of these additional techniques, a standard text such as the one by Melvin and Brooke[16] should be consulted.

Permanent stains

Frequently it is not possible to make a definite identification of the protozoa in direct, temporary wet mounts. Specimens that contain forms resembling protozoa, especially the amebae, and all fresh specimens that cannot be examined immediately should be stained permanently so that critical examinations can be made.

There are a number of permanent stain techniques. The most famous of these in the past has been Heidenhain's iron-hematoxylin method. The original and the modified techniques bring out excellent details of the protozoa. They are somewhat time consuming, however, and it takes a fair amount of practice to produce good results with them. At present, the trichrome stain is the most widely used, and this will be described in some detail. The recently introduced Chlorazol Black E stain, also to be described, is quite suitable for use in hospitals and clinics where fresh fecal specimens are received. It is not suitable for PVA-fixed smears, whereas the iron-hematoxylin and trichrome stain techniques can be used on both fresh fecal smears as well as PVA-fixed smears.

Trichrome stain. The Wheatley[24] modification of Gomori's trichrome stain can be done with a fresh fecal specimen in about 20 minutes and with polyvinyl alcohol–preserved smears in less than 1 hour.

For fresh fecal smears, using an applicator stick, make a thin smear on a clean 1 × 3 inch slide. If necessary, dilute with physiological saline. Immerse the slide into Schaudinn's fixative containing acetic acid. The smear must not be permitted to dry from the time it is made until it is mounted. This is **very important.**

Schaudinn's solution is prepared by mixing one part of 95% ethyl alcohol with two parts of saturated aqueous mercuric chloride (about 14 gm./100 ml. water). Just before use, one part of glacial acetic acid is added to 19 parts of Schaudinn's solution. The fresh smear should be kept in Schaudinn's solution for at least 5 minutes at 50° C. before staining. At room temperature it should be fixed for about one hour. It can be stained days later. The staining sequence is as follows:

1. Place the Schaudinn's fixed smear in 70% alcohol plus iodine — 1 minute
2. In 70% alcohol (1) — 1 minute
3. In 70% alcohol (2) — 1 minute
4. In trichrome stain — 2 to 8 minutes
5. In 90% acidified alcohol (one drop of glacial acetic acid in 10 ml. of alcohol) — 10 to 20 seconds or until stain barely runs from smear
6. In 95% or 100% alcohol — Rinse twice
7. In 100% alcohol or carbol-xylene — 1 minute
8. In xylene — 1 minute or until refraction at smear-xylene interface ends
9. Using Permount or some other mounting fluid, cover stained smear with coverslip

For PVA-preserved smears see the technique for making PVA-preserved fecal smears. The dried PVA slide is processed as follows:

1. Place in 70% alcohol plus iodine — 10 to 20 minutes
2. In 70% alcohol (1) — 3 to 5 minutes
3. In 70% alcohol (2) — 3 to 5 minutes
4. In trichrome stain — 6 to 8 minutes
5. In 90% acidified alcohol (one drop of glacial acetic acid in 10 ml. of alcohol) — 10 to 20 seconds or until stain barely runs from smear
6. In 95% alcohol (1) — Rinse to remove acid; destain
7. In 95% alcohol (2) — 5 minutes
8. In carbol-xylene — 5 to 10 minutes
9. In xylene — 10 minutes
10. Using Permount or other mounting fluid, mount with coverslip

The alcohol-iodine solution (step 1) is prepared by adding enough iodine crystals to 70% ethyl alcohol to make a port-wine color. The carbol-xylene mixture is made up of one volume of carbolic acid (phenol) with three volumes of xylene. Phenol crystals can be liquefied in a water bath. Trichrome stain has the following composition:

Chromotrope 2R	0.60 gm.
Light green SF	0.15 gm.
Fast green FCF	0.15 gm.
Phosphotungstic acid	0.70 gm.
Acetic acid, glacial	1.00 ml.
Distilled water	100.00 ml.

Put the dry stains into a clean flask, add glacial acetic acid, and shake to mix. After 30 minutes add distilled water and mix thoroughly.

The cytoplasm of trophozoites and cysts appear blue-green, tinged with purple. *E. coli* cysts may occasionally appear more purplish than cysts of other species. The nuclear chromatin, chromatoid bodies, ingested red cells, and bacteria stain red or purplish red. Background material usually stains green.

Chlorazol Black E stain. This stain technique, recently introduced by Kohn,[15] and modified by Gleason and Healy[13] fixes and stains fecal smears and tissue sections in a single solution. No destaining is necessary. The method for fecal smears is as follows:

Two stock solutions, the basic solution and the stock stain solution are needed. The undiluted stock stain solution stains the smear rapidly. By diluting with the basic solution, a longer staining time appropriate for the needs of each laboratory can be made.

The basic solution is made up as follows:

90% ethyl alcohol	170 ml.
Methyl alcohol	160ml.
Glacial acetic acid	20 ml.
Liquid phenol	20 ml.
1% phosphotungstic acid	12 ml.
Distilled water	618 ml.

One gm. of phosphotungstic acid crystals is dissolved in 100 ml. distilled water to make the 1% solution. Add alcohols and acids to distilled water and mix thoroughly.

The stock stain solution is made up as follows:

Chlorazol Black E dye	5 gm.
Basic solution (above)	1,000 ml.

Weigh out dye, put in mortar, and grind for at least 3 minutes. Add a small amount of basic solution and grind until a smooth paste is obtained. Add more solution and grind for 5 minutes. Allow particulate matter to settle a few minutes and decant the liquid into a separate dry, clean container. Add more basic solution and continue the grinding and mixing process until all of the dye appears to be in solution. Add any remaining basic solution, bottle the stain, and put aside for 4 to 6 weeks to ripen. A black sediment settles out within a few days, leaving a black-cherry-colored liquid that is the fixative-stain. Filter the stain through Whatman No. 12 filter paper prior to use. Keep the filtered stain in a stoppered bottle protected from moisture, dust, and dirt.

PROCEDURE FOR STAINING. The fecal smear should be prepared in the usual manner by spreading a thin, even layer over approximately one third of the slide. The smear should be placed immediately in the fixative-stain dilution and should not be allowed to dry until mounting.

The following proportions of stock stain solution and basic solution are recommended for optimum dilution and staining.

STOCK STAIN SOLUTION	BASIC SOLUTION	TIME IN HOURS
Undiluted	—	2 to 3
2	1	2 to 4
1	1	2 to 4 to overnight
1	2	2 to overnight
1	3	4 to overnight

A solution producing good overnight staining does not appear to overstain when left for periods of several days. Up to 20 slides can be stained in a 50-ml. Coplin jar of diluted stain. When slides appear visibly red rather than greenish black, discard solution. Process the smear as follows:

1. Place smear in fixative-stain solution — 2 hours to overnight as necessary
2. In 95% ethyl alcohol — 10 to 15 seconds
3. In carbol-xylol or 100% ethyl alcohol — 5 minutes
4. In xylol — 5 minutes
5. Mount in Permount or other suitable media

The carbol-xylol mixture is made by adding one part of liquid phenol to two parts of xylol. Liquid phenol is prepared by placing phenol crystals in a jar and melting them in a water bath. Do not heat over a direct flame.

Protozoa in fresh fecal specimens stain green to gray-green; in older stools organisms are gray to black. Nuclei, chromatoid bodies, karyosomes, and cell membranes stain dark green to black; ingested red cells vary from pink to black. *E. coli* cysts may stain pink or green, and rarely, *E. histolytica* cysts stain faintly pink.

PRESERVATION OF SPECIMENS

Certain parasites in feces degenerate rapidly or continue to develop into stages other than those found in fresh material. Trophozoites of the protozoa and helminth larvae are prime examples of this. If examination of the specimen cannot be carried out right away, the fecal material should be put into an appropriate preservative to retain the morphology of the parasites in recognizable forms when examined at a later time. In the laboratory situation, where competent personnel are not available for immediate parasitological examination, when specimens arrive too late in the day for processing or are collected over weekends, or for specimens that have to be mailed to a central diagnostic laboratory, **preservation is necessary.** Four methods of preservation will be described. The first one, PVA fixative, acts as a preservative as well as an adhesive. It is primarily used for making permanent stains. The other preservatives are formalin, MIF, and PAF solutions.

PVA fixative. PVA fixative is a mixture of the water soluble polymer and a modified Schaudinn's fixative. It is almost an ideal preservative for fecal specimens containing trophozoites and cysts that are to be processed and examined immediately or months later. It is especially suited for mailing fecal specimens to central diagnostic reference and public health laboratories. This fixative can be prepared in the laboratory or purchased commercially. PVA powder* should be of the correct grade.

PREPARATION OF PVA FIXATIVE

1. Dissolve 4.5 gm. of $HgCl_2$ in 31 ml. of 95% ethyl alcohol in a stoppered

---

*Available as Elvanol from E. I. Dupont de Nemours and Company; or as Gelvatol Resin from Shawingan Resins Corporation, Springfield, Mass. The powder as well as the ready-made fixative solution may be purchased from Delkote, Inc., P. O. Box 1335, Wilmington, Del.

50- or 125-ml. flask by swirling at intervals. Add 5 ml. glacial acetic acid, stopper, swirl to mix, and set aside (modified Schaudinn's fixative) until needed.

2. Weigh 5 gm. of PVA powder into a small beaker, add 1.5 ml. of glycerol to it, and mix thoroughly with a glass rod until all particles appear coated with the glycerol. Scrape the mixture into a 125-ml. flask, add 62.5 ml. of distilled water, stopper, and leave at room temperature for at least 3 hours or overnight. Swirl the mixture occasionally.

3. Heat a water bath or large beaker of water to 70° to 75° C and place the loosely stoppered PVA flask in the bath for about 10 minutes, swirling the contents frequently while maintaining the temperature. When the PVA appears to be dissolved, pour in the modified Schaudinn's fixative, restopper the flask, and swirl to mix. Continue to swirl the flask in the water bath for 2 to 3 minutes to dissolve the remaining PVA; allow the bubbles to escape and to clear the solution. Remove the flask from the bath and let cool. Store in a screwcapped or glass-stoppered bottle. This fixative was introduced by Burrows[8] as a modification of the original method of Brooke and Goldman.[3]

USES OF PVA FIXATIVE. Fresh feces or other material may be mixed with PVA fixative on slides or in vials. For the former, a drop of dysenteric stool or other material such as sigmoidoscopic specimens is placed on a microscope slide and mixed thoroughly with three drops of PVA fixative. With an applicator stick the mixture is spread (not smeared like a blood film) over approximately one third of the glass surface, care being taken to extend the smear to the sides of the slide to reduce later peeling. It is allowed to dry thoroughly, preferably overnight, at 36° C. and stained according to the procedure given under the section on permanent staining. Wet mounts, especially sigmoid-oscopic material, can be fixed by removing the coverslip and adding PVA fixative. Dried films remain satisfactory for staining for many months, but for best results they should be stained within 2 months.

In the other method, one part of fresh specimen is mixed thoroughly with three or more parts of PVA fixative in a vial. Dried films for staining can be made on slides from this vial immediately or months later. This procedure is especially useful when many slides are to be made out of the same specimen. In the so-called "two vial method," two screw-capped vials of 20 to 30 ml. capacity, one containing PVA fixative and the other with 10% formalin, are used for mailing specimens.

Formalin. A 5% formalin solution, made by adding 5 ml. commercial formalin to 95 ml. tap or distilled water, is recommended for preserving protozoan cysts, helminth eggs, and larvae. One part of fecal material is mixed with three or more parts of 5% formalin. A 10% formalin solution is also used, particularly for helminth eggs. Since formalin is approximately 40% formaldehyde gas in water, the above concentrations are actually 2% and 4% formaldehyde, respectively. Formalinized specimens can be examined directly as wet mounts, with or without iodine, or after concentration. They are not suitable for making permanent stains.

MIF solution. The MIF solution of Sapero and Lawless[20] can be used as a preservative as well as a stain for direct wet mounts for fresh specimens. The so-called MIFC method[1] is a sedimentation concentration procedure using MIF-preserved material with ether. It is similar to the formalin-ether technique in that the MIF preservative takes the place of formalin solution.

PAF solution.[7] This solution preserves protozoa, including the trophozoites, as well as helminth eggs and larvae. Materials fixed in PAF can be examined as wet mounts, unstained or stained with thionine or azure A, or they can be examined after concentration.

IDENTIFICATION OF INTESTINAL PARASITES
Diagnosis of intestinal protozoa

The four main classes of protozoa are represented in the intestines of man; the **amebae** (Sarcodina), **flagellates** (Mastigophora), **coccidia** (Sporozoa), and **ciliates** (Ciliophora). Only three intestinal protozoa live in the small intestines—the flagellate *Giardia lamblia* and the two *Isospora* (coccidian) species; the rest live in the colon. The isosporae are the only obligate tissue parasites, and they leave the body as oocysts or sporocysts. All the other protozoa occur in the intestine as trophozoites and cysts except the ameba *Dientamoeba fragilis* and the flagellate *Trichomonas hominis,* which do not encyst. Tables 33-1 to 33-4 give the important characteristics of the intestinal protozoa of man. Fig. 33-1 (*A* through *F*) illustrates the more important species.

Of the 15 intestinal protozoa, the clinically important ones are *Entamoeba histolytica, D. fragilis, G. lamblia, I. hominis, I. belli,* and *Balantidium coli.* Clinically, *E. histolytica* is the most important of all the intestinal protozoa of man. *D. fragilis* is incriminated in some persistant diarrhea and nonspecific abdominal complaints. *G. lamblia* is probably the most common of all the protozoa infecting man in this country. The two sporozoa and the ciliate are rarely seen.

The diagnosis of human intestinal protozoan parasites is much more difficult than the diagnosis of the larger helminth eggs and larvae. A great deal of practice and experience with a fairly large variety of material under the supervision of a competent person is needed. A few publications that may be of great help are those by Spencer and Monroe,[22] Burrows,[6] Craig,[9] and the classic work of Dobell and O'Connor.[10] The United States Department of Health, Education, and Welfare publication, *Amebiasis: Laboratory Diagnosis,* developed by the National Communicable Disease Center,[23] is a valuable self-instructional program for beginners. A new color atlas entitled *The Application of*

Comparative Morphology in the Identification of Intestinal Parasites by J. W. Moose published by Charles C Thomas, Springfield, Ill., is also recommended.

Amebae. When they are examined fresh, the motility of the trophozoites should be noted. *E. histolytica* trophozoites have a characteristic progressive or directional motility. The other species are nonprogressive and somewhat sluggish. The cy-

toplasm of the amebae may appear finely granular, coarse, or vacuolated. Bacteria, yeast, and other material may be present. Red blood cells are occasionally seen in the cytoplasm of *E. histolytica* trophozoites. Not only ingested material but details of other structures present in the amebae are best studied in stained slides. The most important structure for the diagnosis of all the amebae, both trophozoites as well as

Text continued on p. 296.

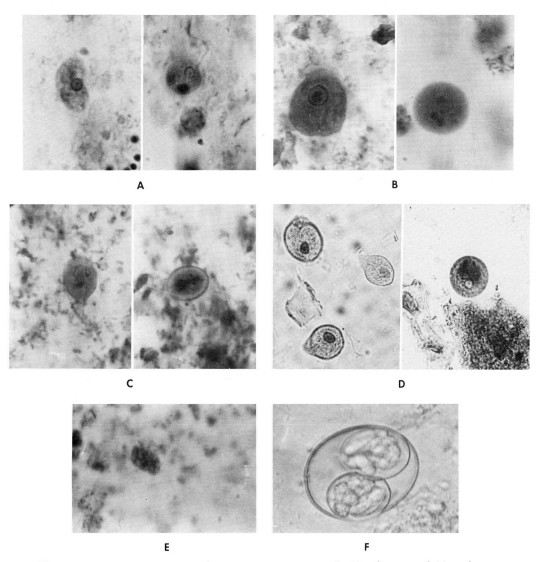

Fig. 33-1. Important intestinal protozoan parasites. **A,** Trophozoite *(left)* and cyst *(right)* of *Entamoeba histolytica.* (900×.) **B,** Trophozoite *(left)* and cyst *(right)* of *Entamoeba coli.* (900×.) **C,** Trophozoite *(left)* and cyst *(right)* of *Giardia lamblia.* (900×.) **D,** Trophozoite *(left)* and cyst *(right)* of *Balantidium coli.* (220×.) **E,** Trophozoite of *Dientamoeba fragilis.* (900×.) **F,** Oocyst of *Isospora* containing two sporocysts. (1200×.)

Table 33-1. Differential morphology of amebae found in stool specimens: trophozoites*

SPECIES	SIZE (DIAMETER OR LENGTH)	MOTILITY	NUMBER
Entamoeba histolytica	10-60 μ Usual range, †15-20 μ — commensal form. ‡Over 20 μ — invasive form.	Progressive with hyaline, finger-like pseudopods.	1 Not visible in unstained preparations.
Entamoeba hartmanni	5-12 μ Usual range, 8-10 μ.	Usually nonprogressive, but may be progressive occasionally.	1 Not visible in unstained preparations.
Entamoeba coli	15-50 μ Usual range, 20-25 μ.	Sluggish, nonprogressive, with blunt pseudopods.	1 Often visible in unstained preparations.
Entamoeba polecki	10-25 μ Usual range, 15-20 μ.	Usually sluggish, similar to E. coli. Occasionally in diarrhetic specimens, motility may be progressive.	1 May be slightly visible in unstained preparations. Occasionally distorted by pressure from vacuoles in cytoplasm.
Endolimax nana	6-12 μ Usual range, 8-10 μ.	Sluggish, usually nonprogressive with blunt pseudopods.	1 Visible occasionally in unstained preparations.
Iodamoeba bütschlii	8-20 μ Usual range, 12-15 μ.	Sluggish, usually nonprogressive.	1 Not usually visible in unstained preparations.
Dientamoeba fragilis	5-15 μ Usual range, 9-12 μ.	Pseudopodia are angular, serrated, or broad lobed and hyaline, almost transparent.	2 (In approximately 20% of organisms only 1 nucleus is present.) Nuclei invisible in unstained preparations.

*From Brooke, M. M., and Melvin, D. M.: Morphology of diagnostic stages of intestinal parasites of man, Atlanta, Ga., 1969.
†Commensal form — usually found in asymptomatic or chronic cases; may contain bacteria.
‡Invasive form — usually found in acute cases; often contain red blood cells.

| NUCLEUS | | CYTOPLASM | |
PERIPHERAL CHROMATIN	KARYOSOMAL CHROMATIN	APPEARANCE	INCLUSIONS
Fine granules. Usually evenly distributed and uniform in size.	Small, discrete. Usually centrally located, but occasionally is eccentric.	Finely granular.	Red blood cells occasionally. Noninvasive organisms may contain bacteria.
Similar to *E. histolytica*.	Small, discrete, often eccentric.	Finely granular.	Bacteria.
Coarse granules, irregular in size and distribution.	Large, discrete, usually eccentric.	Coarse, often vacuolated.	Bacteria, yeasts, other materials.
Usually fine granules evenly distributed. Occasionally, granules may be irregularly arranged. Chromatin sometimes in plaques or crescents.	Small, discrete, eccentric. Occasionally large, diffuse, or irregular.	Coarsely granular, may resemble *E. coli*. Contains numerous vacuoles.	Bacteria, yeasts.
None.	Large, irregularly shaped, blot-like.	Granular, vacuolated.	Bacteria.
None.	Large, usually central. Surrounded by refractile, achromatic granules. These granules are often not distinct even in stained slides.	Coarsely granular, vacuolated.	Bacteria, yeasts, or other material.
None.	Large cluster of 4-8 granules.	Finely granular, vacuolated.	Bacteria.

U. S. Department of Health, Education, and Welfare, Publication No. (HSM) 72-8116.

Table 33-2. Differential morphology of amebae found in stool specimens: cysts*

SPECIES	SIZE	SHAPE	NUMBER
Entamoeba histolytica	10-20 μ Usual range, 12-15 μ.	Usually spherical.	4 in mature cyst. Immature cyst with 1 or 2 occasionally seen.
Entamoeba hartmanni	5-10 μ Usual range, 6-8 μ.	Usually spherical.	4 in mature cyst. Immature cysts with 1 or 2 often seen.
Entamoeba coli	10-35 μ Usual range, 15-25 μ.	Usually spherical. Occasionally oval, triangular or other shapes.	8 in mature cyst. Occasionally supernucleated cysts with 16 or more are seen. Immature cysts with 2 or more occasionally seen.
Entamoeba polecki	9-18 μ Usual range, 11-15 μ.	Spherical or oval.	1 Rarely 2. Occasionally visible in unstained preparations.
Endolimax nana	5-10 μ Usual range, 6-8 μ.	Spherical, ovoidal, or ellipsoidal.	4 in mature cysts. Immature cysts with less than 4 rarely seen.
Iodamoeba bütschlii	5-20 μ Usual range, 10-12 μ.	Ovoidal, ellipsodial, triangular or other shapes.	1 in mature cyst.

*From Brooke, M. M., and Melvin, D. M.: Morphology of diagnostic stages of intestinal parasites of man, Atlanta, Ga., 1969,

| NUCLEUS | | CYTOPLASM | |
PERIPHERAL CHROMATIN	KARYOSOMAL CHROMATIN	CHROMATOID BODIES	GLYCOGEN
Peripheral chromatin present. Fine, uniform granules, evenly distributed.	Small, discrete, usually centrally located.	Present. Elongate bars with bluntly rounded ends.	Usually diffuse. Concentrated mass often present in young cysts. Stains reddish brown with iodine.
Similar to *E. histolytica*.	Similar to *E. histolytica*.	Present. Elongate bars with bluntly rounded ends.	Similar to *E. histolytica*.
Peripheral chromatin present. Coarse granules irregular in size and distribution, but often appear more uniform than in trophozoites.	Large, discrete, usually eccentric but occasionally centrally located.	Present, but less frequently seen than in *E. histolytica*. Usually splinter-like with pointed ends.	Usually diffuse, but occasionally well-defined mass in immature cysts. Stains reddish brown with iodine.
Usually fine granules evenly distributed.	Usually small and eccentric.	Present. Many small bodies with angular or pointed ends, or few large ones. May be oval, rod-like or irregular.	Usually small, diffuse masses. Stain reddish brown with iodine. A dark area called "inclusion mass" (possibly concentrated cytoplasm) is often also present. Mass does not stain with iodine.
None.	Large (blot-like), usually central.	Occasionally granules or small oval masses seen, but bodies as seen in *Entamoeba* spp. are not present.	Usually diffuse. Concentrated mass seen occasionally in young cysts. Stains reddish brown with iodine.
None.	Large, usually eccentric. Refractile, achromatic granules on one side of karyosome. Indistinct in iodine preparations.	Occasionally granules present, but chromatoid bodies as seen in *Entamoeba* spp. are not present.	Compact, well-defined mass. Stains dark brown with iodine.

U. S. Department of Health, Education, and Welfare, Publication No. (HSM) 72-8116.

Table 33-3. Differential morphology of flagellate protozoa found in stool specimens: trophozoites and cysts*

Trophozoites

SPECIES	SIZE (LENGTH)	SHAPE	MOTILITY	NUMBER OF NUCLEI	NUMBER OF FLAGELLA†	OTHER FEATURES
Trichomonas hominis	8-20 μ Usual range, 11-12 μ.	Pear-shaped.	Nervous, jerky.	1 Not visible in unstained mounts.	3-5 anterior. 1 posterior.	Undulating membrane extending length of body.
Chilomastix mesnili	6-24 μ Usual range, 10-15 μ.	Pear-shaped.	Stiff, rotary.	1 Not visible in unstained mounts.	3 anterior. 1 in cytostome.	Prominent cytostome extending $\frac{1}{3}$-$\frac{1}{2}$ length of body. Spiral groove across ventral surface.
Giardia lamblia	10-20 μ Usual range, 12-15 μ.	Pear-shaped.	"Falling leaf."	2 Not visible in unstained mounts.	4 lateral. 2 ventral. 2 caudal.	Sucking disc occupying $\frac{1}{2}$-$\frac{3}{4}$ of ventral surface.
Enteromonas hominis	4-10 μ Usual range, 8-9 μ.	Oval.	Jerky.	1 Not visible in unstained mounts.	3 anterior. 1 posterior.	One side of body flattened. Posterior flagellum extends free posteriorly or laterally.
Retortamonas intestinalis	4-9 μ Usual range, 6-7 μ.	Pear-shaped or oval.	Jerky.	1 Not visible in unstained mounts.	1 anterior. 1 posterior.	Prominent cytostome extending approximately $\frac{1}{2}$ length of body.

Cysts

SPECIES	SIZE	SHAPE	NUMBER OF NUCLEI	OTHER FEATURES
Chilomastix mesnili	6-10 μ Usual range, 8-9 μ.	Lemon shape with anterior hyaline knob or "nipple."	1 Not visible in unstained preparations.	Cytostome with supporting fibrils. Usually visible in stained preparation.
Giardia lamblia	8-9 μ Usual range, 11-12 μ.	Oval or ellipsoidal.	Usually 4. Not distinct in unstained preparations. Usually located at one end.	Fibrils or flagella longitudinally in cyst. Occasionally may be slightly visible in unstained cysts. Deep staining fibers or fibrils may be seen lying laterally or obliquely across fibrils in lower part of cyst. Cytoplasm often retracts from a portion of cell wall.
Enteromonas hominis	4-10 μ Usual range, 6-8 μ.	Elongate or oval.	1-4, usually 2 lying at opposite ends of cyst. Not visible in unstained mounts.	Resembles E. nana cyst. Fibrils or flagella are usually not seen.
Retortamonas intestinalis	4-9 μ Usual range, 4-7 μ.	Pear-shaped or slightly lemon-shaped.	1 Not visible in unstained mounts.	Resembles Chilomastix cyst. Shadow outline of cytostome with supporting fibrils extends above nucleus.

*From Brooke, M. M., and Melvin, D. M.: Morphology of diagnostic stages of intestinal parasites of man, Atlanta, Ga., 1969, U. S. Department of Health, Education, and Welfare, Publication No. (HSM) 72-8116.
†Not a practical feature for identification of species in routine fecal examinations.

Table 33-4. Differential morphology of ciliate protozoa and coccidia found in stool specimens*

SPECIES	SIZE (LENGTH)	SHAPE	MOTILITY	NUMBER OF NUCLEI	OTHER FEATURES
Balantidium coli					
Trophozoite	50-70 μ or more. Usual range, 40-50 μ.	Ovoid with tapering anterior end.	Rotary, boring.	1 large, kidney shaped macronucleus. 1 small subspherical micronucleus immediately adjacent to macronucleus. Macronucleus occasionally visible in unstained preparation as hyaline mass.	Body surface covered by spiral, longitudinal rows of cilia. Contractile vacuoles are present.
Cyst	45-65 μ Usual range, 50-55 μ.	Spherical or oval.		1 large macronucleus visible in unstained preparations as hyaline mass.	Macronucleus and contractile vacuole are visible in young cysts. In older cysts, internal structure appears granular.
Isospora species (*I. belli* and *I. hominis*)	Oocyst. 25-30 μ Usual range, 28-30 μ. Immature oocyst not usually seen in I. hominis.	Ellipsoidal.	Nonmotile.		Mature oocyst contains 2 sporocysts with 4 sporozoites each. *I. belli:* usual diagnostic stage is immature oocyst with single granular mass (zygote) within. *I. hominis:* mature sporocysts, singly or in pairs, are usually passed in feces. Oocyst wall not apparent.
	Sporocyst: *I. belli*-12-14 μ *I. hominis*- 14-16 μ.	Round or oval.			

*From Brooke, M. M., and Melvin, D. M.: Morphology of diagnostic stages of intestinal parasites of man, Atlanta, Ga., 1969, U. S. Department of Health, Education, and Welfare, Publication No. (HSM) 72-8116.

cysts, is the **nucleus.** Both the wet mount stains as well as permanent stains bring out the nuclear details. The nuclei of the genus *Entamoeba* possess a relatively small karyosome and prominent peripheral chromatin granules along the nuclear membranes. The other amebae, *Iodamoeba, Endolimax,* and *Dientamoeba,* have large karyosomes and no chromatin granules along the nuclear membranes. Up to 80% of the trophozoites of *D. fragilis* may have two nuclei. The karyosome of this ameba (which looks large) is actually made up of a cluster of 4 to 8 granules. Sizes of the amebae are useful to some extent. *E. histolytica* and *E. hartmanni* are differentiated primarily by their size.

The nuclei of the cysts of the amebae are identical to those of the trophozoites, but the number of nuclei may vary from one to eight in the mature stages. Whereas the trophozoites have variable shapes, the cysts have rigid shapes, usually spherical, less frequently with other configurations. Cytoplasm of cysts may contain chromatoid bodies and glycogen material. Chromatoid bodies of the cysts of *Entamoeba* species are rod-like or irregular. Glycogen material is usually diffuse in the cytoplasm, but in *I. bütschlii* it is a compact, well-defined mass.

Flagellates. The living trophozoites of the flagellates exhibit characteristic motility. Most of them are pear shaped and possess from two to eight flagella, distributed anteriorly, posteriorly, and laterally, depending upon the species. They all have one nucleus brought out by appropriate stains, except in the case of *G. lamblia,* which has two. The undulating membrane of *T. hominis,* the spiral groove of *Chilomastix mesnili,* and the sucking disc of *G. lamblia* are all distinctive features of these protozoa. Cysts of the flagellates have distinctive shapes and sizes. Nuclear details are not required for the identification of the flagellates.

Ciliates. Only one ciliate, *B. coli,* infects man. It is the largest protozoa that is found in man. Live trophozoites have a rotatory,

boring motion. The body is covered with cilia and there is a large kidney-shaped macronucleus and a small micronucleus adjacent to it in both trophozoites and cysts; these become very prominent in stained preparations. The cyst wall is rather thick. Inside, the cilia may persist for some time in young cysts. Diagnosis of this parasite presents no problem.

Sporozoa. The two coccidian parasites of the intestinal tract are identified by finding the transparent oocysts and sporocysts. *Isopora belli* and *I. hominis* oocysts contain two sporocysts, each sporocyst containing four sporozoites in the mature stage. *I. belli* is usually identified in the immature oocyst stage, but *I. hominis* is found as mature sporocysts, singly or in pairs, while the oocyst wall is not apparent. *Eimeria* species, which are coccidian parasites of lower animals, may sometimes appear in human stools. Mature oocysts of *Eimeria* have four sporocysts, each sporocyst with two sporozoites. Immature oocysts of coccidia can be matured by mixing the fecal specimen with a 2% potassium dichromate solution and allowing it to stand at room temperature for 48 hours. In this way the genus of the coccidia can be determined.

DIAGNOSIS OF INTESTINAL HELMINTHS

The intestinal helminths belong to two phyla: Phylum Nematode, and Phylum Platyhelminthes. The former are better known as **roundworms** because they possess cylindrical bodies, and the latter are known as **flatworms.** The platyhelminthes are subdivided into Class Trematoda (flukes) and Class Cestoidea (tapeworms). Members of a third phylum, the Acanthocephala, also known as the spiny-headed worms, infect many wild and domesticated animals but have very rarely been reported from the intestines of man.

Detection of the presence of most of the intestinal worms is based on finding their eggs in the stool. In a few cases larvae and segments of tapeworms are found. Other exceptions will be indicated in this chapter.

As noted earlier, the identification of

worms is much easier than that of the protozoa, because the eggs and larvae of the helminths are much larger and the shape, size, and color of each is rather unique. Other useful distinctive features are operculum, plugs, spines, mammillated coat, thickness of the shell, and the stage of development of the eggs. No special stains are necessary. Direct wet mounts of fresh or preserved, unconcentrated or con- centrated material will reveal almost all of the diagnostic features. Most textbooks on medical parasitology contain adequate il- lustrations and descriptions for the diag- nosis of intestinal helminths. Of the refer- ences cited in the section of protozoa, those by Spencer and Monroe[22] and Bur- rows[6] can be used for the diagnosis of helminths. See Tables 33-5 to 33-7 and Figs. 33-2 (*A* through *H*), 33-3 (*A* through

Text continued on p. 304.

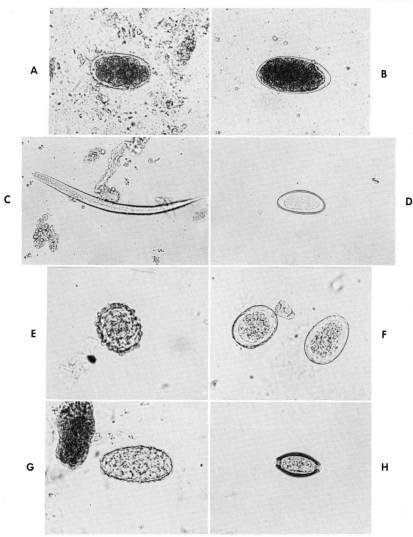

Fig. 33-2.   Important intestinal nematodes. **A,** Egg of hookworm species. (220×.) **B,** Egg of *Trichostrongylus* species. (220×.) **C,** Rhabditiform larva of *Strongyloides stercoralis.* (200×.) **D,** Egg of *Enterobius vermicularis.* (220×.) **E,** Egg of *Ascaris lumbricoides.* (220×.) **F,** Decorticated eggs of *A. lumbricoides.* (220×.) **G,** Infertile egg of *A. lumbricoides.* (220×.) **H,** Egg of *Tricharis trichiura.* (220×.)

Table 33-5. Differential morphology of nematodes: eggs*

SPECIES	SIZE	SHAPE	COLOR
Enterobius vermicularis	55 μ × 26 μ Usual range, 50-60 μ × 20-32 μ.	Elongate, asymmetrical with one side flattened, other side convex.	Colorless
Ascaris lumbricoides Fertile egg	60 μ × 45 μ Usual range, 45-70 μ × 35-50 μ.	Round, or ovoidal with thick shell.	Brown or yellow brown.
Infertile egg	90 μ × 40 μ Usual range, 85-95 μ × 35-45 μ.	Elongate, occasionally triangular, kidney-shaped or other bizarre forms. Shell often very thin.	Brown.
Trichuris trichiura	54 μ × 22 μ Usual range, 49-65 μ × 20-29 μ.	Elongate, barrel-shaped with a polar "plug" at each end.	Yellow to brown. "Plugs" are colorless.
Hookworm *Ancylostoma duodenale*	60 μ × 40 μ Usual range, 57-76 μ × 35-47 μ.	Oval or ellipsoidal with a thin shell.	Colorless with grayish cells.
Nector americanus	65 μ × 40 μ Usual range, 55-79 μ × 35-45 μ.		
Trichostrongylus species	90 μ × 40 μ Usual range, 75-95 μ × 40-50 μ.	Elongate with one (or both) end more pointed than hookworm.	Colorless with grayish cells.

*From Brooke, M. M., and Melvin, D. M.: Morphology of diagnostic stages of intestinal parasites of man, Atlanta, Ga., 1969,

STAGE OF DEVELOPMENT WHEN PASSED	SPECIFIC FEATURES AND VARIATIONS
Embryonated. Contains C shaped or tadpole-like embryo.	Smooth, thin egg shell with one flattened side. Occasionally may contain fully developed larva. (More readily found on anal swabs than in feces.)
1 cell, separated from the shell at both ends.	Mammillated albuminous coat or covering on outer shell. Covering is occasionally lost and the decorticated eggs have a colorless shell with gray or black internal material. Occasionally, the inner cell may be in various stages of cleavage: 2 cells, 4 cells, etc., or a fully developed larva.
Internal material is a mass of irregular globules and granules that fills shell.	Mammillated covering attenuated or missing in many cases.
1 cell or unsegmented.	Polar plugs are distinctive. Occasionally are oriented in a vertical or slanted position and may not be readily recognized. A gentle tap on the coverslip will usually reorient the egg. On rare occassions, atypical eggs lacking polar plugs may be seen.
4- to 8-cell stage.	Occasionally, eggs in advanced cleavage (16 or more cells) or even embryonated may be seen. Rhabditiform larvae may be present if the specimens are old. Species identification cannot be made on eggs alone; therefore, eggs should be reported simply as hookworm.
May be in advanced cleavage or morula stage.	Egg resembles hookworm egg but is larger and more pointed at the ends.

U. S. Department of Health, Education, and Welfare, Publication No. (HSM) 72-8116.

Table 33-6. Differential morphology of cestodes: eggs*

SPECIES	SIZE	SHAPE	COLOR
Taenia saginata *Taenia solium*	35 μ Usual range, 31-43 μ.	Spherical or subspherical with thick striated shell.	Walnut brown.
Hymenolepis nana	47 μ × 37 μ Usual range, 40-60 μ × 30-50 μ.	Oval or subspherical. Shell consists of 2 distinct membranes. On inner membrane are two small "knobs" or poles from which 4 to 8 filaments arise and spread out between the two membranes.	Colorless, almost transparent.
Hymenolepis diminuta†	72 μ Usual range, 70-86 μ × 60-80 μ.	Round or slightly oval. Striated outer membrane and thin inner membrane with slight pole. Space between membranes may appear smooth or faintly granular.	Yellow.
Dipylidium caninum†	35-40 μ Usual range, 31-50 μ × 27-48 μ	Spherical, subspherical or oval. 5-15 eggs (or more) are enclosed in a sac or capsule.	Colorless.
Diphyllobothrium latum	66 μ × 44 μ Usual range, 58-76 μ × 40-51 μ.	Oval or ellipsoidal with an inconspicuous operculum at one end and a small "knob" at the other end.	Yellow to brown.

*From Brooke, M. M., and Melvin, D. M.: Morphology of diagnostic stages of intestinal parasites of man, Atlanta, Ga., 1969.
†Usually found in lower animals, only occasionally found in man.

STAGE OF DEVELOPMENT WHEN PASSED	SPECIFIC FEATURES AND VARIATIONS
Embryonated. 6-hooked oncosphere is present inside a thick shell.	Thick, striated shell. Eggs of *T. solium* and *T. saginata* are indistinguishable and species identification should be made from proglottids or scoleces. *Taenia* spp. should be reported if only eggs are found.
Embryonated. 6-hooked embryo contained in eggs.	Polar filaments.
Embryonated. 6-hooked oncosphere inside shell.	Resembles *H. nana* but lacks polar filaments. Poles are rudimentary and often hard to see.
Embryonated.	Eggs are contained in a sac or capsule which ranges in size from 58 $\mu \times$ 45 μ to 60 $\mu \times$ 170 μ. Occasionally, capsules are ruptured and eggs are free.
Unembryonated. Germinal cell is surrounded by a mass of yolk cells which completely fills inner area of shell.	Egg resembles hookworm egg but has a thicker shell and an operculum.
Germinal cell is usually not visible.	

U. S. Department of Health, Education, and Welfare, Publication No. (HSM) 72-8116.

Table 33-7. Differential morphology of trematodes: eggs*

SPECIES	SIZE	SHAPE	COLOR	STAGE OF DEVELOPMENT WHEN PASSED	SPECIFIC FEATURES AND VARIATIONS
Schistosoma mansoni	155 μ × 66 μ Usual range, 114-118 μ × 45-73 μ.	Elongate with prominent lateral spine near posterior end. Anterior end tapered and slightly curved.	Yellow or yellow brown.	Embryonated. Contains mature miracidium.	Lateral spine. Found in feces; in rare cases, in urine also. Eggs discharged at irregular intervals and may not be found in every stool specimen. Are rare in chronic stages of the infection.
Schistosoma japonicum	90 μ × 70 μ Usual range, 68-106 μ × 45-80 μ.	Oval or subspherical. Small lateral spine is often not seen or may appear as a small hook or "knob" located in a depression in the shell.	Yellow or yellow brown.	Embryonated. Contains mature miracidium.	Found in feces. Often coated with debris and may be overlooked.
Schistosoma haematobium	143 μ × 60 μ Usual range, 112-170 μ × 40-73 μ.	Elongate with rounded anterior end and terminal spine at posterior end.	Yellow brown.	Embryonated. Contains mature miracidium.	Terminal spine. Found in urine, occasionally in feces. Egg often covered with debris.
Clonorchis sinensis	30 μ × 16 μ Usual range, 27-35 μ × 11-20 μ.	Small, ovoidal or elongate with broad, rounded posterior end and a convex operculum resting on "shoulders." A small "knob" may be seen on the posterior end.	Yellow brown.	Embryonated. Contains mature miracidium.	Small size, operculum and "knob" on posterior end. Shell often is covered by adhering debris.
Opisthorchis felineus	30 μ × 12 μ Usual range, 26-30 μ × 11-15 μ.	Elongate, with operculum on anterior end and pointed terminal "knob" on posterior end.	Yellow brown.	Embryonated. Contains mature miracidium.	Lacks prominent shoulders characteristic of *Clonorchis* and has more tapered end.

Heterophyes heterophyes	28 μ × 15 μ Usual range, 26-30 μ × 14-17 μ.	Small, elongate or slightly ovoidal. Operculum. Slight "knob" at posterior end.	Yellow brown.	Embryonated. Contains mature miracidium.	Resembles *Clonorchis* egg but with less distinct shoulders. Operculum is broader than in *Clonorchis*.
Metagonimus yokogawai	28 μ × 17 μ Usual range, 26-35 μ × 15-20 μ.	Small, elongate or ovoidal. Operculum. No "shoulders" at anterior end. Small "knob" often seen on posterior end.	Yellow or yellow brown.	Embryonated. Contains mature miracidium.	Resembles *Clonorchis* and *Heterophyes* eggs. Shell is slightly thinner than *Heterophyes*. Operculum is broader than *Clonorchis*.
Paragonimus westermani	85 μ × 53 μ Usual range, 68-118 μ × 39-67 μ.	Ovoidal or elongate with thick shell. Operculum is slightly flattened and fits into shoulder area of shell. Posterior end is thickened. Egg often asymmetrical with one side slightly flattened.	Yellow brown to dark brown.	Unembryonated. Filled with yolk material in which a germinal cell is imbedded. Cells are irregular in size.	Found in sputum, occasionally in feces. Resembles egg of *D. latum* but is larger, slightly asymmetrical and the operculum is smaller and flatter. The widest part of the *Paragonimus* egg is usually anterior to the center while in *D. latum*, the widest area is about the center.
Fasciola hepatica	150 μ × 80 μ Usual range, 120-182 μ × 63-102 μ.	Ellipsoidal, thin shell. Small, indistinct operculum.	Yellow to light brown.	Unembryonated. Filled with yolk cells in which an indistinct germinal cell is imbedded.	Large size. Broadly oval eggs.
Fasciolopsis buski	140 μ × 80 μ Usual range, 130-159 μ × 78-98 μ.	Ellipsoidal, thin shell. Small, indistinct operculum.	Yellowish brown.	Unembryonated. Filled with yolk cells in which an indistinct germinal cell is imbedded.	Large size. Resembles *F. hepatica* egg and cannot be easily distinguished from *Fasciola*.

*From Brooke, M. M. and Melvin, D. M.: Morphology of diagnostic stages of intestinal parasites of man, Atlanta, Ga., 1969, U. S. Department of Health, Education, and Welfare, Publication No. (HSM) 72-8116.

E), and 33-4 (*A* through *G*) for diagnostic and microscopic features of the helminths eggs.

Nematodes. All six roundworms, listed in Table 33-5, are diagnosed by finding their eggs in the feces. *Strongyloides stercoralis,* also known as threadworm, leaves the human body in the rhabditiform larva stage. As noted at the beginning, this larval stage is very motile. Hookworm eggs in fecal specimens left at room temperature for several days may hatch out, and larvae can be mistaken for *Strongyloides* larvae. Occasionally, filariform stages of *Strongyloides* are also found. The rhabditiform stage of larva of *Strongyloides* has a shorter buccal capsule or cavity as compared with that of the hookworm species. The genital primordium of *Strongyloides* in this stage

is much larger than that of the hookworm. In the filariform or third stage larva, the esophagus of *Strongyloides* is very long and the tip of the tail is notched, whereas hookworm larvae of this stage have a short esophagus and pointed tail. The rhabditiform larva of *Trichostrongylus* has a long buccal cavity, small genital primordium, and a bead-like swelling at the tip of the tail.

The eggs of *Ancylostoma duodenale* and *Necator americanus* cannot be distinguished from each other; hence it is sufficient to report that hookworm eggs were present. They can, of course, be easily differentiated if the adult specimen can be obtained. *Trichostrongylus* eggs are easily mistaken for those of hookworm eggs. The egg of this worm is somewhat larger and

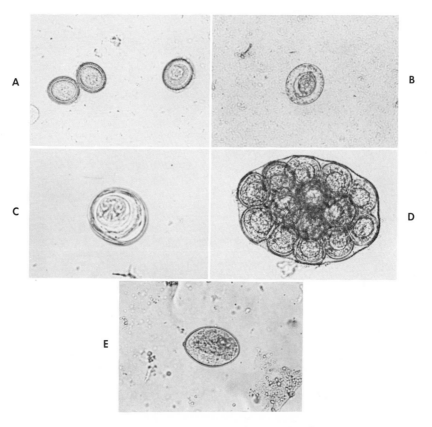

Fig. 33-3. Important intestinal cestodes. **A,** Eggs of *Taenia* species. (220×.) **B,** Egg of *Hymenolepis nana.* (220×.) **C,** Egg of *H. diminuta.* (220×.) **D,** Eggs of *Dipilidium caninum* enclosed in a capsule. (220×.) **E,** Egg of *Diphyllobothrium latum.* (220×.)

one end of the egg is more pointed, compared with that of the hookworm. Also, the number of blastomeres present in a *Trichostrongylus* egg is much larger. Infection with this worm is very rare in this country.

Eggs of the other nematodes are unique and can easily be told apart from one another after some experience with them. The infertile eggs of *Ascaris* should not be missed; although they usually retain the coarse mammillated coat, the shape is usually more elongated than in the fertile forms. The coat can come off, or the egg may break under pressure from the coverslip. Internally, the structures are made up of disorganized irregular granules. Decorticated *Ascaris* eggs present a rather smooth shell. Care should be exercised not to mistake pollens, spores, vegetable cell fragments, and other pseudoparasites for nematode eggs. Vegetable root hairs and

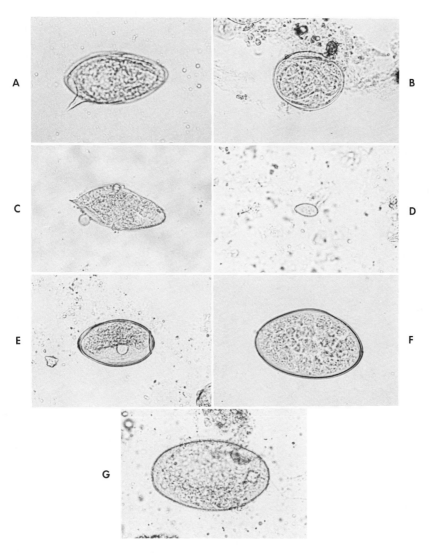

Fig. 33-4. Important intestinal trematodes. **A,** Egg of *Schistosoma mansoni.* (220×.) **B,** Egg of *S. japonicum.* (220×.) **C,** Egg of *S. hematobium.* (220×.) **D,** Egg of *Clonorchis sinensis.* (220×.) **E,** Egg of *Parasonimus westermani.* (220×.) **F,** Egg of *Fasciola hepatica.* (220×.) **G,** Egg of *Fasciolopsis buski.* (220×.)

similar structures can look like dead nematode larvae.

Although the eggs of *Enterobius vermicularis* are sometimes detected in feces, the usual method of diagnosis for pinworm infection is by examination of the so-called **cellophane-tape** preparation (p.307). The adult female pinworm may also be present in this preparation. Although the larvae of *Strongyloides stercoralis* are usually detected in fecal examinations, they can also be found in duodenal washings, sometimes still enclosed inside the egg shells.

Cestodes. Eggs of the tapeworms, with the exception of *Diphyllobothrium latum,* are embryonated when they leave the worm body. The embryo, called an oncosphere or hexacanth, possesses six hooklets.

The embryo of *Taenia saginata* and *T. solium* are enclosed in a thick, radially striated egg shell. Eggs of the two species cannot be told apart. To identify the species, gravid proglottides or the scolices should be examined. *T. solium* proglottides have 7 to 12 main lateral uterine branches, and the scolices possess hooks, whereas *T. saginata* proglottides have 15 to 30 lateral uterine branches, and the scolices are hookless.

Eggs of *Dipilidium caninum,* occasionally found in man, are usually enclosed in packets or capsules of 5 to 15 or more, or they may occur free. Proglottides of this worm are frequently seen; there are two genital pores on each segment. Eggs of *Hymenolepis nana* and *Hymenolepis diminuta* resemble each other, but the latter does not possess polar filaments.

Eggs of the fish tapeworm, *D. latum,* are not embryonated. The undeveloped egg resembles hookworm eggs to some extent, but it has a thicker shell with an operculum, like the eggs of trematodes.

Trematodes. The three schistosomes of man have uniquely shaped eggs: the prominent lateral spine of *Schistosoma mansoni,* the terminal spine of *S. haematobium,* and the small lateral spine in *S. japonicum.* All three eggs are nonoperculated, whereas all the other trematode eggs are operculated. In living eggs, "flame cells" may be seen under subdued light. These eggs in feces hatch out when diluted with fresh water. *S. hematobium* eggs are normally seen in urine but are occasionally found in feces.

The eggs of *Clonorchis, Opisthorchis, Heterophyes,* and *Metagonimus* are quite small and resemble each other closely, making it hard to differentiate the species. *Paragonimus westermani* eggs, found both in sputum or feces, resemble the eggs of *D. latum.* Eggs of *Fasciola hepatica* and *Fasciolopsis buski* are very large and hard to tell apart. It has been noted earlier that trematode eggs are best recovered in sedimentation concentration techniques rather than by zinc sulfate flotation.

Pseudoparasites and other objects. There are certain objects in microscopic fields that superficially resemble the descriptions and illustrations of various parasites. To the beginner, these so-called pseudoparasites pose the problem of differentiation from the parasites, which may result in an incorrect diagnosis and needless therapy. The list of those objects that may be encountered in stool specimens is lengthy. A few of them are seen quite often, and photographs of these are available in some texts.[5,22] Some plant hairs resemble nematode larvae. Oil droplets and yeasts may be mistaken for protozoan cysts. One of the yeast-like organisms, *Blastocystis hominis,* is present in the intestines of many people; it should not be mistaken for *Entamoeba histolytica* cyst. Some pollens and spores resemble helminth eggs or protozoan cysts. Vegetable cells and fibers abound in some fecal specimens.

The presence of certain unmodified or partially modified food particles should be noted and may be reported because these might reflect the overall physiology of the gastrointestinal tract. Among these are undigested muscle fibers, fat globules, and gross vegetable tissues. Abnormal amounts of yeast cells may indicate effects of certain antibiotic therapy. The presence of Charcot-Leyden crystals in a stool specimen is said to be indicative of an allergic response from the host and are sometimes present in *E. histolytica* infections. The presence of erythrocytes and leukocytes in excessive numbers should be noted. These cells, as well as certain mucosal cells, are sometimes mistaken for the trophozoites of the amebae.

ADDITIONAL SPECIMENS AND TESTS

Certain intestinal parasites are found in extraintestinal sites. *E. histolytica,* for example, may be located in the liver, lungs, and other tissues. Some adult parasites, not located in the intestinal tract, lay eggs that find their way into the feces. Some parasites inhabit portions of the upper region of the intestinal tract. Others are intimately associated with the deeper mucosal layers of the intestines. For these reasons, specimens other than fecal material should be examined to diagnose parasites ordinarily associated with the intestinal tract. Cultural methods may also reveal the presence of some parasites, and like other classes of microorganisms, the presence of some of these intestinal parasites are detected by serological procedures.

Sputum. Eggs of the lung fluke, *Paragonimus westermani,* are found in both sputum and feces. Eggs may be scanty and may not be shed regularly; therefore sputum collected over 24 to 48 hours may be preferred. Their presence may be suspected in brown or rust-colored flecks of sputum. Direct wet mounts in saline or after concentration and digestion in 5% NaOH may reveal the large eggs. In pulmonary amebiasis, trophozoites of *E. histolytica* should be differentiated from the mouth commensal, *E. gingivalis,* with one of the wet or permanent stains. Another commensal of the buccal cavity, *Trichomonas tenax,* can be ignored.

Duodenal or biliary drainage. Infections caused by *Strongyloides stercoralis* and *Giardia lamblia* are sometimes detected by duodenal washings when fecal examinations are not helpful. If examined immediately, eggs of *S. stercoralis* may be seen along with the rhabditiform larvae. Trophozoite forms of *G. lamblia* may be detected by this method. Eggs of *Clonorchis sinensis* and *Fasciola hepatica* may be present in biliary drainage specimens. Microscopic study of wet mounts are appropriate.

Sigmoidoscopic material. Aspirated material or mucosal scrapings obtained during sigmoidoscopic examinations of the colon are valuable for the diagnosis of amebiasis and schistosomiasis. Sigmoidoscopy should be done after a recent bowel movement or several hours after purgation. Cotton swabs are not suitable for parasitological studies. A 1-ml. serological pipet fitted with a rubber bulb can be used to aspirate material from suspected lesions and from the mucosa. A curette or Volkmann's spoon can be employed to scrape these areas. These should be examined immediately in wet mounts, or permanent stains should be made and studied. Biopsied tissues taken at sigmoidoscopy, processed with usual histological procedures, can reveal amebae and schistosome eggs in the tissue. The later can be examined directly under the microscope in tissue compressed between two slides.

Aspirated material. Aspirated material from an amebic liver abscess and other lesions submitted to the laboratory should be examined immediately. In the case of liver amebiasis, trophozoites are mostly found at the periphery of the abscess. Wet mount and permanent stain examinations can be done as with fecal specimens.

Perianal swab. Perianal swabs are commonly made with clear cellulose tape for detecting *Enterobius vermicularis* infections.[14] The petrolatum-paraffin swab technique is an alternate method. The perianal area should be swabbed between 10:00 P.M. to midnight, or early in the morning before defecation or bathing. Since the female worm may not migrate every day, a single examination may reveal only 55% of infections. Three consecutive daily examinations will reveal up to 90%

of the total number of cases.[19] Only 5% to 10% of the total cases are revealed by fecal examinations. In addition to eggs of pinworms, eggs of *Taenia,* especially *T. saginata,* can be detected on anal swabs.

Urine, vaginal, and urethral exudates. Although eggs of *Schistosoma hematobium* may occasionally be found in feces, this parasite is best detected by examination of sedimented urine. Urine obtained at noon or shortly after the noon hour gives the best results. Specimens from consecutive days should be made. After centrifugation, wet mounts are examined in the usual manner. *Trichomonas vaginalis* can be found in freshly voided urine, preferably the first portion. Vaginal swabs or scrapings and urethral exudates are examined for *T. vaginalis* with wet mount preparations or permanent stains.

CULTIVATION OF PARASITES

All intestinal protozoa, with the exception of *G. lamblia, B. coli,* and the two *Isospora* can be cultured in vitro; however, *E. histolytica* is the only pathogenic ameba worth the effort. Since cultures are made in diphasic or liquid media, with fecal material inoculated directly into them, each culture is unique, with a mixed flora, and results are therefore not as predictable as with bacteriological cultures. However, if cultures are carried out in addition to the routine methods of examination, a consistently higher number of positive cases are detected. Cultures are not suitable for public health laboratories or where several days will elapse before they can be made. Of the many media and procedures described, three may be mentioned: (1) modified Boeck and Drbohlav's medium, (2) Balamuth's egg yolk infusion medium, and (3) liver infusion agar medium of Cleveland and Collier. The procedures are found in many textbooks of parasitology, including the descriptions in the publication by Melvin and Brooke.[16] A dehydrated base medium with directions is sold by Difco Laboratories as Difco Entamoeba Medium. One of these media can be adopted.

It is sometimes necessary to differentiate the larvae of *Strongyloides* from those of hookworm. In older cultures, eggs of hookworm hatch out into the rhabditiform stage. Cultures can be made to develop the filariform stages of these two worms, using charcoal or sand cultures and test tube cultures with filter paper.

Serodiagnosis. Serological procedures are valuable for several of the parasites closely associated with intestinal infections. This is especially true of those that become intimately involved in the tissue of the host. Three such infections are amebiasis, schistosomiasis, and trichinosis. Details for performing these tests are found in the manual by Blair, Lennette, and Truant.[2]

In amebiasis of the liver and other extraintestinal sites, diagnosis presents a real problem to the physician. *E. histolytica* is now cultured axenically, and the serological tests are highly reliable. Hemagglutination (HA), gel-diffusion, and fluorescent antibody (FA) methods have largely replaced the complement fixation (CF) test of earlier years. For extraintestinal amebiasis the HA test is reported to give 96% to 100% positive reactions. With acute amebic dysentery, figures of 85% to 98% have been reported. Serological tests to diagnose asymptomatic carriers are of less value. Gel-diffusion test results correlate well with those of the HA test. The FA test also compares favorably with these two tests, but the specificity appears to vary.

In addition to fecal examinations, sigmoidoscopy, and biopsy for the diagnosis of schistosomiasis, it is often desirable to employ serological tests. The Center for Disease Control recommends using the CF test along with a bentonite flocculation (BF) test or a cholesterol-lecithin cercarial slide flocculation (CF) test.

Serological tests, in addition to biopsy and skin tests, are available for trichinosis. The BF test is sufficiently sensitive and specific for detecting antibodies after the third week of infection. It can show a rise in titer and may return to normal after 2 to 3 years. The CF test can detect infection a little before the BF test can. The FA test is said to be as sensitive as the BF test.

REFERENCES

1. Blagg, W., Schloegel, E. L., Mansour, N. S., and Khalaf, G. I.: A new concentration technic for the demonstration of protozoa and helminth eggs in feces, Amer. J. Trop. Med. Hyg. **4:**23-28, 1955.
2. Blair, J. E., Lennette, E. H., and Truant, J. P.: Manual of clinical microbiology, Bethesda, Md., 1970, American Society for Microbiology.
3. Brooke, M. M., and Goldman, M.: Polyvinyl alcohol–fixative as a preservative and adhesive for protozoa in dysenteric stools and other liquid materials, J. Lab. Clin. Med. **34**(11):1554-1560, 1949.
4. Brooke, M. M., and Melvin, D. M.: Morphology of diagnostic stages of intestinal parasites of man, Atlanta, Ga., 1969, U. S. Department of Health, Education, and Welfare, Publication No. (HSM) 72-8116.
5. Brown, H. W.: Basic clinical parasitology, ed. 3, New York, 1969, Appleton-Century-Crofts.
6. Burrows, R. B.: Microscopic diagnosis of the parasites of man, New Haven, 1965, Yale University Press.
7. Burrows, R. B.: A new fixative and technics for the diagnosis of intestinal parasites, Amer. J. Clin. Path. **48**(3):342-346, 1967.
8. Burrows, R. B.: Improved preparation of polyvinyl alcohol HgCl₂ fixative used for fecal smears, Stain Tech. **42**(2):93-95, 1967.
9. Craig, F. C.: Laboratory diagnosis of protozoan diseases, ed. 2, Philadelphia, 1948, Lea & Febiger.
10. Dobell, C., and O'Connor, F. W.: The intestinal protozoa of man, London, 1921, John Bale, Sons, and Danielsson, Ltd.
11. Faust, E. C., D'Antoni, J. S., Odom, V., Miller, M. J., Peres, C., Sawitz, W. G., Thomen, L. F., Tobie, J. E., and Walker, J. H.: A critical study of clinical laboratory technics for the diagnosis of protozoan cysts and helminth eggs in feces, Amer. J. Trop. Med., **18:**169-183, 1938.
12. Faust, E. C., and Russell, P. F.: Craig and Faust's clinical parasitology, ed. 7, Philadelphia, 1964, Lea & Febiger.
13. Gleason, N. N., and Healy, G. R.: Modification and evaluation of Kohn's one-step staining technic for intestinal protozoa in feces or tissue, Amer. J. Clin. Path. **43**(5):495-496, 1965.
14. Graham, C. F.: A device for the diagnosis of *Enterobius vermicularis,* Amer. J. Trop. Med. **21:**159-161, 1941.
15. Kohn, J.: A one stage permanent staining method for fecal protozoa, Dapim. Refuiim Med. Quart. Israel **19**(2/3):160-161, 1960.
16. Melvin, D. M., and Brooke, M. M.: Laboratory procedures for the diagnosis of intestinal parasites, Public Health Service Publication No. 1969, Washington, 1969, U. S. Government Printing Office.
17. Nair, C. P.: Rapid staining of intestinal amoebae on wet mounts, Nature **172:**1051, 1953.
18. Ritchie, L. S.: An ether sedimentation technique for routine stool examinations, Bull. U. S. Army Med. Dept. **8:**326, 1948.
19. Sadun, E. H., and Melvin, D. M.: The probability of detecting infections with *Enterobius vermicularis* by successive examination, J. Pediatrics **48:**438-441, 1956.
20. Sapero, J. J., and Lawless, D. K.: The "MIF" stain-preservation technique for the identification of intestinal protozoa, Amer. J. Trop. Med. Hyg. **2:**613-619, 1953.
21. Sawitz, W. G., and Faust, E. C.: The probability of detecting intestinal protozoa by successive stool examinations, Amer. J. Trop. Med. **22:**131-136, 1942.
22. Spencer, F. M., and Monroe, L. S.: The color atlas of intestinal parasites, Springfield, Ill., 1961, Charles C Thomas, Publisher.
23. U. S. Public Health Service: Amebiasis; laboratory diagnosis. Part I. Life cycle of *Entamoeba histolytica.* Part II. Identification of intestinal amoebae. Part III. Laboratory procedures, Public Health Service Publication No. 1187, 1964.
24. Wheatley, W. B.: Rapid stain for intestinal amebae and flagellates, Amer. J. Clin. Path. **21:**990-991, 1951.

PART EIGHT

Antimicrobial susceptibility tests

34 Determination of susceptibility of bacteria to antimicrobial agents

According to Isenberg,[12] the value of the clinical laboratory can be measured only by the significance of the guidance it gives the practicing physician in the treatment of his patients. In no other area of clinical microbiology does this statement become more pertinent than in the testing of clinical isolates for their susceptibility to antimicrobial agents. With the increasing number of these agents at the physician's disposal and the changing pattern of resistance and susceptibility among bacteria—particularly the gram-negative enteric bacilli—the clinician must rely more and more upon sensitivity testing to guide his selection of appropriate drugs or alter an already imposed regimen. Therefore, to a large extent, a laboratory report showing susceptibility or resistance to a particular antimicrobial agent becomes an endorsement of its usefulness or withdrawal.

Since the microbiologist must become an adviser to the physician regarding proper antimicrobial therapy, it follows, therefore, that he must maintain (1) a high level of accuracy in his testing procedures, (2) a high degree of reproducibility for the results, and (3) a good correlation of his results with the clinical response.[12] **Only through close cooperation and exchange of information between the laboratory staff and the clinician can the best possible management of an infectious process be achieved.**

The principal methods presently used by the laboratory to determine susceptibility of a microorganism to an antibiotic include the **dilution tests,** such as the broth tube and agar plate dilution procedures, and the **agar diffusion test,** utilizing antibiotic-impregnated discs. Each method has its advantages and its limitations, and these must be understood and appreciated in order to obtain maximum usefulness of the results. Since there is a place for all of these methods in the clinical laboratory, the procedures and directions for the use of each will be described in detail.

In the interpretation of any in vitro susceptibility tests, it is well to remember that they are essentially **artificial measurements;** the data yielded by them give only the approximate range of effective inhibitory action against the microorganisms. The only absolute criterion of microbial response to antibiotics is the **clinical response** of the patient when adequate dosage of the appropriate antibiotic is administered.

BROTH TUBE DILUTION METHOD FOR DETERMINING SUSCEPTIBILITY TO ANTIBIOTICS

In the broth tube dilution method for determining the susceptibility of an organism to antibiotics, specific amounts of the antibiotic, prepared in decreasing concentration in broth by the serial dilution technique, are inoculated with a culture of the bacterium to be tested. The susceptibility of the organisms is determined, after a suitable period of incubation, by microscopic observation of the presence or absence of growth in the varying concentrations of the antimicrobial agent. This bacteriostatic end-point value is known as the **minimal inhibitory concentration** (MIC). With minor additions, the technique can be adapted to the determination of bactericidal levels of the antibiotic—the **minimal bactericidal concentration** (MBC); this is discussed at length in a further section.

A number of factors must be considered in establishing the procedures and in evaluating the results of these tests.[8] They include the following: (1) the medium in which the tests are performed, (2) the stability of the antibiotic, (3) the size of the inoculum, (4) the rate of growth of the organism, and (5) the period of incubation of the tests. Any variation in one or more of these factors may influence the tests, and the results obtained by one procedure may not agree with those arrived at by a slightly different method.[5] However, if a **standard procedure using only pure cultures** is adopted and strictly adhered to, reproducible results can usually be obtained and the reports from a given laboratory can be readily interpreted by the clinical staff.

The test tube serial dilution method gives a fairly accurate determination of susceptibility to measured amounts (either units or micrograms) of the antibiotic. It is a time-consuming and expensive procedure, however, especially when the clinician wants to know the susceptibility of an organism to a number of antibiotics. For this reason its use may well be restricted to special cases when quantitative results may be of value. In any event, it is strongly recommended that all clinical laboratories should be prepared to offer this service to the clinician, either directly or through a referral laboratory.

The serial dilution method may be recommended for determining the susceptibility of organisms isolated in the following instances: (1) from blood cultures, (2) from patients who fail to respond to apparently adequate therapy, and (3) from patients who relapse while undergoing such therapy. The study of organisms from the third instance usually involves determination of any increase in resistance of subsequent isolations and requires special methods.

Routine procedure for serial dilution tests
Preparation of stock solution of antibiotics

Stock solutions of antibiotics are prepared from concentrated, dehydrated sterile material of known potency that may be obtained from the pharmaceutical manufacturer. Generally, they are prepared in concentrations of 1,000 μg./ml., using phosphate buffer or Mueller-Hinton broth as the diluent and are tubed in 1-ml. amounts in screw-capped vials.

When stored in the frozen state at −20° C., these antibiotics will remain stable for at least 8 weeks*; when refrigerated at 5° C., they show no appreciable loss of potency in 1 week. Any unused thawed solutions of antibiotics should be discarded; each aliquot should be sufficient for 1 day's use only and should not be refrozen.

Table 34-1 is provided as a guide to the preparation of stock solutions of the most frequently used antimicrobial agents.

Selection of media

The fluid media in which the tube dilution sensitivity tests are carried out must be the kind that will support optimal, rapid growth of the test organism in pure culture. A broth medium that will support the growth of pneumococci and streptococci without the addition of serum or blood is preferable, since the addition of such enrichment adds another variable to the test and may influence the results. Trypticase soy broth, or preferably Mueller-Hinton broth† is recommended for sensitivity tests with the following ex-

---

*Ampicillin requires storage at -60° to -70° C. to prevent loss of potency.
†This medium appears to be low in tetracycline and sulfonamide inhibitors and shows good batch-to-batch consistency.

Table 34-1. Procedures for preparing stock solutions*

ANTIBIOTIC	MANUFACTURER	METHOD OF PREPARATION
Ampicillin	Bristol-Myers Co.	Weigh out material and multiply by "activity standard" provided by manufacturer.† Add 0.1 ml. of pH 8.0 phosphate buffer to dissolve; dilute with pH 6.0 phosphate buffer.
Penicillin G	Eli Lilly & Co.	Add 60 ml. of water to a vial containing 1 million units or 600,000 μg.; this makes a stock solution of 10,000 μg./ml.
Methicillin	Bristol-Myers Co.	Weigh out material and multiply by "activity standard" from manufacturer†; dilute with pH 6.0 phosphate buffer.
Oxacillin	Bristol-Myers Co.	Add 16 ml. of water to a vial to give 1,000 μg./ml.
Cephalothin	Eli Lilly & Co.	Weigh out material and multiply by the "activity standard," from the manufacturer†; dilute with pH 6.0 phosphate buffer.
Cephaloridine	Eli Lilly & Co.	Weigh out exactly 30 mg., add 30 ml. of pH 6.0 phosphate buffer to give 1,000 μg./ml.
Carbenicillin	Beecham-Massengill Pharmaceuticals	Add 10 ml. of water to a vial containing 1 gm. of drug, then dilute 1:100 to give 1,000 μg/ml.
Tetracycline	Pfizer Lab.	Add 20 ml. of water to a vial that contains 20 mg. of drug, to give 1,000 μg./ml.
Chloramphenicol	Parke, Davis & Co.	Weigh out exactly 50 mg., add 1 ml. of ethyl alcohol to dissolve drug and sufficient water to 50 ml. This gives a concentration of 1,000 μg./ml.
Erythromycin	Abbott Laboratories	Weigh out material and multiply by the "activity standard" from the manufacturer.† Dissolve in 1-2 ml. alcohol and add water to a final concentration of 1,000 μg./ml.
Lincomycin	The Upjohn Co.	Add 20 ml. water to a vial to make 1,000 μg./ml.
Clindamycin	The Upjohn Co.	Add 15 ml. of water to vial containing 150 mg. of drug, dilute 1:10 to give 1,000 μg./ml.
Kanamycin	Bristol-Myers Co.	Weigh out exactly 30 mg., add 30 ml. water to give 1,000 μg./ml. Agent is unstable in acid range.
Gentamicin	Schering Corp.	Weigh out material and multiply by "activity standard" from manufacturer,† dilute with water to give 1,000 1,000 μg./ml.
Polymyxin B	Burroughs Wellcome & Co.	Add 5 ml. water to a vial containing 50 μg., to give 10,000 μg./ml.; dilute to desired concentration.
Colistin (polymyxin E)	Warner-Chilcott Lab.	Weigh out material and multiply by "activity standard" from manufacturer,† dilute with water to desired concentration.
Bacitracin	The Upjohn Co.	Add 10 ml. of water to a vial containing 10,000 units, to give 1,000 units/ml.
Nitrofurantoin	Eaton Labs., Inc.	Weigh out 120 mg. of drug and transfer to a 50-ml. flask containing 4.0 ml. of dimethyl formamide. Heat in 56° C. water bath with shaking to dissolve. This solution contains 30 mg./ml.
Nalidixic acid	Winthrop Labs.	Weigh out approx. 30 mg., add 2 ml. of 1 N NaOH and allow to stand to dissolve (may require gentle heat). Dilute with sterile water (less 2 ml.) to desired concentration.

*Courtesy of John A. Washington II, Head, Section of Clinical Microbiology, Mayo Clinic.
†Example: 1 mg. = 825 μg. ("activity standard"/μg.). 50 mg. = 50 × 825 = 41,250 μg. Therefore, add 41.2 ml. of diluent to give 1,000 μg./ml.

ceptions: microaerophilic streptococci, which are frequently isolated from cases of subacute bacterial endocarditis, strictly anaerobic streptocci, *Bacteroides* species, and clostridia should be tested in fluid thioglycollate medium enriched with hemin and vitamin K.; some strains may require additional enrichments to support good growth. In the case of all fastidious organisms the growth requirements should be determined **before** sensitivity tests are carried out, in order that the fluid medium supporting the most luxuriant and rapid growth may be selected for the procedure. Hemoglobinophilic organisms, such as *Haemophilus* species, must necessarily be tested in Mueller-Hinton broth containing 1% rabbit blood. The blood may be added to the broth before it is distributed into the test tubes, or it may be added with the inoculum.

Procedure for preparing serial dilutions and determining susceptibility

1. Thaw the frozen stock solution of the antibiotic(s) required and dilute 1:5 with sterile Mueller-Hinton broth. This gives a **working solution** containing 200 μg. or units each. For bacitracin a concentration of 100 units per ml. may be used.
2. Select 10 clear, sterile, cotton-plugged or capped test tubes of small size (13 $\times$ 100 mm.) and mark from 1 to 10.
3. Using aseptic technique, pipette 0.5 ml. of dilution broth into tubes 2 through 10. Do this for each antibiotic to be tested.
4. Add 0.5 ml. of the working solution (200 μg./ml.) of the antibiotic into tubes 1 and 2. Mix contents of the second tube well and transfer 0.5 ml. to tube 3. Mix well and transfer 0.5 ml. to tube 4, continuing this procedure to tube 9. Discard 0.5 ml. from tube 9; the tenth tube receives no antibiotic and serves as the control. Use a **separate** pipet for each transfer to avoid any carry-over.
5. To all tubes add 0.5 ml. of an inoculum containing approximately 10^5 to 10^6 organisms per milliliter. This may may be prepared in most instances by making a 1:1,000 dilution in broth of an overnight (6-hour if a rapidly growing organism) broth culture of the organism to be tested. With slow-growing organisms, such as microaerophilic streptococci, *Bacteroides,* and so forth, it may be necessary to use cultures in thioglycollate medium up to 48 hours old. If numerous antibiotics are to be tested,

Table 34-2. Antibiotic serial dilution—tube setup

TUBE	DILUENT (MEDIUM) ADDED (ML.)	ANTIBIOTIC ADDED	DILUTED CULTURE ADDED (ML.)	FINAL ANTIBIOTIC CONCENTRATION	
				ALL BUT BACITRACIN (UNITS OR μG.)	BACITRACIN (UNITS)
1	None	0.5 ml. working solution	0.5	100	50
2	0.5	0.5 ml. working solution	0.5	50	25
3	0.5	0.5 ml. from tube 2	0.5	25	12.5
4	0.5	0.5 ml. from tube 3	0.5	12.5	6.25
5	0.5	0.5 ml. from tube 4	0.5	6.25	3.125
6	0.5	0.5 ml. from tube 5	0.5	3.125	1.56
7	0.5	0.5 ml. from tube 6	0.5	1.56	0.78
8	0.5	0.5 ml. from tube 7	0.5	0.78	0.39
9	0.5	0.5 ml. from tube 8*	0.5	0.39	0.19
10	0.5	None	0.5	Zero	Zero

*Discard 0.5 ml. from tube 9.

prepare sufficient inoculum in a flask, for uniformity.

The final volume in each tube is now 1 ml., and the antibiotic range covered in the series is from 100 μg. or units to 0.39 μg. or unit per milliliter, in twofold steps (Table 34-2).

For organisms highly susceptible to antibiotics, such as streptococci and pneumococci, a lower range of antibiotic dilutions may be employed by further diluting the working solution (200 μg. or units per milliliter) 1:10. Then, by the serial twofold dilution technique shown in Table 34-2, the final concentrations will be 10, 5, 2.5, 1.25, 0.63, 0.3, 0.15, 0.08 and 0.04 μg. or units per milliliter.

6. Incubate the series at 36° C. and examine macroscopically for **evidence of growth.** Incubate the tubes only as long as it is necessary for the control tube to show turbid growth; usually 12 to 18 hours is the optimal time. The last tube, that is, the lowest concentration of the antibiotic in the series showing no growth, is taken as the MIC of the antibiotic and is expressed as micrograms (or units) per milliliter. It is well to remember that in the serial dilution technique there is a possible error equivalent to one tube dilution, so that the MIC values are not necessarily actual values but are near true values.

To determine the MBC, pipette 0.5 ml. of each tube of the tube dilution set that shows no visible turbidity into 12 ml. of infusion agar, mix, and make a pour plate. Having obtained a colony count of the **initial** inoculum by making a pour plate of the 1:1,000 dilution when setting up the test, one may then calculate the lowest concentration of the antimicrobial that provided a 99.9% and 100% bactericidal activity, by comparing colony counts, after an appropriate incubation period of 48 to 72 hours.

AGAR PLATE DILUTION METHOD FOR DETERMINING SUSCEPTIBILITY TO ANTIBIOTICS

The agar plate dilution method[1] is similar in principle to the tube dilution method, except that a solid medium is used. Mueller-Hinton agar is recommended and is prepared in 100-ml. amounts. Some workers incorporate 5% blood or heated blood in the medium when using it for organisms that require enriched media, such as pneumococci, streptococci, and *Haemophilus.* There appears to be no significant inactivation of the antibiotics by the addition of the blood.

Procedure for preparing serial dilutions

Prepare twofold serial dilutions* of the stock antibiotics, as described in the previous section, using at least ten times the volumes indicated. Stock solutions containing 1,000 μg. per milliliter are most useful, since decimal dilutions are readily prepared from these. For example, to prepare a 10 μg. per milliliter plate, add 1 ml. of the stock to 100 ml. of melted and cooled agar and pour plates of the same.

Preparation of plate dilutions

Melt and cool sufficient screw-capped flasks of agar medium for the number of plates to be prepared (about 20 ml. of medium is required per 90 mm. diameter plate) and allow to equilibrate in a water bath at 50° before adding the antibiotic. Add the required amount of the various antibiotic dilutions to each flask, mix gently by inversion, and pour into plates.† Allow the agar to harden and store in the refrigerator at 5° C. until used, preferably within 24 hours (and not after 1 week) of preparation.‡

---

*Some workers prefer final dilutions to contain 20, 10, 5, 1, 0.1, and 0.01 μg. per milliliter of medium.
†It is **not** recommended that the antibiotic dilutions and culture medium be mixed directly in the plates; this may produce uneven distribution of the antibiotic in the agar.
‡Media containing unstable antibiotics, such as ampicillin, should be prepared twice weekly.

Inoculation of plates

The inoculum size should be adjusted to contain approximately 10^8 organisms per milliliter (equivalent to a McFarland standard of 1 or 2); this will ensure dense, nearly confluent growth on a control plate containing no antibiotic.

Spot inoculation of the plates is made with a 1-mm. loop (approximately 0.001 ml.), a capillary pipet, or, preferably, by using the inocula replicator of Steers and co-workers.[17] In this device, each single manipulation will release thirty-six different cultures from the prongs on a replicator head to the surface of a 100×15 mm. square plastic plate (Falcon) containing agar to a depth of 3.0 mm. Each prong will deliver about 0.001 ml.; thirty-six inoculations can thus be made simultaneously.

In using the Steers replicating device, it is recommended that one space on each plate be allocated to a marking solution (for proper orientation of the plate), one space for testing the viability of the test strain, and two spaces for controls—strains of gram-positive cocci and gram-negative bacilli of known stable MIC's to the antibiotics used. Thus, 32 spaces will then be available per plate for the testing of clinical isolates.

Organisms having a spreading tendency, such as *Proteus* and *Pseudomonas,* may be contained by the use of glass cylinders,* as suggested by Washington.[21]

Incubation and reading of tests

Incubate the plates at 36° C. for 16 to 18 hours and examine for the presence of growth. The lowest concentration of the antibiotic producing **complete inhibition of growth**† is taken as the end point. Partial inhibition can be observed readily by noting the gradual decrease in amount of growth until complete inhibition is ob-

tained. The control cultures on antibiotic-free media should show confluent growth.

STANDARDIZED DISC–AGAR DIFFUSION METHOD FOR DETERMINING SUSCEPTIBILITY TO ANTIBIOTICS

Perhaps the most useful, and certainly the most used, laboratory test for antibiotic susceptibility is the antibiotic disc–agar diffusion procedure, usually called the **disc method.** Its simplicity, speed of performance, economy, and reproducibility (under standardized conditions) makes it ideally suitable for the busy diagnostic laboratory when the more laborious dilution methods previously described may not be practiced.

In this method, as originally described by Bondi and associates,[4] filter paper discs that have been impregnated with various antimicrobial agents of specific concentrations are carefully placed on an agar culture plate that has been inoculated with a culture of the bacterium to be tested. The plate is incubated overnight and observed the following morning for a **zone of growth inhibition** around the disc containing the agent to which the organism is **susceptible,** whereas a **resistant** organism will grow up to (and under) the periphery of the disc.

No attempt will be made here to discuss the complex physicochemical reactions that take place during the diffusion of the antibiotic into the agar gel or the dynamics of bacterial growth on the substrate—the reader is referred to publications by Ericson[6,7] for details concerning these.

In recent years numerous attempts have been made to standardize the disc procedure, including the work of Bauer, Kirby, and co-workers,[3] Ericcson,[7] the World Health Organization (WHO), the Food and Drug Administration,[9] and most recently, the National Committee for Clinical Laboratory Standards (NCCLS). Because the revised *Tentative Standards*[14] recommended by the Subcommittee on Antimicrobial Susceptibility Testing of the NCCLS appears to be the most explicit in

---

*12 × 12 mm. Raschig rings, Scientific Glass Apparatus Co., Bloomfield, N. J.
†A very fine growth or a few visible colonies also may occur when the Steers replicator is used; this may be disregarded in the reading of the test.

methodology, these recommendations are the ones that will be presented herein.

Numerous proficiency testing surveys, including a recent nationwide laboratory evaluation by the Center for Disease Control (CDC)[11] of the disc procedure have revealed that (1) the procedure as practiced is **not** standardized and (2) there are numerous variables that may contribute to these discrepancies. Among those which have been identified are the following:

1. Selection and concentration of antimicrobial discs
2. Selection, volume, and age of plating medium
3. Storage and handling of discs
4. Methodology of testing
5. Criteria used for interpreting results

Selection of antimicrobial discs

The following basic set of drugs* and their concentrations are recommended for routine susceptibility testing:

Ampicillin	10 μg.
Bacitracin	10 U.
Carbenicillin	100 μg.
Cephaloridine	30 μg.
Cephalothin	30 μg.
Chloramphenicol	30 μg.
Clindamycin	2 μg.
Colistin (Polymyxin E)	10 μg.
Doxycycline	30 μg.
Erythromycin	15 μg.
Gentamicin	10 μg.
Kanamycin	30 μg.
Lincomycin	2 μg.
Methicillin	5 μg.
Nafcillin and oxacillin	1 μg.
Nalidixic acid	30 μg.
Neomycin	30 μg.
Nitrofurantoin	300 μg.
Penicillin G	10 U.
Polymyxin B	300 U.
Streptomycin	10 μg.
Sulfonamides	300 μg.
Tetracycline	30 μg.
Vancomycin	30 μg.

A basic set of discs for routine testing against the commonly isolated microorganisms is listed on p. 322.

---

*Available from Baltimore Biological Laboratory, Cockeysville, Md.; Difco Laboratories, Detroit; Pfizer Diagnostics, Flushing, N. Y.; and others.

Selection of plating medium

Although an ideal medium has not yet been perfected for the disc test, the NCCLS Subcommittee considers Mueller-Hinton agar the best compromise for routine susceptibility testing, since it shows good batch-to-batch uniformity and is low in tetracycline and sulfonamide inhibitors. With the addition of 5% defibrinated sheep, horse, or other animal blood, it will support the growth of the more fastidious pathogens that will not grow on the non-enriched medium. When required, the blood-containing medium may be "chocolatized," for testing *Haemophilus* species.

Mueller-Hinton agar* is prepared according to the manufacturer's directions and should be immediately cooled in a 50° C. water bath after removal from the autoclave. This is then poured into sterile dishes (on a level, horizontal surface) to a uniform depth of 4 mm.; this is equivalent to approximately 60 ml. in a 140-mm. (internal diameter) plate, or approximately 25 ml. for 90-mm. plates. After cooling at room temperature, the plates may be used the same day, or stored in the refrigerator at 2° to 8° C. for **not more than 7 days,** unless some method of minimizing water loss from evaporation is taken.† As a sterility control, several plates from each batch of blood-containing Mueller-Hinton agar should be incubated at 36° C. for 24 hours or longer but not used subsequently.

Each batch of Mueller-Hinton agar should be checked for pH when prepared; it should be **pH 7.2 to 7.4** at room temperature. This may be tested by macerating a small amount of the medium in a little distilled water, or by allowing a little of the medium in a small beaker to gel around the pH meter electrode,‡ and reading the pH.

---

*Available from Baltimore Biological Laboratory, Cockeysville, Md.; Difco Laboratories, Detroit; and others.

†Such as wrapping in polystyrene plastic.

‡If available, a surface electrode is desirable.

Just before the medium is used, the plates should be placed in a 36° C. incubator with lids partly ajar, until excess surface moisture has evaporated—usually requiring 10 minutes.

Storage and handling of discs

Antibiotic susceptibility test discs are generally supplied in separate containers with a desiccant* and should be kept under refrigeration (below 10° C.). Discs containing the penicillins (including ampicillin and carbenicillin) and cephalosporin drugs should always be kept frozen (at less than −14° C.) to maintain their potency; a small working supply may be refrigerated for up to **1 week.** For long-term storage, discs are best kept in the frozen state until needed.

As they are required, the unopened containers are removed from the refrigerator or freezer 1 or 2 hours before the discs are to be used and allowed to adjust to room temperature, in order to minimize condensation resulting from warm air reaching the cold containers. If disc dispensers are utilized, they should be equipped with tight covers and supplied with a satisfactory desiccant; when not in use, they should also be refrigerated.

Manufacturer's expiration dates should be noted and listed; discs **must be discarded** on their expiration date.

Preparation of inoculum

It has been shown by various workers that when certified antibiotic discs and a single standard culture medium are used, the greatest factor contributing to reproducibility of the disc test is the control of the inoculum size.

The currently recommended method of preparing a standardized inoculum is as follows:

1. With a wire loop, the tops of four or five isolated colonies of a similar morphological type are transferred to a tube containing 4 to 5 ml. of soybean-casein digest broth.*

2. The broth is incubated at 36° C. until its turbidity exceeds that of the standard (described in step 3). This usually requires 2 to 5 hours' incubation.

3. The turbidity is then adjusted to a **barium sulfate standard** that is prepared by adding 0.5 ml. of 1.175% w/v barium chloride hydrate (Ba $Cl_2 \cdot 2$ H_2O) to 99.5 ml. of 1% v/v (0.36N) sulfuric acid. The standard is distributed in screw-capped tubes of the same size as the ones used in the broth culture, approximately 4 to 6 ml. per tube, which are then tightly sealed and stored at room temperature in the dark. Fresh standards must be prepared at least once every 6 months, although a recent publication suggests that the solution remains stable for a much longer period when heat-sealed and stored in the dark.[22]

4. The barium sulfate standard must be vigorously agitated in a Vortex shaker just before use, and the turbidity of the broth culture is then adjusted visually be adding sterile saline or broth, using an adequate light and comparing the tubes against a white background with a contrasting black line.

Inoculation of the test plates

Within 15 minutes of adjusting the density of the inoculum, a sterile cotton swab on a wooden applicator stick (plastic sticks are not satisfactory) is dipped into the standardized bacterial suspension and the excess fluid is removed by pressing against the inside of the tube above the fluid level. The swab is then used to streak the dried surface of a Mueller-Hinton plate in several planes (by rotating the

---

*Humidity, particularly high humidity, heat, and contamination are important deteriorating factors.[10]

*Trypticase soy broth, Baltimore Biological Laboratory, Cockeysville, Md.; tryptic soy broth, Difco Laboratories, Detroit; and others.

plate approximately 60° each time) to ensure an even distribution of the inoculum.

Allow the inoculated plates to remain on a flat and level surface undisturbed for 3 to 5 minutes to allow for adsorption of excess moisture, then apply the discs, as described in the following section.

Placement of discs

With alcohol-flamed, fine-pointed forceps (cooled before using) or a disc dispenser,* the selected discs are placed on the inoculated plate and pressed firmly into the agar with a sterile forceps or needle, to ensure **complete contact** with the agar. The discs are distributed evenly in such a manner as to be **no closer** than 15 mm. from the edge of the petri dish and so that no two discs are closer than 24 mm. from center to center. Once a disc has been placed, it should not be moved, since some diffusion of the antibiotic occurs almost instantaneously.

An alternative method, using an agar overlay, has been described by Barry and colleagues.[2] This method is useful only for rapidly-growing organisms such as *Staphylococcus aureus*, the enteric bacilli, and *Pseudomonas aeruginosa*, and must be standardized to correspond with results obtained by the cotton swab-streak method already described.

The inoculated and disced plates are inverted and placed in the 36° C. incubator within 15 minutes after application of the discs. Incubation under increased CO_2 tension should **not** be practiced, since the interpretative zone sizes were developed under aerobic conditions; furthermore, CO_2 incubation may significantly alter the zone sizes.

Table 34-3 is presented as a practical guide in the selection of discs for routine susceptibility testing of facultative organisms isolated in clinical practice. Although

not identical to that recommended by the NCCLS,[14] it has proved of value to clinicians at the Wilmington Medical Center, and has served as an aid in reducing the misuse or overuse of antibiotic agents in a large medical complex.

Reading of results

After incubation the relative susceptibility of the organism to the antibiotic is demonstrated by a clear zone of growth inhibition around the disc. This is the result of two processes: (1) diffusion of the antibiotic and (2) growth of the bacteria. As the antibiotic diffuses through the agar medium from the edge of the disc, its concentration progressively diminishes to a point where it is no longer inhibitory for the organism, which then grows freely. The size of this area of suppressed growth, the **zone of inhibition,** is determined by the concentration of the antibiotic present in the area. Therefore, within the limitations of the test, the **diameter of the inhibition zone** denotes the **relative susceptibility** to a particular antibiotic.

After 16 to 18 hours' incubation,* each plate is examined and the diameters of the complete inhibition zones are noted and measured, using reflected light and sliding calipers, a ruler, or a template prepared for this purpose and held on the bottom of the plate. The **end point,** measured to the nearest millimeter, should be taken as the area showing no visible growth that could be detected with the unaided eye. Faint growth or tiny colonies near the edge of the inhibition zones are ignored, as is the veil of swarming occurring in the inhibition zones of some strains of *Proteus* species. With sulfonamides, slight growth (with 80% or more of inhibition) is disregarded, and the margin of heavy growth is measured to determine the zone diameter.

Large colonies growing in an inhibition

---

*Dispensers for both the 90-mm. and 140-mm. Petri plates are available from Baltimore Biological Laboratory, Cockeysville, Md.; Difco Laboratories, Detroit; Pfizer Diagnostics, Flushing, N. Y.; and others.

*Microbial growth should be almost or just confluent; if only isolated colonies are present, the inoculum was too light and the test must be repeated.

Table 34-3. Schema for recommended antimicrobial discs*

ORGANISM	AM 10 μG.	CB 50 μG.	CF 30 μG.	CM 30 μG.	CC 2 μG.	EM 15 μG.	GM 10 μG.	KM 30 μG.	LN 2 μG.	DP 5 μG.	NA 30 μG.	NF 300 μG.	PN 10 U.	PB 300 U.	SM 10 μG.	TE 30 μG.
Gram-negative rods																
Escherichia coli	x	[x]	x	[x]			x	x			(x)	(x)		x	[x]	[x]
Klebsiella-Enterobacter-Serratia	x		x	[x]			x	x			(x)	(x)		x	[x]	[x]
Other enteric bacilli	x		x	[x]			x	x			(x)	(x)		x	[x]	[x]
Proteus species	x	[x]		[x]			x	x			(x)	(x)			[x]	[x]
Pseudomonas aeruginosa, Ps. sp.		x		[x]			x				(x)			x	[x]	[x]
Other nonfermentative bacilli	x	x	x	[x]			x	x			(x)	(x)		x	[x]	[x]
Gram-positive cocci																
Staphylococcus aureus			x	[x]	x	x	[x]		[x]	x			x			[x]
Streptococcus pyogenes (Group A)						[x]			[x]				[x]			[x]
Group D streptococci, including enterococci	x		x	[x]		x	[x]						x		[x]	[x]
Streptococcus pneumoniae			[x]						[x]				[x]			[x]
Other streptococci (alpha, and so forth)	x		x	[x]		x							x			[x]
Miscellaneous groups																
Neisseria meningitidis	x		x	[x]									x			[x]
Haemophilus influenzae	x		x	[x]									x		[x]	[x]
Other facultative organisms	As indicated by identity of isolate															

*From the Section of Microbiology and the Infectious Disease Research Laboratory, Wilmington, Medical Center, Wilmington, Del. Abbreviations: AM, ampicillin; CB, carbenicillin; CF, cephalothin; CM, chloramphenicol; CC, clindamycin; EM, erythromycin; GM, gentamicin; KM, kanamycin; LN, lincomycin; DP, methicillin; NA, nalidixic acid; NF, nitrofurantoin; PN, penicillin G; PB, polymyxin B; SM, streptomycin; TE, tetracycline. x, Recommended for use with the species indicated; (x), recommended for use with urine isolates *only*; [x], recommended for use with blood culture isolates *only*.

Table 34-4. Zone size interpretative chart, Kirby-Bauer method*

ANTIBIOTIC OR CHEMOTHERAPEUTIC AGENT	DISC POTENCY	INHIBITION ZONE DIAMETER TO NEAREST MM.		
		RESISTANT	INTERMEDIATE	SENSITIVE
Ampicillin[1]				
Enterobacteriaceae and enterococci	10 μg.	11 or less	12-13	14 or more
Staphylococci		20 or less	21-28	29 or more
Haemophilus		19 or less	—	20 or more
Bacitracin	10 U.	8 or less	9-12	13 or more
Carbenicillin	100 μg.			
Pseudomonas sp.		13 or less	14-16	17 or more
Proteus and *Escherichia coli*		17 or less	18-22	23 or more
Cephaloridine	30 μg.	14 or less	15-17	18 or more
Cephalothin	30 μg.	14 or less	15-17	18 or more
Chloramphenicol	30 μg.	12 or less	13-17	18 or more
Clindamycin[2]	2 μg.	14 or less	15-16	17 or more
Colistin (polymyxin E)[3]	10 μg.	8 or less	9-10	11 or more
Doxycycline	30 μg.	12 or less	13-15	16 or more
Erythromycin	15 μg.	13 or less	14-17	18 or more
Gentamicin	10 μg.	12 or less	—	13 or more
Kanamycin	30 μg.	13 or less	14-17	18 or more
Lincomycin	2 μg.	9 or less	10-14	15 or more
Methicillin[4]	5 μg.	9 or less	10-13	14 or more
Nafcillin and oxacillin	1 μg.	10 or less	11-12	13 or more
Nalidixic acid[5]	30 μg.	13 or less	14-18	19 or more
Neomycin	30 μg.	12 or less	13-16	17 or more
Nitrofurantoin[5]	300 μg.	14 or less	15-16	17 or more
Penicillin G				
Staphylococci	10 U.	20 or less	21-28	29 or more
Other organisms[6]	10 U.	11 or less	12-21[6]	22 or more
Polymyxin B[3]	300 U.	8 or less	9-11	12 or more
Rifampin (when testing *Neisseria meningitis* susceptibility only)	5 μg.	24 or less	—	25 or more
Streptomycin	10 μg.	11 or less	12-14	15 or more
Sulfonamides[5,7]	300 μg.	12 or less	13-16	17 or more
Tetracycline[8]	30 μg.	14 or less	15-18	19 or more
Trimethoprim/sulfamethoxazole	25 μg.	10 or less	11-15	16 or more
Vancomycin	30 μg.	9 or less	10-11	12 or more

*Courtesy Alfred W. Bauer, Group Medical Center, Seattle, Wash., and John C. Sherris and W. Lawrence Drew, University Hospital, Seattle, Wash. See also reference 3 at end of chapter. Updated and modified by other investigators. See references 9 and 14.

[1]The ampicillin disc is used for testing susceptibility to both ampicillin and hetacillin.

[2]The clindamycin disc is used for testing susceptibility to both clindamycin and lincomycin.

[3]The polymyxins diffuse poorly in agar, and the accuracy of the diffusion method is less than with other antibiotics. Resistance is always significant, but some relatively resistant strains of *Klebsiella* and *Enterobacter* may give zones in the lower end of the sensitive range (up to 15 mm.). When treatment of systemic infections due to susceptible strains is considered, it is wise to confirm the results of a diffusion test with a dilution method.

[4]The methicillin disc is used for testing susceptibility to all penicillin-resistant penicillins: methicillin, cloxacillin, dicloxacillin, oxacillin, and nafcillin. Methicillin-resistant strains of *Staphylococcus aureus* are best detected at 30°C.

[5]Urinary tract infections only.

[6]This category includes some organisms, such as enterococci and gram-negative bacilli, that may cause systemic infections treatable by high doses of penicillin G.

[7]Any of the commercially available 300 or 250 μg. sulfonamide discs can be used with the same standards of zone interpretation.

[8]The tetracycline disc is used for testing susceptibility to all the tetracyclines: chlortetracycline, demeclocycline, doxycycline, methacycline, oxytetracycline, rolitetracycline, minocycline, and tetracycline.

zone may actually be a different bacterial species (a mixed, rather than a pure, culture) and should be subcultured, reidentified, and retested.

Interpretation of zone sizes

The diameters of the inhibition zones are then interpreted by referring to Table 34-4, which represents the NCCLS subcommittee's present recommendations.

The term "susceptible" implies that an infection caused by the strain tested may be expected to respond favorably to the indicated antimicrobial for that type of infection and pathogen. "Resistant" strains, on the other hand, are not inhibited completely by therapeutic concentrations. "Intermediate" implies that the isolant may respond to unusually high concentrations of the agent, due either to high dosage levels or in areas, such as the urinary tract, where the drug is concentrated. In other circumstances, intermediate results might warrant further testing if alternative agents are not available.

Limitations of the test

This modified Bauer-Kirby procedure has been standardized for testing rapidly growing isolants, particularly members of the Enterobacteriaceae, *Staph. aureus,* and *Pseudomonas* species; limited experience also suggests that the interpretative standards hold for *Haemophilus* and streptococci, if blood ("chocolatized" if required) is added to the Mueller-Hinton agar. *Streptococcus pyogenes* and *S. pneumoniae* are generally susceptible to penicillin G and are not routinely tested; however, in those patients hypersensitive to penicillin, the isolant may be tested against erythromycin or lincomycin.

In general, fastidious organisms requiring an increased CO_2 tension or an anaerobic atmosphere, or whose growth rate is unusually slow, do not lend themselves to susceptibility testing by the standardized disc-agar diffusion method; agar plate or broth dilution test procedures are recommended. Likewise, testing of *Neisseria*

gonorrhoeae by the described procedure is not recommended. Because of the interest recently generated by early reports on reliable susceptibility testing of anaerobes using the disc procedure, a subsequent section will consider this technique.

Quality control procedures

It is essential that some form of quality control be carried out in performing the disc procedure, to ensure precision and accuracy of the test results. The NCCLS subcommittee recommends that the tests be monitored **daily** with stock cultures of the Seattle strains of *Staph. aureus* (American Type Culture Collection 25923) and *Escherichia coli* (ATCC 25922), using antibiotic discs representative of those to be used in the testing of clinical isolants.[14] These cultures may be grown on soy-casein digest agar slants and stored under refrigeration (4° to 8° C.), and should be subcultured to fresh slants every 2 weeks.

For testing, the cultures are inoculated to soy-casein digest broth tubes, which are incubated overnight and streaked to agar plates to obtain isolated colonies; these are then picked to broth and tested as described in the preceding sections.

The control strains may be used as long as there is no significant change in the mean inhibition zone diameters not otherwise attributable to technical error. If such changes occur, fresh strains should be obtained from a reference laboratory or other reliable source. Individual values of zone diameters and their permissible differences are indicated in Table 34-5.

Table 34-5 represents a more precise computation, based on standard statistical methods, than previous publications. It is described in NCCLS's *Revised Tentative Standards* (May 1973), free copies of which may be obtained from the Committee.*

Recently, it has been recommended that a well-characterized and confirmed strain

---

*National Committee for Clinical Laboratory Standards, Los Angeles, Calif.

Table 34-5. Maximum acceptable standard deviations and mean zone diameters that should be expected with the *E. coli* (ATCC 25922) and *S. aureus* (ATCC 25923)*

ANTIMICROBIC	DISC CONTENT	MAXIMUM ACCEPTABLE STAND. DEV.	CURRENTLY ACCEPTED TRUE MEAN ZONE DIAMETER (MM)	
			E. COLI	S. AUREUS
Penicillin	10 units	2.9	†	31.5
Ampicillin	10 μg.			
Staphylococci		2.9	†	29.5
Enteric bacilli		1.3	17.5	†
and enterococci				
Methicillin	5 μg.	1.6	†	19.5
Nafcillin				
and oxacillin	1 μg.	1.3	†	—
Cephalothin	30 μg.	1.3	20.5	31.0
Cephaloridine	30 μg.	1.6	—	†
Carbenicillin	50 μg.			
Pseudomonas sp.		1.3	—	†
Proteus and *E. coli*		1.6	—	†
Chloramphenicol	30 μg.	1.9	24.0	22.5
Tetracycline	30 μg.	1.6	21.5	23.5
Erythromycin	15 μg.	1.6	11.0	26.0
Lincomycin	2 μg.	1.9	†	—
Clindamycin	2 μg.	1.6	†	—
Kanamycin	30 μg.	1.6	21.0	22.5
Neomycin	30 μg.	1.6	20.0	22.0
Streptomycin	10 μg.	1.3	16.0	18.0
Gentamicin	10 μg.	1.3	22.5	23.0
Sulfonamides	300 μg.	1.6	—	†
Nitrofurantoin	300 μg.	1.6	—	†
Nalidixic acid	30 μg.	1.9	—	†
Polymyxin B	300 units	1.3	14.0	†
Vancomycin	30 μg.	1.3	†	17.0

*NCCLS Subcommittee on Antimicrobial Susceptibility Testing: Performance standards for antimicrobial disc susceptibility tests as used in clinical laboratories, revised tentative standards, May 1973. In Balows, A., editor: Current techniques for antibiotic susceptibility testing, Springfield, Ill., 1974, Charles C Thomas, Publisher.
†Data not relevant; —data not yet established.

of *Pseudomonas aeruginosa* be added to the quality control system.[11] Apparently, some lots of Mueller-Hinton agar may contain increased concentrations of Ca^{++} and Mg^{++}. Since *Staph. aureus* and *E. coli* are not effected by these ions, they will demonstrate no changes in zone sizes, but growth of *Ps. aeruginosa* is enhanced and will therefore demonstrate **smaller** zones. Thus, the effect of increased concentrations of these cations would influence the interpretation of susceptibility and should be predetermined.

Susceptibility testing of anaerobes

With the diverse spectrum of activity of various antimicrobial agents against clinically significant anaerobes, it is clearly apparent that a simple, rapid, reliable test for their susceptibility is in demand. Attempts to adapt the Bauer-Kirby disc-agar diffusion technique for predicting sensitivity of anaerobes has been found to be generally unsatisfactory,[19] but a number of workers in the field of anaerobic bacteriology are attempting to correlate the zone diameters obtained by the disc test, with the MIC's obtained by either broth or agar dilution techniques, with varying degrees of success. Chief among these has been the work of Sutter and colleagues,[18] who have obtained a statistically good correlation with most of the antibiotics tested against a variety of known strains of anaerobes. The authors point out, however, that if predictions of the antibiotic sensitivity of **unidentified** isolants are to be made by the disc-diffusion procedure, it may be necessary to establish separate criteria for organisms that have different growth rates —slow, moderately rapid, or rapid—since this is one of the major variables that affect zone sizes.

Therefore, we feel that until a standardized, reproducible, and clinically correlated disc-diffusion technique for predicting antibiotic susceptibility of significant anaerobes becomes available, the reader is best directed to the employment of methodologies utilizing broth or agar dilution procedures.[13,14]

As a guide to the microbiologist and clinician, Table 34-6 from a recent publication by Finegold and co-workers* is presented. This is based on their correlation of in vitro laboratory findings with an evaluation of clinical effectiveness.

Other uses for antibiotic discs

One unexpected advantage of performing disc-sensitivity tests on primary plates inoculated with clinical specimens likely to contain more than a single pathogenic species (such as sputum, throat swabs, or urine) is the likelihood of **uncovering** organisms overgrown by other bacterial species. For example, an agar plate inoculated with a mixed culture of staphylococci and streptococci and "disced" with two different antibiotics may result in two distinct patterns of inhibition zones. One zone may show an inhibition of the staphylococci, whereas colonies of the streptococci may be growing within that zone.

*In Kagan, B. M., editor: Antimicrobial therapy, ed. 2, Philadelphia, 1973, W. B. Saunders Co.

Table 34-6. Susceptibility of anaerobes to antimicrobial agents*

ANTIMICROBIAL AGENT	MICROAEROPHILIC AND ANAEROBIC COCCI	BACTEROIDES FRAGILIS	BACTEROIDES MELANINOGENICUS	FUSOBACTERIUM VARIUM	OTHER FUSOBACTERIUM SPECIES	EUBACTERIUM AND ACTINOMYCES	CLOSTRIDIUM PERFRINGENS	OTHER CLOSTRIDIA
Penicillin G	+++	+	++++	++++†	++++	++++	++++†	++ to +++
Lincomycin	+++	+ to ++	+++	++	+++	++ to +++	+ to ++	+
Clindamycin	+++	+++	+++	++	+++	+++‡	++++‡	++
Metronidazole	++	+++	+++	+++	+++	?	+++	?+++
Chloramphenicol	+++	+ to ++	+++	+++	+++	+++	++++†	++++
Tetracycline	++	+	++	+	+	++	++	++
Erythromycin	++ to +++	+	++ to +++	+	++ to +++	++ to +++	+	++ to +++§
Vancomycin	++ to +++	+	+		+	?+++	+++	++++†

*Chart courtesy of Sydney M. Finegold. From Kagan, B. M., editor: Antimicrobial therapy, ed. 2, Philadelphia, 1973, W. B. Saunders Co. ++++, Drug of choice; +++ good, activity; ++, moderate activity; +, poor or inconsistant activity.
†A few strains are resistant.
‡Rare strains resistant.
§Based on old studies (resistance may have developed subsequently).

Around the other disc there may be only colonies of staphylococci, the streptococci having been inhibited.

Since antibiotic discs demonstrate distinctly different inhibitory capacities, they may be used for the sole purpose of **selectively isolating** various microorganisms. Following Vera's suggestions,[20] it has been our practice to place discs containing penicillin (10 units), neomycin (30 μg.), and bacitracin (10 units) on all **primary plates inoculated with a potentially mixed flora.** The inhibition zones around the neomycin discs have been particularly useful in exposing colonies of Group A beta hemolytic streptococci, pneumococci, enterococci, and other streptococci. *Haemophilus influenzae* has been isolated with ease from within the zones surrounding the 10-unit bacitracin disc on blood agar plates inoculated with sputum, and from material from the throat and nasopharynx. Penicillin discs (10 units) have aided in unmasking colonies of coliform bacilli, pseudomonads and species of *Proteus, Candida albicans,* and others. A 10-unit penicillin disc also is useful in revealing colonies of *Bordetella pertussis* on Bordet-Gengou plates of nasopharyngeal cultures. Kanamycin discs (30 ug.) have been reported to be helpful in separating *Bacteroides* and *Clostridium* species from other wound bacteria when the plates are incubated anaerobically.[20]

DETERMINATION OF ANTIBACTERIAL LEVEL OF SERUM DURING ANTIBIOTIC THERAPY

A **direct** method for determining the antibacterial potency of serum was first described by Schlichter and associates.[15,16] We have evaluated the Schlichter method in many cases of acute infections (particularly subacute bacterial endocarditis, staphylococcal septicemia, enterococal endocarditis, and others) and can recommend it as a valuable and practical guide to the antibiotic therapy of severe or complicated bacterial infections.

Technique of Schlichter test*

1. Subculture a recent isolate of the organism to infusion agar or blood agar slants and store in the refrigerator until the test is run, then subculture to a tube of broth early on the day of the test.

2. Obtain the first blood specimen **before** therapy, if possible; this serves as a control. Allow a 24-hour period to elapse after initiation of therapy in order to permit stabilization of absorption and excretion rates. Then take blood samples at any desired interval, at the low point of the blood concentration curve, if the patient is on intermittent dosage. Collect 10 ml. of the patient's blood in a sterile tube. On receipt in the laboratory, the clot is separated and the serum is obtained by centrifugation. The serum is then transferred to a sterile, rubber-stoppered tube; it may be titrated at that time (or within 2 to 3 hours if refrigerated) or frozen immediately in a slanting position and stored at $-20°$ C., a temperature at which it remains stable for several days.

3. Prepare serial twofold dilutions of serum in 1 ml. amounts in Mueller-Hinton broth,† using eight sterile, gauze-stoppered Kahn tubes **(use a separate pipet for each dilution).** The first tube contains only **undiluted** serum, and a ninth tube contains only broth and serves as a culture **control.** The serum dilutions range from undiluted through 1:128. Very sensitive organisms, such as alpha hemolytic streptococci, may require dilutions up to 1:2,048.

4. To each tube of the series add 0.05 ml. of a 1:1,000 dilution of a 6-hour broth culture of the organism isolat-

---

*Modified by John A. Washington II, Head, Section of Clinical Microbiology, Mayo Clinic.[21]
†Use soy-casein digest broth or others, such as brain-heart infusion or Levinthal's broth, for organisms not growing in Mueller-Hinton medium.

ed from the patient. Also prepare a pour plate using 1 ml. of the inoculum.

5. Incubate the test at 36° C. for 18 to 24 hours and examine. The **bacteriostatic** end point is taken as the highest dilution in which no visible growth occurs. Because of the inherent turbidity of some sera, it is recommended that subcultures be made from each tube to a sector of a blood agar plate. To determine **bactericidal** end points, transfer 0.05 ml. from each tube showing no growth to a tube of thioglycollate medium. Mix and incubate at 36° C. for 72 hours. The tube showing no growth in thioglycollate is taken as the end point. Good growth should be evident in the control tube.

6. In cases where the organism grows slowly, a loopful of an overnight broth culture may be used as the inoculum. When microaerophilic or anaerobic bacteria have been isolated, the tubes should be incubated anaerobically.

7. Since thioglycollate medium contains sufficient agar to permit the growth of discrete colonies, one can determine by inspection the number of colonies growing out, and thus the degree of killing. It is thus recommended that one report the results as follows:
   a. **Complete killing** at serum dilution_____, with no growth in subculture.
   b. Partial inhibition at serum dilution_____, with 1 to 2+ growth in subculture.
   c. Record and report the presence of less than 10 colonies at the serum dilution observed.

8. If the **percentage of bactericidal activity** is required, the following may be carried out:
   a. Pipette 0.5 ml. of each tube showing no gross turbidity into tubes containing 12 ml. of melt-

ed and cooled (45° to 50° C.) brain-heart infusion agar, mix, and make pour plates.
   b. Incubate for 72 hours at 36° C.
   c. On the basis of the inoculum colony count (step 4), calculate the lowest titers of serum dilution showing 99.9% and 100% bactericidal activity, and report accordingly.

9. Schlichter indicated that optimal antibiotic dosage (either single or combined drugs) had been achieved when a bactericidal level of 1:2 (complete inhibition in the first two tubes) had been demonstrated; others, however, believe that a bactericidal level of 1:8 should be the minimum effective level in problem cases. Dosage is adjusted according to results of the test and is maintained for the duration of the infection.

REFERENCES

1. Anderson, T. G.: Testing of susceptibility to antimicrobial agents and assay of antimicrobial agents in body fluids. In Blair, J. E., Lennette, E. H., and Truant, J. P., editors: Manual of clinical microbiology, Bethesda, Md., 1970, American Society for Microbiologists.
2. Barry, A. L., Garcia, F., and Thrupp, L. D.: An improved single-disk method for testing the antibiotic susceptibility of rapidly-growing pathogens, Amer. J. Clin. Path. **53:**149-158, 1970.
3. Bauer, A. W., Kirby, W. W. M., Sherris, J. C., and Turck, M.: Antibiotic susceptibility testing by a standardized single disc method, Amer. J. Clin. Path. **45:**493-496, 1966.
4. Bondi, A., Spaulding, E. H., Smith, E. D., and Dietz, C. C.: A routine method for the rapid determination of susceptibility to penicillin and other antibiotics, Amer. J. Med. Sci. **214:**221-225, 1947.
5. Branch, A., Starkey, D. H., and Power, E. E.: Diversifications in the tube dilution test for antibiotic sensitivity of microorganisms, Appl. Microbiol. **13:**469-472, 1965.
6. Ericsson, H.: Rational use of antibiotics in hospitals, Scando. J. Clin. Lab. Invest. **12:**1-59, 1960.
7. Ericsson, H.: The paper disc method in quantitative determination of bacterial sensitivity to antibiotics, Stockholm, 1961, Karolinska Sjukhuset.
8. Fink, F. C.: Special features of the tube dilution method of antibiotic susceptibility testing, Presented at the Interscience Conference on Anti-

microbial Agents and Chemotherapy, American Society of Microbiologists, Chicago, 1962.

9. Food and Drug Administration: Standardized disc susceptibility test, Federal Register, **37**(191):20527-20529, September 30, 1972.

10. Griffith, L. J., and Mullins, C. G.: Drug resistance as influenced by inactivated sensitivity discs, Appl. Microbiol. **16**:656-658, 1968.

11. Hall, C. T., and Webb, C. D., Jr.: Proficiency testing—bacteriology III and IV (July and Oct.) 1972, Atlanta, Ga., 1973, Center for Disease Control.

12. Isenberg, H. D.: A comparison of nationwide microbial susceptibility testing using standardized discs, Health Lab. Sci. **1**:185-256, 1964.

13. Martin, W. J., Gardner, M., and Washington, J. A. II: In vitro antimicrobial susceptibility of anaerobic bacteria isolated from clinical specimens, Antimicrob. Ag. Chemother. **1**:148-158, 1972.

14. NCCLS Subcommittee on Antimicrobial Susceptibility Testing: Performance standards for antimicrobial disc susceptibility tests, as used in clinical laboratories, revised tentative standards, August, 1974. In Balows, A., editor: Current techniques for antibiotic susceptibility testing, Springfield, Ill., 1974, Charles C Thomas, Publisher.

15. Schlichter, J. G., and MacLean, H.: A method of determining the effective therapeutic level in the treatment of subacute bacterial endocarditis with penicillin, Amer. Heart J. **34**:209-211, 1947.

16. Schlichter, J. G., MacLean, H., and Milzer, A.: Effective penicillin therapy in subacute bacterial endocarditis and other chronic infections, Amer. J. Med. Sci. **217**:600-608, 1949.

17. Steers, E., Foltz, E. L., and Graves, B. S.: An inocula replicating apparatus for routine testing of bacterial susceptibility to antibiotics, Antibiot. Chemother. (N.Y.) **9**:307-311, 1959.

18. Sutter, V. L., Kwok, Y.-Y., and Finegold, S. M.: In vitro susceptibility testing of anaerobes; standardization of a single disc test. In Balows, A., editor: Current techniques of antibiotic susceptibility testing. Springfield, Ill., 1974, Charles C Thomas, Publisher.

19. Thornton, G. F., and Cramer, J. A.: Antibiotic susceptibility of Bacteroides species, Antimicrob. Ag. Chemother. **10**:509-513, 1970.

20. Vera, H. D.: Sensitivity plate tests. In BBL Manual of Products and Laboratory Procedures, Cockeysville, Md., 1968, Baltimore Biological Laboratory.

21. Washington, J. A. II: Personal communication (E. G. S.), 1973.

22. Washington, J. A. II, Warren, E., and Karlson, A. G.: Stability of barium sulfate turbidity standards, Appl. Microbiol. **24**:1013, 1972.

PART NINE
Serological methods in diagnosis

35 Serological identification of microorganisms

GROUPING AND TYPING OF BETA HEMOLYTIC STREPTOCOCCI BY PRECIPITIN TEST

Both the group and the type of a beta streptococcus may be determined by the Lancefield precipitin procedure using the same antigen. Typing should be carried out soon after isolation because the M type specific protein substance, which determines the type, is lost on laboratory cultivation. Frozen fresh isolates can be typed successfully later.

LANCEFIELD PROCEDURE
Materials needed

1. Two sizes of capillary tubing are required: (a) 1.2 to 1.5 mm., outside diameter for grouping, and (b) 0.7 to 1 mm. outside diameter for typing.* Tube lengths of 7.5 cm. should be used.
2. Wooden blocks 12 inches in length containing Plasticine are used for holding the capillary tubes upright after they are filled with antigen and antiserum.
3. Todd-Hewitt broth at pH 7.8 to 8, tubed in 40-ml. amounts and inoculated 24 hours previously with the culture to be tested.
4. Beta streptococcus grouping and typing serum.†

---

*Kimble Glass Co., Vineland, N. J.
†Difco Laboratories, Detroit; Baltimore Biological Laboratory, Cockeysville, Md.; and others.

5. N/5 hydrochloric acid—1 ml. 12 N HCl plus 59 ml. of 0.85% saline.
6. Buffer solution—N/5 sodium hydroxide in M/15 phosphate buffer solution at pH 7. To prepare, dissolve 1 gm. of anhydrous acid sodium phosphate in 100 ml. of N/5 sodium hydroxide.
7. Phenol red—0.01%. To prepare, add 0.01 gm. of phenol red to 60 ml. of alcohol and 40 ml. of water.
8. Thymol blue—0.01%. To prepare, add 0.01 gm. of thymol blue to 60 ml. of alcohol and 40 ml. of water.

Preparation of antigen

1. Grow the organisms to be tested for 18 to 24 hours in 40 ml. of Todd-Hewitt broth.
2. Centrifuge the culture for 30 minutes and **remove all of the supernate.**
3. Resuspend the sediment in 0.4 ml. of N/5 hydrochloric acid pH 2.0 to 2.4.
4. Mix well with a wooden applicator stick.
5. Add 1 drop of thymol blue indicator. A peach color will result.
6. Transfer to a 15-ml. conical bottom centrifuge tube and heat in a boiling water bath for 3 to 10 minutes, shaking occasionally.
7. Cool in a refrigerator or a cold water bath for 10 minutes.

8. Centrifuge at 3,000 r.p.m. for 10 minutes.

9. Decant the clear supernatant fluid into a clean test tube and add 1 drop of phenol red indicator. The solution will be a distinct yellow.

10. Add the buffer solution drop by drop until a pale pink color develops (pH 7.4 to 7.8). Do not make too strongly alkaline.

11. Centrifuge as described previously and remove the supernatant fluid. This must be crystal clear for use as the antigen.*

12. The extract may be stored at 4° C. for several days.

Technique

For grouping beta streptococci. Clean the outside of a piece of 1.2 to 1.5 mm. capillary tubing with tissue paper or lens paper. Dip the capillary tube into Group A streptococcus antiserum and permit the serum to rise one third of the length of the tube, equivalent to a 2-cm. column. Wipe the outside of the tube to remove excess serum and to prevent adulteration of the antigen. Dip the capillary tube into the prepared antigen and draw up an equal amount of the extract. Wipe the outside of the tube with tissue and invert the tube until there is an air space both above and below the column of liquid. Place the tube upright in the Plasticine in the wooden block.

Repeat the procedure, using Groups B, C, D, and G antisera, respectively. Immediately after the tests are set up, examine the capillary tubes with a hand lens by strong light in front of a black background. The tubes should be perfectly clear if the test has been correctly per-

formed. Leave the tubes at room temperature and reexamine after 15 to 30 minutes. In the tube containing antigen and homologous antiserum a **milky ring** will be formed at the interface of the reactants. A **positive** reaction such as this is sharp and definite and appears in 5 to 10 minutes. After an hour or two the ring disappears and a heavy white precipitate settles to the bottom of the liquid. The **negative** tubes should remain perfectly clear.

If the culture tested proves to be Group A streptococcus and typing of the strain is desired, the same extract can be used for the antigen, which can be kept in a refrigerator for at least a week and still give satisfactory results. The procedure for the typing of beta streptococci follows.

For typing beta streptococci. Clean the outside of a piece of 0.7 to 1 mm. capillary tubing with tissue or lens paper. Use the same technique as just described, substituting beta streptococcus **typing serum** for the grouping serum, and insert the capillary tubing upright into the Plasticine. Incubate the tests for 2 hours, refrigerate overnight, and read as described earlier. Sometimes it is possible to read the tubes after incubation; in other cases it is necessary to refrigerate them in order to obtain a reaction.

Cross-reactions may occur in the typing tests. In such cases the cross-reaction can be eliminated by repeating the test, using the sera in which precipitation has occurred and diluting such sera one half, one fourth, or one eighth with isotonic salt solution. Incubate for 2 hours, refrigerate overnight, and read as in the regular test.[6]

The reader's attention also is directed to the typing of Group A streptococci by the T-agglutination technique of Griffith.[2] The reproducibility of the procedure has been well established internationally; studies have shown an agreement of 86% in the T-typing of 355 Group A strains by two separate laboratories.[8] Moody and coworkers also were able to T-type 88% of over 1,300 strains referred to the Center for Disease Control (CDC), compared

---

*A method for preparing streptococcal extracts by use of a proteolytic enzyme from *Streptomyces albus* has been described by Maxted.[4] It appears to be more rapid and simpler than the Lancefield method but is not suitable for some Group D strains. The prepared enzyme may be obtained commercially from Baltimore Biological Laboratory, Cockeysville, Md.; Difco Laboratories, Detroit.

with only 47% typable by the M-precipitin technique.[5] The methods used were described in the report.

QUELLUNG METHOD FOR TYPING PNEUMOCOCCI

Pneumococcus typing by the **quellung reaction** was formerly one of the most important procedures in a medical diagnostic laboratory. With the advent of antimicrobial therapy for the treatment of pneumococcal infections, serological typing of pneumococci is no longer necessary as a guide to therapy. However, the quellung test remains the most rapid and satisfactory method for the identification of a pneumococcus directly from clinical material shortly after the specimen has been received in the laboratory. For this reason it is still an important laboratory procedure and will be discussed here.

Clinical materials

The quellung test is a useful procedure with the following clinical specimens:

1. Sputum in which organisms resembling pneumococci are readily demonstrated in direct smears.
2. Broth cultures (containing a few drops of blood) of nasopharyngeal or throat swabs that show pneumococci in a stained smear after 6 to 8 hours' incubation.
3. Cerebrospinal fluid sediment that contains organisms suggestive of pneumococci (successful typing of the organisms definitely and immediately establishes their identity).
4. Suspected fresh pneumococcus colonies from blood agar plates, suspended in a few drops of broth.
5. Positive blood culture bottles, empyema fluid, and other specimens that show organisms resembling pneumococci on direct examination.

Technique

1. Spread a small loopful of the specimen in a thin film on a clean glass slide and **allow to air dry.** Make six

of these preparations. If stained smears of the material reveal more than 15 to 25 organisms per oil-immersion field, **dilute** the specimen, since too many organisms tend to agglutinate in the antiserum and make it difficult to see the capsular reaction.

2. Place a large loopful of typing serum on a coverglass and add a small loopful of methylene blue stain. Invert the coverglass over one of the dried films made previously. When available, the typing sera are individually prepared for pneumococcus types **1** through **34,** inclusive, and as alphabetically identified mixtures of these, they are labeled **A** through **F.*** Typing is carried out first with the mixtures, until a positive quellung is obtained and then with antisera for the types making up that mixture. Pools and individual sera are produced by the CDC and are available to Public Health Laboratories and government research agencies. As many as three coverglasses may be used on one slide. A polyvalent diagnostic antiserum, **Omniserum,** is available.† This gives capsular swelling with the 82 recognized types of pneumococci. Also available from the same source are nine pools (A through I) covering these types and 46 monovalent antisera. All contain methylene blue stain.

3. Examine each preparation under the microscope, using the oil-immersion lens with **reduced** illumination. A **positive** reaction is indicated by the appearance of a definitely outlined and refractile capsule surrounding the blue-stained pneumococcus. There is some variation in the reaction among strains; type 3 pneumococci, for example, have very large

---

*Pools A through F are available from Difco Laboratories, Detroit.
†From Dr. Erna Lund, Statenserum-Institut, Copenhagen, Denmark.

capsules. In general, however, it is the **sharpness** of the capsular outline rather than the size of the capsule that indicates a positive reaction. In **negative** preparations the capsules may be visible without swelling. In some negative reactions a thin halo with a definite outline may confuse the inexperienced worker. By focusing above and below such organisms nothing will be observed; in positive reactions the capsules are readily visible in these focal planes. Capsular reactions usually take place in several minutes; negative preparations should be reexamined after 1 hour before discarding. Avoid letting the preparations dry out.

QUELLUNG TEST ON SPINAL FLUID FOR DIAGNOSIS OF HAEMOPHILUS INFLUENZAE MENINGITIS

A rapid diagnosis of meningitis caused by *Haemophilus influenzae* can be made by performing a quellung test with the spinal fluid. Gram-stained smears should first be made of the material and observed carefully for the presence of small gram-negative rods, and if not observed the fluid should be centrifuged and gram-stained smears prepared from the sediment. If gram-negative bacilli resembling *H. influenzae* are seen in the smears from either source, quellung tests should be set up, using the appropriate specimen. Since almost all cases of *Haemophilus* meningitis are caused by serological **type b** organisms, the homologous antiserum* should be used. The technique is as described for typing of pneumococci. If influenza bacilli of serological type b are present, a typical quellung reaction will be observed.

QUELLUNG TEST FOR IDENTIFICATION OF KLEBSIELLA TYPES (FRIEDLÄNDER'S BACILLI)

Julianelle[3] studied the serological relationship of Friedländer's bacilli and divid-

ed the organisms into four types: A, B, C, and a heterogeneous Group X. Seventy-two capsular types are now recognized. Types 1, 2, and 3 of the new series correspond to A, B, and C of the former series. These sera may be used for the performance of a quellung test.* The technique is the same as that described for the typing of pneumococci.

SEROLOGICAL GROUPING OF NEISSERIA MENINGITIDIS

Four serological groups of *Neisseria meningitidis* are currently recognized.[1] These are Groups A, B, C, and D (Group D is not prevalent.)† The grouping of meningococci is of epidemiological significance only, but if serological identification is desired, agglutination tests can be carried out as follows:

1. Prepare twofold serial dilutions of the antisera.‡
2. Place 0.1 ml. of each serial dilution in a very small test tube.
3. Suspend the organisms from a young (5- to 24-hour) culture on solid medium using physiological saline containing 0.05% potassium cyanide to promote smoothness of bacterial suspension and also to kill organisms.
4. Screen the suspension by pipetting through a wisp of nonabsorbent cotton or centrifuge for 2 minutes at low speed to remove coarse particles. Clumps will not settle out spontaneously.
5. Dilute the antigen to approximately No. 8 McFarland standard (Chapter 41).
6. Add 0.1 ml. of the antigen to each serum dilution and shake the rack for 3 minutes at room temperature.

---

*Difco Laboratories, Detroit; Hyland Laboratories, Los Angeles.

*Capsular antisera for types 1 through 6 and a pooled antiserum are available from Difco Laboratories, Detroit.
†A fifth serological group of encapsulated *Neisseria meningitidis,* tentatively classified as Group E, has been described.[7]
‡These antisera may be obtained from the Central Public Health Laboratories, London, and from Difco Laboratories, Detroit.

7. Add 0.8 ml. of physiological saline to facilitate the reading of the reactions.
8. Read for macroscopic clumping of the organisms.

Typing can also be done by the quellung reaction, using specific antisera, as previously described. No quellung reaction is obtained with group B, because no morphological capsule exists.

SLIDE AGGLUTINATION TEST IN SEROLOGICAL IDENTIFICATION

Final type identification of the various members of the Enterobacteriaceae is dependent on serological analysis. The technique is used primarily for the numerous types within the *Salmonella* and *Shigella* genera and enteropathogenic *Escherichia coli* (EPEC), but it may be applied to other organisms. It is possible, with the aid of commercially available diagnostic antisera, to identify the more common salmonellae and EPEC and to group the shigellae. An exact antigenic analysis, however, requires the use of absorbed sera, which are available at some state health laboratories and at reference centers, such as the CDC, Atlanta, Ga., and the Laboratory Center for Disease Control, Ottawa, Canada, on this continent.

Technique

Mark off a number of squares ($^3/_4$ inch) on a clean glass slide with a grease pencil (Blaisdell, red 169T). Prepare a milky concentration of cells in saline in a tube from the growth on a TSI slant, from an agar slant, or from the colonies on an agar plate. Place 1 loopful or a small drop of each antiserum to be tested per square and leave 1 square blank. Add a drop of 0.85% saline to the blank square. With a Pasteur pipet add a small drop of the cell suspension to each square containing serum and to the blank square (as a control). Tilt the slide back and forth for 1 minute to mix, then observe for agglutination macroscopically. Agglutination is recognized by the **prompt** formation of fine granules or large aggregates. The control and any

negative tests should remain homogeneous.

K antigens, which mask the heat-stable somatic complex, are found in *Salmonella, Shigella,* and *Escherichia.* Should the cells fail to agglutinate in O antiserum, the suspension should be heated at 100° C. for 10 to 30 minutes, cooled, and retested in the appropriate O antisera. Suspensions of live cells will agglutinate in antisera that contain K antibodies, for example, *Salmonella* Vi and *Escherichia coli* OB.

The following list of diagnostic antisera* is provided for the reader's convenience.

Salmonella diagnostic sera†

Salmonella polyvalent (O antigens, 1, 2, 3, 4, 5, 6, 7, 8, 9, 10, 15, 19, Vi)
Salmonella Group A (O antigens 1, 2, 12)
Salmonella Group B (O antigens 4, 5, 12)
Salmonella Group C_1 (O antigens 6, 7)
Salmonella Group C_2 (O antigens 6, 8)
Salmonella Group D (O antigens 9, 12)
Salmonella Group E (E_1, E_2) (O antigens 3, 10, 15)
Salmonella Group F
Salmonella Group G
Salmonella Group H
Salmonella Group I
Salmonella Vi
Salmonella (flagellar a antigen)
Salmonella (flagellar b antigen)
Salmonella (flagellar c antigen)
Salmonella (flagellar d antigen)
Salmonella (flagellar i antigen)
Salmonella (flagellar v antigen)
Salmonella (flagellar y antigen)
Salmonella (flagellar 1, 2, 3, 5, 6, and 7 antigens)

Shigella grouping sera

Shigella grouping serum, Group A *(Sh. dysenteriae)*
Shigella grouping serum, Group B *(Sh. flexneri)*
Shigella grouping serum, Group C *(Sh. boydii)*
Shigella grouping serum, Group D *(Sh. sonnei)*

Enteropathogenic *Escherichia coli* diagnostic sera

Escherichia coli diagnostic serum, O18:B20
Escherichia coli diagnostic serum, O18:B21

---

*Lederle Laboratories, Pearl River, N. Y.; Baltimore Biological Laboratory, Cockeysville, Md.; Difco Laboratories, Detroit.

†The term antibody should be substituted for antigen in the *Salmonella* sera. The latter are used in testing for the presence of the antigens indicated.

Escherichia coli diagnostic serum, O20:B7
Escherichia coli diagnostic serum, O26:B6
Escherichia coli diagnostic serum, O28:B18
Escherichia coli diagnostic serum, O55:B5
Escherichia coli diagnostic serum, O86:B7
Escherichia coli diagnostic serum, O111:B4
Escherichia coli diagnostic serum, O112:B11
Escherichia coli diagnostic serum, O112:B13
Escherichia coli diagnostic serum, O119:B14
Escherichia coli diagnostic serum, O124:B17
Escherichia coli diagnostic serum, O125:B15
Escherichia coli diagnostic serum, O126:B16
Escherichia coli diagnostic serum, O127:B8
Escherichia coli diagnostic serum, O128:B12

REFERENCES

1. Breed, R. S., Murray, E. G. D., and Smith, N. R.: Bergey's manual of determinative bacteriology, ed. 7, Baltimore, 1957, The Williams & Wilkins Co.
2. Griffith, F.: The serological classification of *Streptococcus pyogenes,* J. Hyg. **34:**542-584, 1934.
3. Julianelle, L. A.: A biological classification of *Encapsulatus pneumoniae* (Friedländer's bacillus), J. Exp. Med. **44:**113, 1926.
4. Maxted, W. R.: Preparation of streptococcal extracts for Lancefield grouping, Lancet **2:**255-256, 1948.
5. Moody, M. D., Padula, J., Lizana, D., and Hall, C. T.: Epidemiologic characterization of group A streptococci by T-agglutination and M-precipitation tests in the public health laboratory, Health Lab. Sci. **2:**149-162, 1965.
6. Swift, H. F., Wilson, A. T., and Lancefield, R. C.: Typing group A hemolytic streptococci by M precipitin reactions in capillary pipettes, J. Exp. Med. **78:**127-133, 1943.
7. Vedros, N. A., and Culver, G.: A new serological group (E) of *Neisseria meningitidis,* J. Bact. **95:**1300-1304, 1968.
8. Wilson, E., Zimmerman, R. A., and Moody, M. D.: Value of T-agglutination typing of group A streptococci in epidemiologic investigations, Health Lab. Sci. **5:**199-207, 1968.

36 Antigen-antibody determinations on patients' sera

AGGLUTINATION TEST FOR DIAGNOSIS OF PNEUMONIA CAUSED BY MYCOPLASMA PNEUMONIAE (PRIMARY ATYPICAL PNEUMONIA)

The differential diagnosis of *Mycoplasma* pneumonia may be clinically difficult and usually necessitates laboratory confirmation. Serological evidence can be obtained by the use of the so-called cold hemagglutination test. The procedure for this test and its interpretation follows.*

Cold hemagglutination test

The development of cold hemagglutinins in the serum of patients with primary atypical pneumonia was first reported by Peterson and co-workers in 1943.[5] They observed that these antibodies (now considered macroglobulins) caused human erythrocytes to form visible clumps when incubated at 0° to 10° C. but not at 37° C.

Cold agglutinins are found in normal sera in low titers (less than 1:16) but are present in titers of 1:40 to 1:2,048 in a high proportion of patients with *Mycoplasma pneumoniae* infection. The titer rises during the course of the illness, usually reaching a maximum during the third or fourth week, followed by its rapid disappearance thereafter. The cold agglutinin response is generally related directly to the severity and duration of the illness,

*Bennett's excellent text[1] on clinical serology is recommended for additional reading.

although mild cases may also develop a significant titer.[2]

In obtaining serum for the cold hemagglutination test, it is essential that the drawn blood **not** be refrigerated before separation of the serum. This could result in the absorption by the red cells of most or all of the cold agglutinins present, resulting in their removal and thus a valueless test. It should also be noted that prolonged refrigeration of the serum will usually result in the disappearance of the cold agglutinins. Inactivation of the serum by heat to destroy complement, however, will not affect the titer.

Technique

Prepare twofold dilutions of the patient's serum as described in Table 36-1. A 2% suspension of Group O human red cells or patient's own red cells is used. The latter is prepared by adding 5 ml. of 0.85% sodium chloride solution to the clot, suspending as many cells as possible, and removing the clot. Centrifuge the suspension for 5 minutes, decant the supernate, and repeat the process until the supernate is clear. Prepare a 2% suspension by diluting 0.1 ml. of packed cells with 5 ml. of saline. Add 0.5 ml. of the suspension to each tube, including the control tube. The final serum dilutions will range from 1:10 to 1:2,560, and the final concentration of red cells is 1%. After mixing the antigen and antiserum by shaking the rack of

Table 36-1. Protocol for single agglutination test by serial dilution system

TUBE	1	2	3	4	5	6	7	8	9	10
Amount of saline (ml.)	0.8	0.5	0.5	0.5	0.5	0.5	0.5	0.5	0.5	0.5
Amount of serum (ml.)	0.2 (Mix)	0.5 of tube 1	0.5 of tube 2	0.5 of tube 3	0.5 of tube 4	0.5 of tube 5	0.5 of tube 6	0.5 of tube 7	0.5 of tube 8	(Discard 0.5 from tube 9)
Initial serum dilution	1/5	1/10	1/20	1/40	1/80	1/160	1/320	1/640	1/1280	(Control)
Amount of antigen (ml.)	0.5	0.5	0.5	0.5	0.5	0.5	0.5	0.5	0.5	0.5
Final serrrum dilution	1/10	1/20	1/40	1/80	1/160	1/320	1/640	1/1280	1/2560	—

tubes, place the tubes in a refrigerator (4° C.) overnight. Read **immediately** for agglutination after refrigeration; do not allow the tubes to stand around at room temperature before reading. The titer is the highest dilution of serum showing a 1+ (least amount of visible clumping) or greater agglutination. After reading the test, place the tubes in a 37° C. water bath for 2 hours, and reread the test; agglutination due to cold agglutinins will disappear. A titer of 1:32 or greater is considered significant, although it should be pointed out that the demonstration of a fourfold increase in titer in **paired** (acute and convalescent) sera is of greater diagnostic significance.

DETERMINATION OF ANTISTREPTOLYSIN O TITERS

A significant number of patients suffering from a recent infection with Group A streptococci develop an antibody to streptolysin O, a specific hemolysin of these strains (and an occasional Group C and G strain). The antibody, found in patients' sera, is a globulin, and it will combine with and neutralize streptolysin O in vitro, thereby inhibiting its hemolytic activity on erythrocytes. By a serial tube dilution procedure using the patient's serum and a prestandardized fixed amount of streptolysin O with a red blood cell indicator system, the level of antistreptolysin O can be measured. Since the occurrence of this an-

tibody in the patient's serum is dependent on the production of the streptolysin O by the infecting streptococcus, a **rising** antistreptolysin O (ASO) titer aids in the diagnosis of rheumatic fever, acute hemorrhagic glomerulonephritis, and other Group A streptococcal infections.

Because the reagents required for the determination of the ASO titer are readily available commercially,* accompanied by complete directions, the test will not be described in detail here. In brief, the test is set up by preparing a series of dilutions of the patient's serum to which is added a constant volume of streptolysin O reagent. After 15 to 45 minutes' incubation at 37° C., a constant volume of Group O human or rabbit erythrocytes is added to each serial dilution and the test reincubated. The last tube of the series showing **no hemolysis** is the ASO titer, which is expressed as the reciprocal of that dilution and given in Todd units. For example, if the highest dilution showing no hemolysis is 1/250, the ASO titer would be 250 Todd units. A titer of 300 Todd units is considered significant because most normal adults show titers of up to 200 Todd units. An elevated titer appears from 1 to 3 weeks after onset; a rising titer on repeated weekly specimens is of diagnostic sig-

---

*Difco Laboratories, Detroit; Baltimore Biological Laboratory, Cockeysville, Md.; Sylvana Co., Millburn, N. J.; and others.

nificance. Antibiotic and corticosteroid therapy may inhibit the production of streptolysin O.

DETERMINATION OF C-REACTIVE PROTEIN

C-reactive protein (CRP) is an abnormal alpha globulin that appears rapidly in the serum of patients who have suffered from inflammatory conditions of either infectious or noninfectious origin and is absent in serum from normal persons.

This protein has the capacity for precipitating the somatic C carbohydrate of pneumococci, and its presence was first determined by mixing patient's serum and the purified pneumococcal C-polysaccharide. It was subsequently demonstrated that by injecting animals with the C-reactive protein, a specific antibody reacting with the protein could be produced. It is this anti-CRP serum that is used as a more sensitive agent in this precipitin test. The test has proved useful in the diagnosis and prognosis of rheumatic fever, since it disappears when the inflammation subsides, reappearing only when the process becomes reactivated.

The reagents for the test, complete with directions, are readily available commercially,* and it is unnecessary, therefore, to discuss it here.

PRECIPITIN TEST ON CEREBROSPINAL FLUID

In this test, cerebrospinal fluid and a specific antibacterial serum are allowed to react in a capillary tube. This procedure has proved useful in making an etiological diagnosis of meningitis, when negative cultures are obtained. The technique recommended is that of H. D. Isenberg,[3] of the Long Island Jewish Hospital. Various commercial antisera† are individually

taken up in capillary tubes, followed by spinal fluid. The mixtures are incubated for 2 hours at 37° C., refrigerated overnight, and read. A positive test is indicated by the formation of a precipitin ring at the interface of the reactants.

SLIDE AGGLUTINATION TESTS

A simple, rapid, quantitative slide agglutination test for the detection of serum agglutinins developed during certain febrile infections has become a useful diagnostic procedure in many laboratories. A number of investigators have found this technique as informative as that obtained by the tube agglutination procedure, and may be recommended.

The antigens are standardized and include the following*:

Brucella abortus antigen
Proteus OX19 antigen
Salmonella Group A (O antigens 1, 2, 12)
Salmonella Group B (O antigens 4, 5, 12)
Salmonella Group C (C_1 and C_2) (O antigens 6, 7, 8, Vi)
Salmonella Group D (O antigens 1, 9, 12, Vi)
Salmonella Group E (E_1, E_2, E_3, E_4) (O antigens 1, 3, 15, 19, 34)
Paratyphoid A antigen (flagellar a)
Paratyphoid B antigen (flagellar b, 1, 2)
Paratyphoid C antigen (flagellar c, 1, 5)
Typhoid H antigen (flagellar d)

Technique

1. Using a glass slide 9 × 14 inches ruled in $1\frac{1}{2}$-inch squares,† deliver 0.08, 0.04, 0.02, 0.01, and 0.005 ml. quantities of serum with a 0.2-ml. pipet graduated in 10^{-3} ml. to the squares of one row, from left to right. Repeat this for as many rows as there are antigens to be used (usually six).
2. Shake the antigen vials so that the

---

*Difco Laboratories, Detroit; Baltimore Biological Laboratory, Cockeysville, Md.; Sylvana Co., Millburn, N. J.; and others.
†Difco Laboratories, Detroit; Baltimore Biological Laboratory, Cockeysville, Md.

---

*Febrile Antigens are available from Lederle Laboratories, Pearl River, N. Y.; Baltimore Biological Laboratories, Cockeysville, Md.; Hyland Laboratories, Los Angeles; and others.
†Permanently ruled glass slides are available from Arthur H. Thomas Co., Philadelphia. These are the Perma Slides, 20 ring 14 mm. I.D., No. 6690-M10, or equivalent.

contents are well mixed. By means of the standardized dropper provided with each antigen vial, deliver 1 drop of antigen (0.03 ml.) on each volume of serum in each row, from left to right. When this quantity of antigen is mixed with the quantities of serum indicated, the result will be approximately equivalent to a dilution series of: 1:20, 1:40, 1:80, 1:160, and 1:320 in a tube test using diluted antigen. Further dilutions may be prepared by using a 1:10 dilution of the serum in physiological saline solution and the volumes described in step 1.

3. Using applicator sticks or toothpicks, mix the serum and antigen, proceeding in each row from right to left to minimize the carry-over of serum from the low to the high dilutions.

4. Rotate the slide over a surface that is lighted for maximum visibility for a period of 3 minutes.

5. Read and record the degree of agglutination as follows:

4+	Complete agglutination
3+	75% agglutination
2+	50% agglutination
1+ or less	25% or less agglutination

6. The smallest quantity (highest dilution) of serum giving a 2+ agglutination is considered the end point, or **titer.** Therefore, if a serum specimen shows the following pattern:

Serum	Equivalent dilution	Antigen A	Antigen B	Antigen C
0.08 ml.	1:20	4+	4+	3+
0.04 ml.	1:40	4+	3+	2+
0.02 ml.	1:80	4+	2+	2+
0.01 ml.	1:160	2+	+/−	−
0.005 ml.	1:320	−	−	−

Report it as the following serum titers:

Antigen A = 1:160
Antigen B = 1:80
Antigen C = 1:80

Interpretation of test results is similar to that described for tube-agglutination tests, described elsewhere in this chapter. In general, agglutinin titers must be considered as **indirect** aids to clinical diagnosis rather than as specific evidence of the suspected disease. The possibility of nonspecific reactions must be considered.

TUBE AGGLUTINATION TESTS

As indicated previously, the examination of a patient's serum for the detection of agglutinins against various organisms is an important diagnostic procedure. In many infections, such as **tularemia** and **brucellosis,** the agglutination test may be the only means of laboratory diagnosis available and, as such, may be of considerable value. Also, in cases of enteric fever, this test, properly interpreted, may present valuable information concerning the etiology of the disease. Furthermore, the agglutination test may offer an explanation of a fever of unknown origin by demonstrating the presence of agglutinins, or it may rule out the existence of a suspected infection.

Since agglutinins of variable titer may occur in normal sera, a positive agglutination test on a single specimen of serum has little or no significance. In order for results to be meaningful, a **rise in titer** must be demonstrated during the course of the infection. This can be determined only by comparing the titers of two or more samples of serum, one during the acute phase and one during convalescence.

When serum is sent to the laboratory for an agglutination test, the clinician should provide the pertinent case history to give some indication to the laboratory personnel as to the suspected etiology. In cases of suspected **typhoid fever,** the serum should be set up against both the H and the O antigens of *Salmonella typhi.* Both H and O antigens should also be used for the detection of agglutinins against other salmonellae. For the serological diagnosis of **rickettsial infections,** antigens prepared from *Proteus* OX19, *Proteus* OX2, and *Proteus* OXK should be used, since the typhus fevers usually give a higher titer with *Proteus* OX19, whereas the **spotted fevers** may give a higher titer with *Proteus* OX2 antigens. *Proteus* OXK aids in the diagnosis of tsutsugamushi rickettsioses.

Preparation of antigens

Standardized suspensions of most bacteria for use as antigens are available commercially.* It has been found that a formalinized broth suspension provides a more stable H antigen than a formalinized saline suspension of cells harvested from agar.

Preparation of Salmonella H antigens. Select a motile, smooth strain of *Salmonella* that is recommended for H antigen production. Monophasic variants of diphasic types such as *Salmonella paratyphi B* and *Salmonella paratyphi C* are available.† Enhance the motility of the culture by passage on 0.3% to 0.4% agar. Small tubes of infusion broth or trypticase soy broth may be inoculated to provide young seed cultures that are then added to larger volumes of broth. Incubate the bulk suspension at room temperature for 24 to 48 hours. Add formalin (40% formaldehyde) to give a concentration of 0.3% by volume and place in a refrigerator for 48 hours. A sterility test should be performed for safety. Dilute the sterile stock suspension with a phosphate buffer solution (pH 6.8 to 7) containing 0.3% formalin to give a cell density approximately equal to No. 3 McFarland standard. The adjusted suspension should be allowed to stabilize in a refrigerator for about 2 weeks before use.

Preparation of O antigens. O antigens should be prepared for *Salmonella typhi, Salmonella paratyphi A, Salmonella paratyphi B, Salmonella paratyphi C, Proteus* OX2, *Proteus* OX19, and *Proteus* OXK for

detecting the presence of O agglutinins against these organisms. There are two accepted methods of preparing bacterial suspensions for use as O antigens.

USING A NONMOTILE STRAIN. Nonmotile strains of most of the aforementioned cultures are available.* An O antigen suspension may be prepared from such a strain by using a technique similar to that used in the preparation of H antigen but eliminating preliminary passage on semisolid medium.

USING A MOTILE STRAIN. Prepare seed cultures in broth from a smooth colony on agar. Inoculate Roux bottles or large Petri dishes containing 2% infusion or trypticase soy agar and remove surplus inoculum with a pipet. Incubate the Roux bottles or dishes in an inverted position at 36° C. for 18 to 24 hours. Harvest the growth with an appropriate amount of saline and collect in centrifuge bottles. To the concentrated cell suspension add 95% alcohol in the ratio of 4 volumes of alcohol to 1 volume of suspension. Shake in a shaking machine for 30 minutes, then place at room temperature in the dark for 48 hours. Centrifuge to pack the cells, discard the supernate, and wash twice with 0.85% saline. Resuspend the cells in phosphate buffer pH 6.8 to 7 containing 0.2% formalin by volume. This concentrate may be maintained as a stock suspension, and diluted with saline to No. 3 McFarland standard density for use when required.

Technique

Two methods of preparing serial dilutions are available. The first method is employed if a single test is required, and the protocol for this is given in Table 36-1. When there are a number of tests to be set up, such as in the Widal test for typhoid and paratyphoid fevers, the **parallel dilution system** is employed for accuracy

---

*Excellent standardized antigens are available from Central Public Health Laboratories, London.
†International Salmonella Center, Copenhagen; Center for Disease Control, Alanta, Ga.; National Salmonella Center, Ottawa; American Type Culture Collection, Rockville, Md.

---

*International Salmonella Center, Copenhagen; Center for Disease Control, Atlanta, Ga.; National Salmonella Center, Ottawa; American Type Culture Collection, Rockville, Md.

Table 36-2. Protocol for agglutination test by parallel dilution system

TUBE	1	2	3	4	5	6	7	8	9
					PREPARATION OF MASTER DILUTIONS				
Amount of saline (ml.)	8	5	5	5	5	5	5	5	5
Amount of serum (ml.)	2	(Mix and transfer 5 ml. from tube 1 to tube 2. Continue mixing and transferring 5 ml. serially through to tube 9, discarding 5 ml. from tube 9.)							
Serum dilution	1/5	1/10	1/20	1/40	1/80	1/160	1/320	1/640	1/1280
Volume of serum dilution (ml.)	5	5	5	5	5	5	5	5	5

PROCEDURE FOR PARALLEL AGGLUTINATION TESTS

(Commencing with the highest dilution in tube 9 of the master dilution series, pipette 0.5 ml. of this dilution into the corresponding agglutinating tube and proceed in reverse until tube 1 is reached, as indicated, using the same pipet.)

AGGLUTINATION TUBE	1	2	3	4	5	6	7	8	9	10
Amount of master dilution (ml.)	0.5 of dilution 1	0.5 of dilution 2	0.5 of dilution 3	0.5 of dilution 4	0.5 of dilution 5	0.5 of dilution 6	0.5 of dilution 7	0.5 of dilution 8	0.5 of dilution 9	(0.5 of saline)
Amount of antigen (ml.)	0.5	0.5	0.5	0.5	0.5	0.5	0.5	0.5	0.5	0.5
Final serum dilution	1/10	1/20	1/40	1/80	1/160	1/320	1/640	1/1280	1/2560	Control

and speed. The protocol for the latter is shown in Table 36-2.

Single serial dilution test. Set up ten clean clear agglutinating tubes (10 × 100 mm.) in a rack. Using physiological saline (0.85%) as the diluent, add to the tubes in the amounts shown in Table 36-1. Add the patient's undiluted serum with a 1-ml. serological pipet to the first tube, mix, and transfer as indicated. Discard 0.5 ml. from tube 9. Tube 10 serves as the control and contains only saline. Add 0.5 ml. of the appropriate antigen suspension to each tube of the series.

Parallel dilution method*. Set up nine clean 18- or 20-mm. test tubes in a rack. Add saline to each tube of this series, as indicated in Table 36-2. Add undiluted patient's serum to tube 1 as shown, and, after thorough mixing, serial dilution is effected by transferring 5 ml. from tube to tube until nine master dilutions have been established. With a 5-ml. pipet, transfer 0.5 ml. of the highest dilution (tube 9) to the corresponding agglutination tube in each of the test series set up in agglutinating racks. By starting with the highest dilution and working backward, the same pipet may be used throughout the procedure with little concern for carrying over any appreciable amount of antibody. With the system shown, at least nine parallel series could be established for agglutination tests with different antigens. Adjustments may be made either way in the master series to accommodate the number of agglutination tests planned. The anti-

---

*Bailey, W. R.: Unpublished data.

gen suspensions are added to the racks, respectively. One control should be set up in agglutinating tube 10 for each test, as shown in Table 36-2.

Since the antigens used in agglutination tests are subject to variations, it is recommended that **control tests,** using positive antisera of known titer, be carried out occasionally as a check on the agglutinability of the test antigens. Such tests are performed according to the technique used in the single serial dilution test.

Period of incubation

The times and temperatures used for the incubation of tube agglutination tests will vary from one laboratory to another. Each laboratory selects an adequate yet convenient system according to its particular needs, the number of antisera received daily, and the pressure of laboratory routine.

Higher temperatures hasten the reaction and allow for an earlier reading. Centrifugation, following short incubation periods, facilitates reading in some cases. Overnight refrigeration may help to clarify final readings. Incubation temperatures vary from 36° to 56° C., depending on the laboratory. Table 36-3 shows some of the recommended conditions.

Reading agglutination tests

Observe every tube for clearing of the supernatant fluid and the amount and character of the sediment and agglutinated particles. The pattern of the sediment can be more accurately observed if the tubes are held over a concave mirror. In the **control** and **negative** test, the antigen will

Table 36-3. Time and temperature of incubation for agglutination tests

ANTIGEN	TIME AND TEMPERATURE	READING
Salmonella H	1 to 2 hours at 50° C.	At end of incubation
Salmonella O	18 to 24 hours at 50° C.	At end of incubation
Proteus OX19 *Proteus* OX2	2 hours at 36°C.	After overnight refrigeration
Brucella *Francisella tularensis*	24 hours at 36°C.	At end of incubation

settle out in the bottom of the tube in a small round disc with smooth edges. In the **positive** tubes the cells will settle out in a larger area and may even extend up the sides of the tube. The pattern of this sediment will be most irregular and will vary with the extent of the agglutination. After examining the pattern of the sediment, shake the tube gently. H agglutinins produce large floccular aggregates, which are easily broken up, whereas O agglutinins produce granular or small flaky aggregates. Complete agglutination with complete clearing of the supernatant fluid indicates a 4+ reaction. Decreasing amounts of agglutination and increasing cloudiness of the supernatant fluid are read as 3+, 2+ and 1+ reactions.

Interpretation of results

It is almost impossible to assign positive or negative values to arbitrary titers in any agglutination test because of a group of variable factors, such as past history of infection, vaccination, time at which the specimen was taken, and naturally occurring agglutinins. As has been mentioned in the introduction to this section, only a **rise in titer** over a period of time is truly significant. However, the following suggestions may prove helpful in interpreting results.

Negative results. Negative results may be due to either of the following: (1) taking the sample of blood before the appearance of agglutinins in the serum or (2) incorrect diagnosis (patient not having infection for which the tests were prescribed).

Negative results are of value for comparison with titers that may possibly be obtained with later samples of serum. Any increase in titer during the course of an infection is usually significant.

Positive results. The interpretation of positive results will vary with the disease involved and is given only brief mention in this chapter.

1. In **typhoid fever:** (a) Indication of current infection—titer of 1:160 with O antigen—titer rising in subsequent sera; (b) indication of past infection, recent vaccination, or an anamnestic reaction—titer of 1:80 to 1:160 with H antigen only.
2. In **brucellosis:** A titer of 1:160 or over usually suggests infection, past or present.
3. In **tularemia:** A titer of 1:80 to 1:160 or over usually suggests definite infection.
4. In **Rocky Mountain** and **typhus fevers:** Variable results are obtained with *Proteus* OX19 and *Proteus* OX2. In general, typhus fever gives a higher titer with OX19, and Rocky Mountain spotted fever and tick bite fever may give a higher titer with OX2. In both instances, diagnostic titers are high, over 1:320, and only a rising titer is conclusive. Agglutination with *Proteus* OXK may be diagnostic for tsutsugamushi fever, but it may also be indicative of the spirochetal disease, relapsing fever.

REFERENCES

1. Bennett, C. W.: Clinical serology, rev. ed., Springfield, Ill., 1968, Charles C Thomas, Publisher.
2. Hayflick, L., and Chanock, R. M.: *Mycoplasma* species of man, Bact. Rev. **29:**185-221, 1965.
3. Isenberg, H. D.: Personal communication.
4. Mirick, G. S., et al.: Studies on a nonhemolytic streptococcus isolated from the respiratory tract of human beings, J. Exp. Med. **80:**391-406, 1944.
5. Peterson, O. L., Ham, T. H., and Finland, M.: Cold agglutinins (autohemagglutinins) in primary atypical pneumonias, Science **97:**167, 1943.

37 Fluorescence microscopy in diagnostic procedures

The fluorescent antibody (FA) method, more properly known as the **immunofluorescence method,** has reached such a level of scientific maturity that it is no longer a laboratory curiosity. It is a rapid, reproducible, and reliable aid in the hands of experienced workers for the identification of microorganisms or the antibodies they engender.

Space does not permit a complete review of the vast literature about FA; the interested reader is referred to the excellent monographs and technical manuals on the method, including those of Coons,[4] Cherry and associates,[3] Beutner,[1] and the various manufacturers of reagents.

The FA method is essentially a more sophisticated technique for **demonstrating antigen-antibody reactions.** In this technique a film preparation or tissue section is treated with an appropriate solution of an immune globulin (antibody) that has been labeled (conjugated) with a fluorescent dye (such as fluorescein isothiocyanate and others). This preparation is then examined against a dark background (dark field) lighted by a very bright light source rich in the near-ultraviolet spectrum. This light causes the antigen-antibody complex to become **fluorescent** and appear as a brightly glowing **yellow-green** object against the dark background when stained with fluorescein isothiocyanate.

The diagram in Fig. 37-1 represents the basic optical system utilized in fluorescence microscopy and consists of a high-pressure mercury vapor lamp (Osram HBO 200) that provides light of a very high intensity. This is passed through Schott BG-14 and BG-22 heat-absorbing filters and excited in wavelengths of 350 to 450 mμ by passage through a BG-12 filter, the range in which the fluorescein dye is most brilliant. Since this wavelength is in the ultraviolet spectrum, it must be removed by the insertion of a barrier filter in the eyepiece (usually Schott OG-1) for protection of the observer's eyes. This orange filter holds back wavelengths below 500 mμ and transmits visible wavelengths of light emited by the specimen. Other equivalent Corning or Wratten filter systems also may be used, and newer optics and lighting systems are available.

The choice of the particular FA technique to be used depends largely on the information desired. Generally, the **direct** staining procedure is used in identifying an unknown antigen, such as Group A streptococci in throat swabs. The **indirect** method is employed chiefly in the detection of antibody, as in the fluorescent treponemal antibody (FTA-ABS) test for syphilis. Other uses of these basic principles are described in the references previously cited.

To date, a number of FA diagnostic procedures have been tested and compared, and thus a number of standardized commercially prepared conjugates have be-

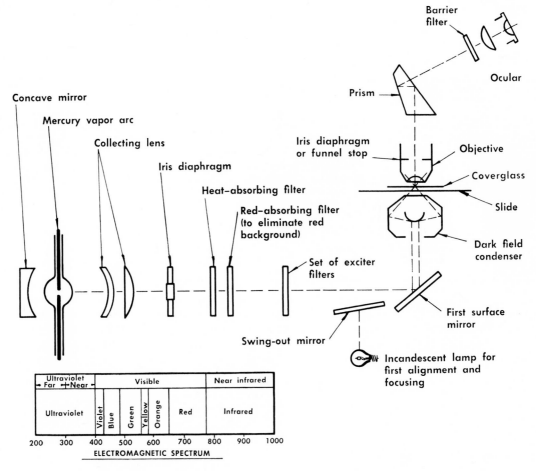

Fig. 37-1. Schematic representation of equipment for fluorescence microscopy. (Prepared with suggestions by Dr. Peter Bartals, E. Leitz, Inc., New York.)

come available.* These procedures are in regular use in many public health and hospital microbiology laboratories at the present time and include the following: rapid identification of Group A streptococci; the rapid and specific diagnosis of rabies in tissue; identification of enteroviruses and respiroviruses in isolates and clinical material, herpes simplex, arboviruses, rickettsiae, and so forth; identification of enteropathogenic *Escherichia coli* in fecal material; the highly specific FTA-ABS test

for syphilis; the rapid identification of *Neisseria gonorrhoeae* in exudates; and other tests mentioned in various sections of this text. For a critical review with an extensive bibliography of the FA techniques used in diagnostic bacteriology, the reader is referred to the excellent monograph by Cherry and Moody.[2]

A list of FA tests useful in diagnostic bacteriology and mycology are presented in Tables 37-1 and 37-2.

There are numerous pitfalls in the application of immunofluorescence to diagnostic bacteriology; the procedures must be carried out by workers well trained in a meticulous technique and with a background of methodology in serology and

*Baltimore Biological Laboratory, Cockeysville, Md.; Difco Laboratories, Detroit; Sylvana Co., Millburn, N.J.; Clinical Sciences, Inc., Whippany, N.J.; and others.

Table 37-1. List of fluorescent antibody tests most useful in diagnostic microbiology*

TEST	STATUS
I. Bacteriology	
A. Group A streptococci. Direct test.	Most highly evaluated and massively use of all FA tests.
B. Serogrouping of enteropathogenic *Escherichia coli* (EEC). Direct test.	Highly effective for rapid screening of specimens in children up to 2 years of age. Eliminates necessity of culturing negatives.
C. Fluorescent treponemal antibody test (FTA-ABS). Indirect test.	Highly specific and sensitive for detection of true antibody to *T. pallidum*. Replacing treponemal immobilization test.
D. Identification of the gonococcus. Direct test. (An indirect FA technique has been developed for detecting gonococcal antibodies in the sera of females with uncomplicated disease.[7] It shows promise.)	A good test. Receiving widespread evaluation and showing great promise for facilitating diagnosis of gonorrhea (p. 398, No. 7).
E. Identification of the diphtheria bacillus. Direct test.	Good for use on nasopharyngeal specimens from patients suspected of having diphtheria. Not recommended for carrier surveys, contacts, and so forth. Specificity needs improvement.
F. *Bordetella pertussis.* Direct test.	A rapid, specific, and sensitive test as compared with conventional culture techniques.

G. Incitants of bacterial meningitis. Direct test on spinal fluids
    1. *Haemophilus influenzae*
    2. *Neisseria meningitidis*
    3. *Streptococcus pneumoniae*
	A rapid, highly efficient test to apply to CSF sediments. Good coverage of pneumococci; laborious due to large number of type conjugates required. No problem with *H. influenzae* or with meningococci.

H. *Listeria monocytogenes.* Direct test.	Excellent for specific detection of organisms in formalin-fixed, paraffin embedded sections, CSF, impression smears of tissue, and so forth.
I. *Brucella* (three species.) Direct test.	Sensitive, rapid, and genus specific for detection of smooth strains of *Brucella* in tissues. No advantage over agglutination tests for detection of antibody.
J. *Yersinia pestis.* Direct test.	Excellent for rapid diagnosis of plague when used on blood, stomach contents of vectors, stored animal tissues, and phagocytic cell preparations. Direct or indirect staining of impression smears, frozen or freeze-dried material may be used. FA is method of choice for rapid and specific identification of plague bacillus in tissues and in culture.
K. *Francisella tularensis.* Tularemia. Direct test.	Rapid, sensitive, and specific for detection of organisms in culture, impression smears, frozen sections, formalin-fixed, paraffin embedded sections of specimens from human cases, experimental animals, or from air.
L. *Bacillus anthracis.* Direct test.	Rapid and sensitive for detection of anthrax organisms in impression smears or formalin-fixed, paraffin-embedded tissue, also from human cutaneous vesicles, and tissues of experimental or naturally infected animals. Test not entirely specific.
M. Leptospirae.	Has been used successfully to detect leptospirae in urine sediments, frozen sections, impression smears, and in formalin-fixed frozen sections. Not reliable for typing.
N. *Salmonella typhi.* Direct test. Using sorbed Vi conjugate.	Approximately the same sensitivity and specificity as culture on bismuth sulfite agar but very rapid. Excellent when used for detection of chronic typhoid carriers. Culture is more reliable for use on cases and convalescents. *Continued.*

*Courtesy William B. Cherry, Chief, Bacterial Chemistry Unit, Bacteriology Section. From the Laboratory Program, Center for Disease Control, Atlanta, Ga.

Table 37-1. List of fluorescent antibody tests most useful in diagnostic microbiology — cont'd

TEST	STATUS
Bacteriology — cont'd	
O. *Shigella*. Direct test. *S. sonnei* *S. flexneri*	Sensitive and specific for screening fecal smears. Well evaluated. Appears promising for screening of fecal smears. Needs further evaluation.
P. Clostridia. Direct test.	Toxigenic types of *C. botulinum* can be differentiated with absorbed conjugates. Staining of smears from tissues or body fluids is very helpful in diagnosis of animal diseases due to clostridia.
Q. *Treponema pallidum*. Direct test.	Procedure may be applied to specimens from lesions transported either in capillary tubes or on dried slides and mailed to the laboratory[6].

Table 37-2. Value and limitations of fluorescent antibody reagents in medical mycology*

CONJUGATE FOR	PROCEDURE	STAINING REACTION WITH FUNGI IN		
		CULTURES	TISSUE SMEARS, PUS, EXUDATES	TISSUE SECTIONS
Actinomyces israelii	Direct staining	Specific identification	Specific identification	Not developed
Actinomyces naeslundii	Direct staining	Specific identification	Specific identification	Not developed
Arachnia (Actinomyces) propionica	Direct staining	Specific identification	Specific identification	Specific identification
Aspergillus species	Direct staining	—	Conjugate available for differentiating *Aspergillus* sp. from *Candida* sp. and phycomycetes[5]	Genus identification[5]
Blastomyces dermatitidis	Direct staining	Specific identification of yeast-form; no value for identification of mycelial form	Specific (tissue form) identification	Specific identification
Candida species	Direct staining	No specific reactivity for any *Candida* sp. but a good rapid screening agent		
Coccidioides immitis, tissue form	Direct staining	No value for identification of mycelial form	Specific identification	Specific identification
Cryptococcus neoformans	Direct staining	Specific identification; does not stain some strains	Specific identification	Specific identification
Histoplasma capsulatum	Direct staining in conjunction with *B. dermatitidis* conjugate	Specific identification; no value for identification of mycelial form	Specific identification	Not optimal
Sporotrichum schenckii	Direct staining	Specific identification of yeast-form	Specific identification	Specific identification
Trichophyton sp.	Direct staining	Differentiation of *T. mentagrophytes* from *T. rubrum* only.		

*Courtesy Leo Kaufman, Chief, Fungus Immunology Unit, Mycology Section, Center for Disease Control, Atlanta, Ga.

microscopy. The procedures must be monitored at all times by the inclusion of positive and negative controls; reagents must be standardized and their specificity must be checked regularly. In short, the FA method should not be used exclusively as a substitute for the conventional cultural procedures; rather, it should be utilized as an adjunct to these tests. Its great potential in the field of diagnostic microbiology has not yet been fully realized.

REFERENCES

1. Beutner, E. H.: Immunofluorescent staining; the fluorescent antibody method, Bact. Rev. **25**:49-76, 1961.
2. Cherry, W. B., and Moody, M.D.: Fluorescent antibody techniques in diagnostic bacteriology, Bact. Rev. **29**:222-250, 1965.
3. Cherry, W. B., Goldman, M., Carski, T. R., and Moody, M.D.: Fluorescent antibody techniques in the diagnosis of communicable diseases, Public Health Service Pub. No. 729, Washington, D. C., 1960, U. S. Government Printing Office.
4. Coons, A. H.: The diagnostic application of fluorescent antibodies, Schweiz. Path. Bakt. **22**:700-723, 1959.
5. Kaplan, W.: Direct fluorescent antibody tests for the diagnosis of mycotic diseases, Ann. Clin. Lab. Sci. **3**:25-29, 1973.
6. Kellogg, Jr., D. S.: The detection of Treponema pallidum by a rapid, direct fluorescent antibody darkfield (DFATP) procedure, Health Lab. Sci. **7**:34-41, 1970.
7. Welch, B. G., and O'Reilly, R. J.: An indirect fluorescent technique for study of uncomplicated gonorrhea, I. Methodology, J. Infect. Dis. **127**:69-76, 1973.

PART TEN
Quality control

38 Quality control in the microbiology laboratory

Although the concept of quality control has come only recently to the clinical microbiology laboratory, this is no excuse for its absence. In the present and continuing reexamination of health care delivery by the medical and allied health professions, it is incumbent upon the microbiologist to ensure that his laboratory contributes to the highest quality of patient care by its expertise, its accuracy, and the prompt reporting of its findings to the attending physician. It is only by constant self-evaluation of the laboratory's performance through a quality control program that such a level of excellence can be developed and maintained.

It will not be the function of this section to detail the methodology for carrying out a quality control program—these are presently available in excellent texts and publications by Bartlett[2,3] and others [6,7,8]—but to suggest a basic format that can be tailored to individual situations or requirements.

CONTROL OF EQUIPMENT

Generally, equipment is the simplest control to monitor and consists chiefly of the daily, weekly, or monthly observation and recording of the temperatures of incubators (optimal, 35° C.), refrigerators, water baths, hot air ovens, freezers, autoclaves (by temperature-time cycle charts), and so forth. A range of $\pm$ 1° C. is generally acceptable for heating devices, with a greater range acceptable for refrigerating devices; both should be measured with thermometers that have been calibrated against a Bureau of Standards instrument.* Autoclaves should also be checked at monthly intervals with biological devices such as the Kilit or Attest spore test ampules or strips.†

In addition to their temperature recordings, carbon dioxide incubators should also have their CO_2 concentration measured by the portable Fyrite CO_2 gas analyzer,‡ which determines the CO_2 level by absorption with change in level of the absorbent in a graduated tube. Tank pressure also must be monitored, and an extra tank of gas should be kept on hand at all times.

Every laboratory handling material for mycobacteriological examination should possess a **biological safety hood,** which also requires regular recording of the negative pressure within the hood, the interval between the HEPA filter changes, the effective output of the UV Sterilamp (measured by a photometric cell§), and so forth. Directions for these tests are includ-

---

*Thermometers should be in tubes of water wherever possible.

†Available from Baltimore Biological Laboratory, Cockeysville, Md.

‡Arthur H. Thomas, Co., Philadelphia, catalog No. 5566-C 10.

§Westinghouse SM-600 meter, from Westinghouse Electric Corp., Lamp Division, Bloomfield, N. J.

ed in the hood operation manual, or from the Biological Hazards Control Officer of the Center for Disease Control (CDC), Atlanta, Georgia.

Other miscellaneous items requiring inspection, calibration, or replacement include the standard (0.01 ml.) inoculating loops, chipped or cracked glassware, and the presence of residual detergent in them; mechanical devices, such as centrifuges, vacuum pumps, or pipetting machines require frequent oiling and adjustments. Microscopes and balances should be cleaned an a semiannual schedule; pH meters and other measuring devices such as thermometers, graduated cylinders, pipets, and so forth also should be calibrated periodically.

CONTROL OF CULTURE MEDIA

Most clinical bacteriology laboratories today utilize ready-made, commercially available solid and fluid culture media. These are generally of uniformly high quality and good batch-to-batch consistency, and are readily obtainable. If, on the other hand, a laboratorian has the good fortune to include on his staff an experienced and dedicated media chef, he is assured of a product that is superior in many ways to that available in the marketplace. In this situation, however, strict attention to details of preparation must be adhered to, such as following exact directions of the manufacturer regarding methods of preparation and sterilization (overheating is the most frequent source of error), properly sealing and dating dehydrated media, and so forth. With certain media, such as decarboxylase media or Mueller-Hinton agar, it is also necessary to check the pH of the final product.

The foregoing obviously implies the use of a quality control program that is best carried out by **performance testing** for the desired reaction with a number of stock cultures of microorganisms of known stability. At the Wilmington Medical Center, a culture collection of 26 strains (identified only by number) are dispersed in groups at weekly or monthly intervals to the several hospital laboratories comprising the microbiology section. The technologists are then requested to carry out the reactions indicated on a file card accompanying the cultures (submitted on half sections of a blood agar plate) and return the results along with the original plate to the quality control laboratory, where the results are recorded in a continuous master log book. Any discrepancies noted between expected and submitted results are then rechecked by the quality control technologist, using the original plate cultures and a sample of the medium from the central media kitchen. After this retesting, a follow-up evaluation is held with members of the staff, with a view toward correction of the error, either in performance of the medium or in the method of testing (usually the latter is at fault).

The four sets of stock cultures (A, B, C, D), referred to in Table 38-1, are submitted on a scheduled basis, designed to avoid overburdening the technologists during a work period. The stock cultures are preserved in the frozen state at approximately −70° C. on glass beads, according to the method of Nagel and Kunz[5], where they remain viable for many months. When required, a single glass bead is removed from its reservoir and placed in a tube of trypticase soy broth and incubated until growth is visible. This is then streaked on one half a blood agar plate to obtain isolated colonies, and subsequently submitted for media evaluation, as indicated.

CONTROL OF REAGENTS AND ANTISERA

The various chemical reagents used in biochemical tests are best monitored, after preparation, by integrating into the daily laboratory workload through the use of a master schedule calendar.[8] Reagents should be dated on preparation and stored in light-proof, tightly stoppered bottles either at room temperature or in the refrigerator.

The solutions are checked by testing each new batch for the expected reactions, using known positive and negative culture

Table 38-1. Recommended cultures for monitoring routine media*

MEDIUM	POSITIVE (OR GROWTH)	NEGATIVE (OR NO GROWTH)
	SET A (MONTHLY)	
Lysine decarboxylase	*Edwardsiella tarda*	*Citrobacter freundii*
Arginine dihydrolase	*Enterobacter cloacae*	*Proteus vulgaris*
Ornithine decarboxylase	*Enterobacter aerogenes*	*Citrobacter freundii*
Indole	*Escherichia coli*	*Enterobacter cloacae*
Methyl red	*Escherichia coli*	*Enterobacter cloacae*
VP reaction	*Enterobacter cloacae*	*Escherichia coli*
Citrate (Simmond)	*Enterobacter aerogenes*	*Edwardsiella tarda*
Phenylalamine deaminase	*Proteus rettgeri*	*Citrobacter freundii*
Glucose OF		
Oxidative	*Pseudomonas aeruginosa*	
Fermentative	*Escherichia coli*	
Nonfermentative	*Acinetobacter lwoffii*†	
Cetrimide agar	*Pseudomonas aeruginosa* (pigment)	*Acinetobacter lwoffii*
Nitrate reduction	*Pseudomonas aeruginosa*	*Acinetobacter lwoffii*
10% lactose	*Acinetobacter anitratum*†	*Pseudomonas aeruginosa*
Oxidase	*Pseudomonas aeruginosa*	*Acinetobacter lwoffii*
SF broth	*Streptococcus faecalis*	*Streptococcus bovis*
Bile esculin agar	*Streptococcus faecalis*	*Streptococcus pyogenes*
Catalase	*Staphylococcus aureus*	*Streptococcus pyogenes*
	SET B (MONTHLY)	
Dextrose ascitic fluid, semisolid	*Haemophilus influenzae* *Neisseria meningitidis*	
DNase	*Serratia marcescens*	*Staphylococcus epidermidis*
KCN broth	*Enterobacter aerogenes*	*Edwardsiella tarda*
Malonate	*Enterobacter aerogenes*	*Edwardsiella tarda*
Urea agar	*Proteus rettgeri* (rapid) *Klebsiella pneumoniae* (slow)	*Escherichia coli*
Motility	*Escherichia coli*	*Klebsiella pneumoniae*
	SET C (WEEKLY)	
Chocolate agar	*Haemophilus influenzae* *Neisseria meningitidis*	
Thayer-Martin agar	*Neisseria gonorrhoeae* *Neisseria meningitidis*	*Escherichia coli*
CNA agar	*Streptococcus pyogenes*	*Proteus vulgaris*
Carbohydrate broth	As indicated by carbohydrate used	
SS agar	*Salmonella typhimurium*	*Escherichia coli*
TSI	*Salmonella typhimurium* (K/AG, H_2S^+) *Shigella sonnei* (K/A, H_2S^-)	*Escherichia coli* (A/AG, H_2S^-)
	SET D (WEEKLY)	
EMB agar	*Escherichia coli* (lactose$^+$) *Salmonella typhimurium* (lactose$^-$)	*Staphylococcus aureus*
XLD agar, Hektoen agar	*Escherichia coli* (lactose$^+$) *Salmonella typhimurium* (lactose$^-$)	
SS agar	*Salmonella typhimurium*	*Escherichia coli*
Sheep blood agar (hemolysis)	*Streptococcus pyogenes* (beta) *Streptococcus salivarius* (alpha)	
Sheep blood agar (disc)	*Streptococcus pyogenes* (bacitracin-susc.) *Streptococcus* Group B (bacitracin-resist.)	
Sheep blood agar (disc)	*Streptococcus pneumoniae* (optochin-susc.) *Streptococcus salivarius* (optochin-resist.)	
Thioglycollate medium	*Bacteroides fragilis*	
Sabouraud	*Candida albicans*	
TB media	*Mycobacterium tuberculosis*	

*These stable stocks have been collected from the Center for Disease Control reference strains, from the American Type Culture Collection, Rockville, Md. (obtained from Bact-Chek, Roche Diagnostics, New York) and from patient isolants confirmed by the Center for Disease Control or the Wilmington Medical Center, Wilmington, Del.

†See footnote, p.165.

controls. These would include tests for indole and acetylmethylcarbinol, methyl red reaction, the deamination of phenylalanine, reduction of nitrate, the oxidase reaction, ONPG tablets, the rapid test strips, and so forth.

Gram-staining reagents are best monitored by staining and examining slides prepared from a suspension of *E. coli* and *S. aureus* that has been grown in broth and sterilized by autoclaving; this procedure should be a part of each worker's **daily** staining schedule. Acid-fast staining reagents can be similarly checked with a suspension of killed mycobacteria.

Antisera that have been obtained from reliable sources are dated on receipt and diluted to the manufacturer's specifications. They are then tested with standard stock cultures for sensitivity and specificity on each new lot, and every 3 months thereafter. The manufacturer's directions for storage and test performance likewise should be explicitly followed. In a similar manner, bacterial antigens for agglutination tests should be monitored with known positive and negative antisera, if available.

PROCEDURE BOOK

The procedure book delineates the current standard procedures employed in the microbiology laboratory and should be comprehensive enough to include all techniques from the simple preparation of reagents to the methods for identifying isolates of the common genera. It should be easily interpreted by the least experienced bench worker and should not consist of lengthy descriptions more readily found in a standard microbiology test. It should be reviewed at regular intervals, with obsolete material deleted and new methods inserted, and it should be the **working reference standard** for the entire staff.

QUALITY CONTROL IN SPECIMEN COLLECTION

Until recently, one of the least monitored areas of quality control in the microbiology laboratory has been in the quality of the clinical specimen itself. Is it representative of the suspected infection site? Has it been collected correctly in a proper container? Was it submitted promptly and cultured without undue delay? These are questions that must be answered positively, so that the laboratory results are meaningful and not misleading to the attending physician.

The microbiologist tends to be too lenient in his acceptance of improperly collected specimens, particularly those from the respiratory or urinary tract, or from grossly infected ulcers and wounds. Too frequently treatment has been initiated on the basis of a report of potential pathogens, including their antibiograms, when in reality the isolates represent only colonization of the patient's oropharynx or superficial skin debris.

Recently, Bartlett,[1] at the Hartford Hospital, has instituted a program for evaluating the quality of the sputum sample based on an examination of the Gram stain of a direct smear. The decision to culture the specimen is based on the numbers of squamous epithelial cells (minus values) and mucous and pus cells (plus values) observed under $10\times$ magnification. If such a specimen appears to be of oropharyngeal origin, a repeat specimen is requested immediately. Along with these and other requirements, Bartlett has been able to reduce unnecessary speciation and antibiotic susceptibility testing of sputum cultures by more than 25%, with a resultant savings in labor and a reduction in noncontributory or misleading findings.

While it is well recognized that a count of 10^5 bacteria or greater in a clean-voided urine specimen promptly cultured is associated with urinary tract infection, it should also be noted that counts in the range of 10^3 to 10^5 per ml. also may represent infection, especially in patients under therapy; such isolates should be speciated and their antibiotic susceptibilities determined. Contaminated specimens usually can be detected when 3 or more bacterial species are present in numbers greater than 10^3 per ml., and a repeat specimen should be requested under these circumstances.

Reference has already been made to the prime importance of proper specimen collection in the isolation of anaerobic microorganisms (see Chapter 13). The use of "gassed out" tubes, prereduced media, and proper anaerobic techniques is directly related to the successful isolation of these bacteria.

MISCELLANEOUS MEASURES

The role of "blind unknowns," that is, simulated clinical specimens (urine, wound, fluids, and so forth) seeded with known bacteria, is well established as an internal quality control procedure. Two such specimens may be introduced each week into the regular laboratory routine and are identified in such a way that the worker is not aware of the unknown, or more important, that a fictitious report is not routed to the ward or physician indicated on the laboratory request slip. When the report is completed, the microbiologist discusses the results with the technologist performing the culture, with recommendations for improvement, if indicated. Quality control procedures in antibiotic susceptibility testing has already been described (Chapter 34).

Proficiency testing specimens, such as those disseminated from state laboratories, the CDC, the College of American Pathologists, and so forth, are not as valuable as quality control measuring devices as are "blind unknowns," in that realistic laboratory conditions are not applicable. The worker is aware of the source, there is usually a prolonged time limit, and the specimen is given a laboratory workup exceeding that normally carried out. However, the proficiency test unknowns can serve an exceedingly useful purpose as an ongoing educational program for the staff, particularly when accompanied by the excellent critiques published regularly by the Proficiency Testing Section, Licensure and Development Branch, Center for Disease Control.[4]

Other miscellaneous methods of quality control are self-evident: an ongoing educational program within the laboratory; weekly departmental conferences with the microbiologist regarding problems, changes, or personnel relations; encouragement and provision for staff members to attend medical conferences, workshops, and seminars; and so forth. It is also necessary to constantly review and correct the performances of the night and weekend staff, particularly in plating procedures, relatively simple identification techniques, and especially in the carrying out and proper interpretation of a "stat" Gram stain.

In summary, good quality control resides essentially in the conscience of the individual worker, and it is the duty of the microbiologist, by his own leadership and knowledge, to encourage and nourish this desire for superiority and service at all times.

REFERENCES

1. Bartlett, R. C.: Personal communication (E. G. S.), 1973.
2. Bartlett, R. C.: Medical microbiology; quality, cost and clinical relevance, New York, 1974, Wiley-Interscience, Division of John Wiley & Sons, Inc.
3. Bartlett, R. C., Irving, W. R., Jr., and Rutz, C.: Quality control in clinical microbiology, Chicago, 1968, Amer. Soc. Clin. Path., Commiss. Contin. Educ.
4. Hall, C. T., and Webb, C. D., Jr.: Proficiency testing—Bacteriology III and IV (July and Oct.) 1972, Atlanta, Ga., 1973, Center for Disease Control.
5. Nagel, J. G., and Kunz, L. J.: Simplified storage and retrieval of stock cultures, Appl. Microbiol. **23:**837-839, 1972.
6. Russell, R. L., Yoshimori, M. A., Rhodes, R. F., Reynolds, J. W., and Jennings, E. R.: A quality control program for clinical microbiology, Amer. J. Clin. Path. **39:**489-494, 1969.
7. Vera, H. D.: Quality control in diagnostic microbiology, Health Lab. Sci. **8:**176-189, 1971.
8. Woods, D., and Byers, J. F.: Quality control recording methods in microbiology, Amer. J. Med. Technol. **30:**79-85, 1973.

PART ELEVEN

Culture media, stains, reagents, and tests

39 Formulae and preparation of culture media

The principles governing the selection, preparation, sterilization, and storage of laboratory culture media have been discussed in Chapter 1; thus the ensuing chapter will include the media to which reference is made throughout the text, their formulae, and any specific directions that may be needed. Complete directions for the preparation of media from the dehydrated products obtained commercially are adequately described by the manufacturers.

1. Beef extract broth*

Beef extract	3 gm.
Peptone	10 gm.
Sodium chloride	5 gm.
Distilled water	1,000 ml.

Dissolve ingredients by boiling, adjust to pH 7.2, tube in 10-ml. amounts in 18 × 150 mm. tubes, and autoclave at 121° C. for 15 minutes. For use as a fermentation base broth, add 100 ml. of a 0.04% aqueous solution of bromthymol blue before tubing and sterilizing.

The medium is used for differentiation of *Candida* species by carbohydrate fermentation tests.

2. Beef extract agar*

Add 2.5% agar to beef extract broth; dissolve by boiling; adjust to pH 7.6; tube and autoclave at 121° C. for 15 minutes; and slant before solidifying.

The medium is used for pure culture preparation of *Candida* species prior to carrying out fermentation tests.

3. Bile esculin agar*

Beef extract		3.0 gm.
Peptone		5.0 gm.
Oxgall		40.0 gm.
Esculin		1.0 gm.
Ferric citrate		0.5 gm.
Agar		15.0 gm.
Distilled water	1,000	ml.

a. Suspend 64 gm. of the dehydrated medium in water, heat to boiling to dissolve, and tube in screw-capped tubes.

b. Sterilize by autoclaving at 121° C. for 15 minutes.

c. Cool to 55° C. and add aseptically 50 ml. of filter-sterilized horse serum (optional). Mix well.

d. Dispense in sterile tubes and cool in slanted position.

This medium is useful in the selective detection of Group D streptococci. After inoculation, these organisms will form brownish black colonies, surrounded by a black zone. *Listeria monocytogenes* also reacts positively, as do some other organisms. Other forms of agar and broth media also are available.*

---

*Baltimore Biological Laboratory, Cockeysville, Md.; Difco Laboratories, Detroit.

*Pfizer Diagnostics, Flushing, N. Y.; Difco Laboratories, Detroit; Baltimore Biological Laboratory, Cockeysville, Md.

4. Bismuth sulfite agar*

Beef extract	5	gm.
Peptone	10	gm.
Dextrose	5	gm.
Disodium phosphate	4	gm.
Ferrous sulfate	0.3	gm.
Bismuth sulfite indicator	8	gm.
Agar	20	gm.
Brilliant green	0.025	gm.
Distilled water	1,000	ml.

Bismuth sulfite agar medium is highly recommended for the isolation of *Salmonella typhi*. It may be used for both streak and pour plates. **Read's modification** of the medium (through the addition of 1% sodium chloride and 1% mannose, the elimination of the brilliant green, and adjustment of the final pH to 9.2) is recommended for the isolation of *Vibrio cholerae*.

5. Blood agar plates for streaking

Blood agar base medium	500 ml.
Sterile blood	20 ml.

Sterilize the base medium (trypticase soy agar [BBL] or tryptic soy agar [Difco] are recommended). Cool to 48° or 50° C. and add 5% to 10% sterile defibrinated sheep, horse, or rabbit blood, aseptically. Rotate to mix thoroughly and pour into sterile Petri dishes in approximately 12-ml. amounts. When agar is set, invert and incubate 1 or 2 plates overnight to test for sterility. Discard the test plates. These plates may be stored for a week at refrigeration temperature without deterioration, but they should be packaged to minimize water loss, which could amount to 7% per week if unprotected.

The pouring should be done carefully to avoid air bubbles. If these appear in the poured plate, pass a Bunsen flame over the agar before it sets. This will break the bubbles. An alternate method for avoiding air bubbles, if automated pouring equipment† is not available, is as follows: prepare the agar base in an Erlenmeyer

flask and cover the mouth of the flask, first with a single layer of cheesecloth, which may be secured with a rubber band, and then with a single sheet of fine filter paper, which is also secured with a rubber band. Trim with scissors. Autoclave the flask and contents, cool to the desired temperature, paint the filter paper with tincture of iodine, and add the blood aseptically with a needle and syringe by injecting through the paper and cheesecloth. Mix well, remove and discard the filter paper, and pour the medium **through** the cheesecloth.

6. Blood agar pour plates
(Brown: Monograph No. 9, The Rockefeller Institute for Medical Research, 1919)

Heat 12-ml. tubes of infusion agar in boiling water until the agar is thoroughly melted. Cool the agar to 48° C. by standing the tubes in a container of warm water or a 48° C. water bath. With a sterile pipet add about 0.7 to 1 ml. of sterile blood to each tube. Inoculate the fluid blood agar with a proper dilution of culture or the original material. Twirl the tube to mix the blood, inoculum, and agar, taking care not to form bubbles, and pour into a sterile Petri dish.

7. Blood cystine dextrose agar*
(Francis: J.A.M.A. **91:**1155, 1928; Rhamy: Amer. J. Clin. Path. **3:**121, 1933)

Beef heart infusion	500 gm.
Proteose peptone	10 gm.
Glucose	10 gm.
Sodium chloride	5 gm.
Cystine	1 gm.
Agar	15 gm.
Distilled water	1,000 ml.

Blood cystine dextrose agar medium may be used satisfactorily for the cultivation of *Francisella tularensis*.

To prepare this medium for the cultivation of *Fran. tularensis*, dissolve 16.8 gm. of the dehydrated product in 300 ml. of

---

*Baltimore Biological Laboratory, Cockeysville, Md.; Difco Laboratories, Detroit.
†Brewer pipetting machine, BBL.

*Available in dehydrated form by Difco Laboratories, Detroit, and Baltimore Biological Laboratory, Cockeysville, Md.

distilled water. Adjust the reaction to pH 7.3 and autoclave for 20 minutes at 15 pounds' pressure. Cool to a temperature of 60° to 70° C., and add 18 ml. of whole rabbit blood or dehydrated hemoglobin, mix well, and distribute into test tubes aseptically. Cool in slanting position.

8. Bordet-Gengou medium*

Potatoes, infusion from	125	gm.
Sodium chloride	5.5	gm.
Agar	20	gm.
Distilled water	1,000	ml.

Sterilize at 15 pounds for 15 minutes. Cool to 50° C. and add 15% to 20% sterile sheep, rabbit, or human blood aseptically.

The medium is recommended for cultivation of *Bordetella pertussis.*

9. Brain-heart infusion blood agar

Brain-heart infusion agar†	26	gm.
Agar, powdered	2.5	gm.
Distilled water	500	ml.
(Final pH approximately 7.4)		

Suspend the ingredients in the water and dissolve by boiling. Autoclave for 15 minutes at 121° C. Cool to about 45° C. and add aseptically 30 ml. of defibrinated animal blood. Mix well and distribute aseptically in approximately 20-ml. amounts in cotton-plugged and sterilized 25 × 150 mm. Pyrex test tubes (without lips). Slant, allow to harden, and store in a refrigerator.

If antibiotics are to be added, this may be done after heating the medium and before autoclaving. The following amounts are recommended (Center for Disease Control):

Cycloheximide‡ (0.5 mg. per milliliter)	250 mg.
Chloramphenicol§ (0.05 mg. per milliliter)	25 mg.

---

*Available as dehydrated base from Difco Laboratories, Detroit; Baltimore Biological Laboratory, Cockeysville, Md.

†Baltimore Biological Laboratory, Cockeysville, Md.; Difco Laboratories, Detroit.

‡Cycloheximide is available in 4-gm. amounts from Upjohn Co., Kalamazoo, Mich., as Acti-dione.

§Chloramphenicol is available from Parke-Davis & Co., Detroit, as Chloromycetin.

Add the chloramphenicol, dissolved in 5 ml. of 95% ethanol, and the cycloheximide, dissolved in 5 ml. of acetone. Mix well and autoclave at 118° C. for not longer than 10 minutes. Cool to approximately 45° C., add the blood, and dispense as indicated above.

This medium is used for the isolation of fastidious fungi.

10. Brain-heart infusion broth*

Calf brains, infusion	200	gm.
Beef heart, infusion	250	gm.
Proteose peptone	10	gm.
Dextrose	2	gm.
Sodium chloride	5	gm.
Disodium phosphate	2.5	gm.
Distilled water	1,000	ml.
(Reaction of medium pH 7.4)		

Recommended for cultivating the pneumococcus for the bile solubility test.

11. Brilliant green agar*

Yeast extract	3	gm.
Proteose peptone No. 3	10	gm.
Sodium chloride	5	gm.
Lactose	10	gm.
Saccharose	10	gm.
Phenol red	0.08	gm.
Brilliant green	0.0125	gm.
Agar	20	gm.
Distilled water	1,000	ml.
(Final pH 6.9)		

Brilliant green agar is a selective medium recommended for the isolation of the salmonellae other than *Salmonella typhi*. It is not recommended for the isolation of the shigellae.

12. Brucella-menadione blood agar

a. Menadione solution:

Add 2 ml. of menadione† (Synkavite-Roche, 10 mg./ml.) to 388 ml. of sterile distilled water and store at 4° C. Filter sterilize as needed.

b. Preparation of base:

---

*Available in dehydrated form from Difco Laboratories, Detroit, and Baltimore Biological Laboratory, Cockeysville, Md.

†Some workers prefer vitamin K_1 (10 μg./ml. final concentration in medium).

Brucella agar	43 gm.
Agar	2.5 gm.
Distilled water	1,000 ml.

Sterilize by autoclaving at 121° C. for 15 minutes, cool to 50° C.

c. Add aseptically:

| Defibrinated or laked sheep blood | 50 ml. |
| Menadione solution | 10 ml. |

Mix well and pour plates, using approximately 20 ml. per plate. Store at room temperature.

The medium is used for the selective isolation of fastidious anaerobes.

13. Carbohydrate media for fermentation tests

When some of the carbohydrates are sterilized by heating in alkaline broth, they are more or less broken down into simpler carbohydrates. It has been found definitely advantageous by many investigators to sterilize sugar solutions by filtration through Seitz or membrane filters and to add these aseptically to the broth base in the required amounts. Although several carbohydrates can stand autoclave temperature and pressure, one is advised to sterilize the following by **filtration:** xylose, lactose, sucrose, arabinose, trehalose, rhamnose, and salicin. These may be prepared as 5% or 10% solutions, depending on solubility, then sterilized by filtration, and added aseptically to the base containing an indicator, to give 0.5% to 1% final concentration.

Other less heat-susceptible carbohydrates may be added to the broth base containing indicator (bromcresol purple, for example) before autoclaving at 116° to 118° C. (10 to 12 pounds' pressure) for 15 minutes. Ten Broeck, according to a report made in 1920, found that unheated serum contains an enzyme that hydrolyzes maltose to glucose. Serum to be added to maltose broth should therefore be heated for 1 hour at 60° C. to inactivate the enzyme. Incubate the carbohydrate broth to test the sterility.

14. Casein medium for separation of Nocardia and Streptomyces
(Center for Disease Control)

Prepare separately:

a. Skimmed milk
| (dehydrated or instant nonfat milk) | 10 gm. |
| Distilled water | 100 ml. |

Autoclave at 121° C. for 20 minutes.

b.
| Distilled water | 100 ml. |
| Agar | 2 gm. |

Autoclave as above.

Cool both solutions to approximately 45° C., mix, and pour into sterile Petri dishes.

Test for hydrolysis. Streak or make point inoculations of each culture on casein plates, using half a plate for each organism. Incubate at 25° C. (or 36° C. if it does not grow at room temperature). Observe for clearing of casein in 7 and 14 days.

Nocardia asteroides does not hydrolyze casein.

Nocardia brasiliensis and *Streptomyces* species hydrolyze casein.

15. Cetrimide agar*
(Lowbury and Collins: J. Clin. Path. **8:**47, 1955)

Peptone	20 gm.
Magnesium chloride	1.4 gm.
Potassium sulfate	10.0 gm.
Agar (dried)	·13.6 gm.
Cetrimide†	0.3 gm.
Distilled water	1000 ml.
(Final pH 7.2±)	

Suspend powder in water, add 10 ml. of glycerol, heat with frequent agitation, and boil for 1 minute.

Dispense in 5-ml. amounts in 15 × 125 mm. screw-capped tubes, autoclave at 118° to 121° C. for 15 minutes, and slant to give a generous slant.

This medium is used for the selective identification of *Pseudomonas aeruginosa,*

---

*Pseudosel agar, Baltimore Biological Laboratory, Cockeysville, Md.

†Cetyl trimethyl ammonium bromide.

whose growth is not inhibited by the cetrimide. Other members of the genus (except *Ps. fluorescens*) and related nonfermentative organism are inhibited.

16. Chlamydospore agar

Chlamydospore agar*	18.5 gm.
Distilled water	500 ml.

Suspend the dehydrated product in a 1-liter Erlenmeyer flask; dissolve by heating; and tube in approximately 15-ml. amounts in screw-capped, 20 × 150 mm. test tubes. Autoclave at 121° C. for 15 minutes with caps loosened. When cool, tighten caps and store at room temperature. When needed, melt a tube in a water bath, pour into a sterile Petri plate, and allow to solidify.

17. Chocolate agar

Chocolate agar is agar to which blood or hemoglobin has been added and heated until the medium becomes brown or chocolate in color. The recommended method of preparing this is as follows:

Suspend sufficient proteose No. 3 agar,† Eugonagar,‡ or GC agar base§ in distilled water to make a double-strength base. Mix thoroughly and heat to boiling for 1 minute, with frequent agitations; autoclave at 121° C. for 15 minutes. At the same time, autoclave an equal volume of 2% hemoglobin,§ made by the gradual addition of distilled water to the dehydrated hemoglobin, to obtain a **smooth** suspension. Cool both solutions to approximately 50° C., add the supplement (Iso-Vitalex enrichment‡ or Supplement B or C† are recommended) and combine with bacteriological precautions; then pour into sterile, disposable Petri plates. Best results will be obtained if plates are **freshly** prepared.

*Available in dehydrated form from Difco Laboratories, Detroit; Baltimore Biological Laboratory, Cockeysville, Md.
†Difco Laboratories, Detroit.
‡Baltimore Biological Laboratories, Cockeysville, Md.
§Baltimore Biological Laboratory, Cockeysville, Md.; Difco Laboratories, Detroit.

One or two plates should be incubated at 36° C. overnight to determine sterility, and later discarded.

Thayer-Martin agar for the selective isolation of *Neisseria gonorrhoeae* and *N. meningitidis* may be prepared from the aforementioned chocolate agar by adding an antibiotic inhibitor (V-C-N, BBL, or Difco). (The use of Thayer-Martin medium results in a higher recovery of neisseriae from clinical specimens likely to be contaminated with other bacteria.)

Chocolate agar slants may be prepared individually by the same method just described, slanting the medium after tubing in 5- to 10-ml. amounts in sterile screw-capped tubes. To prepare a large number of slants, use a flask containing 50 to 100 ml. of agar, proceeding as before.

18. Chopped meat glucose (CMG)

Ground beef, lean, fat-free	500 gm.
Distilled water	1,000 ml.
Sodium hydroxide, 1 N	25 ml.

a. Mix ingredients, bring to boil, and simmer with frequent stirring for 20 minutes.
b. Cool to room temperature, skim off fat, and filter through three layers of gauze; squeezing out gauze, retain both meat particles and filtrate (filtered through coarse and fine paper).
c. Restore filtrate to 1 liter with distilled water, and add:

Trypticase	30 gm.
Yeast extract	5 gm.
Dipotassium phosphate	5 gm.
Resazurin solution (25 mg./100 ml. water)	4 ml.

d. Boil, cool, adjust to pH 7.8, and add 0.5% glucose and 0.5 gm. cystine.
e. Dispense 6- to 7-ml. amounts into tubes containing meat particles (step b), 1 part meat to 4 or 5 parts of broth, and autoclave at 121° C. for 20 minutes.

This medium is recommended for the primary isolation of anaerobes and also for growing pure cultures for gas-liquid chromatographic analysis. It may also be obtained as a dehydrated medium—cooked meat phytone.*

19. Chopped-meat medium†

To 1 pound of finely ground beef heart and other muscle (fat free) add 500 ml. of boiling sodium hydroxide (N/15 to N/20); boil for 20 minutes. Cool, strain off fat, and filter through muslin. Adjust fluid to pH 7.5 and add 1% peptone. Add approximately 2 inches of the meat and a small ball of steel wool to each tube and add enough broth to overlay the meat by about 1 inch. Heat tubes for 30 minutes in boiling water; sterilize by autoclaving at 121° C. for 15 minutes. The medium should be boiled for a few minutes to drive off dissolved oxygen if not used the same day. Incubation in an anaerobic jar or under a petrolatum seal may be required for cultivation or maintenance of certain *Clostridium* species.

20. Citrate agar‡
(Simmons: J. Infect. Dis. **39:**209, 1926)

Distilled water	1,000	ml.
Agar	20	gm.
Sodium chloride	5	gm.
Magnesium sulfate	0.2	gm.
Ammonium dihydrogen phosphate	1	gm.
Dipotassium phosphate	1	gm.
Sodium citrate	2	gm.
Bromthymol blue	0.08	gm.
(Final pH 6.9±)		

Citrate agar is used to determine the utilization of citrate as the sole carbon source.

21. Columbia CNA agar*

Polypeptone peptone	10.0 gm.
Biosate peptone	10.0 gm.
Myosate peptone	3.0 gm.
Corn starch	1.0 gm.
Sodium chloride	5.0 gm.
Agar (dried)	13.5 gm.
Colistin	10 mg.
Nalidixic acid	15 mg.
Distilled water	1,000 ml.
(Final pH 7.3±)	

a. Suspend 42.5 gm. of the dehydrated medium in water, heat with frequent agitation, and boil for 1 minute.

b. Sterilize by autoclaving at 121° C. for 15 minutes.

c. Cool to 50° C., add 5% defibrinated sheep blood and pour plates.

This medium is excellent for the selective growth of gram-positive cocci, when gram-negative bacilli, especially *Proteus* species, tend to overgrow on conventional blood agar plates.

22. Cornmeal agar*

Yellow cornmeal	125 gm.
Distilled water	3,000 ml.

Heat cornmeal in water at 60° C. for 1 hour, filter through paper, make up to volume, and add 50 gm. of agar. Expose to flowing steam for 1 hour, filter through absorbent cotton, dispense in tubes, and autoclave at 121° C. for 30 minutes. Cornmeal agar suppresses vegetative growth of many fungi while stimulating sporulation.

With the addition of 1% Tween 80, it is useful in stimulating production of chlamydospores of *Candida albicans.*

23. Cystine tellurite blood agar
(Frobisher: J. Infect. Dis. **60:**99, 1937)

a. Melt 100 ml. of sterile 2% infusion agar in a flask and cool to 45° to 50° C. Care should be taken to maintain this temperature throughout the following steps in the preparation of the medium.

b. Add aseptically 15 ml. of sterile 0.3% solution of potassium tellurite in distilled water. The tellurite solution may be sterilized by autoclaving.

---

*Baltimore Biological Laboratory, Cockeysville, Md.
†Available in dehydrated form from the Baltimore Biological Laboratory, Cockeysville, Md.
‡Baltimore Biological Laboratory, Cockeysville, Md.; Difco Laboratories, Detroit.

---

*Available in dehydrated form from the Baltimore Biological Laboratory, Cockeysville, Md., and others. Some workers find that "homemade" media give more consistent results.

c. Add aseptically 5 ml. of sterile blood and mix well.

d. Add 3 to 5 mg. of cystine. The dry powder is used and need not be sterilized. Since different lots of cystine vary, the optimal amount necessary to produce the best growth of *Corynebacterium diphtheriae* may vary from 3 to 5 mg.%.

e. Mix the medium well and pour into sterile Petri dishes. Since the cystine does not go entirely into solution, shake the flask frequently while pouring the plates.

This medium is used for the isolation of *Corynebacterium diphtheriae.*

24. Cystine trypticase agar (CTA)*
(Vera: J. Bact. **55**:531, 1948)

Cystine	0.5	gm.
Trypticase	20	gm.
Agar	3.5	gm.
Sodium chloride	5	gm.
Sodium sulfite	0.5	gm.
Phenol red	0.017	gm.
Distilled water	1,000	ml.

(Final pH, 7.3±)

CTA is an excellent medium for the growth of pathogenic organisms. It can be used for the maintenance of cultures, including fastidious organisms (held at 25° C.), the determinination of motility, and with the addition of carbohydrates, for determining fermentation reactions of the fastidious organisms, including *Neisseria.* Sterilize at 115° to 118° C. (not over 12 pounds' pressure) for 15 minutes.

25. Decarboxylase test media
(Moeller: Acta Path. Microbiol. Scand. **36**:161, 1955)

Basal medium† :		
Peptone (Orthana special)	5	gm.
Beef extract	5	mg.
Bromcresol purple (1.6%)	0.625	ml.
Cresol red (0.2%)	2.5	ml.
Glucose	0.5	gm.

Pyridoxal	5	mg.
Distilled water	1,000	ml.

(Adjust to pH 6)

Divide the basal medium into four 250-ml. portions; tube one portion without addition of the amino acid (control). To one of the portions add 1% of L-lysine dihydrochloride,* to the second portion add 1% of L-arginine monohydrochloride, and to the third portion add 1% of L-ornithine dihydrochloride. Adjust the portion to which the ornithine was added to pH 6 with N/1 NaOH before sterilization. The media are tubed in 3- to 4-ml. amounts in small (13 × 100 mm.) screw-capped tubes. Autoclave at 121° C. for 10 minutes.

Inoculate all four tubes from an agar slant culture and overlay each tube with 4 to 5 mm. of sterile mineral oil.† If oil is not added, the reactions are not valid after 24 hours. Some workers have recommended the addition of 0.3% agar to the basal medium in place of an oil overlay.

Incubate all four tubes at 36° C. and read daily for not more than 4 days (most positive reactions with the enteric bacilli will occur in 1 to 2 days). A **positive** reaction is indicated by alkalinization of the medium with a change in color from yellow (due to the initial fermentation of glucose) to **violet** (due to the decarboxylation of the amino acid). Therefore, a **yellow** color after several days' incubation indicates a **negative** test, which shows the absence of the enzymes decarboxylase or dihydrolase. All positive tests should be compared with the control, which remains yellow.

These media are used for differentiating members of the Enterobacteriaceae.

26. Desoxycholate agar‡
(Leifson: J. Path. Bact. **40**:581, 1935)

Peptone	10	gm.
Lactose	10	gm.

---

*Available in dehydrated form from the Baltimore Biological Laboratory, Cockeysville, Md.
†Baltimore Biological Laboratory, Cockeysville, Md.; Difco Laboratories, Detroit.

---

*Amino acids are available from Nutritional Biochemicals Corp., Cleveland, Ohio.
†Sterilize in test tubes at 121° C. for 45 minutes.
‡Baltimore Biological Laboratory, Cockeysville, Md.; Difco Laboratories, Detroit.

Sodium citrate	1	gm.
Ferric citrate	1	gm.
Sodium chloride	5	gm.
Dipotassium phosphate	2	gm.
Sodium desoxycholate	1	gm.
Agar	16	gm.
Neutral red	0.033	gm.
Distilled water	1,000	ml.

(Final pH 7.2±)

Desoxycholate agar is used for the isolation of gram-negative enteric bacilli and the differentiation of lactose-fermenting and nonlactose-fermenting species.

27. Desoxycholate citrate agar*
(Leifson: J. Path. Bact. **40**:581, 1935)

Meat, infusion from	350	gm.
Peptone	10	gm.
Lactose	10	gm.
Sodium citrate	20	gm.
Ferric citrate	1	gm.
Sodium desoxycholate	5	gm.
Agar	17	gm.
Neutral red	0.02	gm.
Distilled water	1,000	ml.

(Final pH 7.3±)

Desoxycholate citrate agar is used for the isolation of salmonellae and shigellae from stool and other specimens.

28. Desoxyribonuclease (DNAse) test medium*

Desoxyribonucleic acid	2 gm.
Phytone	5 gm.
Sodium chloride	5 gm.
Trypticase	15 gm.
Agar	15 gm.
Distilled water	1,000 ml.

(Approximate final pH 7.3)

The medium may be sterilized in the autoclave at 15 pounds' pressure for 15 minutes at 121° C., cooled, and poured into Petri dishes. The dry surface may be streaked in several places ($\frac{1}{2}$-inch streaks) with strains of staphylococci to be tested for DNAse activity. Flooding of the grown plate after 24 hours with N/1 hydrochloric acid will reveal **clear zones** around the streaks of the DNAse-positive strains.

---

*Baltimore Biological Laboratory, Cockeysville, Md.; Difco Laboratories, Detroit.

Some workers prefer the flooding of the plate with 0.1% toluidine blue; a bright **rose pink** color is produced around the growth of DNAse producers.

29. Dextrose ascitic fluid semisolid agar for spinal fluid cultures

Phenol red broth base	4	gm.
Agar, powdered (weigh accurately)	0.5	gm.
Distilled water	250	ml.

Suspend ingredients in a 1-liter Erlenmeyer flask, plug with cotton, and autoclave at 121° C. for 15 minutes. At the same time sterilize a rack full of 18 × 125 mm. screw-capped test tubes and a wrapped Cornwall automatic pipet.* Cool medium and add the following aseptically:

Ascitic fluid†	50 ml.
Dextrose sterile 20% solution	15 ml.

Using the sterile pipet, tube in 5-ml. amounts in screw-capped tubes, incubate overnight for sterility, and store at room temperature.

30. Drug-containing media for susceptibility testing of mycobacteria

a. Prepare the 7H 10–OADC agar as described in the section on culture media in this text and divide into as many parts as are desired for the various drug concentrations and controls.

b. Cool the medium to 56° C., and add sterilized stock solutions to give the desired final concentrations. The drugs and their concentrations in Table 39-1 are recommended.

c. Mix the drug thoroughly with the medium, dispense aseptically in the indicated quadrants of disposable Felsen dishes,‡ and allow medium to harden. The drug-containing media should be dispensed within 1 hour following sterilization and kept away

---

*Becton-Dickinson's Cornwall Pipetting Unit with 5 ml. syringe.

†Bacto-Ascitic Fluid, 10 ml. ampules, Difco Laboratories, Detroit.

‡Falcon Plastics, Rutherford, N. J.

Table 39-1. Drugs and their concentrations for drug-containing media*

DRUG	CONCEN-TRATION μG./ML.	AMOUNT OF DRUG OR DYE TO ADD TO 200 ML. OF MEDIUM
Control	—	
Isoniazid	0.2	0.4 ml. of 100 μg./ml. †
Streptomycin	2	0.4 ml. of 1,000 μg./ml. ‡
PAS	2	0.4 ml. of 1,000 μg./ml. † Add dilute NaOH to dissolve‡
Ethionamide	5	1 ml. of 1,000 μg./ml.§
Kanamycin	5	1 ml. of 1,000 μg./ml.‡
Viomycin	5	1 ml. of 1,000 μg./ml.‡
Cycloserine	50	1 ml. of 10,000 μg./ml.‡
Ethambutol	5	1 ml. of 1,000 μg./ml. †
Rifampin	1	0.2 ml. of 1,000 μg./ml. †

*In vitro studies on drug susceptibility to pyrazinamide are not consistently satisfactory because many strains of *Mycobacterium tuberculosis* will not grow on 7H 10 agar at pH 5.8, at which level the drug is most active.
†Sterilized by autoclaving.
‡Prepared from sterile vials or ampules.
§Available as ethionamide HCl; soluble; dissolve in 95% ethanol; self-sterilizing.

from direct light. It is, therefore, necessary to have on hand all reagents, plates, dispensing apparatus, and other equipment prior to the cooling of the base medium.

d. Incubate plates overnight to check for sterility; store in refrigerator at 4° C.

e. All drug-containing media should be discarded if not used within 1 month.

f. Each batch of media should be checked with cultures of tubercle bacilli of known susceptibility; the H37 Rv strain of *Mycobacterium tuberculosis* and a known drug-resistant strain should be used. These cultures may be obtained by writing to the Trudeau Laboratory, Saranac Lake, N. Y.

31. Egg yolk agar

Scrub the shells of fresh hens' eggs and sterilize with diluted mercuric chloride. Aspirate the egg white and discard; aspirate the yolks to sterile tubes or a flask. Add an equal volume of sterile 0.85% sodium chloride solution and mix by gentle shaking. Test the sterility by making 1-ml. pour plates; if sterile, store the yolks at 4° C. for up to 2 weeks.*

Add 10 ml. of yolk suspension to 90 ml. of melted and cooled blood agar base and pour the plates. The modified base formula of McClung and Toabe (J. Bact. **53:**139-147, 1947) is also recommended.

Proteose peptone No. 2	40	gm.
Sodium orthophosphate (Na$_2$HPO$_4$)	5	gm.
Potassium dihydrophosphate (KH$_2$PO$_4$)	1	gm.
Sodium chloride	2	gm.
Magnesium sulfate	0.1	gm.
Glucose	2	gm.
Agar	25	gm.
Distilled water	1,000	ml.

(Adjust to pH 7.6; autoclave at 121° C. for 15 minutes)

The medium is used for the isolation and identification of clostridia.

32. Eosin-methylene blue agar (EMB)†

Peptone	10	gm.
Lactose	5	gm.
Sucrose	5	gm.
Dipotassium phosphate	2	gm.
Agar	13.5	gm.
Eosin Y	0.4	gm.
Methylene blue	0.065	gm.
Distilled water	1,000	ml.

The agar concentration may be increased to 5% (use an additional 3.65 gm. of agar) to inhibit the spreading of *Proteus.* The Levine EMB agar, preferred by some investigators does not contain sucrose.

Lactose-fermenting, gram-negative bacteria may be distinguished from nonlactose-fermenting types by their appearance. Colonies of *Escherichia coli* have a characteristic metallic sheen. If the sucrose-containing medium is used, *Proteus* colonies will also show this characteristic if they are inhibited from spreading by the higher agar concentration.

---

*A 30% egg yolk emulsion in saline is available from Baltimore Biological Laboratory, Cockeysville, Md.
†Available in dehydrated form from Baltimore Biological Laboratory, Cockeysville, Md.; Difco Laboratories, Detroit.

33. Fermentation broth for Listeria

Proteose peptone No. 3*	10	gm.
Beef extract	1	gm.
Sodium chloride	5	gm.
Bromcresol purple	0.1	gm.
Distilled water	1,000	ml.

Dissolve with the aid of heat and sterilize at 121° C. for 20 minutes. Add aseptically 0.5% of the following carbohydrates: glucose, salicin, rhamnose, dulcitol, and raffinose.

These carbohydrate broths are recommended for biochemical studies on *Listeria*.

34. Fermentation media for neisseriae

a. Suspend 4.3 gm. of dehydrated CTA medium (see no. 24) in 150 ml. of distilled water and heat to boiling with frequent agitation.
b. Adjust to pH 7.4 to 7.6 with 1 N NaOH, dispense in 50-ml. amounts in 250-ml. Erlenmeyer flasks, and sterilize at not over 118° C. (12 pounds) for 15 minutes.
c. Prepare 20% solutions of glucose, lactose, maltose, and sucrose in distilled water; tube and sterilize by membrane filtration.
d. Add 2.5 ml. of each carbohydrate to separate flasks of the 50—ml. cooled CTA medium.
e. Mix and dispense in 2.0-ml amounts in sterile screw-capped tubes (13 × 100 mm.) and store in refrigerator at 4° C.

35. Fildes enrichment agar*

Sodium chloride, 0.85% solution	150	ml.
Hyrochloric acid	6	ml.
Defibrinated sheep blood	50	ml.
Pepsin, granular	1	gm.

a. Mix well and heat in 56° C. water bath for 2 to 24 hours, with occasional agitation.
b. Adjust to pH 7.5 with 20% NaOH (approximately 12 ml.)
c. Readjust to pH 7.0 to 7.2 with HCl.

d. Add 0.25% chloroform, mix thoroughly, stopper tightly, and store in refrigerator at 4° C.
e. After heating 4 to 10 ml. of this in a sterile container in a 56° C. water bath for 30 minutes to get rid of the chloroform, add to 200 ml. melted nutrient agar cooled to 56° C., and pour into Petri dishes.

This is an excellent medium for the isolation of *Haemophilus influenzae*.

36. Fletcher's semisolid medium for Leptospira*

a. Sterilize 1.76 liters of distilled water by autoclaving at 121° C. for 30 minutes.
b. Cool to room temperature and add 240 ml. of sterile normal rabbit serum.
c. Inactivate by incubation at 56° C. for 40 minutes.
d. Add 120 ml. of melted and cooled (not greater than 56° C.) 2.5% meat extract agar at pH 7.4.
e. Dispense 5-ml. amounts in sterile, 16 × 130 mm., screw-capped test tubes and 15-ml. amounts in sterile, 25-ml., diaphragm-type, rubber-stoppered vaccine bottles.
f. Inactivate at 56° C. for 60 minutes on 2 successive days.

37. GN broth*
(Hajna: Pub. Health Lab. **13**:83, 1955)

Peptone	20	gm.
Dextrose	1	gm.
D-mannitol	2	gm.
Sodium citrate	5	gm.
Sodium desoxycholate	0.5	gm.
Dipotassium phosphate	4	gm.
Monopotassium phosphate	1.5	gm.
Sodium chloride	5	gm.
Distilled water	1,000	ml.
(Final pH 7±)		

Dissolve the dehydrated medium in distilled water and autoclave at 116° C. (10

---

*Difco Laboratories, Detroit.

*Difco Laboratories, Detroit; Baltimore Biological Laboratory, Cockeysville, Md.

pounds' stream pressure) for 15 minutes or steam for 30 minutes at 100° C.

The medium is used as an enrichment medium for salmonellae and shigellae in fecal specimens.

38. Gelatin medium*

To extract or infusion broth add 15% gelatin. Heat in an Arnold sterilizer or in a double boiler until the gelatin is thoroughly dissolved. Adjust the reaction to pH 7. Tube and autoclave at 121° C. (15 pounds' pressure) for 15 minutes. Sodium thioglycollate, 0.05%, may be added for cultivation of the clostridia in an external aerobic environment.

39. Gelatin medium (dilute) for differentiation of Nocardia and Streptomyces
(Center for Disease Control)

Gelatin	4 gm.
Distilled water	1 liter
(Adjust to pH 7)	

Dispense in tubes, approximately 5 ml. per tube. Autoclave at 121° C. for 5 minutes. Inoculate with a small fragment of growth from Sabouraud dextrose agar slant. Incubate at room temperature for 21 to 25 days. Examine for quantity and type of growth.

Nocardia asteroides exhibits no growth or very sparse, thin, flaky growth. *Nocardia brasiliensis* shows good growth and round compact colonies. *Streptomyces* species show poor to good growth (stringy or flaky).

40. Hektoen enteric agar*

Proteose peptone	12	gm.
Bile salts	9	gm.
Yeast extract	3	gm.
Lactrose	12	gm.
Salicin	2	gm.
Sucrose	12	gm.
Sodium chloride	5	gm.
Sodium thiosulfate	5	gm.
Ferric ammonium citrate	1.5	gm.
Agar	14	gm.

Acid fuschsin	0.1	gm.
Bromthymol blue	0.065	gm.
Distilled water	1,000	ml.

Suspend 76 gm. of dehydrated medium, if using the commercial product, in 1,000 ml. distilled water. Heat to boiling and continue until the medium dissolves. **Do not autoclave.** Cool to 50° C. and pour into Petri dishes.

This medium is used for the isolation and differentiation of the gram-negative enteric pathogens. The coliforms can be distinguished from other enteric organisms. The former are salmon to orange in color, while the salmonellae and shigellae are bluish green.

41. Kanamycin-vancomycin blood agar

1. Menadione-hemin solution:
   Dissolve 200 mg. of recrystallized hemin in 8 ml. of 1 N NaOH. To this add 388 ml. of sterile distilled water and 2 ml. of menadione* (Synkarite-Roche, 10 mg./ml.).

2. Kanamycin-vancomycin + menadione-hemin:
   Add 2 gm. of kanamycin or 6 ml. (Bristol—1 gm./3 ml. vial) and 150 mg. of vancomycin or 3.3 ml. (Lilly—500 mg./10 ml. vial) to 206 ml. of menadione-hemin solution (see above). Store at 4° C.

   Note: The kanamycin-vancomycin solution may be stored at 4° C. and appropriate amounts filter-sterilized as needed.

3. Preparation of base:

Trypticase soy agar	40	gm.
Agar	2.5	gm.
Distilled water	1,000	ml.

   Sterilize by autoclaving at 121° C. for 15 minutes and cool to 50° C.

4. Add aseptically:

Defibrinated sheep blood	50 ml.
Kanamycin-vancomycin + menadione-hemin	10 ml.

5. Mix well, pour plates (approximately

---

*Difco Laboratories, Detroit; Baltimore Biological Laboratory, Cockeysville, Md.

*Some workers prefer vitamin K.₁ (Konakion-Roche, 10 mg./ml.).

20 ml. per plate), and store at room temperature.

Used for primary inoculation of clinical specimens for selective isolation of anaerobes.

42. Lactose agar (10%) in phenol red broth base
(Chilton and Fulton: J. Lab. Clin. Med. **31:**824, 1946)

Phenol red broth base*	7.5 gm.
Lactose	50 gm.
Agar	10 gm.
Distilled water	500 gm.

Add agar to the distilled water. Heat with frequent stirrings until the agar is melted. Cool to a temperature of 50° to 60° C. Add the lactose and the phenol red broth base. Mix well. Tube in 5- to 6-ml. amounts. Autoclave at 118° C. (12 pounds' pressure) for 10 to 12 minutes. As soon as the autoclave pressure is reduced, remove the medium from the autoclave and place in a slanting position.

43. Loeffler coagulated serum slants
(Centralbl. Bakt. **2:**105, 1887)

Add 3 volumes of beef, hog, or horse serum (collected as cleanly as possible) to 1 volume of glucose infusion broth (1% glucose). Tube, slant, and inspissate for about 2 hours at 75° to 85° C. on each of 3 successive days. The medium may also be sterilized in a slanted position in a horizontal autoclave. Close the autoclave tightly and, without allowing the air to escape, autoclave at 15 pounds' pressure for 15 minutes to coagulate the medium. Then allow the air to escape slowly, and admit steam so that the pressure is maintained at 15 pounds. When an atmosphere of pure steam has been obtained, close the outlet valve tightly and sterilize the medium with steam under 15 pounds' pressure (121° C.) for 15 or 20 minutes.

---

*Available in dehydrated form from Difco Laboratories, Detroit; Baltimore Biological Laboratory, Cockeysville, Md.

44. Lowenstein-Jensen medium*
(Holm and Lester: Public Health Rep. **62:**847, 1947, abst.)

Salt		solution
Monopotassium phosphate	2.4	gm.
Magnesium sulfate ($7H_2O$)	0.24	gm.
Magnesium citrate	0.6	gm.
Asparagin	3.6	gm.
Glycerol, reagent grade	12	ml.
Distilled water	600	ml.
Potato flour	30	gm.
Homogenized whole eggs	1,000	ml.
Malachite green, 2% aqueous	20	ml.

Add the potato flour to the flask of salt solution and autoclave the mixture at 121° C. for 30 minutes. Clean fresh eggs, not more than 1 week old, by vigorous scrubbing in 5% soap solution and allow to remain in it for 30 minutes. Place these in running cold water until it becomes clear. Immerse the washed eggs in 70% alcohol for 15 minutes, remove and break into a sterile flask; homogenize completely by shaking with glass beads. Then filter the homogenate through four layers of sterile gauze.

Add one liter of the homogenized egg suspension to the flask of cooled potato flour–salt solution. Add the malachite green to this emulsion and mix thoroughly. Dispense the medium aseptically with a sterile aspirator bottle in 6-ml. amounts into sterile, screw-capped 150-mm. glass tubes. Inspissate the tubes at 85° C. for 50 minutes and check for sterility by incubating at 36° C. for 48 hours. Store in a refrigerator, where the medium will keep for at least 1 month if the tubes are tightly sealed to prevent loss of moisture.

This medium is used for the cultivation of *Mycobacterium tuberculosis*.

45. MacConkey agar†

Peptone	17	gm.
Proteose peptone	3	gm.

---

*Prepared tubed media are available from Difco Laboratories, Detroit; Baltimore Biological Laboratory, Cockeysville, Md.

†Available in dehydrated form from Baltimore Biological Laboratory, Cockeysville, Md.; Difco Laboratories, Detroit.

Lactose	10	gm.
Bile salts	1.5	gm.
Sodium chloride	5	gm.
Agar	13.5	gm.
Neutral red	0.03	gm.
Crystal violet	0.001	gm.
Distilled water	1,000	ml.

The agar concentration may be increased to 5% (use an additional 3.65 gm. of agar) to inhibit the spreading of *Proteus.* The medium is inhibitory and differential rather than selective. Coliforms (lactose fermenting) produce red colonies on the medium while nonlactose fermenters produce colorless colonies.

46. Mannitol salt agar*

Beef extract	1	gm.
Proteose peptone No. 3	10	gm.
Sodium chloride	75	gm.
Mannitol	10	gm.
Agar	15	gm.
Phenol red	0.025	gm.
Distilled water	1,000	ml.

Sterilize the medium at 15 pounds' pressure for 15 minutes, cool to 48° C., and pour into sterile Petri dishes. Use 12 to 15 ml. per plate.

This medium is recommended for the selective isolation of pathogenic staphylococci, since most other bacteria are inhibited by the high salt concentration. Colonies of pathogenic staphylococci are surrounded by a yellow halo, indicating manitol fermentation.

47. McBride medium, modified
(J. Lab. Clin. Med. **55:**153, 1960)

Phenylethanol agar*	35.5	gm.
Glycine, anhydride	10	gm.
Lithium chloride	0.5	gm.
Distilled water	1,000	ml.

Dissolve with the aid of heat and sterilize at 121° C. for 20 minutes.

The medium is recommended for the cultivation of *Listeria.*

---

*Available in dehydrated form from Baltimore Biological Laboratory, Cockeysville, Md.; Difco Laboratories, Detroit.

48. Methyl red–Voges-Proskauer medium (MR-VP) (Clark and Lubs medium)*

Buffered peptone	7 gm.
Dextrose	5 gm.
Dipotassium phosphate	5 gm.
Distilled water	1,000 ml.

(Final pH 6.9 ±)

The medium is used for the methyl red test and the Voges-Proskauer test (production of acetylmethylcarbinol).

49. Middlebrook 7H 10 agar with OADC enrichment

The preparation of 7H 10 oleic acid–albumin agar as originally described is extremely tedious and complicated. Its preparation may be simplified by the use of a combination of six stock solutions prepared in advance, as described in the handbook *Tuberculosis Laboratory Methods* of the Veterans Administration.† However, for the average diagnostic microbiology laboratory, it is strongly recommended that **commercially prepared** 7H 10–OADC complete medium be used whenever possible.‡ For purposes of orientation the ingredients of each medium will be tabulated, but the directions are intended solely for preparation from dehydrated medium and prepared enrichment.

a. Middlebrook 7H 10 agar‡

Ammonium with	0.5	gm.
d-Glutamic acid	0.5	gm.
Sodium citrate	0.4	gm.
Disodium phosphate	1.5	gm.
Monopotassium phosphate	1.5	gm.
Ferric ammonium phosphate	0.04	gm.
Magnesium sulfate	0.05	gm.
Pyridoxine	0.001	gm.
Biotin	0.0005	gm.
Malachite green	0.001	gm.
Agar	15	gm.

b. Middlebrook OADC enrichment‡

Oleic acid	0.5	gm.
Bovine albumin, fraction V	50	gm.

---

*Available in dehydrated form from Baltimore Biological Laboratory, Cockeysville, Md.; Difco Laboratories, Detroit.

†Obtainable from the Superintendent of Documents, U. S. Government Printing Office, Washington, D. C.

‡Baltimore Biological Laboratory, Cockeysville, Md.; Difco Laboratories, Detroit.

Dextrose	20	gm.
Beef catalase	0.04	gm.
Sodium chloride	8.5	gm.
Distilled water	1,000	ml.

To rehydrate the 7H 10 agar, suspend 20 gm. in 1 liter of cold distilled water containing 0.5% reagent grade glycerol and heat to boiling to dissolve completely. Distribute in 180-ml. amounts in flasks; sterilize by autoclaving at 121° C. for 10 minutes; cool to 50° to 55° C.; add 20 ml. of OADC enrichment aseptically to each flask; and dispense in sterile plastic Petri dishes. Store at 4° C. in the dark for not longer than 2 months in a plastic bag to prevent dehydration. If the medium is to be used for drug susceptibility testing, the various therapeutic agents may be added to the melted and cooled medium.

50. Milk media

a. Skimmed milk

Fresh, clean, skimmed cow's milk may be used. This is dispensed in tubes and sterilized either by tyndallization (flowing steam for 30 minutes on 3 successive days) or in the autoclave at 10 pounds' pressure for 10 minutes.

Commercial dehydrated skimmed milk is also completely satisfactory. The concentration used is 10% in distilled water. The method of sterilization is as before.

b. Litmus milk

Skimmed milk powder	100 gm.
Litmus	5 gm.
Distilled water	1,000 ml.

Autoclave at 10 pounds' pressure for 10 minutes.

Reactions in litmus milk. Pink color of litmus indicates an acid reaction caused by fermentation of lactose. A purple or blue color (alkaline) indicates no fermentation of lactose. White color (reduction) results when litmus serves as an electron acceptor and is reduced to its leuco base. Coagulation (clot) is caused by precipitation of the casein by the acid produced from lactose. Coagulation may also be caused by the conversion of casein to paracasein by the enzyme rennin.

Peptonization (dissolution of the clot) indicates digestion of the curd or milk proteins by proteolytic enzymes.

c. Milk medium with a reducing agent*

Skimmed milk powder	100	gm.
Peptone	10	gm.
Sodium thioglycollate	0.5	gm.
Litmus	5	gm.
Distilled water	1,000	ml.

d. Methylene blue milk

Skimmed milk powder	10	gm.
Methylene blue (1% aqueous)	10	ml.
Distilled water	90	ml.

Methylene blue milk is used in the identification of the enterococci.

51. Moeller cyanide broth†
(Acta Path. Microbiol. Scand. **34:**115, 1954)

Peptone (Orthana special‡)	10	gm.
Sodium chloride	5	gm.
Monobasic potassium phosphate	0.225	gm.
Dibasic sodium phosphate ($2H_2O$)	5.64	gm.
Distilled water	1,000	ml.
(Adjust reaction to pH 7.6)		

Autoclave the medium in flasks. Dissolve 0.5 gm. of potassium cyanide in 100 ml. of **cold** sterile broth. Place 1 ml. of the cyanide broth in tubes 12 × 100 mm., stopper with paraffined corks **immediately,** and store at 4° to 5° C. The medium remains stable for 2 to 3 weeks.

Inoculate the medium with one small loopful of a 24-hour broth culture of the organisms. Incubate at 36° C. and observe daily for 2 days for **growth,** which is recognized by turbidity of the medium.

52. Motility test medium§

| Beef extract | 3 gm. |
| Gelysate peptone | 10 gm. |

*This medium has been found satisfactory for the cultivation of *Clostridium* species and will allow observation of their reactions in litmus milk. Autoclave at 10 pounds' pressure for 10 minutes; stopper tightly.

†The base medium is available from Baltimore Biological Laboratory, Cockeysville, Md.; Difco Laboratories, Detroit.

‡Proteose peptone No. 3, 0.3%, has been found to serve as a satisfactory substitute.

§Baltimore Biological Laboratory, Cockeysville, Md.

Sodium chloride	5 gm.
Agar	4 gm.
Distilled water	1,000 ml.

(Final pH 7.3±)

a. Suspend 22 gm. of dehydrated medium in water, add 0.05 gm. triphenyltetrazolium. Heat with frequent agitation; boil 1 minute to dissolve ingredients.

b. Dispense in screw-capped tubes and sterilize by autoclaving at 121° C. for 15 minutes.

c. Tighten caps when cool; store at room temperature.

This medium is stabbed once and read after 1 to 2 days' incubation at 36° C. Motile organisms spread out from line of inoculation; nonmotile organisms grow only along stab.

53. Motility test semisolid agar for Listeria

Bacto-tryptose*	10 gm.
Sodium chloride	5 gm.
Agar	5 gm.
Dextrose	1 gm.
Distilled water	1,000 ml.

Dissolve with the aid of heat, distribute in tubes, and sterilize at 121° C. for 20 minutes.

Inoculate by stabbing with a 24-hour broth culture. Incubate the tubes at 36° C. and observe for growth from the stab line (motile).

The medium is used for demonstrating the motility of *Listeria.*

54. Mueller-Hinton agar†
(Proc. Soc. Exp. Biol. Med. **48:**330, 1941)

Beef, infusion from	300 gm.
Peptone	17.5 gm.
Starch	1.5 gm.
Agar	17 gm.

(Final pH 7.4±)‡

Suspend the medium in distilled water, mix thoroughly, and heat with frequent

agitation. Boil for about 1 minute and dispense and sterilize by autoclaving at 116° to 121° C. (12 to 15 pounds' steam pressure) for not more than 15 minutes, and cool by placing immediately in a 50° C. water bath before pouring.

The medium is used in the disc-agar diffusion method of testing for antibiotic susceptibility of microorganisms.

Five percent defibrinated animal blood may be added as enrichment; this may be chocolatized in the testing of *Haemophilus* species.

55. Mycoplasma (PPLO) isolation media
a. Mycoplasma isolation agar (PPLO agar)*

Beef heart, infusion from	50 gm. fresh tissue
Peptone	10 gm.
Sodium chloride	5 gm.
Agar	14 gm.

(Final pH 7.8)

1. Suspend 34 gm. of dry medium in 1,000 ml. distilled water, mix, and boil for 1 minute. Dispense in 70- to 75-ml. amounts. Autoclave at 121° C. for 15 minutes.

2. Cool to approximately 50° C., and add aseptically:

> 20 ml. horse serum†
> 10 ml. 25% yeast extract (see preparation C)
> 2 ml. Penicillin G solution (100,000 units/ml.)

3. Mix gently, pour into sterile Petri dishes (100 × 15 mm.) approximately 20 ml. per plate. The smaller (60 × 15 mm.) dishes require a lesser amount (8 ml.) and are used for subcultures.

4. If bacterial contamination is excessive, add 5 ml. of a 1:2,000 dilution of thallium acetate; reduce fungal contamination by adding 0.7 ml. of a 0.05% solution of amphotericin B.

5. Seal plates with parafilm and store in the refrigerator.

---

*Difco Laboratories, Detroit.
†Baltimore Biological Laboratory, Cockeysville, Md.; Difco Laboratories, Detroit.
‡Check by using a surface electrode after gelling, if available.

---

*Available from Difco, Detroit, and Baltimore Biological Laboratory, Cockeysville, Md.
†Available from Baltimore Biological Laboratory, Cockeysville, Md.

b. Mycoplasma isolation broth*

This fluid medium is identical to the agar medium above with the agar omitted. It is prepared by dissolving 21 gm. of the dry medium in 1,000 ml. distilled water, dispensing in 70-ml. volumes and sterilizing in the autoclave at 121° C. for 15 minutes. After cooling to about 50° C., the horse serum, yeast extract, and penicillin are added as described previously.

c. Preparation of 25% yeast extract

1. Suspend 125 gm. of Fleischmann's pure dry yeast (Type 20-40)† in 500 ml. distilled/deionized water in a 1-liter beaker. Boil for 2 minutes with constant stirring.
2. Dispense in centrifuge tubes and refrigerate overnight to settle the yeast cells.
3. Centrifuge at 3,000 r.p.m. for 30 minutes and carefully remove the supernatant.
4. Adjust to pH 8.0 with 1 N NaOH (if used for T-strains, adjust to pH 6.0 with 1 N HCl).
5. Dispense in 2.5-ml. amounts in screw-capped tubes or small bottles; autoclave at 121° C. for 15 minutes.
6. Store in frozen state at −20° C.; use within 3 months. When thawing out, the extract becomes quite turbid, but clears on standing at room temperature.

d. Biphasic isolation medium for M. pneumoniae

1. Prepare basal broth medium as described previously.
2. Add 1 ml. of a 0.1% phenol red solution and 2 ml. of a 0.1% methylene blue solution to 70 ml. of broth.
3. Sterilize by autoclaving at 121° C. for 15 minutes.
4. Cool to approximately 50° C. and add

aseptically the horse serum, yeast extract, and penicillin as described previously.

5. Add aseptically about 3 ml. to tubes of the isolation agar (2-ml. amounts in screw-capped tubes) and seal tightly.

Growth of *M. pneumoniae* is indicated by a change in color of the indicator.

e. Differential medium for T-strains (Shepard's medium A-6)
(Shepard and Howard: Ann. N. Y. Acad. Sci. **174:**809-818, 1970)

Trypticase soy broth*	4.8	gm.
Manganous sulfate		
(MnSO$_4$·H$_2$O, ACS)	0.06	gm.
Distilled/deionized water	163	ml.

1. Dissolve and adjust to pH 5.5 with 2 N HCl.
2. Add 1.734 gm. Ionagar No. 2 (Oxoid),† and **without** heating to dissolve agar autoclave at 121° C. for 15 minutes.
3. On removal from autoclave, rotate flask on a flat surface for 10 to 15 seconds to thoroughly dispense the agar but avoid air bubbles. Cool to 50° to 55° C. in a water bath.
4. Without delay, add the following:

Sterile horse serum	40 ml.
Sterile 25% yeast extract	2 ml.
Sterile 10% urea solution	2 ml.
(filter sterilized)	
Penicillin G stock solution	2 ml.
(100,000 units/ml.)	

5. Mix contents thoroughly as described before and pour approximately 20 ml. per plate into 100 × 15 mm. Petri plates.
6. After hardening, invert plates and leave at room temperature overnight. Store in the refrigerator for not more than one week.

This medium is designed as a **primary** solid medium for the isolation of T-strain

---

*Available from Difco, Detroit, and Baltimore Biological Laboratory, Cockeysville, Md.
†Available from Standard Brands, Inc., New York, N. Y.

---

*Baltimore Biological Laboratory, Cockeysville, Md.
†Colab Laboratories, Inc., Glenwood, Ill., Oxoid Ltd., London.

mycoplasmas from clinical material. In either pure or mixed cultures, T-strains produce intense, dark, golden brown colonies. Colonies of other mycoplasma species appear normal or slightly yellow or amber when viewed through the bottom of the dish.

56. Nitrate broth for nitrate reduction test

Tryptone	5	gm.
Neopeptone	5	gm.
Distilled water	1,000	ml.
Potassium nitrate (reagent grade)	1	gm.
Glucose	0.1	gm.

Before adding the potassium nitrate and glucose, boil the other ingredients in the water and adjust the pH to 7.3 to 7.4. Dispense 5 ml. per tube and sterilize at 15 pounds' pressure for 15 minutes. The reagents and tests for nitrate reduction will be listed in Chapter 41.

57. ONPG (beta-galactosidase) test

The ability of certain gram-negative bacilli to ferment lactose is a useful criterion for identifying certain members of the family Enterobacteriaceae. Since some coliform organisms ferment this carbohydrate slowly or not at all, they have been confused with salmonellae (for example, *Citrobacter*).

Lactose fermentation is dependent on permease, which allows the lactose to enter the bacterial cell, and beta-galactosidase, which breaks the lactose into glucose and galactose. The slow lactose fermenters are deficient in permease, and the demonstration of beta-galactosidase in such organisms permits rapid identification as a lactose fermenter. This enzyme can be detected by the use of a tablet of ortho-nitrophenyl-beta-galactopyranoside (ONPG)* dissolved in distilled water and inoculated with a heavy suspension of the organism to be tested. The test is incubated at 36° C. for 6 hours; hydrolysis of the ONPG is detected by the liberation of ortho-nitrophenol, with a characteristic

yellow color, often within 30 minutes. Thus a **positive** ONPG test indicates that the organism contains lactose-fermenting enzymes and may be classed as a lactose fermenter.

58. Oxidative-fermentative (OF) basal medium (Hugh and Leifson)*
(J. Bact. **66:**24-26, 1953)

Trypticase or tryptone	2	gm.
Sodium chloride	5	gm.
Dipotassium phosphate	0.3	gm.
Agar (dried)	2.5	gm.
Bromthymol blue	0.03	gm.
(3 ml. of a 1% aqueous [not alcoholic] solution)		
Distilled water	1,000	ml.
(Final pH 7.1 ± 0.2)		

Suspend the material in 1 liter of distilled water and heat to dissolve with frequent agitation; adjust the reaction to a pH of approximately 7.1. Sterilize by autoclaving at 121° C. for 15 minutes. To 100 ml. of sterile base, add 10 ml. sterile 10% dextrose solution (10% lactose, mannitol, and sucrose also may be added to separate 100 ml. portions if required). The completed medium should be a **green** color.

OF basal medium is used for differentiating *Acinetobacter, Alcaligenes,* and *Pseudomonas* from members of the Enterobacteriaceae.

59. Phenol red broth base†

Trypticase	10	gm.
Sodium chloride	5	gm.
Phenol red	0.018	gm.
(Final pH 7.4±)		

Phenol red broth base can be used as a base for determining carbohydrate fer-

---

*Key Scientific Products Co., Los Angeles.

*Available in dehydrated form from Baltimore Biological Laboratory, Cockeysville, Md.; Difco Laboratories, Detroit.

†Available in dehydrated form from Baltimore Biological Laboratory, Cockeysville, Md. Similar media with bromcresol purple indicator can be obtained either from Baltimore Biological Laboratory or from Difco Laboratories, Detroit. If bromcresol purple is preferred to phenol red indicator, purple broth base is also available in dehydrated form from these laboratories.

mentation reactions. Carbohydrates can be added to the medium in 0.5% to 1% concentrations before tubing and sterilization. Durham tubes should be inserted into the tubed medium for the detection of gas formation.

60. Phenylalanine agar*
(Ewing et al.: Pub. Health Lab. **15**:153, 1957)

Yeast extract	3 gm.
DL-Phenylalanine	2 gm.
(or L-phenylalanine)	1 gm.
Disodium phosphate	1 gm.
Sodium chloride	5 gm.
Agar	12 gm.
Distilled water	1,000 ml.

Tube 3-ml. amounts in small tubes and sterilize at 121° C. for 10 minutes. Allow to solidify in a slanted position.

This medium is used to test for the deamination of phenylalanine to phenylpyruvic acid by members of the Enterobacteriaceae.

After incubation of the culture for 18 to 24 hours at 36° C., allow 4 or 5 drops of 10% ferric chloride solution to run down over the growth on the slant. If the acid has been formed, a **green** color immediately develops on the slant and in the fluid at the base of the slant.

61. Phenylethyl alcohol agar* (Brewer and Lilley)
(J. Amer. Pharm. Assoc. [Scient. Ed.] **42**:6-8, 1953)

Trypticase	15	gm.
Phytone	5	gm.
Sodium chloride	5	gm.
β-phenylethyl alcohol	2.5	gm.
Agar	15	gm.
Distilled water	1,000	ml.
(Final pH 7.3±)		

Suspend powder in water, mix thoroughly, heat with agitation, and boil 1 minute. Dispense as butts in 16 × 150 mm. screw-capped tubes. Sterilize at 118° to 121° C. (12 to 15 pounds' pressure) for 15 minutes. Five percent blood may be

added to the cooled medium (45° to 50° C.), if desired, before pouring plates.

Phenylethyl alcohol agar is used for the isolation of gram-positive cocci and the inhibition of gram-negative bacilli (particularly *Proteus*) when these are in mixed culture.

62. Potato-carrot agar

Carrots, peeled	20 gm.
Potatoes, peeled	20 gm.
Distilled water	1,000 ml.

Wash potatoes and carrots, mash both, and place in distilled water for 1 hour. Boil for 5 minutes, filter through paper, make up to 1,000 ml. of volume, and add 15 gm. of agar and 5 ml. of Tween 80. Dispense in tubes, autoclave at 121° C. for 20 minutes, and slant and cool. Potato-carrot agar is an excellent medium for demonstrating color characteristics of a fungal colony.

63. Potato-dextrose agar*

Potatoes, infusion from	200 gm.
Dextrose	20 gm.
Agar	20 gm.
Distilled water	1,000 ml.

Boil potatoes in water for 15 minutes, filter through cotton, and make up to volume with water. Add dry ingredients and dissolve agar with heat. No pH adjustment is required. Dispense as desired and autoclave at 121° C. for 10 minutes. The agar is used to stimulate spore production of fungi.

64. Rice grain medium

White rice	8 gm.
Distilled water	25 ml.

Place in a 125-ml. Erlenmeyer flask and autoclave at 121° C. for 15 minutes.

Rice grain medium is used for differentiation of *Microsporum* species. *M. canis* and *M. gypseum* grow and sporulate well in this medium; *M. audouini* grows poor-

---

*Difco Laboratories, Detroit; Baltimore Biological Laboratory, Cockeysville, Md.

*Available in prepared tubes from Baltimore Biological Laboratory, Cockeysville, Md.; Difco Laboratories, Detroit.

ly. Conidial formation of some of the *Trichophyton* species is stimulated.

65. Sabouraud dextrose agar

Sabouraud dextrose agar*	32.5	gm.
Agar, powdered	2.5	gm.
Distilled water	500	ml.

Suspend the ingredients in the water; dissolve by heating to boiling; tube in approximately 20-ml. amounts in cotton-plugged, 25 × 150 mm. Pyrex test tubes (without lips). If antibiotics are to be added, this may be done after heating of the medium and before autoclaving. The following amounts are recommended (Center for Disease Control):

Cycloheximide† (0.5 mg. per milliliter)	250 mg.
Chloramphenicol‡ (0.05 mg. per milliliter)	25 mg.

Add the chloramphenicol dissolved in 5 ml. of 95% ethanol and the cycloheximide dissolved in 5 ml. of acetone. Mix well and distribute into tubes as indicated. Autoclave at 118° C. for not longer than 10 minutes. Slant, allow to harden, and store in refrigerator. This selective medium is used for the isolation of fungi when other contaminating microorganisms may be present.

66. Sabouraud dextrose broth

Dextrose	40	gm.
Peptone	10	gm.
Distilled water	1,000	ml.

Dissolve ingredients, tube in 10-ml. amounts in 18 × 150 mm. tubes, and autoclave at 121° C. for 10 minutes. Sabouraud dextrose broth is used for differentiation of *Candida* species.

67. Salmonella-Shigella (SS) agar*

Beef extract	5	gm.
Peptone	5	gm.
Lactose	10	gm.
Bile salts mixture	8.5	gm.
Sodium citrate	8.5	gm.
Sodium thiosulfate	8.5	gm.
Ferric citrate	1	gm.
Agar	13.5	gm.
Brilliant green	0.33	gm.
Neutral red	0.025	gm.
Distilled water	1,000	ml.

Dissolve by boiling. **Do not autoclave.**

Salmonella-Shigella agar is used for isolation of the pathogenic, intestinal, gram-negative bacilli, inhibition of coliform bacilli, and differentiation of lactose-fermenting and nonlactose-fermenting strains.

68. Scott's modified Castaneda media for blood culture

a. *Agar bottle—slant:*		
Trypticase soy agar*	40	gm.
Agar, granulated	15	gm.
Water, distilled	1,000	ml.

Heat in autoclave at 121° C. for 5 minutes to melt agar, or heat on a hot plate with a magnetic stirring bar.

b. *Agar bottle—broth:*		
Trypticase soy broth*	30	gm.
Sodium citrate, A.R.	5	gm.
Water, distilled	1,000	ml.
c. Thioglycollate medium (135-C.)†	30	gm.
Sucrose	100	gm.
Sodium polyanethol sulfonate (SPS) ("Grobax," 5% sterile solution SPS)‡	2	ml.
Distilled water	1,000	ml.
(pH of all media after autoclaving should be 7.2 ± 0.2)		

a. Dispense agar medium (a) in the melted state in approximately 20-ml. amounts into each of fifty clean,

---

*This medium and also media containing the antibiotics are available from Baltimore Biological Laboratory, Cockeysville, Md.; Difco Laboratories, Detroit.
†Cycloheximide is available in 4-gm. amounts from Upjohn Co., Kalamazoo, Mich., as Acti-dione.
‡Chloramphenicol is available from Parke, Davis & Co., Detroit, as Chloromycetin.

---

*Available in dehydrated form from Difco Laboratories, Detroit; Baltimore Biological Laboratory, Cockeysville, Md.
†Baltimore Biological Laboratory, Cockeysville, Md.
‡Roche Diagnostics, N. Y.

nonchipped 4-ounce, square, clear glass, screw-capped bottles (10 ml. in 2-ounce bottles).*

b. Insert the disposable rubber diaphragms in the screw caps, which are then applied loosely on the bottle tops.

c. Dispense thioglycollate medium (c) into fifty clean bottles, approximately 75 ml. per bottle (30 ml. in 2-ounce bottles) and apply the screw caps as noted earlier.

d. Make up trypticase broth (b) in a 2-liter flask and plug. Along with this prepare a dispensing buret (a 300-ml. Salvarsan tube† or satisfactory substitute) to the end of which are attached a 2-foot length of rubber tubing, a needle holder, and a 1½-inch, 21-gauge needle, in that order. Insert the attached needle, together with a portion of the rubber tubing, in the top of the buret and make fast by plugging, thus assuring a closed unit during sterilization.

e. Autoclave all media and equipment at 118° to 121° C. for 12 to 15 minutes.‡

f. As soon as the pressure reaches zero, open the autoclave, remove all bottles rapidly, and tightly fasten the screw caps. The hands must be protected from the hot bottles by asbestos gloves.

g. Place the bottles containing the agar on their sides on a cool table top to permit hardening of the agar layer. After about an hour place the bottles upright and sterilize the tops with al-

cohol sponges. (The agar has not become detached from the side of the bottles in the many thousands made.)

h. Clamp the sterile dispensing buret to a ring stand. By means of aseptic technique, fill the buret with sterile trypticase broth.

i. To each agar slant bottle then add 60 ml. of broth (25 ml. in 2-ounce bottles) by puncturing the rubber diaphragm with the sterile needle. The partial vacuum in the bottle will easily permit addition of this amount.

j. Incubate the agar slant bottles at 36° C. for 48 hours to ensure sterility. The thioglycollate bottles do not need incubation for a sterility check.

k. Inspect the bottles, label, and store at room temperature.

69. Selenite-F enrichment medium*
(Leifson: Amer. J. Hyg. **24:**423, 1936)

Sodium hydrogen selenite (anhydrous)	0.4%
Sodium phosphate (anhydrous)	1 %
Peptone	0.5%
Lactose	0.4%
(Final pH 7)	

Dissolve the ingredients in distilled water. Sterilize gently; 30 minutes in flowing steam in an autoclave is sufficient. It is important to note that the medium should **not** be autoclaved. This medium is used selectively in the isolation of *Salmonella* and *Shigella*.

70. Sellers differential agar†
(Bact. Proc. 1963, p. 65.)

Yeast extract	1	gm.
Peptone	20	gm.
L-Arginine	1	gm.
D-Mannitol	2	gm.
Bromthymol blue	0.04	gm.
Phenol red	0.008	gm.

---

*Blood culture bottles (St. Louis Health Dept. type) with special screw caps and disposable rubber diaphragms, 4-ounce and 2-ounce sizes, from Curtin Scientific Co., Rockville, Md., and other addresses.
†Arthur H. Thomas Co., Philadelphia.
‡At the Wilmington Medical Center, we have found that the loose application of a piece of steam autoclave tape (No. 1222-3M) to the bottle tops, which is then fastened down after sterilization, obviates disinfection of the diaphragm when filling with broth (step h above) or when collecting the blood culture.

---

*Available in dehydrated form from Baltimore Biological Laboratory, Cockeysville, Md; Difco Laboratories, Detroit.
†Difco Laboratories, Detroit; Baltimore Biological Laboratory, Cockeysville, Md.

Sodium chloride	2	gm.
Sodium nitrate	1	gm.
Sodium nitrite	0.35	gm.
Magnesium sulfate	1.5	gm.
Dipotassium phosphate	1	gm.
Agar	15	gm.

To rehydrate the medium, suspend 45 gm. in 1,000 ml. of cold distilled water and heat to boiling to dissolve the medium completely. Dispense into test tubes and stopper with cotton plugs or loosely fitting caps. Sterilize in the autoclave for 10 minutes at 15 pounds' pressure (121° C.). Allow the tubes to cool in the slanted position to give approximately 1½-inch butts and 3-inch slants. Immediately **before** inoculating, add 2 large drops or 0.15 ml. of a sterile 50% bacto-dextrose solution to each tube by letting it run down the **side of the tube opposite the slant.** Inoculate the tubes by deep stab into the butt and by streaking the slant. Incubate for 24 hours at 36° C. The final reaction of the medium will be pH 6.7 at 25° C.

Sellers differential agar is recommended for differentiating and identifying nonfermentative gram-negative bacilli that produce an alkaline reaction on TSI agar. The dehydrated medium is prepared according to the formula of Sellers and is recommended. This medium is particularly useful in differentiating *Pseudomonas aeruginosa, Acinetobacter anitratum, Acinetobacter lwoffi,* and *Alcaligenes faecalis.*

The reactions produced by the nonfermentative gram-negative bacilli are given in Table 39-2.

71. SF (Streptococcus faecalis) broth (SF medium)*

Tryptone	20	gm.
Dextrose	5	gm.
Dipotassium phosphate	4	gm.
Monopotassium phosphate	1.5	gm.
Sodium azide	0.5	gm.
Sodium chloride	5	gm.
Bromcresol purple	0.032	gm.
Distilled water	1,000	ml.

(Final pH 6.9)

SF medium is a selective broth medium for enterococci, since growth of all other cocci is inhibited.

72. Sodium chloride broth (6.5%)

Heart infusion broth†	100 ml.
Sodium chloride	6 gm.

Brain-heart infusion broth contains 0.5% sodium chloride. Thus, by adding an additional 6%, the desired concentration is obtained. The medium is selective for enterococci.

73. Sodium hippurate broth
(Ayers and Rupp: J. Infect. Dis. 30:388, 1922)

Sodium hippurate	10 gm.
Infusion broth	1,000 ml.

Dissolve the salt in the broth first. Tube in 3-ml. amounts in small tubes and steri-

---

*Available in dehydrated form from Difco Laboratories, Detroit; Baltimore Biological Laboratory, Cockeysville, Md. Prepare as directed on the label.
†Available in dehydrated form from Baltimore Biological Laboratory, Cockeysville, Md.; Difco Laboratories, Detroit.

Table 39-2. Reactions produced on Sellers medium by nonfermentative gram negative bacilli

ORGANISM	SLANT COLOR	BUTT COLOR	BAND COLOR	FLUORESCENT SLANT	NITROGEN GAS
Pseudomonas aeruginosa	Green	Blue or no change	Sometimes blue	Yellow Green	Produced
*Acinetobacter anitratum**	Blue	No change	Yellow	Absent	Absent
*Acinetobacter lwoffii**	Blue	No change	Absent	Absent	Absent
Alcaligenes faecalis	Blue	Blue or no change	Absent	Absent	Produced

*See footnote, p. 165.

lize at 15 pounds' pressure for 15 minutes.

This medium is used to test for the hydrolysis of sodium hippurate by streptococci. The reagent and test are described in Chapter 41.

74. Sodium malonate broth

(Leifson: J. Bact. **26:**329, 1933; Ewing et al.: Pub. Health Lab. **15:**153, 1957)

Yeast extract	1	gm.
Ammonium sulfate	2	gm.
Dipotassium sulfate	0.6	gm.
Monopotassium phosphate	0.4	gm.
Sodium chloride	2	gm.
Sodium malonate	3	gm.
Glucose	0.25	gm.
Bromythymol blue	0.025	gm.
Distilled water	1,000	ml.

(Final pH 6.7±)

Sterilize in small tubes in 3-ml. amounts at 121° C. for 15 minutes. Inoculate from TSI slants or broth cultures; incubate cultures at 36° C. for 48 hours.

The medium is used to test for the utilization of sodium malonate by members of the Enterobacteriaceae. A **positive test** is shown by the color change of the indicator from green to a **Prussian blue.**

75. Starch agar medium

Bacto-agar	20 gm.
Bacto-peptone	5 gm.
Beef extract	3 gm.
Sodium chloride	5 gm.
Soluble starch	20 gm.
Distilled water	1,000 ml.

(Final pH 7.2±)

a. Dissolve the agar in 300 ml. of water with heat.
b. Dissolve the beef extract and peptone in 200 ml. of water.
c. Mix the solutions in steps a and b and make up to 1,000 ml. volume.
d. To this mixture add the starch, dissolve, and autoclave at 121° C. for 15 minutes.

The medium may be dispensed in tubes (15- to 20-ml. amounts) or flasks convenient for pouring plates, and it should be stored in a refrigerator. For the pouring of plates, melt the tubes as required. If **poured plates of starch agar are refrigerated, the medium becomes opaque.**

Starch agar is useful in developing smooth cultures by streaking borderline rough strains on the surface of the medium. It may also be used for testing cultures for starch hydrolytic activity.

76. Sucrose (5%) broth or agar

Prepare a 50% aqueous solution of sucrose, sterilize in an autoclave at 10 pounds' pressure for 10 minutes, and store in a refrigerator.

To prepare broth, add aseptically 0.5 ml. of the stock sucrose solution to 5 ml. of tubed sterile infusion broth.

To prepare agar plates, add aseptically 1.5 ml. of the stock sucrose solution to 15 ml. of sterile melted agar. Mix well and pour into a sterile Petri dish.

77. Tellurite reduction test medium
Medium:

a. Suspend 4.7 gm. of Middlebrook 7H 9 dehydrated base* in 900 ml. of distilled water and add 0.5 ml. of Tween 80.
b. Autoclave at 121° C. for 15 minutes, cool to 55° C., and add aseptically 100 ml. ACD enrichment.*
c. Dispense aseptically in 5-ml. amounts in 20 × 150 mm. screw-capped tubes, check for sterility, and store in the refrigerator.

Tellurite solution:
a. Dissolve 0.2 gm. of potassium tellurite in 100 ml. of distilled water.
b. Dispense in 2- to 5-ml. amounts and sterilize by autoclaving at 121° C. for 10 minutes.

This medium is used to test the ability of certain mycobacteria of Runyon Group III nonphotochromogens to reduce the tellurite rapidly to the black, metallic tellurium.

---

*Difco Laboratories, Detroit; Baltimore Biological Laboratory, Cockeysville, Md.

78. Tetrathionate broth*

Proteose peptone†	5 gm.
Bile salts	1 gm.
Calcium carbonate	10 gm.
Sodium thiosulfate	30 gm.
Distilled water	1,000 ml.

Tube the medium in 10-ml. amounts and heat to boiling. **Before use,** add 0.2 ml. of iodine solution (6 gm. of iodine crystals, 5 gm. of potassium iodide in 20 ml. of water) to each tube.

This is a selective liquid enrichment medium for use in the isolation of *Salmonella,* except *S. typhi.* The sterile base **without** iodine may be stored indefinitely in a refrigerator.

79. Thayer-Martin agar

GC agar base* (double strength)	72 gm.
Agar	10 gm.
Distilled water	1,000 ml.

a. Suspend dehydrated medium in water, mix well, and heat with agitation; boil for 1 minute.

b. Sterilize by autoclaving at 121° C. for 15 minutes.

c. At the same time, autoclave a suspension of 20 gm. of dehydrated hemoglobin in 1,000 ml. water for 15 minutes (suspension must be **smooth** before sterilizing).

d. Cool both to 50° C., mix aseptically, then add 20 ml. IsoVitalex enrichment (BBL) and 20 ml. of V-C-N inhibitor (BBL) containing antibiotics; pour plates using 20 ml. per plate. Store in refrigerator.

This medium is recommended for the isolation of *Neisseria gonorrhoeae* from all sites, as well as the recovery of *N. meningitidis* from nasopharyngeal and throat cultures.

---

*Available in dehydrated form from Difco Laboratories, Detroit; Baltimore Biological Laboratory, Cockeysville, Md.

†Difco Laboratories, Detroit.

80. Thioglycollate medium without indicator*
(Brewer: J. Bact. **39**:10, 1940; and **46**:395, 1943)

Peptone	20	gm.
L-Cystine	0.25	gm.
Glucose	6	gm.
Sodium chloride	2.5	gm.
Sodium thioglycollate	0.5	gm.
Sodium sulfite	0.1	gm.
Agar	0.7	gm.
Distilled water	1,000	gm.

(Final pH 7.2±)

Tube the medium in 15-ml. amounts in 6 × ³/₄-inch test tubes, making a column of medium 7 cm. high. Autoclave for 15 minutes at 121° C. **Store at room temperature.**

81. Thionine or basic fuchsin agar
(Huddleson, et al.: Brucellosis in man and animals, New York, 1939, The Commonwealth Fund)

Trypticase soy agar may be used as a base for differential media containing thionine and basic fuchsin.

Prepare the dyes, thionine and basic fuchsin, in 0.1% stock solutions in sterile distilled water. These stock solutions may be stored indefinitely. Before adding to the media, heat the dye solutions in flowing steam in an autoclave for 20 minutes, shake well, and while still hot, add to melted agar. In trypticase soy agar the final concentration of the dye should be 1:100, 000 (10 ml. per liter of medium). Thoroughly mix the dyes (added individually) and the melted agar and pour immediately into Petri dishes, one set containing thionine and one containing basic fuchsin. Place plates in a 36° C. incubator until the water of condensation disappears, at which time they are ready for

---

*Available in dehydrated form from Baltimore Biological Laboratory, Cockeysville, Md. (No. 11720); Difco Laboratories, Detroit (No. 0430). It is also available with indicator. This may be enriched by the addition of 10% normal rabbit or horse serum when cool.

use. Inoculate the plates within 24 hours of preparation.

Streak the surface of the plates with a heavy suspension of *Brucella* prepared from a 48- to 72-hour trypticase soy agar slant culture. It is advisable to streak plates in duplicate, incubating one set aerobically and the other in 10% carbon dioxide. Incubate the plates for 72 hours and observe for **inhibition of growth** by either thionine or basic fuchsin or by both of the dyes.

82. Todd-Hewitt broth, modified*
(J. Path. Bact. **35:**973, 1932)

Beef heart infusion	1,000 ml.
Neopeptone	20 gm.

Adjust to pH 7 with normal sodium hydroxide and add the following:

Sodium chloride	2	gm.
Sodium bicarbonate	2	gm.
Disodium phosphate	0.4	gm.
Glucose	2	gm.
(Final pH 7.8±)		

Mix chemicals in broth and bring to a slow boil. Boil for 15 minutes, filter through paper, dispense in tubes, and autoclave at 115° C. for 10 minutes.

Modified Todd-Hewitt broth is used for growing streptococci for serological identification.

83. Triple sugar iron agar—TSI agar*
(Hajna: J. Bact. **49:**516, 1945)

Peptone	20	gm.
Sodium chloride	5	gm.
Lactose	10	gm.
Sucrose	10	gm.
Glucose	1	gm.
Ferrous ammonium sulfate	0.2	gm.
Sodium thiosulfate	0.2	gm.
Phenol red	0.025	gm.
Agar	13	gm.
Distilled water	1,000	gm.
(Final pH 7.3±)		

Triple sugar iron agar is used for determining carbohydrate fermentation and hydrogen sulfide production as a first step in the identification of gram-negative bacilli.

84. Trypticase dextrose agar*

Trypticase	20	gm.
Dextrose	5	gm.
Agar	3.5	gm.
Bromthymol blue	0.01	gm.
Distilled water	1,000	ml.
(Final pH 7.3±)		

Trypticase dextrose agar can be used to determine motility and dextrose fermentation of aerobic and anaerobic organisms.

Prepare according to directions on label.

85. Trypticase lactose iron agar*

Trypticase	20	gm.
Lactose	10	gm.
Ferrous sulfate	0.2	gm.
Agar	3.5	gm.
Sodium sulfite	0.4	gm.
Sodium thiosulfate	0.08	gm.
Phenol red	0.02	gm.
Distilled water	1,000	ml.
(Final pH 7.3±)		

The medium can be used for the determination of motility, lactose fermentation, and production of hydrogen sulfide by aerobes and anaerobes.

86. Trypticase sucrose agar*

Trypticase	20	gm.
Sucrose	10	gm.
Agar	3.5	gm.
Phenol red	0.02	gm.
Distilled water	1,000	ml.
(Final pH 7.2±)		

The medium can be used to determine motility and sucrose fermentation by aerobes and anaerobes.

87. Trypticase nitrate broth*

Trypticase	20	gm.
Disodium phosphate	2	gm.
Glucose	1	gm.
Agar	1	gm.
Potassium nitrate	1	gm.
Distilled water	1,000	ml.
(pH 7.2)		

---

*Available in dehydrated form from Baltimore Biological Laboratory, Cockeysville, Md.; Difco Laboratories, Detroit.

*Available in dehydrated form from Baltimore Biological Laboratory, Cockeysville, Md.

The medium is used to demonstrate indole production and nitrate reduction by aerobes and anaerobes.

88. Trypticase soy agar*

Trypticase	15 gm.
Phytone	5 gm.
Sodium chloride	5 gm.
Agar	15 gm.
Distilled water	1,000 ml.

(Final pH 7.3±)

Trypticase soy agar is an excellent **blood agar base** and can be used for the isolation and maintenance of all organisms except those with special nutritional requirements.

89. Trypticase soy broth*

Trypticase	17 gm.
Phytone	3 gm.
Sodium chloride	5 gm.
Dipotassium phosphate	2.5 gm.
Glucose	2.5 gm.
Distilled water	1,000 ml.

(Final pH 7.3±)

Trypticase soy broth is excellent for the rapid (6 to 8 hours) growth of most organisms and will support the growth of pneumococci and streptococci without the addition of blood or serum. It will also support the growth of *Brucella.* However, because of the fermentation of the glucose present and the resulting drop in pH, acid-sensitive organisms, particularly pneumococci, may die in 18 to 24 hours.

90. Tryptophane broth

Tryptophane broth is a more popular medium for the detection of indole production. Baltimore Biological Laboratory's trypticase and Difco Laboratories' tryptone are recommended, in 1% aqueous solution.

91. Urea agar—urease test medium†
(Christensen: J. Bact. **52**:461, 1946)

Peptone	1	gm.
Glucose	1	gm.
Sodium chloride	5	gm.
Monopotassium phosphate	2	gm.
Phenol red	0.012	gm.
Agar	20	gm.
Distilled water	1,000	ml.

(Final pH 6.8 to 6.9)

Prepare the agar base and sterilize in the autoclave at 121° C. for 15 minutes in flasks containing 100- to 200-ml. amounts. Store until needed. Prepare a 29% solution of urea. Sterilize by filtering through a sterile bacteriological filter. Add the sterile urea solution in a final concentration of 10% to a flask of the agar base that has been melted and cooled to a temperature of 50° C. Mix well and distribute aseptically into sterile small tubes in amounts of 2 to 3 ml. Allow the medium to solidify in a slanting position in such a way as to obtain an agar butt of $1/_2$ inch and an agar slant of 1 inch.

Urea agar can be used to demonstrate **urease production** by species of *Proteus.* It will also detect the smaller amounts of urease produced by some coliform and other enteric bacilli, thus differentiating them from urease-negative *Salmonella* and *Shigella.* It may also be used to detect urease production by *Cryptococcus* species.

92. XLD agar*
(Taylor: Amer. J. Clin. Path. **44**:471, 1965)

The medium may be prepared by using the dehydrated xylose lysine agar base* and adding the sodium thiosulfate, ferric ammonium citrate, and sodium desoxycholate (procedure recommended by some workers) or by utilizing the complete xylose lysine desoxycholate (XLD) agar.

Xylose	3.5	gm.
L-Lysine	5	gm.
Lactose	7.5	gm.
Sucrose	7.5	gm.
Sodium chloride	5	gm.
Yeast extract	3	gm.
Phenol red	0.08	gm.

---

*Available in dehydrated form from Baltimore Biological Laboratory, Cockeysville, Md.
†Available in dehydrated form from Baltimore Biological Laboratory, Cockeysville, Md.; Difco Laboratories, Detroit.

---

*Baltimore Biological Laboratory, Cockeysville, Md.; Difco Laboratories, Detroit.

Agar (dried)	13.5	gm.
Sodium desoxycholate	2.5	gm.
Sodium thiosulfate	6.8	gm.
Ferric ammonium citrate	0.8	gm.
Distilled water	1,000	ml.

(Final pH 7.4±)

Suspend the medium in distilled water and heat with frequent agitation just to the boiling point. **Do not boil.** Transfer immediately to a 50° C. water bath and pour plates as soon as the medium has cooled. The medium should be of an orange color and clear, or nearly so. Excessive heating or prolonged holding at 50° C. causes a precipitation, which may cause some differences in colony morphology.

The medium is used for the isolation of enteric pathogens, especially shigellae.

40 Staining formulae and procedures

STAINS

For the reader's convenience the solubilities of the more widely used stains and dyes, in water and in alcohol, are shown in Table 40-1.

1. Acid-fast stain

Kinyoun carbolfuchsin method
(Kinyoun: Amer. J. Pub. Health **5**:867, 1915)

Basic fuchsin	4	gm.
Phenol	8	ml.
Alcohol, 95%	20	ml.
Distilled water	100	ml.

Dissolve the basic fuchsin in the alcohol and add the water slowly while shaking. Melt the phenol in a 56° C. water bath and add 8 ml. to the stain using a pipet with a rubber bulb.

Stain the fixed smear for 3 to 5 minutes (no heat necessary) and continue as with Ziehl-Neelsen stain.

By the addition of a detergent or wetting agent the staining of acid-fast organisms may be accelerated (Muller and Chermock: J. Lab. Clin. Med. **30**:169, 1945). Tergitol No. 7* may be used. Add 1 drop of Tergitol No. 7 to every 30 to 40 ml. of the Kinyoun carbolfuchsin stain. Stain the smears for 1 minute, decolorize, and counterstain as described in the following section.

*Carbide and Carbon Chemical Corporation, New York.

The technique and interpretation of the acid-fast stain are given in Chapter 3.

Ziehl-Neelsen method

a. *Carbolfuchsin stain:*

Basic fuchsin	0.3	gm.
Ethanol, 95%	10	ml.

Mix these with following:

Phenol, melted crystals	5	ml.
Distilled water	95	ml.

b. *Acid alcohol:**

Hydrochloric acid, concentrated	3	ml.
Ethanol, 95%	97	ml.

c. *Counterstain:*

Methylene blue	0.3	gm.
Distilled water	100	ml.

Some workers may prefer 0.5% aqueous brilliant green or a saturated solution of picric acid as a counterstain; the latter is pale and does not selectively stain cellular material.

a. Prepare a smear of appropriate thickness; dry and fix as described previously.

b. Place a strip of filter paper slightly smaller than the slide over the smear.

c. Flood the slide with carbolfuchsin stain; heat to steaming with a low Bunsen flame or electrically heated

*Use a 1% aqueous solution of sulfuric acid as a decolorizer when staining smears of suspected acid-fast *Nocardia,* such as *N. asteroides.*

Table 40-1. Solubility of stains*

| | PERCENT SOLUBLE AT 26°C. | |
STAIN	IN WATER	IN 95% ETHANOL
Bismarck brown	1.36	1.08
Congo red	0	0.19
Crystal violet (chloride)	1.68	13.87
Eosin Y	44.2	2.18
Fuchsin, basic (chloride)	0.26	5.93
Malachite green (oxalate)	7.60	7.52
Methylene blue (chloride)	3.55	1.48
Neutral red (chloride)	5.64	2.45
Safranin O	5.45	3.41
Thionin	0.25	0.25

*Based on data from Conn, J. G.: Biological stains, Commission on Standardization of Biological Stains, Geneva, N. Y., 1928, W. F. Humphrey Press.

slide warmer. **Do not boil** and do not allow to dry out.

d. Allow to stand 5 minutes without further heating; then remove the paper and wash the slide in running water.

e. Decolorize to a faint pink with acid alcohol while continuously agitating the slide until no more stain comes off in the washings (approximately 1 minute for films of average thickness). Thoroughness in decolorization is essential in order to prevent the possibility of a false positive reading.

f. Wash with water; counterstain with methylene blue for 20 to 30 seconds.

g. Wash with water, dry in air, and examine under the oil-immersion lens.

2. Capsule stain

The principles of capsule stains are discussed in Chapter 3.

Anthony method

a. Make a thin even smear of a culture in skimmed milk or litmus milk by spreading with a glass slide or an inoculating needle bent at a right angle. If it is not a milk culture, a loopful of the material may be mixed with a loopful of skimmed milk and then spread to give a uniform background.

b. Air dry. Do not fix with heat.

c. Stain with 1% aqueous crystal violet for 2 minutes.

d. Wash with a solution of 20% copper sulfate.

e. Air dry in a vertical position and examine under the oil-immersion lens. The capsule is unstained against a purple background; the cells are deeply stained.

Muir method

Muir mordant:

Tannic acid, 20% aqueous solution	2 parts
Saturated aqueous solution of mercuric chloride	2 parts
Saturated aqueous solution of potassium alum	5 parts

a. Prepare a thin even film of the bacteria; allow to dry in the air.

b. Cover the film with a piece of filter paper the size of the smear and flood the slide with Ziehl-Neelsen carbolfuchsin.

c. Heat to steaming with a low Bunsen flame for 30 seconds.

d. Rinse gently with 95% ethanol and then with water.

e. Add the mordant for 15 to 30 seconds; wash well with water.

f. Decolorize with ethanol to a faint pink; wash with water.

g. Counterstain with 0.3% methylene blue for 30 seconds.

h. Air dry and examine under the oil-immersion lens. The cells are stained red and the capsules blue.

India ink method*

In the India ink method, the capsule displaces the colloidal carbon particles of the ink and appears as a clear halo around the microorganism. The procedure is

---

*Not all India inks are suitable. Pelikan India ink made by Gunther Wagner of Hanover, Germany, is recommended; add about 0.3% tricresol as a preservative.

especially recommended for demonstrating the capsules of **pathogenic cryptococci.**

    a. To a small loopful of saline, water, or broth on a clean slide, add a **minute** amount of growth from a young agar culture, using an inoculating needle.

    b. Mix well; then add a small loopful of India ink and immediately cover with a thin coverglass, allowing the fluid to spread as a thin film beneath the coverglass.

    c. Examine immediately under the oil-immersion objective, reducing the light considerably by lowering the condenser. Capsules, when present, stand out as **clear halos** against a dark background.

3. Flagella stain
(Gray: J. Bact. **12:**273, 1926)

Mordant:

Potassium alum, saturated aqueous solution	5 ml.
Tannic acid, 20% aqueous solution	2 ml.
Mercuric chloride, saturated aqueous solution	2 ml.

Mix and add 0.4 ml. of a saturated alcoholic solution of basic fuchsin. Make up mordant freshly for use each day.

Gray method

    a. Using a grease-free slide that has been flamed and cooled, spread a drop of distilled water on the slide to cover an area of approximately 2 sq. cm.

    b. Fish part of a colony from a young agar culture or take a small amount of growth from a slant with an inoculating needle and **touch gently** into the drop of water at several places on the slide; then gently rotate the slide.

    c. Allow to **air dry. Do not heat.**

    d. Add the mordant and allow to act for 10 minutes.

    e. Wash gently with distilled water or clean tap water.

    f. Add Ziehl-Neelsen carbolfuchsin and leave on for 5 to 10 minutes.

    g. Wash with tap water, air dry, and examine under oil.

4. Fluorochrome stain (Truant method)[1]

By staining the smear with fluorescent dyes, such as auramine and rhodamine, and examining by fluorescence microscopy using an ultraviolet light source, acid-fast bacilli, when present, will appear to glow with an orange to red color. These are visible under lower magnifications of the microscope; thus, a stained smear can be examined in much less time than is required by conventional methods. Numerous modifications of the procedure have been introduced; that reported by Truant and coworkers[1] is recommended.

Auramine O (Allied Chemical Corp., C.I. No. 41000)	1.5	gm.
Rhodamine B (Allied Chemical Corp., C.I. No. 749)	0.75	gm.
Glycerol	75	ml.
Phenol	10	ml.
Distilled water	50	ml.

Combine solutions, mix well (magnetic stirring device for 24 hours or heat until warm and stir vigorously for 5 minutes), filter through glass wool, and store in a glass-stoppered bottle at 4° C. The stain is stable for several months under refrigeration.

    a. Heat fix on a slide warmer* for 2 hours at 65° C. or overnight.

    b. Cover smear with auramine-rhodamine solution.

    c. Stain for 15 minutes at room temperature or at 36° C.

    d. Rinse off with distilled water.

    e. Decolorize with 0.5% hydrochloric acid in 70% ethanol for 2 to 3 minutes, then rinse thoroughly with distilled water.

    f. Flood smear with counterstain, a 0.5% solution of potassium perman-

---

*Micro-slide staining and drying bath are available from Scientific Products Corp. American Hospital Supply Corp., Flushing, N. Y.

ganate (filter and store in amber bottle), for 2 to 4 minutes (no longer—excessive exposure results in loss of brilliance).

g. Rinse with distilled water, dry, and examine.

The smears are examined under a binocular microscope using an ultraviolet light source. The Leitz, Zeiss, and Reichert fluorescent microscopy units are highly recommended and are equipped with Osram HBO 200 maximum pressure mercury vapor lamps as light sources, BG-12 (3 or 4 mm.) or C-5113 (2 mm.) violet exciter filters, and OG-1 deep yellow barrier filters in the eyepieces. This recommended filter combination results in bright **yellow-orange** staining bacilli against a dark background; nonspecific background debris fluoresces a pale yellow, quite distinct from the yellow-orange bacilli.

It is suggested that a drop of immersion oil be placed on the dark-field condenser, the slide inserted, and the microscope first focused under bright light until a clear central area is seen on the slide, then switched over to the ultraviolet source. Smears may be rapidly examined under low- or high-power objective (25× or 40×) with a 10× eyepiece; after a little practice all smears may be examined at these magnifications or lower in a matter of seconds. Occasionally, it may be necessary to switch to the oil-immersion objective to confirm typical morphological characteristics, such as beading or cording. It is recommended that the microscopic examination be carried out in a darkened room for maximum efficiency.

All positive smears should be confirmed with a Kinyoun or Ziehl-Neelsen stain. This may be done without removing the auramine-rhodamine stain, but the reverse of this procedure is not satisfactory.

5. Gram stain (Hucker modification)

Reagents:
a. *Stock crystal violet:*

Crystal violet, 85% dye	20 gm.
Ethanol, 95%	100 ml.

b. *Stock oxalate solution:*

Ammonium oxalate	1 gm.
Distilled water	100 ml.

Working solution: Dilute the stock crystal violet solution 1:10 with distilled water and mix with 4 volumes of stock oxalate solution. Store in a glass-stoppered bottle.

c. *Gram iodine solution:*

Iodine crystals	1 gm.
Potassium iodide	2 gm.

Dissolve these completely in 5 ml. of distilled water; then add:

Distilled water	240 ml.
Sodium bicarbonate 5% aqueous solution	60 ml.

Mix well; store in an amber glass bottle.

d. *Decolorizer:*

Ethanol, 95%	250 ml.
Acetone	250 ml.

Mix; store in a glass-stoppered bottle.

e. *Counterstain:*

Stock safranin

Safranin O	2.5 gm.
Ethanol, 95%	100 ml.

Working solution: Dilute stock safranin 1:5 or 1:10 with distilled water; store in a glass-stoppered bottle.

The principles of the Gram stain are discussed in Chapter 3. The following procedure is for the **rapid** method:

a. Prepare a thin film of the material to be examined; dry and fix as previously described.

b. Flood the slide with crystal violet stain and allow to remain on the slide for 10 seconds.

c. Pour off the stain and wash off the remainder with the iodine solution.

d. Flood with iodine solution and allow to mordant for 10 seconds.

e. Rinse off with running water. Shake off the excess.

f. Decolorize with alcohol-acetone solution or 95% alcohol (an alcohol-acetone solution may prove to be too rapid) until the solvent flows colorlessly from the slide. This usually takes from 10 to 20 seconds, depending on the thickness of the smear. Care should be taken not to overdecolorize the film, which may result in an incorrect reading.

g. Counterstain with safranin for 10 seconds; then wash off with water.

h. Blot between clean sheets of bibulous paper and examine under oil immersion.

6. Metachromatic granule stain

Albert stain

The Albert stain is a differential stain and is recommended for its simplicity in staining *Corynebacterium diphtheriae.*

a. Prepare the smear and fix with heat.

b. Flood the smear with Albert stain for 3 to 5 minutes.

c. Wash in tap water and drain off excess.

d. Flood with Gram's iodine. Allow to react for 1 minute.

e. Wash, blot dry, and examine.

Granules appear blue-black, the bands appear blue to blue-green, and the cytoplasm appears green.

Methylene blue stain

Methylene blue	0.3 gm.
Ethyl alcohol, 95%	30 ml.

When dissolved add

Distilled water	100 ml.

Cover the fixed smear with the staining solution and stain for 1 minute. Wash with water and blot dry with blotting paper.

Loeffler methylene blue stain, as formerly used, was prepared by adding alkali to the foregoing solution. Modern purified samples of methylene blue do not require the addition of alkali. The older preparations contained acid impurities.

It is a simple stain. Prepare a smear of the organism and fix with heat. Flood the smear with Loeffler methylene blue and allow to react for 1 minute. Wash and blot dry. The granules take up the dye readily and appear deep blue in color. Overstaining lessens contrast.

7. PPLO (Mycoplasma) stain

(Dienes and Weinberger: Bact. Rev. **15:**245, 1951)

Reagents:

Methylene blue	2.5	gm.
Azure II	1.25	gm.
Maltose	10.0	gm.
Sodium carbonate	0.25	gm.
Distilled water	100	ml.

Dissolve the ingredients in the water.

a. Spread out a drop of the stain on a grease-free coverglass and allow to dry.

b. With a sterile scalpel, cut out a small block of the agar medium containing a few *Mycoplasma* colonies and place this, colonies up, on a clean glass slide.

c. Lay the coverglass, stain side down, on the agar block carefully, without rubbing.

d. Seal the preparation with a mixture of 3 parts petrolatum and 1 part of paraffin to prevent drying.

e. Examine under the low-power objective of the microscope.

f. *Mycoplasma* colonies are quite distinct with dense blue-stained centers and light blue peripheries.

8. Relief staining (Dorner)

Reagents:

Nigrosin	10	gm.
Distilled water	100	ml.

Boil for 30 minutes, add 0.5 ml. of formalin when cool, filter through paper, and store in 2-ml. amounts in sterile corked tubes.

a. Place a loopful of the bacterial suspension on a grease-free slide and immediately add a loopful of the nigrosin solution.

b. Spread out in a thin film.

c. Dry the slide in air or hasten drying with gentle heat.

d. Examine under oil-immersion lens.

e. Cells are unstained against the dark background.

9. Staining of rickettsiae

Castañeda stain

Reagents:
Solution A:

Potassium phosphate (KH_2PO_4), 1% aqueous	100 ml.
Sodium phosphate (Na_2HPO_4 $12H_2O$), 25% aqueous	100 ml.

Mix and add 1 ml. of formalin.

Solution B:

Methyl alcohol	100 ml.
Methylene blue	1 gm.

Mix 20 ml. of solution A with 0.15 ml. of solution B and add 1 ml. of formalin.

Solution C (counterstain):

Safranin O, 0.2% aqueous	25 ml.
Acetic acid, 0.1%	75 ml.

a. Prepare a homogeneous film and dry in air.
b. Cover film with stain (mixture of solutions A and B).
c. Drain off the stain. Do not wash.
d. Counterstain with safranin O (solution C) for 1 to 4 seconds.
e. Wash with tap water. Blot dry.
f. Examine under oil-immersion lens.
g. The rickettsiae stain blue whereas the cellular elements stain red.

Giemsa method*

a. Prepare a homogeneous film on a clean glass slide. Allow to dry in air.
b. Flood with methyl alcohol for 1 minute.
c. Drain off the alcohol and allow to dry.
d. Cover film with Giemsa stain (15 drops) and allow to react for 1 minute.
e. Add distilled water (30 drops) and continue staining for 5 minutes. Drain off.
f. Wash with distilled water.
g. Place slide on end and allow to dry in air.
h. Examine under oil-immersion lens.
i. The rickettsiae stain a bluish purple.

---

*This stain may also be used for the spirochetes, which stain blue.

10. Spore stain

Dorner method

a. Make a heavy suspension of the organisms in distilled water in a test tube and add an equal volume of freshly filtered carbolfuchsin.
b. Place tube in boiling water bath for 5 to 10 minutes.
c. Mix a loopful of the aforementioned combination with a loopful of a boiled and filtered 10% aqueous solution of nigrosin on a clean slide.
d. Spread out and dry film quickly with gentle heat.
e. Examine under oil. The spores stain red, and the bacterial cells are almost colorless against a dark gray background.

A modification of the Dorner method may be employed whereby a smear of the culture is prepared and fixed. The smear is then covered with a strip of filter, to which the carbolfuchsin is added. The dye is heated to steaming for 5 to 7 minutes with a Bunsen burner, and the filter paper is removed. Wash with water, blot dry, and cover with a thin film of nigrosin using a second slide or a needle. The appearance of the cells will be as described previously.

11. Wayson stain for smears of pus

Dissolve 0.2 gm. of basic fuchsin and 0.75 gm. of methylene blue in 20 ml. of absolute ethanol. Add the dye solution to 200 ml. of a 5% solution of phenol in distilled water. Filter. Stain smears for a few seconds. Wash, blot, and dry. This stain is useful in polar staining.

12. Wright-Giemsa method for staining conjunctival scrapings

The Wright-Giemsa stain, along with a Gram stain of the scrapings, gives immediate information to the opthalmologist regarding (a) conjunctivitis of bacterial origin, (b) inclusion body conjunctivitis and trachoma, or (c) eosinophilia of allergic conjunctivitis.

a. Two slide preparations of the scrappings are made—one is stained by the Gram method and one by the Wright-Giemsa technique.

b. To carry out staining by the Wright-Giemsa technique, apply Wright stain to the slide for 1 minute. Add an equal volume of neutral distilled water and stain for 4 minutes.

c. Shake off the stain; then apply dilute Giemsa stain (1 drop to 1 ml. of neutral distilled water), and allow to stain for 15 minutes.

d. Shake off, decolorize lightly with ethanol, and air dry (do not blot).

MOUNTING FLUIDS
1. Lactophenol cotton blue fluid

Phenol crystal	20 gm.
Lactic acid	20 gm.
Glycerin	40 gm.
Distilled water	20 ml.

Dissolve these ingredients by heating gently over a steam bath. Add 0.05 gm. of cotton blue dye (Poirrier's blue).

This may be used for yeasts as well as fungi and serves as both a mounting fluid and a stain.

a. Place a drop of this fluid on a clean slide.

b. Place a small amount of culture in this drop. If the culture is on agar, remove a piece of the medium with the embedded growth.

c. Cover with a coverglass and press down gently to flatten.

d. Warm gently to remove air bubbles if necessary.

e. Examine under the microscope with high dry or oil-immersion objectives.

f. Ring edges of coverglass with nail polish if a permanent mount is required.

2. Chloral lactophenol

Chloral lactophenol, recommended by Dr. F. Blank of Temple University School of Medicine, is used in place of 10% potassium hydroxide, and in the same manner.

Chloral hydrate	2 parts
Phenol crystals	1 part
Lactic acid	1 part

Dissolve the ingredients by gentle heating over a steam bath.

3. Sodium hydroxide-glycerin

Glycerin	10 ml.
Sodium hydroxide*	20 gm.
Distilled water	90 ml.

This mounting fluid is used in moist preparations when examining clinical material for fungi.

---

*Potassium hydroxide may be substituted for sodium hydroxide.

REFERENCE

1. Truant, J. P., Brett, W. A., and Thomas, W., Jr.: Fluorescence microscopy of tubercle bacillus stained with auramine and rhodamine, Henry Ford Hosp. Med. Bull. **10:**287-296, 1962.

41 Reagents and tests

1. Alpha hemolysin test for virulent staphylococci

a. Inoculate a tube of tryptose phosphate broth with *Staphylococcus* and incubate at 36° C. in a 25% to 30% carbon dioxide atmosphere for 48 hours.

b. Centrifuge the culture and sterilize the supernate by filtration.

c. Set up ten tubes (13 × 100 mm.) in a rack. Add 1 ml. of saline to tubes 2 through 8.

d. Place 1 ml. of supernate in tube 1 and 1 ml. in tube 2. Mix the contents of tube 2 by aspirating with the pipet and transfer 1 ml. of this to tube 3. Continue through tube 8 to establish a dilution series of 1:1 to 1:128 in 1 ml. volumes. (Discard the 1 ml. surplus from tube 8.)

e. Tube 9 should contain only saline. Place in tube 10, 1 ml. of an effective dilution of a known alpha hemolysin. These will serve as negative and positive controls respectively.

f. Add to each tube of the ten-tube series 1 ml. of a 5% suspension of washed rabbit erythrocytes. Final dilutions will now be 1:2 to 1:256.

g. Incubate in a 37° C. water bath for 2 hours and place in a refrigerator overnight.

h. The highest dilution of the supernate that gives 50% hemolysis is recorded as the **alpha hemolysin titer.**

2. Arysulfatase test reagent

a. *Stock substrate:*
Dissolve 2.6 gm. of tripotassium phenolphthalein* in 50 ml. of distilled water (0.08 M); sterilize by filtration and store under refrigeration.

b. *Stock solution of substrate:*
Prepare a 0.001 M substrate solution by adding 2.5 ml. of the 0.08 M stock substrate to 200 ml. of Dubos Tween-albumin broth.† Dispense aseptically in 2-ml. amounts in 16 × 125 ml. screw-capped tubes.

3. Arylsulfatase color standards

a. Prepare a stock solution of 0.1 gm. of phenolphthalein in 10 ml. of ethyl alcohol.

b. Prepare a 2 N sodium carbonate solution: 10.6 gm. of sodium carbonate in 100 ml. of distilled water.

c. Prepare standards according to the chart shown.

d. Mix each dilution and dispense in 2-ml. quantities in 16 × 125 mm. screw-capped tubes.

e. Add 6 drops of 2 N sodium carbonate to each tube.

f. Solutions without added sodium carbonate may be stored in the refriger-

---

*Nutritional Biochemicals Corp., Cleveland, Ohio, or L. Light & Co., Colinbrook, Bucks, England.
†Baltimore Biological Laboratory, Cockeysville, Md.; Difco Laboratories, Detroit.

Arylsulfatase color chart

Tube	Phenolphthalein	Distilled water	Amount	2 N Na₂CO₃	Reading
1	1 ml. of stock	50 ml.	2 ml.	6 drops	5+
2	5 ml. of tube 1	25 ml.	2 ml.	6 drops	4+
3	2 ml. of tube 1	25 ml.	2 ml.	6 drops	3+
4	1.5 ml. of tube 1	50 ml.	2 ml.	6 drops	2+
5	0.5 ml. of tube 1	50 ml.	2 ml.	6 drops	1+
6	0.5 ml. of tube 1	100 ml.	2 ml.	6 drops	±

ator for several months; tubes containing sodium carbonate will fade in 2 to 4 weeks and must be freshly prepared as indicated.

g. A more stable set of standards, valid for 6 months, can be prepared using M/15 Na_2HPO_4 as reagent and 0.1% phenol red as color indicator.*

4. Buffer solutions

Buffered glycerol-saline solution
(Teague and Clurman, 1916; modified by Sachs, 1939).

Sodium chloride	4.2 gm.
Dipotassium phosphate, anhydrous	3.1 gm.
Monopotassium phosphate, anhydrous	1 gm.
Glycerol	300 ml.
Distilled water	700 ml.

Dispense in bottles with tightly fitting screw caps in approximately 10-ml. amounts; autoclave for 15 minutes at 116° C. Add sufficient phenol red to give a distinct red color; if the solution becomes yellow (acid), it should be discarded. The solution is used for preserving fecal specimens.

Sorensen pH buffer solutions

Buffer solutions may be added to culture media to prevent a significant change in hydrogen ion concentration. Sorensen buffers, prepared from potassium and sodium phosphates, are readily prepared from the anhydrous salts, or they may be purchased from commercial sources.

---

*Vestal, A. L.: Procedures for the isolation and identification of mycobacteria, Public Health Service Publication No. 1995, 1969.

Reagents:
Solution A:
  M/15 Na_2PO_4

Dissolve 9.464 gm. of the anhydrous salt, previously dried at 130° C. in distilled water, to make 1 liter of solution.

Solution B:
  M/15 KH_2PO_4

Dissolve 9.073 gm. of the anhydrous salt, previously dried at 110° C. in distilled water, to make 1 liter of solution. Mix solutions A and B as indicated.

pH	Solution A	Solution B
5.29	0.25 ml.	9.75 ml.
5.59	0.5 ml.	9.5 ml.
5.91	1 ml.	9 ml.
6.24	2 ml.	8 ml.
6.47	3 ml.	7 ml.
6.64	4 ml.	6 ml.
6.81	5 ml.	5 ml.
6.98	6 ml.	4 ml.
7.17	7 ml.	3 ml.
7.38	8 ml.	2 ml.
7.73	9 ml.	1 ml.
8.04	9.5 ml.	0.5 ml.

5. Catalase test

Use an 18- to 24-hour agar slant culture* incubated at 36° C. Pour 1 ml. of a 3% solution of hydrogen peroxide over the growth and set the tube in an inclined position. The reaction is **positive** if there is a rapid ebullition of gas. Micrococci and staphylococci are catalase positive; streptococci and pneumococci are catalase negative; *Bacillus* species are catalase positive.

The test may also be carried out with a

---

*The agar slant should be inoculated quite heavily. An old slant culture will not give a proper test.

24- to 48-hour culture in broth or thiogly-collate medium (microaerophiles and anaerobes). Add approximately 1 ml. of the hydrogen peroxide to the culture and observe for gas as before.

The test, however, **cannot** be applied to cultures grown on blood agar because of the catalase present in the red blood cells.

6. Digesting and decontaminating solutions for culturing of sputum for Mycobacterium tuberculosis

N-acetyl-L-cystine method
(Kubica et al.: Amer. Rev. Resp. Dis. **89:**284, 1964)

a. Prepare the necessary volume of digestant as shown in Table 41-1. The solution is self-sterilizing but should be used within 24 hours, since it deteriorates on standing.

b. Under a well-ventilated safety cabinet transfer no more than 10 ml. of sputum to a sterile 50-ml. screw-capped, aerosol-free centrifuge tube. Smaller volumes may be used in smaller tubes, but in no case should the volume of sputum exceed one fifth of the volume of the tube.

c. Add an equivalent volume of the acetyl cysteine–sodium hydroxide digestant (Table 41-1) to the specimen; mix well in a Vortex mixer. Digestion is generally effected in 5 to 30 seconds. Avoid extreme agitation, which may inactivate the acetyl cysteine by oxidation. Carry on as described on p.199.

Trisodium phosphate–benzalkonium chloride method

a. Dissolve 1,000 gm. of trisodium phosphate .12H_2O in 4,000 ml. of hot distilled water. Add 7.4 ml. of 17% aqueous benzalkonium chloride concentrate.*

b. M/15 phosphate buffer, pH 6.6

(1) Sodium monohydrogen   9.47 gm.
    phosphate (anhydrous)

Distilled water   1,000   ml.
(2) Potassium dihydro-   9.08 gm.
    phosphate
    Distilled water   1,000   ml.

Mix 625 ml. of (2) with 375 ml. of (1). Check the reaction and adjust the pH to 6.6 if required; dispense in small volumes in appropriate containers, and sterilize at 121° C. for 15 minutes.

7. Direct immunofluorescence procedure for Neisseria gonorrhoeae

1. Prepare a **thin** film of a suspected colony from a Thayer-Martin plate (may be made up to 15 minutes after performing the oxidase test) on a slide containing a 6-mm. diameter etched circle.

2. Thoroughly dry the film in air.

3. Overlay with adsorbed (with anti-meningococcus Group B serum) fluorescin-labeled *N. gonorrhoeae* antiserum,* keeping within the 6-mm. diameter circle.

4. Incubate the slide at 36° C. in a moist chamber for 30 minutes (alternately for 5 minutes at room temperature).

5. Rinse with pH 7.2 phosphate buffer, dry, mount in buffered glycerine with a cover glass, and examine under fluorescence microscopy. Gonococci appear as **yellow-green** diplococci of typical size and shape.

6. Include a positive urethral smear or one prepared from a known **fresh** isolate of *N. gonorrhoeae,* along with one of a boiled suspension of *Enterobacter cloacae,* as positive and nonspecific staining controls.

8. Ferric chloride test for detecting hydrolysis of sodium hippurate

(Ayers and Rupp: J. Infect. Dis. **30:**388, 1922)

Reagent:

Dissolve 12 gm. of ferric chloride ($FeCl_3$ .6H_2O) in 100 ml. of 2% aqueous hydrochloric acid.

When sodium hippurate broth is

---

*17% Zephiran chloride is available from Winthrop Laboratories, New York.

*Available from Sylvana Co., Millburn, N. J.; Difco Laboratories, Detroit; and others.

Table 41-1. Preparation of acetyl cysteine–sodium hydroxide digestant

REAGENT	VOLUME OF DIGESTANT NEEDED				
	50 ml.	**100**	**200**	**500**	**1,000**
1 N (4%) sodium hydroxide	25 ml.	50	100	250	500
0.1 M (2.94%) sodium citrate · 2H$_2$O	25 ml.	50	100	250	500
N-acetyl-L-cysteine powder*	0.25 gm.	0.5 gm.	1 gm.	2.5 gm.	5 gm.

*Powdered N-acetyl-L-cysteine is available from Mead Johnson Laboratories, Evansville, Ind.; Sigma Chemical Co., St. Louis; Baltimore Biological Laboratory, Cockeysville, Md.

prepared, the level of the medium in each tube should be marked with a wax pencil. During storage and incubation of the medium, some evaporation occurs and the concentration of sodium hippurate in the medium is increased. Since the concentration of sodium hippurate must be exactly 1%, make up the loss due to evaporation by adding distilled water up to the wax pencil mark after incubation and before the ferric chloride test is performed. If this precaution is not taken, a false positive reaction may be obtained.

Transfer 0.8 ml. of culture in sodium hippurate broth to a small test tube (Wassermann tube) and add 0.2 ml. of the reagent. Mix immediately and observe after 10 to 15 minutes. A **permanent precipitate** indicates the presence of benzoic acid **(positive hydrolysis).**

Since sodium hippurate is first precipitated and later redissolved by the amount of reagent specified and since benzoic acid is also redissolved by a greater excess of the reagent, it is necessary to have the reagent and the medium balanced and to measure accurately the amounts used in the test. A **control test** of the sterile medium should always be made. If the culture is quite turbid, thus causing confusion in reading the result, it should be centrifuged and the clear supernatant fluid used for the test.

9. Gastric mucin (5%)
(Strauss and Klegman: J. Infect. Dis. **88:**151, 1951)

Emulsify 5 gm. of gastric hog mucin (granular type)* in 95 ml. of distilled

water in a blender for 5 minutes. Autoclave for 15 minutes at 121° C. Cool to room temperature; adjust to pH 7.3 with sterile sodium hydroxide. Check for sterility and store in a refrigerator.

Mix equal parts of 5% gastric mucin and the fungus suspension; inject 1 ml. intraperitoneally into the appropriate laboratory animal.

10. Gluconate oxidation test
(Haynes: J. Gen. Microbiol. **5:**939, 1951)

Pseudomonas aeruginosa is able to oxidize dextrose or gluconate to ketogluconate, which in turn is detected by the reduction of copper salts, as found in Benedict's solution. This test is helpful in identifying nonpigmented strains of *Ps. aeruginosa.*

The use of gluconate substrate tablets* is recommended for this test. A single tablet is added to 1 ml. of distilled water, which is then heavily inoculated with the test organism. After 12 to 18 hours' incubation at 36° C., test the culture for reducing substances with Benedict's solution.† A positive test is indicated by a color change from blue to green-yellow.

11. Hydrogen sulfide production
Lead acetate paper test

Saturate filter paper strips (5 × 1 cm.) with 5% lead acetate solution. Air dry, then autoclave at 15 pounds' pressure for 15 minutes.

Inoculate a sulfur-containing liquid medium and insert a lead acetate strip be-

---

*Wilson Laboratories, Chicago, Ill.

*Key Scientific Products Co., Los Angeles.
†A Clinitest tablet from Ames Co., Elkhart, Ind., may be substituted.

tween the plug and inner wall of tube and above the liquid. Hydrogen sulfide production is evidenced by the **blackening** of the lower portion of the strip.

A negative test may be checked by adding a small amount of 2 N hydrochloric acid to the tube and closing the tube as before. Any dissolved sulfide will be liberated and will combine with the lead in the strip to form the black lead sulfide.

Note: The lead acetate paper test may be positive when the butt reaction in TSI agar is negative or only weakly positive. It is more sensitive.

Triple sugar iron (TSI) agar method

The butt of this medium is stabbed with the culture. Hydrogen sulfide production is detected by the blackening of the butt (p. 139). Lead acetate paper may also be used by inserting a strip between the loosened cap and inner wall of the TSI tube.

12. Indicators for anaerobiosis
Fildes and McIntosh indicator

Prepare the following solutions:
a. 6% aqueous glucose (add a small crystal of thymol as a preservative).
b. One-tenth normal sodium hydroxide; 6 ml. to 94 ml. of distilled water.
c. Aqueous methylene blue (0.5%); 3 ml. to 100 ml. of distilled water.

For use, mix 1 ml. of each solution in a test tube, boil the mixture until colorless, then place in a loaded anaerobic jar before sealing. A blue color at the end of incubation indicates that anaerobiosis was **not** achieved.

Smith modified methylene blue indicator
(Smith, L. DS., ASM meeting, Washington, D. C., 1964)

Mix thoroughly 1 pound of sodium bicarbonate (commercial grade is satisfactory) with 50 gm. of glucose and 20 mg. of methylene blue. For use, add about 1 inch to a 16 × 100 mm. test tube, half fill with tap water, invert to mix, and place within the jar. The solution will slowly become colorless during incubation at 36°

C.; if more rapid decolorization is required, the solution may be heated to boiling and cooled rapidly immediately before placing in the jar. The indicator should be **colorless** at the end of incubation if anaerobiosis was maintained.

13. McFarland nephelometer standards

a. Set up ten test tubes or ampules of equal size and of good quality. Use new tubes that have been thoroughly cleaned and rinsed.
b. Prepare 1% chemically pure sulfuric acid.
c. Prepare 1% aqueous solution of chemically pure barium chloride.
d. Add the designated amounts of the two solutions to the tubes as shown in Table 41-2 to make a total of 10 ml. per tube.
e. Seal the tubes or ampules. The suspended barium sulfate precipitate corresponds approximately to homogeneous *Escherichia coli* cell densities per milliliter throughout the range of standards, as shown in Table 41-2.

14. Methyl red test
(Clark and Lubs.: J. Infect. Dis. **17:**160, 1915)

To 5 ml. of the culture in MR-VP broth, add 5 drops of methyl red solution. A positive reaction is indicated by a distinct **red** color, showing the presence of acid (pH less than 4.5). A negative reaction is indicated by a **yellow** color. *Escherichia coli* and other methyl red–positive organisms produce a high acidity from the dextrose in this medium within 48 hours, which turns the indicator red.

The solution of methyl red is prepared by dissolving 0.1 gm. of the indicator in 300 ml. of 95% alcohol and diluting to 500 ml. with distilled water.

15. Nitrate reduction test
Reagents:
Solution A:
Sulfanilic acid   8 gm.
Acetic acid (5N)   1,000 ml.

Table 41-2. McFarland nephelometer standards

TUBE NUMBER	1	2	3	4	5	6	7	8	9	10
Barium chloride (ml.)	0.1	0.2	0.3	0.4	0.5	0.6	0.7	0.8	0.9	1
Sulfuric acid (ml.)	9.9	9.8	9.7	9.6	9.5	9.4	9.3	9.2	9.1	9
Approx. cell density ($\times 10^8$/ml.)	3	6	9	12	15	18	21	24	27	30

Solution B:

Alphanaphthylamine	5 gm.
Acetic acid (5N)	1,000 ml.

Add 5 drops of each reagent to the tube. A **positive** test for nitrites is revealed by the development of a **red** color in 1 to 2 minutes.

Some organisms can reduce nitrate beyond the nitrite stage to nitrogen or ammonia. **A negative test for nitrite, therefore, should not be construed necessarily as a negative nitrate reduction test without first testing for the presence of unreduced nitrate:**

Add a very small amount of zinc dust to the broth medium, which has shown a negative reduction test with the foregoing reagents. The presence of unreduced nitrate is revealed by the development of a red color, thus confirming a negative nitrate reduction test.

The test for nitrate reduction is carried out after 24 to 48 hours of incubation at 36° C.

16. Nitrate reduction test for mycobacteria (nitrite standards)

a. Prepare a M/100 sodium nitrite solution by dissolving 0.14 gm. of sodium nitrite in 200 ml. of distilled water.

b. Carry out twofold serial dilutions of the foregoing in 2 ml.-volumes in thirteen marked tubes.

c. To tubes 6, 7, 8, 10, 11, and 13 add 1 drop of 1:2 dilution of hydrochloric acid, 2 drops of 0.2% aqueous solution of sulfanilamide, and 2 drops of 0.1% aqueous solution of N-naphthylethylenediamine dihydrochloride.

d. Nitrite standards **fade rapidly** and must be freshly prepared each time.

Tube	Dilution	Reading
6	1:64	5+
7	1:128	4+
8	1:256	3+
10	1:1024	2+
11	1:2048	1+
13	1:8192	+/−

17. Oxidase test for detecting colonies of Neisseria

(Gordon and McLeod: J. Path. Bact. **31:**185, 1928)

For the oxidase test a 1% solution (0.1 gm. in 10 ml.) of para-aminodimethylaniline monohydrochloride* is used. Add the dye to distilled water and let stand for 15 minutes before using or until a definite purple color has developed. The solution should be used immediately and then discarded, the dye being active only 1 to 2 hours after the solution has been prepared.

Place a few drops of the dye solution on portions of the plate containing colonies suspected of being *Neisseria*. Colonies producing indophenol oxidase become **pink**, progressing to maroon, dark **red**, and finally to **black**. Organisms from pink colonies are usually viable, but after the colonies have become black the *Neisseria* organisms are dead. The dye does not interfere with the Gram reaction.

A 1% aqueous solution of tetramethyl-para-phenylenediamine dihydrochloride may also be used. It is less toxic and gives the colonies a lavender color that eventually turns purple. The reagent sometimes colors the surrounding medium, and it

---

*N, N-Dimethyl-*p*-phenylenediamine monohydrochloride, from Eastman Organic Chemicals, Rochester, N. Y.

costs considerably more than the dimethyl reagent.

18. Oxidase test for Pseudomonas
(Kovacs: Nature **178**:703, 1956)

Reagent:

Tetramethyl-para-phenylenediamine dihydrochloride	0.1 gm.
Distilled water	10 ml.

Add the dye to the water and allow to stand for 15 minutes before using. Prepare freshly for use each time. The dye loses its activity after 2 hours.

Place 2 or 3 drops of the reagent on a piece of Whatman No. 1 filter paper (6 sq. cm.). Remove a suspected colony* from the plate, or some growth from an agar slant, and smear with a loop on the reagent-saturated paper. A positive reaction (oxidase-positive), recognized by a **dark purple** color, develops in 5 to 10 seconds.

19. Preservation of fungal cultures†

Obtain a good grade of **heavy** mineral oil‡ and dispense in approximately 125-ml. amounts in 250-ml. Erlenmeyer flasks (cotton plugged). Autoclave at 121° C. for 45 minutes. Using sterile technique, pour the oil over a small but actively growing fungus culture on a short slant of Sabouraud agar. It is essential that the oil cover not only the fungus colony but also the whole agar surface (about 1 inch above the top of the slant); otherwise the exposed agar or colony will act as a wick and in time cause the medium to dry out.

The stock culture is stored upright at room temperature and will remain viable without further attention for several years.

In transferring from oiled cultures, remove a bit of the fungus with a long inoculating needle, drain off the oil by touching it on the inside of the tube, and inoculate to a fresh slant. Rinse off the needle in xylol before flaming.

---

*Use a platinum wire loop—nichrome may cause a false-positive reaction.
†Recommended by the Mycology Unit of the National Center for Disease Control, Atlanta, Ga.
‡Available from Parke-Davis Co., Detroit.

20. Ringer solution

Sodium chloride	8.50 gm.
Potassium chloride	0.20 gm.
Calcium chloride .2H$_2$O	0.20 gm.
Sodium carbonate	0.01 gm.
Distilled water	1,000 ml.

(pH 7)

Sterilize by autoclaving at 121° C. for 15 minutes. To use in dissolving calcium alginate swabs, prepare in one-quarter strength and add 1% sodium hexametaphosphate. Approximately 10 minutes of shaking usually suffices for complete solution of the swab.

21. Tests for indole
Ehrlich indole test
(Modification of Bohme: Centralbl. Bakt., Orig. **40**:129, 1906)

Reagent:

Paradimethylaminobenzaldehyde	2 gm.
Ethyl alcohol (95%)	190 ml.
Hydrochloric acid (concentrated)	40 ml.

Add 1 ml. of xylene to a 48-hour culture of organisms in tryptone or trypticase broth or other appropriate medium. Shake well and allow to stand for a few minutes until the solvent rises to the surface.

Gently add about 0.5 ml. of the reagent down the sides of the tube so that it forms a ring between the medium and the solvent. If indole has been produced by the organisms, it will, being soluble in solvent, be concentrated in the solvent layer, and on the addition of the reagent, a brilliant **red ring** will develop just below the solvent layer. If no indole is produced, no color will develop.

Kovacs indole test

Reagent:

Pure amyl or isoamyl alcohol	150 ml.
Paradimethylaminobenzaldehyde	10 gm.
Concentrated hydrochloric acid (A.R.)	50 ml.

Dissolve the aldehyde in the alcohol and add the acid slowly. Prepare in small quantities and store in the refrigerator when not in use.

Inoculate tryptophane broth and incubate for 48 hours at 36° C. Add 5 drops of Kovacs reagent. A **deep red** color indicates the presence of indole.

22. Tween 80 hydrolysis substrate

M/15 phosphate buffer, pH 7	100 ml.
Tween 80	0.5 ml.
Neutral red (0.1% aqueous solution)	2.0 ml.

a. Mix the foregoing solutions, dispense in 16 × 125 mm. screw-capped tubes in 2-ml. amounts.

b. Sterilize by autoclaving at 121° C. for 10 minutes.

c. Check for sterility by incubating overnight at 36° C. Final color is amber or straw.

d. Store in refrigerator in the dark for not more than 2 weeks.

This substrate is used to test the ability of certain strains of mycobacteria to rapidly degrade the Tween 80 to oleic acid, which is detected by a change in color of the indicator.

23. Voges-Proskauer test for acetyl-methylcarbinol or acetoin

(Coblentz: Amer. J. Pub. Health **33:**315, 1943)

Reagents:

Alpha-naphthol (5%) in absolute ethyl alcohol
Potassium hydroxide (40%) containing 0.3% creatine

a. Pipet 1 ml. of a 48-hour culture grown in MR-VP broth into a clean Wassermann tube.

b. Add 0.6 ml. of 5% alpha-naphthol in absolute ethyl alcohol.

c. Add 0.2 ml. of 40% potassium hydroxide–creatine solution.

d. Shake well and allow to stand for 10 to 20 minutes. If acetylmethylcarbinol has been produced, a bright **orange-red** color will develop at the surface of the medium and will gradually extend throughout the broth.

Index